I0056557

Wildlife Habitats: Ecology, Environment and Conservation

Wildlife Habitats: Ecology, Environment and Conservation

Edited by Kason Hurst

SYRAWOOD
PUBLISHING HOUSE
New York

Published by Syrawood Publishing House,
750 Third Avenue, 9th Floor,
New York, NY 10017, USA
www.syrawoodpublishinghouse.com

Wildlife Habitats: Ecology, Environment and Conservation
Edited by Kason Hurst

© 2018 Syrawood Publishing House

International Standard Book Number: 978-1-68286-538-5 (Hardback)

This book contains information obtained from authentic and highly regarded sources. Copyright for all individual chapters remain with the respective authors as indicated. All chapters are published with permission under the Creative Commons Attribution License or equivalent. A wide variety of references are listed. Permission and sources are indicated; for detailed attributions, please refer to the permissions page and list of contributors. Reasonable efforts have been made to publish reliable data and information, but the authors, editors and publisher cannot assume any responsibility for the validity of all materials or the consequences of their use.

Trademark Notice: Registered trademark of products or corporate names are used only for explanation and identification without intent to infringe.

Cataloging-in-Publication Data

Wildlife habitats : ecology, environment and conservation / edited by Kason Hurst.
    p. cm.
Includes bibliographical references and index.
ISBN 978-1-68286-538-5
1. Habitat (Ecology). 2. Wildlife habitat improvement. 3. Wildlife conservation. 4. Animal ecology.
5. Wildlife management. I. Hurst, Kason.
QL82 .W55 2018
591.7--dc23

TABLE OF CONTENTS

Preface..IX

Chapter 1   **Tropical Fishes Dominate Temperate Reef Fish Communities within**
             Western Japan... 1
             Yohei Nakamura, David A. Feary, Masaru Kanda, Kosaku Yamaoka

Chapter 2   **A Flexible Approach for Assessing Functional Landscape Connectivity, with**
             Application to Greater Sage-Grouse (*Centrocercus urophasianus*)........................ 9
             Seth M. Harju, Chad V. Olson, Matthew R. Dzialak, James P. Mudd,
             Jeff B. Winstead

Chapter 3   **Seasonal Dynamics of Atlantic Herring (*Clupea harengus* L.) Populations**
             Spawning in the Vicinity of Marginal Habitats...20
             Florian Eggers, Aril Slotte, Lísa Anne Libungan, Arne Johannessen,
             Cecilie Kvamme, Even Moland, Esben M. Olsen, Richard D. M. Nash

Chapter 4   **Spatio-Temporal Patterns of Major Bacterial Groups in Alpine Waters**..........33
             Remo Freimann, Helmut Bürgmann, Stuart E. G. Findlay, Christopher T. Robinson

Chapter 5   **Patterns and Variability of Projected Bioclimatic Habitat for *Pinus albicaulis***
             in the Greater Yellowstone Area... 41
             Tony Chang, Andrew J. Hansen, Nathan Piekielek

Chapter 6   **Bathymetric Variation in Recruitment and Relative Importance of Pre- and**
             Post-Settlement Processes in Coral Assemblages at Lyudao (Green Island),
             Taiwan...55
             Yoko Nozawa, Che-Hung Lin, Ai-Chi Chung

Chapter 7   **Catchment-Scale Conservation Units Identified for the Threatened Yarra Pygmy**
             Perch (*Nannoperca obscura*) in Highly Modified River Systems.......................62
             Chris J. Brauer, Peter J. Unmack, Michael P. Hammer, Mark Adams,
             Luciano B. Beheregaray

Chapter 8   **Projected Polar Bear Sea Ice Habitat in the Canadian Arctic Archipelago**.......75
             Stephen G. Hamilton, Laura Castro de la Guardia, Andrew E. Derocher,
             Vicki Sahanatien, Bruno Tremblay, David Huard

Chapter 9   **Conservation Investment for Rare Plants in Urban Environments**....................82
             Mark W. Schwartz, Lacy M. Smith, Zachary L. Steel

Chapter 10   **Defining Landscape Resistance Values in Least-Cost Connectivity Models for**
             the Invasive Grey Squirrel: A Comparison of Approaches Using Expert-Opinion
             and Habitat Suitability Modelling...91
             Claire D. Stevenson-Holt, Kevin Watts, Chloe C. Bellamy, Owen T. Nevin,
             Andrew D. Ramsey

Chapter 11   **Thermal Carrying Capacity for a Thermally-Sensitive Species at the Warmest Edge of its Range**.. 102
Daniel Ayllón, Graciela G. Nicola, Benigno Elvira, Irene Parra, Ana Almodóvar

Chapter 12   **Similar Processes but Different Environmental Filters for Soil Bacterial and Fungal Community Composition Turnover on a Broad Spatial Scale**........................... 113
Nicolas Chemidlin Prévost-Bouré, Samuel Dequiedt, Jean Thioulouse, Mélanie Leliévre, Nicolas P. A. Saby, Claudy Jolivet, Dominique Arrouays, Pierre Plassart, Philippe Lemanceau, Lionel Ranjard

Chapter 13   **Primates Living Outside Protected Habitats are More Stressed: The Case of Black Howler Monkeys in the Yucatán Peninsula**... 124
Ariadna Rangel-Negrín, Alejandro Coyohua-Fuentes, Roberto Chavira, Domingo Canales-Espinosa, Pedro Américo D. Dias

Chapter 14   **Predation Limits Spread of *Didemnum vexillum* into Natural Habitats from Refuges on Anthropogenic Structures**... 132
Barrie M. Forrest, Lauren M. Fletcher, Javier Atalah, Richard F. Piola, Grant A. Hopkins

Chapter 15   **Great Apes and Biodiversity Offset Projects in Africa: The Case for National Offset Strategies**... 144
Rebecca Kormos, Cyril F. Kormos, Tatyana Humle, Annette Lanjouw, Helga Rainer, Ray Victurine, Russell A. Mittermeier, Mamadou S. Diallo, Anthony B. Rylands, Elizabeth A. Williamson

Chapter 16   **Patch Size and Isolation Predict Plant Species Density in a Naturally Fragmented Forest**... 158
Miguel A. Munguía-Rosas, Salvador Montiel

Chapter 17   **Contrasting Regeneration Strategies in Climax and Long-Lived Pioneer Tree Species in a Subtropical Forest**.. 165
Haiyang Wang, Hui Feng, Yanru Zhang, Hong Chen

Chapter 18   **Habitat Loss, not Fragmentation, Drives Occurrence Patterns of Canada Lynx at the Southern Range Periphery**.. 172
Megan L. Hornseth, Aaron A. Walpole, Lyle R. Walton, Jeff Bowman, Justina C. Ray, Marie-Josée Fortin, Dennis L. Murray

Chapter 19   **Global Drivers and Tradeoffs of Three Urban Vegetation Ecosystem Services**............................ 183
Cynnamon Dobbs, Craig R. Nitschke, Dave Kendal

Chapter 20   **Fast Growing, Healthy and Resident Green Turtles (*Chelonia mydas*) at Two Neritic Sites in the Central and Northern Coast of Peru: Implications for Conservation**.. 192
Ximena Velez-Zuazo, Javier Quiñones, Aldo S. Pacheco, Luciana Klinge, Evelyn Paredes, Sixto Quispe, Shaleyla Kelez

Chapter 21   **Evidence for Frozen-Niche Variation in a Cosmopolitan Parthenogenetic Soil Mite Species (Acari, Oribatida)**.. 204
Helge von Saltzwedel, Mark Maraun, Stefan Scheu, Ina Schaefer

Chapter 22   **Habitat Capacity for Cougar Recolonization in the Upper Great Lakes Region**...........................214
Shawn T. O'Neil, Kasey C. Rahn, Joseph K. Bump

Permissions

List of Contributors

Index

PREFACE

Many endangered species rely on conservation for survival and growth. Habitat management seeks to conserve wildlife habitats for both plants and animals. Habitat fragmentation, reduction and extinction are some of the greatest threats that are faced by many species of flora and fauna. Methods involved in such conservation include spatial modeling tools, creating habitat corridors and maintaining sustainable access to resources. Biodiversity hotspots and sustainable land development are other methods that aim to sustain biodiversity. Different approaches, evaluations, methodologies and advanced studies on wildlife habitats have been included in this book. Scientists and students actively engaged in this field will find this book full of crucial and unexplored concepts.

Various studies have approached the subject by analyzing it with a single perspective, but the present book provides diverse methodologies and techniques to address this field. This book contains theories and applications needed for understanding the subject from different perspectives. The aim is to keep the readers informed about the progresses in the field; therefore, the contributions were carefully examined to compile novel researches by specialists from across the globe.

Indeed, the job of the editor is the most crucial and challenging in compiling all chapters into a single book. In the end, I would extend my sincere thanks to the chapter authors for their profound work. I am also thankful for the support provided by my family and colleagues during the compilation of this book.

Editor

Tropical Fishes Dominate Temperate Reef Fish Communities within Western Japan

Yohei Nakamura[1]*, David A. Feary[2], Masaru Kanda[1,3], Kosaku Yamaoka[1]

1 Graduate School of Kuroshio Science, Kochi University, Nankoku, Kochi, Japan, **2** School of the Environment, University of Technology, Sydney, New South Wales, Australia, **3** Kuroshio Zikkan Center, Otsuki, Kochi, Japan

Abstract

Climate change is resulting in rapid poleward shifts in the geographical distribution of tropical and subtropical fish species. We can expect that such range shifts are likely to be limited by species-specific resource requirements, with temperate rocky reefs potentially lacking a range of settlement substrates or specific dietary components important in structuring the settlement and success of tropical and subtropical fish species. We examined the importance of resource use in structuring the distribution patterns of range shifting tropical and subtropical fishes, comparing this with resident temperate fish species within western Japan (Tosa Bay); the abundance, diversity, size class, functional structure and latitudinal range of reef fishes utilizing both coral reef and adjacent rocky reef habitat were quantified over a 2 year period (2008–2010). This region has undergone rapid poleward expansion of reef-building corals in response to increasing coastal water temperatures, and forms one of the global hotspots for rapid coastal changes. Despite the temperate latitude surveyed (33°N, 133°E) the fish assemblage was both numerically, and in terms of richness, dominated by tropical fishes. Such tropical faunal dominance was apparent within both coral, and rocky reef habitats. The size structure of the assemblage suggested that a relatively large number of tropical species are overwintering within both coral and rocky habitats, with a subset of these species being potentially reproductively active. The relatively high abundance and richness of tropical species with obligate associations with live coral resources (i.e., obligate corallivores) shows that this region holds the most well developed temperate-located tropical fish fauna globally. We argue that future tropicalisation of the fish fauna in western Japan, associated with increasing coral habitat development and reported increasing shifts in coastal water temperatures, may have considerable positive economic impacts to the local tourism industry and bring qualitative changes to both local and regional fisheries resources.

Editor: Athanassios C. Tsikliras, Aristotle University of Thessaloniki, Greece

Funding: This research was funded by the Ministry of Education, Culture, Sports, Science, and Technology of Japan. DA Feary was supported by the University of Technology, Sydney under the Chancellors Postdoctoral Research Fellowship scheme. The funders had no role in study design, data collection and analysis, decision to publish, or preparation of the manuscript.

Competing Interests: The authors have declared that no competing interests exist.

* E-mail: ynakamura@kochi-u.ac.jp

Introduction

The world's oceans have substantially warmed since 1955 [1], and there is increasing evidence that the geographic range of marine organisms has shifted in accordance with this warming [2–4]. One of the most well documented shifts in marine communities has been the poleward movement of habitat-forming species [3]. For example, within Japan over the last 30 years, tropical habitat-forming macroalgal *Sargassum* (Fucales) species have substantially increased in cover within temperate regions, while temperate *Sargassum* species have significantly decreased in cover [5]. In addition, in response to rising seawater temperatures, the range of four hermatypic (reef-building) coral species, including two common tropical *Acropora* species, have been expanding throughout temperate Japanese coasts since 1930, with the speed of these expansions reaching ~14 km/year [6].

There is increasing evidence that tropical fish populations may be expanding into temperate regions, but there is still little information on the factors which may facilitate or constrain such expansion [4,7,8]. Tropical range shifts are likely to be limited by species-specific resource requirements [9,10]. In particular, for

tropical fishes, temperate reefs may lack a range of settlement substrates, settlement cues or specific tropical dietary components [11]. As approximately 10% of coral reef fishes are classified as coral dependent throughout some part of their life stage [12], for these species a reliance on live coral resources is likely to constrain shifts into temperate reef habitats [13–15]. For example, previous studies suggests that highly specialised trophic groups (e.g., obligate coral-feeding butterflyfishes) will be unlikely to recruit and survive in habitats devoid of extensive cover of preferred coral species [12,16]. We can therefore expect then that the availability of specific coral reef resources may strongly constrain range extensions for coral reef fishes. Within regions devoid of high live coral cover, fish communities are expected to comprise species with little reliance on coral habitats and the resources they provide [4], independent of increased coastal temperatures [9].

There is increasing evidence for the expansion of reef-building corals into temperate regions globally, which has been closely associated with the proximity of regions to western boundary currents [6,17,18]. For example, Tosa Bay (western Japan, Fig. 1) has shown globally significant increases in coastal water temperatures and concomitant increases in coral development [19,20].

This bay is strongly influenced by the offshore Kuroshio Current [20]. Winter sea surface temperature (SST) offshore of the bay is higher than average for Japanese temperate zones, rising approximately 1.51°C from 1902 to 2012 (Fig.2), double the average rate of global ocean warming [21]. In particular, mean winter SST (January - March) in central Tosa Bay has increased by approximately 1.7°C (from 15.8°C to 17.5°C) over the past 30 years [22]. With increased SSTs hermatypic coral habitats have been rapidly expanding within areas of the bay since the 1990s [19]. At present, 136 coral species belonging to 50 genera (16 families) have been surveyed in the bay [19,20]. Such coral communities have shown extensive increases in species richness within the last 20 years (Fig. 1), with the tropical corals *Acropora muricata* and *A. latistella* having now developed into stable and permanent coral communities within the bay [19].

Although the expansion of tropical habitats (i.e., hermatypic corals and tropical algal beds) into temperate regions may be expected to facilitate the settlement and survival of tropical and subtropical fish species, for a range of resident temperate fish species such expansion may result in reduced cover of preferred temperate habitats, and concomitant reductions in the abundance of species closely associated with these habitats [22]. In addition, rapid changes in coastal water temperatures may also have a substantial mediating influence on species ability to compete for resources. For example, recent work (Beck et al. In review) suggests that competitive interactions between tropical vagrants and temperate residents are predominantly associated with temperature fluctuations; within warmer waters vagrants may numerically dominate fish populations, whereas with cooler waters such numerical dominance substantially reduces. Despite this, there is still little empirical evidence to determine the role of competition for resources in structuring patterns in tropical range shifts [4]. As space limitation is important in determining juvenile abundance in numerous site attached tropical fish species [23–26], we can expect that species specific differences in competition for temperate benthic resources both during and following settlement may have

substantial consequences for the abundance and diversity of tropical and subtropical range-shifting species within temperate ecosystems.

The aim of this work was to determine the importance of tropical benthic habitats (i.e., hermatypic coral) and temperate rocky reef habitat in determining the abundance, diversity, size structure and functional composition of tropical and subtropical range shifting species, and temperate resident fish populations. Therefore, this work examined the composition of fish communities, within bimonthly surveys over 2 years, associated with adjacent coral – and rocky reef habitats within the temperate coastline of western Japan. We hypothesized that the abundance and diversity of tropical and subtropical fishes would be constrained by the availability of live coral reefs, correlated with species coral resource specialization. In parallel, the structure and function of existing temperate fish communities were expected to be substantially associated with temperate habitat availability, with little utilization of tropical habitats.

Materials and Methods

Ethics Statement

This study involved no capture or handling of fishes. This research was approved by local fisheries cooperative and the Kochi Prefecture.

Study sites

This study was conducted in Tosa Bay, western Japan (33°N, 133°E) (Fig. 1). Two locations were selected in Tosa Bay: the Yokonami Peninsula (YK) and Kashiwajima (KA) (Fig. 1). Within each location 2 separate study sites were designated according to the dominant benthic community composition: coral habitats were designated as areas where live coral cover exceeded >60% (at the transect level), whereas rocky reef sites were composed of 100% rock habitat (Fig. 3).

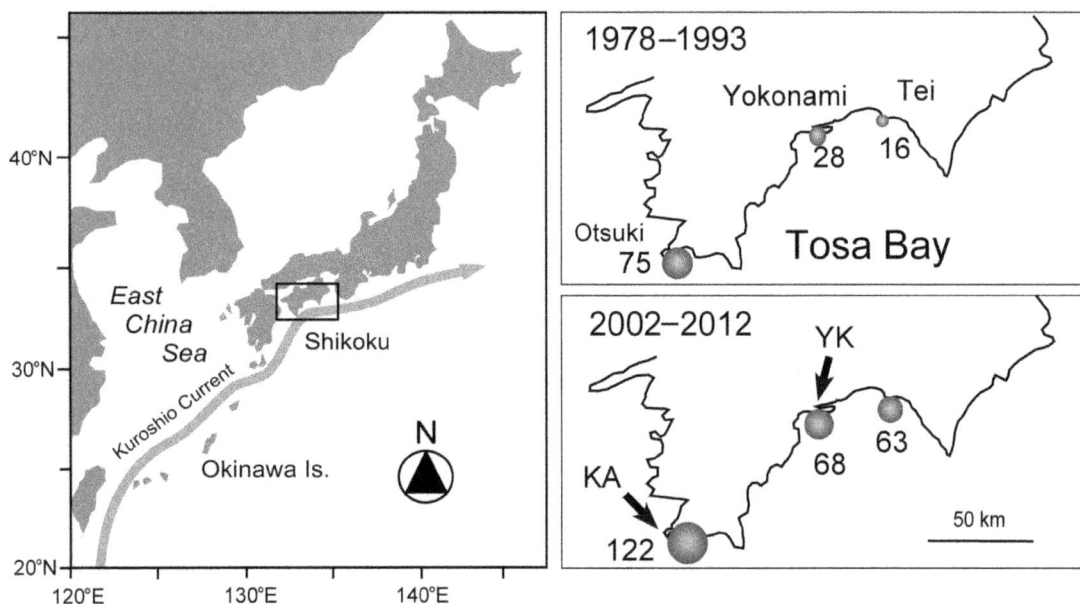

Figure 1. Map of the study site showing the location of Tosa Bay, western Japan. Figures on the right show the species richness of hermatypic corals observed at Otsuki, Yokonami, and Tei along the coast of Tosa Bay during the 1978–1993 and 2002–2012 survey periods, respectively. The numbers indicate coral species richness, and the gray circle indicates the degree of coral species richness. Arrows indicate the study sites (YK, Yokonami; KA, Kashiwajima). The coral species richness map was modified from that of Mezaki and Kubota [19].

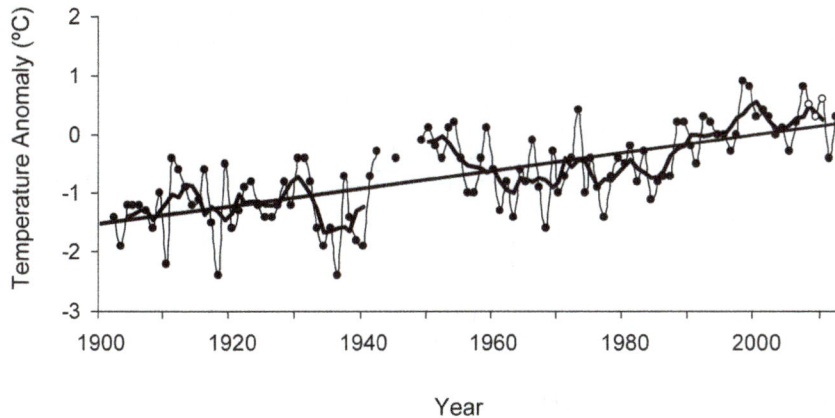

Figure 2. Time-series representation of winter surface seawater temperature annual anomalies at around offshore of Tosa Bay. Time series plot of annual averaged winter (Jan–Mar) surface seawater temperature anomalies (●) at around offshore of Tosa Bay for the period from 1902 to 2012. The 1981–2010 average is used as the normal. (○); study period (2008–2010). The bold line shows the five-year running mean, and the straight line indicates the long-term linear trend ($y = 0.0151x - 1.5287$). Data were provided by the Japan Meteorological Agency.

YK is a small peninsula located in central Tosa Bay, with both study sites located on the Pacific Ocean side of the peninsula. One site was dominated by a relatively shallow (2–8 m) and large (3 ha) coral reef, predominantly composed of branching and tabular *Acropora* (e.g., 50%–90% live coral cover dominated by *A. muricata*, *A. solitaryensis*, and *A. hyacinthus*, [27]). The second study site at YK was 100 m from the coral dominated site and encompassed 100% temperate rocky reef habitat (2–7 m depth, 4.5 ha). The second location (KA) is a small island (3.9 km circumference, with an area of 0.57 km^2) located in Otsuki, the southwestern end of Tosa Bay. This location holds 122 coral species [19], with the coral-dominated site surveyed (within 2–10 m depth) holding patchily distributed coral reefs, dominated by tabular *Acropora* and *Pavona decussata* (>50% live coral cover); this sites was ~100 m away from the rocky reef dominated site.

Temperature loggers (YK and Ohtoshima [15 km from KA, but having similar water temperature to KA]) deployed at 5 m recorded SST from Dec 1, 2008 to December 30, 2010, while salinity at each location was measured *in situ* using a portable salinometer in August and December each year.

Survey protocols

Fish assemblages were surveyed bimonthly at each site using underwater visual belt transect surveys from December 2008 to December 2010 (52 surveys in total). During each bimonthly survey, five 20×1 m haphazardly laid transects, separated by at least 10 m, were utilised and all fishes within transects counted. All individuals were taxonomically identified to species (using [28]),

and their total length estimated to 1 cm (roving fishes) and 0.5 cm (resident fishes). Fishes were categorized by latitudinal group (tropical, subtropical, and temperate) and primary feeding guild (i.e., benthivore, planktivore, corallivore, herbivore, omnivore, detritivore, piscivore, and parasite cleaner) based on FishBase [29]. All surveys were undertaken between 0900 and 1500 h.

Data analysis

Fish species richness and abundance (20 m^2) were compared between habitat types (coral reef, rocky reef; n = 5 in each habitat) and sampling months (total 13 months) within each location using mixed-model two-way analysis of variance (ANOVA). In this analysis, habitat type was considered to be a fixed factor and sampling month a random factor. Before analyses, all data were square root transformed (X+0.5). In addition, using separate mixed-model two-way analyses of variance, the abundance of fishes between latitudinal ranges and functional groups were examined between habitat types and sampling months within each location. Species richness and abundance were expressed per 20 m^2. All statistical analyses were conducted using SPSS (14.0J).

Results

Fish assemblage structure

The majority of species surveyed within both coral and rocky habitats were tropical in origin (comprising 83.3% of total species diversity within each site), with relatively low species richness of subtropical or temperate fish species observed (comprising 10.1% and 6.4%: Table 1). In parallel the numerical abundance of individuals throughout both coral and rocky habitats was dominated by tropical fishes (85.6% of the total number of individuals surveyed over the 2 years), while both the subtropical fauna (9.7%) and temperate fauna (4.6%) held comparatively low abundances (Table S1)

The coral habitat within both locations held significantly higher species richness and abundance than the rocky habitat (YK Species richness: $F_{1,104} = 6.5$, $p < 0.05$, YK species abundance: $F_{1,104} = 19.5$, $p < 0.01$; KA species richness: $F_{1,104} = 267.4$, $p < 0.01$, species abundance: $F_{1,104} = 154.5$, $p < 0.01$; Fig. 4). The high diversity and abundance of fishes within the coral habitat was primarily tropical in origin, with significantly higher tropical species richness and abundance in the coral than rocky habitat at

Figure 3. Underwater view of Yokonami, central Tosa Bay. (a) rocky and (b) coral habitats.

Table 1. Species richness of each range group in rocky and coral habitats at Yokonami and Kashiwajima during the study period.

Range group	Yokonami			Kashiwajima			
	Rock	Coral	Total	Rock	Coral	Total	Total
Tropical fish	40	61	80	107	174	199	221
Subtropical fish	15	11	17	22	22	27	27
Temperate fish	5	4	5	11	14	16	17

both the locations (two-way ANOVA: YK, $F_{1,104} = 28.5$ and 31.1, $p < 0.001$ for species richness and abundance; KA, $F_{1,104} = 291.5$ and 108.5, $p < 0.001$ for species richness and abundance; Fig. 5). There were varying patterns in the richness and diversity of subtropical fauna between habitats at both locations. There was little difference in the richness or abundance of subtropical species between habitats at YK (two-way ANOVA: $F_{1,104} = 3.6$, $p = 0.08$ for species richness and $F_{1,104} = 0.9$, $p = 0.36$ for abundance), while within KA the coral habitat held significantly higher species richness and abundance of subtropical fishes than within the rocky habitat (two-way ANOVA: $F_{1,104} = 52.7$, $p < 0.001$ for species richness and $F_{1,104} = 21.4$, $p < 0.01$ for abundance). With YK there was significantly higher temperate faunal richness and abundance in the rocky habitat than the coral habitat (two-way ANOVA: $F_{1,104} = 25.3$, $p < 0.001$ for species richness and $F_{1,104} = 19.0$, $p < 0.01$ for abundance), whereas there were no significant differences in temperate species richness or abundance between habitats at KA (two-way ANOVA: $F_{1,104} = 0.7$, $p = 0.43$ for species richness and $F_{1,104} = 0.2$, $p = 0.91$ for abundance).

YK and KA species numbers within coral habitats was dominated by tropical butterflyfish (YK: 19.7%, KA: 10%) and tropical wrasses (YK: 11.8%, KA: 16.2%), while the species numbers associated with the rocky habitat were dominated by tropical damsels (15%) and subtropical wrasses (10%) at YK, and tropical wrasses (15%) at KA (Table S1). The numerical composition of fish assemblages at both locations within coral habitats were dominated by tropical damsels (YK: 30.1%, KA: 34.9%), while tropical butterflyfish (45.1%) in YK and tropical cardinalfish (41.1%) in KA were also numerically dominant in the coral habitats (Table S1). The fish communities within the rocky habitats at both locations were relatively dissimilar, with KA numerically dominated by tropical damsels (54.7%) and tropical cardinalfish (13%), while YK assemblages were numerically dominated by tropical damsels (37.9%), with lower abundances of subtropical wrasses (12.7%) and temperate wrasses (21.2%).

Seasonal patterns in fish assemblage structure

There was a substantial decline in assemblage richness and abundance approximately 1–2 months following seasonal reductions in SST, with significantly more abundant and rich assemblages during middle boreal summer to early boreal winter (Aug to Dec), than middle boreal winter to early boreal summer (Feb to Jun) (two-way ANOVA: YK Species richness: $F_{12,104} = 3.0$, $p < 0.05$, YK species abundance: $F_{12,104} = 2.9$, $p < 0.05$; KA species richness: $F_{12,104} = 6.1$, $p < 0.01$; KA species abundance: $F_{12,104} = 3.0$, $p < 0.05$; Fig. 4).

There were clear differences in the effect of SST changes on fish assemblage structure, dependent on species latitudinal distribution range. Tropical species showed a clear increase and then decrease in both abundance and richness closely matched with seasonal changes in SST (two-way ANOVA: YK Species richness: $F_{12,104} = 4.1$, $p < 0.05$, YK species abundance: $F_{12,104} = 1.7$, $p = 0.18$; KA species richness: $F_{12,104} = 6.3$, $p < 0.01$, KA species abundance: $F_{12,104} = 2.5$, $p = 0.06$), (Fig. 5). In addition, within this fauna, despite a substantial decrease in abundance after December, individuals of 40 species were surveyed during the middle of winter (February) and following early spring (April) (Table S2). For several species there was also evidence to suggest that overwintering of breeding sized individuals within coral reef habitats is occurring (i.e., *Chaetodon auripes*: ≥ 12 cmTL, *Dascyllus reticulatus*: ≥ 6 cmTL, *Pomacentrus coelestis*: ≥ 6 cmTL, *P. nagasakiensis*: ≥ 8 cmTL ([30], Y Nakamura unpublished database) (Fig. S1). In comparison to the tropical fauna, there was no significant seasonal trend in both subtropical and temperate fishes richness or abundance (two-way ANOVA: $F_{12,104} = 0.5-1.7$, $p > 0.05$ for species richness and abundance, excluding species richness of temperate fishes at KA [2-way ANOVA: $F_{12,104} = 3.1$, $p < 0.05$]).

Functional composition of fish assemblages

The functional composition of both the subtropical and temperate fauna was a subset of the functional diversity found within the tropical fauna (Fig. 6). However, all three latitudinal groupings were dominated (both in mean richness and abundance) by benthivores, while planktivores were relatively abundant and diverse within the tropical fauna. Between habitats (coral and rocky reef) there were significant differences in functional groups between latitudinal groupings. Within the tropical fauna, the species richness and abundance of planktivores, corallivores, and parasite cleaners were significantly higher in the coral than rocky habitats at both the locations (two-way ANOVA: YK Species richness: $F_{1,104} = 10.6$, 58.1 and 15.6, $p < 0.01$, YK species abundance: $F_{1,104} = 5.6$, 29.9 and 17.7, $p < 0.05$; KA species richness: $F_{1,104} = 73.3$, 45.7 and 30.5, $p < 0.01$, KA species abundance: $F_{1,104} = 40.8$, 29.9 and 25.0, $p < 0.01$), while significantly higher abundances of benthivores, herbivores, omnivores, and piscivores were surveyed in the coral than rocky habitat in KA (two-way ANOVA: Species richness: $F_{1,104} = 81.9$, 91.0, 49.0 and 15.2, $p < 0.01$, species abundance: $F_{1,104} = 71.7$, 21.8, 14.1 and 23.3, $p < 0.01$; Fig. 6). Within the subtropical fauna both benthivores and planktivores showed significantly higher abundance and richness in the coral than rocky habitat at KA (two-way ANOVA: Species richness: $F_{1,104} = 98.3$ and 14.0, $p < 0.01$, species abundance: $F_{1,104} = 30.4$ and 9.2, $p < 0.05$), while a significantly higher abundance and richness of herbivores were found in the rocky habitat at YK (two-way ANOVA: Species richness: $F_{1,104} = 8.0$, $p < 0.05$, species abundance: $F_{1,104} = 6.2$, $p < 0.05$; Fig. 6). The temperate fauna showed significantly higher richness and abundance of benthivores and omnivores in the rocky, than coral habitat at YK (two-way ANOVA: Species richness: $F_{1,104} = 20.1$ and 16.8, $p < 0.01$, species abundance: $F_{1,104} = 14.4$ and 11.7, $p < 0.01$; Fig. 6).

Environmental variables

There were distinct seasonal changes in average SST throughout both years of surveys, with the highest average SST across both locations within August ($27.7°C \pm 1.0$), and the lowest average SST in February ($17°C \pm 1.2$). However, both locations recorded SST lows of $15.3°C$ and $15.1°C$ in February, for KA and YK, respectively, while maximum SSTs of approximately $30°C$ (encompassing late August - early September) were recorded at

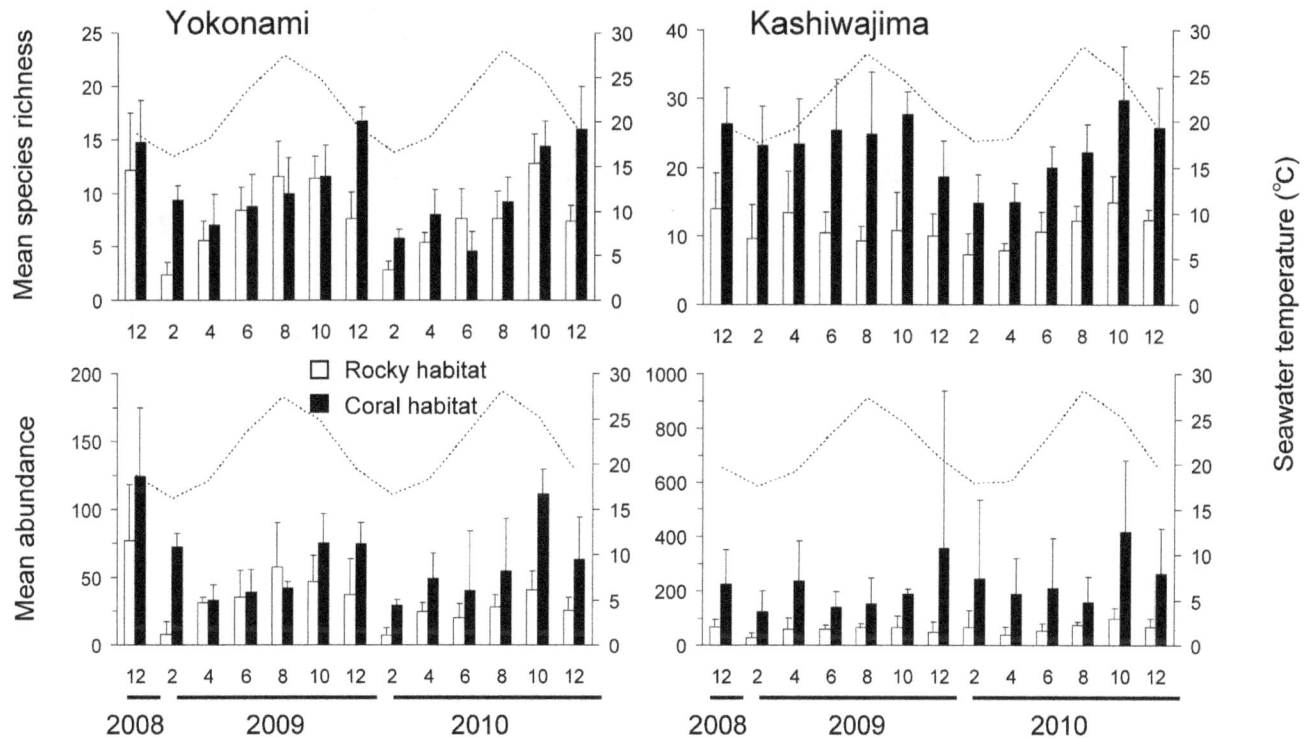

Figure 4. Seasonal patterns in species richness and abundance of fishes in the rocky and coral habitats. Bars indicate mean species richness and abundance [+standard deviation (SD)] of fishes per transect (n = 5, 1 ×20 m) in the rocky habitat and coral habitat. Months are labeled by numbers. The dotted line indicates the monthly average seawater temperature (depth, 5 m) around each site. Please note differences in y-axis scale between graphs.

both locations (Fig. 4). Salinity was similar at each location and season (approximately 34–35).

Discussion

Despite both surveyed locations being temperate in climate (33°N, 133°E) the fish fauna was both numerically, and in terms of richness, dominated by fishes with a tropical origin. This is the first study worldwide to show that a temperate reef habitat is numerically dominated by a tropical fauna [4]. Within the Pacific coastline of Japan tropical fish juveniles are routinely observed throughout summer and autumn, most of these larvae presumably being transported from southern subtropical regions (e.g., Ryukyu Islands) by the northern flowing Kuroshio Current [31,32]; historical records show that such vagrant incursions have been observed in western Japan from before the mid-20th century [33,34], with several cold tolerant tropical species (e.g. *Siganus fuscescens*, *Calotomus japonicus* and *Kyphosus* sp.) already resident within western Japan centuries ago [35]. For example, within Kamohara's catalog [fish list based on a collection of fishes made in Kochi Prefecture (Tosa Bay) from the late 1920's to the early 1960's] [34], 153 out of 244 tropical and subtropical species identified in the present study (63%) have been reported from Tosa Bay (Table S1). However, the mean winter SST of Tosa Bay before the 1940's was below 16°C (Fig. 2), which crosses the minimum average winter temperature threshold for most tropical reef fish survival (16°C–18°C) [7,36]. In comparison, the mean winter SST of Tosa Bay has substantially increased over the last 30 years, and has remained relatively high (~17°C) since the late 1980's (Fig. 2) [21,22]. Such high winter SSTs have been associated with a well-acknowledged and permanent development

of tropical benthic species (i.e., hermatypic coral and *Sargassum* spp.) within western Japan [5,6,19]. Therefore, although data is lacking on the temporal development of tropical and subtropical reef fish populations within this region, we assume that an increase in the rate of larval incursion into Tosa Bay and the development of permanent fish populations have occurred in the same time frame as apparent for tropical benthic communities.

Range shifts in tropical and subtropical fishes into temperate regions are expected to be limited by species specific resource requirements [4,9,10]. In particular, we would expect that temperate reefs may lack the range of resources important in the development of non-native assemblages, including specific settlement substrates, settlement cues, or dietary components found on tropical and subtropical reefs [11]. However, the present work has shown that for some tropical and subtropical reef fish species the availability of live coral resources may not necessarily determine their distribution patterns between coral and rocky-dominated habitats. Such generalist habitat use is expected to be predominantly exhibited by species that can effectively utilise one of more resource (e.g., habitat, trophic) available within both the habitat types. For example, the vast majority of tropical species found in both the coral and rocky habitats had no strong association with coral trophic resources, with these fish's predominantly comprising benthivores (benthic invertebrate feeders) and herbivores (which included browsers and scrapers; Table S1); trophic resources appropriate for both functional groups were available throughout coral and rocky reef habitats. In comparison, there was a significantly higher abundance and diversity of obligate corallivores and obligate coral settlers/dwellers in the coral than rocky habitats at both locations; some of these species numerically dominated the tropical assemblage within sites (e.g. *Chaetodon*

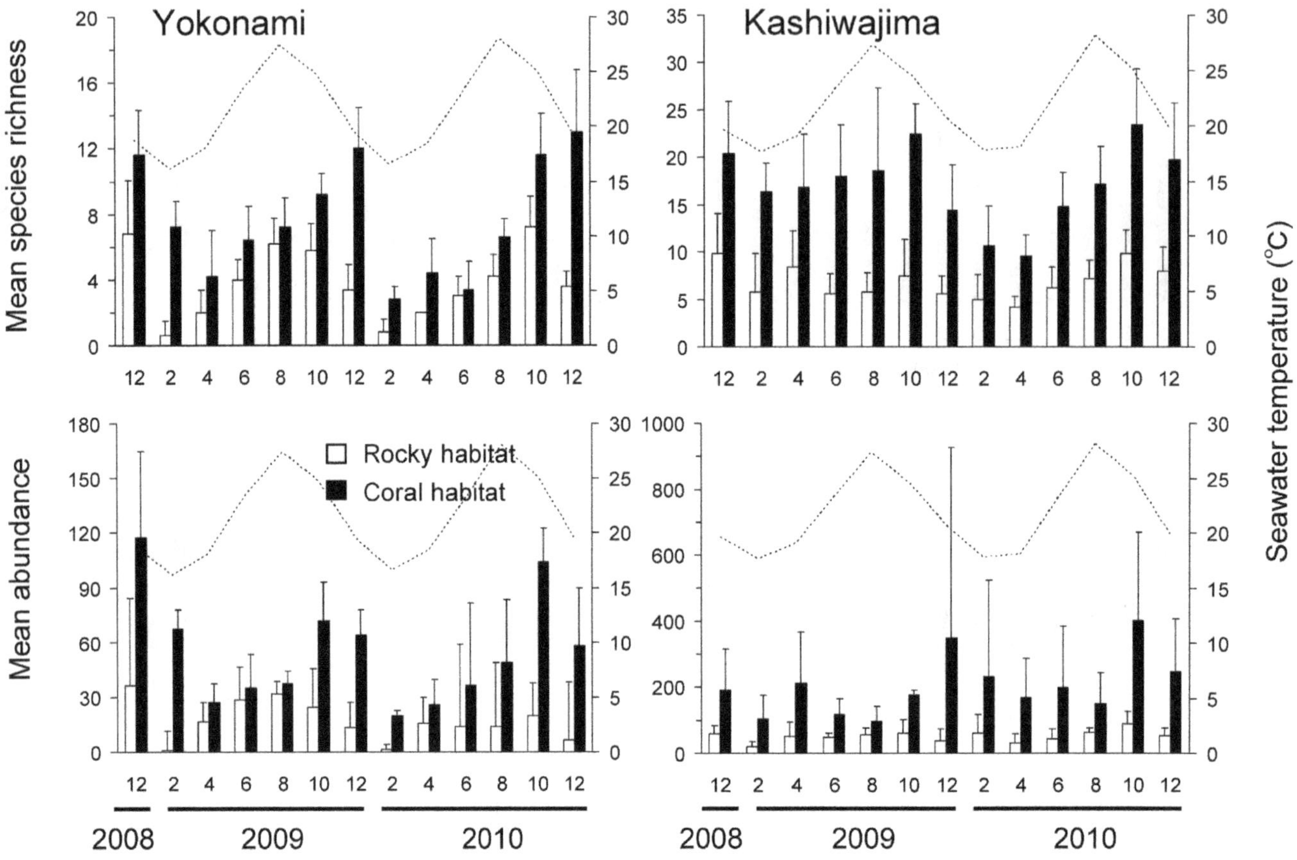

Figure 5. Seasonal patterns in species richness and abundance of tropical fishes in the rocky and coral habitats. Bars indicate mean species richness and abundance [+standard deviation (SD)] of fishes per transect (n = 5, 1×20 m) in the rocky habitat and coral habitat. Months are labeled by numbers. The dotted line indicates the monthly average seawater temperature (depth, 5 m) around each site. Please note differences in y-axis scale between graphs.

speculum and *C. lunulatus*). This pattern in assemblage structure suggests that the availability of coral resources may likely constrain the latitudinal shift of a range of tropical species that are obligately associated with coral reef resources [4,37]. In fact, recent evidence has suggested that tropical species that are highly associated with live coral may show little success within temperate environments [38]. For example, the overwhelming majority of butterflyfishes surveyed over 12 years within south eastern Australia, where there has been little hermatypic coral shift into temperate environments, were non-coral or facultative coral feeders, despite obligate coral feeding butterflyfishes being relatively abundant in the southern Great Barrier Reef [39]. In Tosa Bay, although corallivorous butterflyfishes had been collected until the early 1960's (Table S1), the previous work has reported them as rare or very rare within assemblages [34]. Therefore, we argue that the expansion of hermatypic coral habitats coupled with increasing coastal SSTs over the last 30 years has led to a substantial shift in the species richness and abundance of coral-associated tropical fishes in this temperate region.

There is evidence to suggest that a relatively large number of tropical fish species survived throughout the two winter seasons in Tosa Bay, with a small subset of these species overwintering that are potentially reproductively active. The establishment of permanent (i.e., breeding) populations at high latitudes is a key indicator of successful geographic range shifts in tropical fishes [4,40]. However, there has been little prior evidence of tropical fish species forming reproductively active populations in temperate

regions ([4,41] but see [42]). There is broad evidence to show that tropical fishes are able to effectively withstand water temperatures much lower than predominantly found in tropical latitudes (i.e., within temperate regions: [7,36,41]). Moreover, there is evidence to suggest that high latitude populations of tropical fishes may compensate for a shorter reproductive season and lower water temperature by producing gametes at higher rates than is the case at lower latitudes [43,44]. However, within western Japan, rapid increases in coastal water temperatures associated with a strengthening of the Kuroshio Current [45] may have reduced the oceanographic variables constraining reproduction in several tropical fishes, resulting in viable breeding populations developing within this region. Therefore, although winter is still a key bottleneck for survival and population establishment of tropical fishes within western Japan [27,46], we can predict that substantial changes in winter SST within regions associated with the increasing strength of the Kuroshio Current [45], amid fluctuations in natural warming cycles (associated with the Pacific Decadal Oscillation) may result in the successful breeding, and therefore establishment of permanent populations of tropical fish communities within western Japanese regions.

Changes in the structure and function of temperate fish communities within Japan, associated with rapid shifts in climate, are expected to have flow-on economic impacts to the local tourism industry and bring qualitative changes to local and regional fisheries resources. Scuba diving and glass-bottomed boat tourism attract large numbers of tourists and generate important

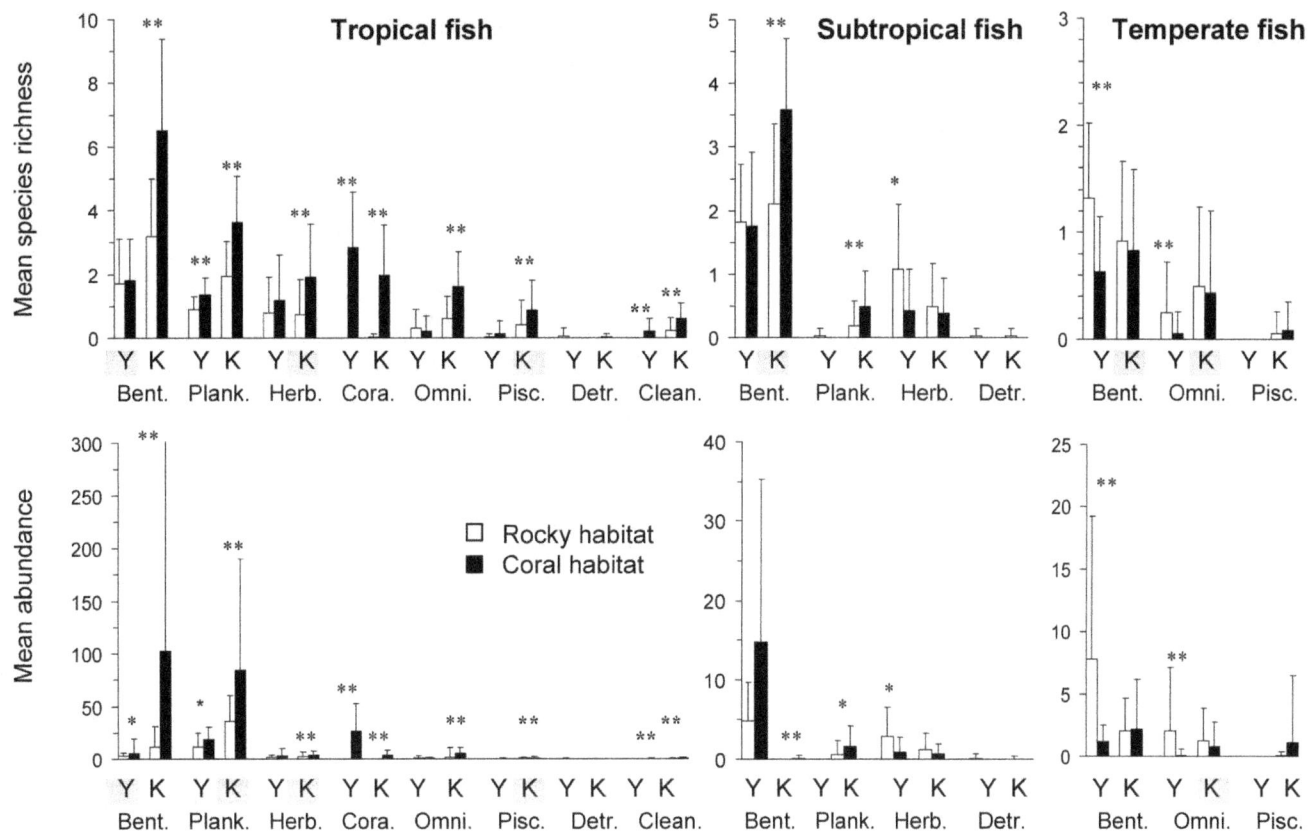

Figure 6. Functional composition of tropical, subtropical and temperate fish assemblages in the rocky and coral habitats. Bars indicate mean species richness and abundance [+standard deviation (SD)] of each feeding group of tropical, subtropical, and temperate fishes per transect in the rocky and coral habitats at Yokonami (Y) and Kashiwajima (K) during the study period (n = 65 in each habitat, 1 ×20 m). *p<0.05 and **p<0.01 between coral and rocky habitats by two-way analysis of variance (ANOVA) in abundance at each site. Shaded site symbols indicate significant seasonal differences at p<0.05 by two-way ANOVA in abundance. Bent. (Benthivore), Plank. (Planktivore), Herb. (Herbivore), Cora. (Corallivore), Omni. (Omnivore), Pisc. (Piscivore), Detr. (Detritivore), Clean. (Parasite cleaner).

tourism revenue for this region [20]. In Kashiwajima, for example, the scuba diving industry began in 1992, and now brings up to 10,000–15,000 Japanese and international divers per year to this area [47]. This industry contributes substantially to the local economy, both directly (e.g., through scuba diving packages) and indirectly (e.g., accommodation, food services and other local tourism ventures). With predictions of increasing warming occurring in this region [48], we can expect the cover and diversity of tropical coral reefs (and associated fish communities) to become more dominant. Such future 'tropicalisation' of Tosa Bay, and western Japan is expected to then increase the economic value of dive-based tourism within the region, associated with warmer summer waters and more tropical and subtropical dominated reef communities. In addition to local changes in reef-based tourism, shifts in the structure of fish communities within temperate Japan are expected to have substantial impacts on the fisheries sector. As this work found a large range of families important in tropical fisheries within both locations (including grouper (F. Serranidae), parrotfish (F. Scaridae), snapper (F. Lutjanidae), and emperor fish (F. Lethrinidae)), we may predict with increasing coastal waters that local and regional fishing productivity may hold increasing proportions of such tropical fish resources [8,48,49].

Supporting Information

Figure S1 Seasonal size distribution of the most abundant 10 possible overwintering species. Total 10 transects in each month at Yokonami (YK) and Kashiwajima (KA). Months are labeled by numbers. The dotted line indicates the monthly average seawater temperature (depth, 5 m) around each location. For the most abundant 10 possible overwintering species, see Table S2.

Table S1 Total abundance of each fish spceis across all the transects in rocky and coral habitats at Yokonami and Kashiwajima during the study period. n = 65 in each habitat in each location. *Recorded in Kamohara (1964)[34].

Table S2 List of possible overwintering tropical species; i.e. species observed in both middle winter (February) and following early spring (April). Numbers indicate total abundance of each tropical spceis across all the transects in rocky and coral habitats at Yokonami and Kashiwajima during February and April in 2009 and 2010.

Acknowledgments

We are grateful to T. Kawasaki and R. Ito for assistance with fieldwork and Ikenoura Fishery Cooperatives for logistic support. SST logger data were provided by Kochi Prefectural Fisheries Experiment Station. This study is a contribution from the Yokonami Rinkai Experimental Station.

Author Contributions

Conceived and designed the experiments: YN. Performed the experiments: YN MK KY. Analyzed the data: YN DAF. Wrote the paper: YN DAF.

References

1. Belkin LM (2009) Rapid warming of large marine ecosystems. Prog Oceanogr 81: 207–213.
2. Harley CDG, Hughes AR, Hultgren KM, Miner BG, Sorte CJB, et al. (2006) The impacts of climate change in coastal marine systems. Ecol Lett 9: 228–241.
3. Hoegh-Guldberg O, Bruno JF (2010) The impact of climate change on the world's marine ecosystems. Science 328: 1523–1528.
4. Feary DA, Pratchett M, Emslie M, Fowler A, Figueira W, et al. (2013) Latitudinal shifts in coral reef fishes: why some species do and others don't shift. Fish Fish DOI: 10.1111/faf.12036
5. Tanaka K, Taino S, Haraguchi H, Prendergast G, Hiraoka M (2012) Warming off southwestern Japan linked to distributional shifts of subtidal canopy-forming seaweeds. Ecol Evol 2: 2854–2865.
6. Yamano H, Sugihara K, Nomura K (2011) Rapid poleward range expansion of tropical reef corals in response to rising sea surface temperatures. Geophys Res Lett 38: L04601
7. Figueira WF, Booth DJ (2010) Increasing ocean temperature allow tropical fishes to survive overwinter in temperate waters. Glob Change Biol 16: 506–516.
8. Madin EMP, Ban NC, Doubleday ZA, Holmes TH, Pecl GT, et al. (2012) Socio-economic and management implications of range-shifting species in marine systems. Glob Ecol Biogeogr 22: 137–146.
9. Munday PL, Jones GP, Pratchett MS, Williams AJ (2008) Climate change and the future for coral reef fishes. Fish Fish 9: 261–285.
10. Cheung WWL, Lam VWY, Sarmiento JL, Kearney K, Watson R, et al. (2009) Projecting global marine biodiversity impacts under climate change scenarios. Fish Fish 10: 235–251.
11. Harriott VJ, Banks SA (2002) Latitudinal variation in coral communities in eastern Australia: a qualitative biophysical model of factors regulating coral reefs. Coral Reefs 21: 83–94.
12. Pratchett MS, Munday PL, Wilson SK, Graham NAJ, Cinner JE, et al. (2008) Effects of climate-induced coral bleaching on coral-reef fishes-ecological and economic consequences. Oceanogr Mar Biol 46: 251–296.
13. Munday PL, Jones GP, Caley MJ (1997) Habitat specialisation and the distribution and abundance of coral-dwelling gobies. Mar Ecol Prog Ser 152: 227–239.
14. Gardiner NM, Jones GP (2005) Habitat specialisation and overlap in a guild of coral reef cardinalfishes (Apogonidae). Mar Ecol Prog Ser 305: 163–175.
15. Feary DA, Almany G, Mccormick MI, Jones G (2007) Habitat choice, recruitment and the response of coral reef fishes to coral degradation. Oecologia 153: 727–737.
16. Pratchett MS (2005) Dietary overlap among coral-feeding butterflyfishes (Chaetodontidae) at Lizard Island, northern Great Barrier Reef. Mar Biol 148: 373–382.
17. Precht WF, Aronson RB (2004) Climate flickers and range shits of reef corals. Front Ecol Environ 2:307–314.
18. Baird AH, Sommer B, Madin JS (2012) Pole-ward range expansion of Acropora spp. along the east coast of Australia. Coral Reefs 31: 1063.
19. Mezaki T, Kubota S (2012) Changes of hermatypic coral community in coastal sea area of Kochi, high-latitude, Japan. Aquabiology 201: 332–337 (in Japanese with English abstract)
20. Ministry of the Environment and Japanese Coral Reef Society (2004) Coral Reefs of Japan. Tokyo, Ministry of the Environment.
21. Japan Metrological Agency (2013) Weather statistics information: global environment and climate. Available: http://www.data.kishou.go.jp/climate/index.html. Accessed 2013 Aug.
22. Terazono Y, Nakamura Y, Imoto Z, Hiraoka M (2012) Fish response to expanding tropical Sargassum beds on the temperate coasts of Japan. Mar Ecol Prog Ser 464: 209–220.
23. Hixon MA, Beets JP (1989) Shelter characteristics and Caribbean fish assemblages: experiments with artificial reefs. Bull Mar Sci 44: 666–680.
24. Munday PL, Jones GP, Caley MJ (2001) Interspecific competition and coexistence in a guild of coral-dwelling fishes. Ecology 82: 2177–2189.
25. Holbrook SJ, Schmitt RJ (2002) Competition for shelter space causes density-dependent predation mortality in damselfishes. Ecology 83: 2855–2868.
26. Bonin MC, Srinivasan M, Almany GR, Jones GP (2009) Interactive effects of interspecific competition and microhabitat on early post-settlement survival in a coral reef fish. Coral Reefs 28: 265–274.
27. Hirata T, Oguri S, Hirata S, Fukami H, Nakamura Y, et al. (2011) Seasonal changes in fish assemblages in an area of hermatypic corals in Yokonami, Tosa Bay, Japan. Japan J Ichthyol 58: 49–64.
28. Nakabo T (Ed) (2002) Fishes of Japan with pictorial keys to the species, English edn. Tokyo, Tokai University Press.
29. Froese R, Pauly D (Eds) (2012) FishBase. World Wide Web electronic publication. Available: http://www.fishbase.org. Accessed 2012 April.
30. Kobayashi K, Suzuki K, Hioki S (2007) Early gonadal formation in Chaetodon auripes and hermaphroditism in sixteen Japanese butterfly fishes (Chaetodontidae). Japan J Ichthyol 54: 21–40.
31. Briggs JC (1974) Marine zoogeography. McGraw-Hill Book Company, New York.
32. Soeparno, Nakamura Y, Shibuno T, Yamaoka K (2012) Relationship between pelagic larval duration and abundance of tropical fishes on temperate coasts of Japan. J Fish Biol 80: 346–357.
33. Kamohara T (1950) Description of the fishes from the provinces of Tosa and Kishu, Japan. Kochi bunkyo kyoukai, Kochi.
34. Kamohara T (1964) Revised catalogue of fishes of Kochi Prefecture, Japan. Rep Usa Mar Biol Sta 11:1–99.
35. Niwa S (Ed)(1735–1738) Flora, fauna, and crops of the Japan Islands in Yedo Era (Shokoku-sanbutsuchou).
36. Emme J, Bennett WA (2008) Low temperature as a limiting factor for introduction and distribution of Indo-Pacific damselfishes in the eastern United States. J Therm Biol 33: 62–66.
37. Soeparno, Nakamura Y, Yamaoka K (2013) Habitat choice and recruitment of tropical fishes on temperate coasts of Japan. Environ Biol Fish 96: 1101–1109.
38. Booth DJ, Figueira WF, Gregson MA, Brown GL, Beretta G (2007) Occurrence of tropical fishes in temperate southeastern Australia: Role of the East Australian Current. Estuar Coast Shelf Sci 72: 102–114.
39. Emslie MJ, Pratchett MS, Cheal AJ, Osborne K (2010) Great Barrier Reef butterflyfish community structure: the role of shelf position and benthic community type. Coral Reefs 29: 705–715.
40. Booth DJ, Bond N, Macreadie P (2011) Detecting range shifts among Australian fishes in response to climate change. Mar Freshw Res 62: 1027–1042.
41. Figueira WF, Biro P, Booth D, Valenzuela V (2009) Performance of tropical fish recruiting to temperate habitats: role of ambient temperature and implications of climate change. Mar Ecol Prog Ser 384: 231–239.
42. Ochi H (1989) Mating behaviour and sex change of the anemonefish, Amphiprion clarkii, in the temperate waters of southern Japan. Environ Biol Fish 26: 257–275.
43. Conover DO (1992) Seasonality and the scheduling of life history at different latitudes. J Fish Biol 41 (suppl.B): 161–178.
44. Kokita T (2004) Latitudinal compensation in female reproductive rate of a geographically widespread reef fish. Environ Biol Fish 71: 213–224.
45. Wu L, Cai W, Zang L, Nakamura H, Timmermann A, et al. (2012) Enhanced warming over the global subtropical western boundary currents. Nature Clim Change 2: 161–166.
46. Nakazono A (2002) Fate of tropical reef fish juveniles that settle to a temperate habitat. Fish Sci 68 (suppl. 1): 127–130.
47. Shinbo T (2008) Rethinking the sea as commons-from a case of the "Coral Sea" of Kashiwajima Island, Kochi, Japan. Kuroshio Sci 2: 77–83.
48. Kuwahara H, Akeda S, Kobayashi S, Takeshita A, Yamashita Y, et al. (2006) Predicted changes on the distribution areas of marine organisms around Japan caused by the global warming. Global Environ Res 10: 189–199.
49. Cheung WWL, Lam VWY, Sarmiento JL, Kearney K, Watson R, et al. (2010) Large-scale redistribution of maximum fisheries catch potential in the global ocean under climate change. Glob Change Biol 16: 24–35.

A Flexible Approach for Assessing Functional Landscape Connectivity, with Application to Greater Sage-Grouse (*Centrocercus urophasianus*)

Seth M. Harju*, Chad V. Olson, Matthew R. Dzialak, James P. Mudd, Jeff B. Winstead

Hayden-Wing Associates LLC, Natural Resource Consultants, Laramie, Wyoming, United States of America

Abstract

Connectivity of animal populations is an increasingly prominent concern in fragmented landscapes, yet existing methodological and conceptual approaches implicitly assume the presence of, or need for, discrete corridors. We tested this assumption by developing a flexible conceptual approach that does not assume, but allows for, the presence of discrete movement corridors. We quantified functional connectivity habitat for greater sage-grouse (*Centrocercus urophasianus*) across a large landscape in central western North America. We assigned sample locations to a movement state (encamped, traveling and relocating), and used Global Positioning System (GPS) location data and conditional logistic regression to estimate state-specific resource selection functions. Patterns of resource selection during different movement states reflected selection for sagebrush and general avoidance of rough topography and anthropogenic features. Distinct connectivity corridors were not common in the 5,625 km^2 study area. Rather, broad areas functioned as generally high or low quality connectivity habitat. A comprehensive map predicting the quality of connectivity habitat across the study area validated well based on a set of GPS locations from independent greater sage-grouse. The functional relationship between greater sage-grouse and the landscape did not always conform to the idea of a discrete corridor. A more flexible consideration of landscape connectivity may improve the efficacy of management actions by aligning those actions with the spatial patterns by which animals interact with the landscape.

Editor: Brock Fenton, University of Western Ontario, Canada

Funding: This study was funded by ConocoPhillips, EnCana Corporation and Noble Energy. No additional external funding was received for this study. The funders had no role in study design, data collection and analysis, decision to publish, or preparation of the manuscript.

Competing Interests: The authors have the following interests. This study was funded by ConocoPhillips, EnCana Corporation and Noble Energy. Hayden-Wing Associates LLC provided in-kind contributions to support travel costs associated with dissemination of this work and the time and materials associated with data collection and analysis including: Radiotelemetry equipment, GPS devices and statistical analysis software. All authors are employed by Hayden-Wing Associates LLC. There are no patents, products in development or marketed products to declare.

* E-mail: seth@haydenwing.com

Introduction

Maintaining connectivity of landscapes for animal populations is a primary challenge for conservation and land managers [1], [2]. This challenge arises from two general sources: methodological and implementation. Methodological challenges include assumptions made about the nature of the system under study, and often include decisions implicitly based on our expectation of how animals should interact with the landscape, rather than modeling the ways in which animals exhibit functional interactions with the landscape. For example, previous methodological limitations have included basing connectivity models on expert opinion, treating the landscape as a binary classification of habitat vs. non-habitat, modeling structural connectivity of individual landscape components, and assuming animal pseudo-presence along straightline movement paths connecting known animal locations [3], [4], [5], [6], [7], [8], [9]. In particular, recent work [10] has overcome many of these challenges by looking at functional animal-environment relationships in a complex multivariate landscape based on known animal locations. However, a persistent methodological challenge that is not addressed in recent connectivity literature is that investigators (often unwittingly) use analytical methods that make the implicit assumption that connectivity is best achieved via delineation of corridors. Addressing this methodological challenge may help overcome implementational challenges in applied connectivity management.

One of the primary challenges in implementation is that connectivity management is often addressed long after landscapes are heavily developed. For this reason, most connectivity work has focused on a corridor approach to maintaining connectivity. This approach has its own set of challenges (e.g., potentially ineffective design and cost-benefit tradeoffs [8], [11]), yet a more basic challenge is that animals often move through landscapes without regard for human-designated corridors [5]. Most methods for modeling connectivity implicitly assume the presence of discrete corridors (e.g., looking for the least-cost path between patches of habitat), essentially treating the landscape as patches of habitat in a sea of non-habitat [12]. This view is accurate in many conservation situations, such as landscapes that are already heavily developed (e.g., urban areas) or where a species' ecology dictates a strict distinction between habitat and non-habitat (e.g., butterflies inhabiting meadows surrounded by willow thickets and coniferous forest, [3]). However, delineating discrete corridors may be unnecessarily limiting or unrealistic in other situations. Habitat generalists or highly mobile species may not necessarily use distinct corridors. For example, recent work [13] has documented altered resource selection of migrating mule deer (*Odocoileus hemionus*) in developed areas, although mule deer maintained high fidelity to a multitude of historic routes spread across the landscape. Even strict habitat obligates may not need corridors in landscapes with low to moderate levels of human modification. A generalized approach to modeling connectivity may also inform management of invasive species (often habitat generalists), a situation often considered outside of the realm of conservation-based connectivity research but one that may strongly benefit from applied management to reduce connectivity [14]. Stakeholders may resist implementing measures to maintain connectivity in a take-it-or-leave-it corridor network. Conceptualizing and investigating landscape connectivity as a generalized or potentially diffuse process, rather than as a series of rigid structural features such as corridors, would establish a more flexible framework from which to identify and conserve functional connectivity habitat. Addressing the methodological challenge of imposing spatial constraints on connectivity habitat may help overcome the largest challenges associated with implementation of connectivity management.

We used simple analytical techniques to test whether a species' functional relationship with the landscape justified imposing a priori spatial constraints on connectivity habitat using a generalized connectivity modeling method that does not assume the existence of or need for discrete corridors. However, the approach we used will delineate discrete movement corridors if they are a natural pattern resulting from the functional relationship between animals and the landscape. We used the association between greater sage-grouse *Centrocercus urophasianus* (hereafter sage-grouse) and shrub-steppe habitat in western North America as a model system. Sage-grouse populations have experienced long term declines between 17 and 47% [15], currently occupy approximately 56% of their historic range [16], and have been designated warranted (but precluded due to higher priority species) for listing as threatened or endangered under the United States federal Endangered Species Act [17]. Connectivity is thought to be an important part of conservation and management in this system

[18], [19]. Sage-grouse are a good case study for a generalized approach to modeling connectivity because they occur across a variety of shrub-steppe habitats, are considered a potential umbrella species for sagebrush steppe conservation [20] and exhibit both non-migratory and migratory movement between seasonal use areas within populations [21]. Energy development in the Intermountain West of the United States is occurring over large spatial scales and is currently the predominant expanding human use of the study area. Energy development, like other types of widespread human activity, has been shown to affect resource selection and population dynamics in sage-grouse [22], [23], [24] and through landscape-level modification of habitat may function to reduce connectivity of sage-grouse habitat via human activity associated with such development. Specific objectives of this work were to 1) use sage-grouse occurrence locations to infer latent movement states, 2) develop resource selection functions (RSF) for both sexes within each movement state, 3) generate maps predicting probability of occurrence across the landscape during moderate to long distance movement states (e.g., a connectivity map) and 4) validate the connectivity map using occurrence locations from independent sage-grouse in moderate to long-distance movement states.

Materials and Methods

Study Area

The 5,625 km^2 study area included portions of the Wind River Basin in central Wyoming, USA (Fig. 1). Topography is variable with gently sloping flats, cut banks, dry washes, steep forested slopes and rocky canyons ranging in elevation from 1478–2776 m. In general, the southern half of the study area is a geographic basin and the northern half is characterized by increasingly steep slopes, valleys, and ridges. Minimum and maximum temperatures for each year during the study period were −29.3 and 36.7°C in 2008, −34.6 and 34.6°C in 2009, and −27.0 and 34.2°C in 2010; average total precipitation from three weather stations across the study site was 5.42 cm in 2008, 20 cm in 2009, and 17.9 cm in 2010 (C.V. Olson, unpublished data). Dominant plant species at lower elevation included Wyoming big sagebrush (*Artemisia tridentata wyomingensis*), basin big sagebrush (*A. t. tridentata*), black greasewood (*Sarcobatus vermiculatus*), winterfat (*Ceratoides lanata*) and

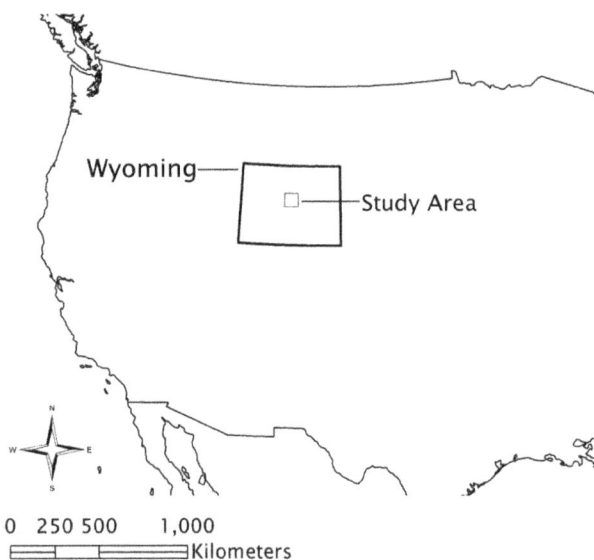

Figure 1. Map of study area boundary in central Wyoming, USA.

Figure 2. Female greater sage-grouse (*Centrocercus urophasianus*) in central Wyoming, USA, wearing rump-mount GPS unit. Photo credit T. Dorval.

shadscale (*Atriplex confertifolia*). At higher elevation, mountain big sagebrush (*A. t. vaseyana*), limber pine (*Pinus flexilis*), Douglas fir (*Pseudotsuga menziesii*) and rocky mountain juniper (*Juniperus scopulorum*) were present. The study area encompassed historic and ongoing development of energy resources. Oil and natural gas development was initiated in the 1920 s; gas development accelerated in the 1990 s. In 2010 there were 1,085 wells associated with oil and gas development in the study area.

Field Procedures and Location Data

From 2008–2010, we captured 42 male and 98 female sage-grouse by spotlighting [25] on and around leks dispersed throughout the study area, mostly during spring. Some captures occurred in summer and fall; in such cases capture effort was based on known location of other grouse to which GPS units were affixed. Permission to capture sage-grouse was granted by the Wyoming Game and Fish Department (Chapter 33 Permit #649), the relevant regulatory body concerned with protection of wildlife. The work was not approved by an Institutional Animal Care and Use Committee as all study design, animal capture and handling, and subsequent data collection and analysis was performed by employees of a private consulting firm. However, all animal capture and handling protocols were approved and conducted under a permit issued by Wyoming Game and Fish Department. We determined age and sex and fitted sage-grouse with 30-g ARGOS/GPS Solar PTTs (PTT–100, Microwave Telemetry Inc., Columbia, MD, USA). GPS units were attached using a rump-mount technique [26] (Fig. 2). GPS units had Ultra High Frequency (UHF) beacons for ground tracking and detection of mortality and had a 3-year operational life. Collars were programmed to record location information during 15 Feb–14 May every 3 h from 0700–2200, during 15 May–15 July every 1 hr from 0700–2100. During 16 July–31 Oct collars recorded location information every 3 hr from 1000–2200 and during 1 Nov–14 Feb every 6 h from 1000–2200. We did not include locations from female sage-grouse during the period when the individual was

incubating eggs or caring for broods because in these cases movement behavior was constrained by distance from the nest or by slow movement capabilities of chicks. We had detailed field-based and GPS data on nest initiation, nest failure, hatch dates and brood fates for each adult female to determine excluded dates (C.V. Olson, unpublished data). We did not delineate the analysis by season because we viewed resource selection during long-distance movement as important during all seasons. Additionally, practical land management decisions would address comprehensive connectivity habitat needs (rather than only managing for fall connectivity habitat at the exclusion of spring connectivity habitat). Because we systematically collected and subsampled the GPS data throughout the course of each year, any seasonal differences were averaged out.

Assigning Movement State

We focused on three movement states for this analysis: encamped, traveling and relocating. We used 10 am locations for this analysis to remove diel variation in resource selection, to better capture long-distance movements, to model resource selection in relation to disturbance during the day time when human activity was most prominent and because preliminary analysis showed that distance between locations (steplength) was correlated with time between locations (thus requiring a consistent time interval between locations). Each location was assigned to a movement state based on distance from the 10 am location 24 hrs previous and distance to the next 10 am location 24 hrs hence. We discarded the first and last location from each individual because we did not have a previous or successive location from which to assign a movement state. Sophisticated methods to delineate movement states (e.g., [27], [28], [29]) did not work with our data set (S.M. Harju, unpublished data), perhaps because of differences in species ecology, temporal scales of data collection or the lack of distinct processes underlying separate movement states in sage-grouse [30], [31]. To approximate the ecological relationships between sage-grouse and resources in different latent movement

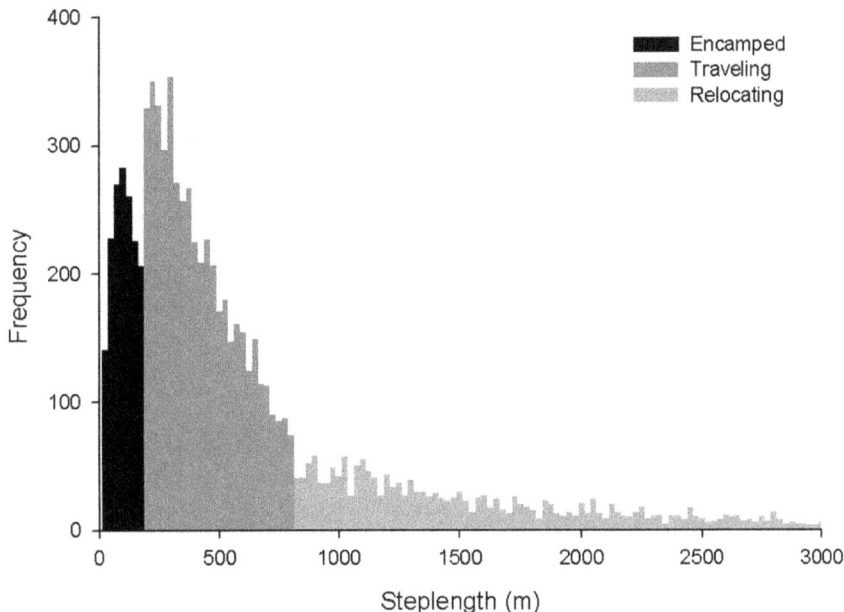

Figure 3. Histogram of 24-hr steplengths by greater sage-grouse in central Wyoming, USA, 2008–2010. Figure was right-truncated for display; longest 24-h steplength was 17,852 m. The movement states encamped, traveling and relocating were assigned to the shortest 25%, middle 50% and longest 25% of 24-hr steplengths, respectively.

Table 1. Sample sizes for data used in model building and validation of the predictive connectivity map (travel + relocate) in central Wyoming, USA, 2008–2010.

	Sex	Movement state	# locations	# birds	Median # locns/bird	Min # locns/bird	Max # locns/bird
Model-building	Female	Encamped	913	73	8	1	54
		Travel	2909	78	25	1	136
		Relocate	1303	75	12	1	79
	Male	Encamped	397	33	6	1	79
		Travel	1244	34	27	1	107
		Relocate	610	31	17	1	61
Validation	Female	Travel	587	13	36	2	125
		Relocate	259	12	19	1	52
	Male	Travel	254	6	29	2	114
		Relocate	111	5	30	1	47

states, we used a simple objective cutoff. We divided steplengths into three categories (Fig. 3) based on the distribution of distances between successive 24 hr locations (Males: mean = 732.44 m, median = 375.31 m, max = 17200.7 m; Females: mean = 738.49 m, median = 371.43 m, max = 19975.72 m). We assigned the shortest 25% of steplengths to the encamped movement state (<177.25 m Males; <168.81 m Females), the middle 50% of steplengths to the traveling movement state (>177.25 m and <800.38 m Males; >168.81 m and <798.87 m Females) and the longest 25% of steplengths to the relocating movement state (>800.38 m Males; >798.87 m Females).

Locations of sage-grouse are samples taken from unobserved movement states. To ensure confidence in the estimated movement state from which a location was sampled we retained only those locations with a consistent movement state immediately before and after that location. For example, location l_1 from a male sage-grouse was assigned to the movement state 'relocating' if the steplengths from l_0 to l_1 (24 hrs previous) and l_1 to l_2 (successive 24 hrs) were both >800.38 m. Although this resulted in discarding

some location data (when we were unsure as to the movement state from which a given location was sampled), it provided a basis from which to objectively establish movement states from location data.

Covariate Calculation

Using a Geographic Information System (GIS; ArcGIS® 9.3, ESRI, Redlands CA) and a 30 m grid cell size, we calculated covariates depicting landscape features that, based on field observation and previous research, influenced behavior of sage-grouse [21], [22], [32], [33]. We generated 7 covariates depicting predominant human modifications of the landscape (road density within 1 km and oil and natural gas well density within 1 km), landscape vegetation (mean percent coverage of sagebrush within an 800 m window, the standard deviation of the percent coverage of sagebrush across grid cells within an 800 m window, and the number of mesic grid cells within a 2 km window) and topographic features of the landscape (slope and terrain roughness [standard deviation of elevation] within 800 m). Raster images for oil or gas

Table 2. Odds ratios of selection for predictor variables during different movement states by greater sage-grouse in central Wyoming, USA, 2008–2010.

		Predictor variable						
Sex	Movement state	Mesic areas[a]	Slope[b]	Topographic roughness[c]	Road Density[d]	Well Density[e]	Sagebrush[f]	Patchiness of sagebrush[g]
Female	Encamped	1.000	**0.929**	**0.942**	1.360	1.116	1.002	1.178
	Traveling	1.000	**0.939**	**0.906**	1.063	**0.868**	**1.147**	**1.123**
	Relocating	**0.999**	0.947	**0.952**	1.059	**0.749**	**1.115**	**1.173**
Male	Encamped	1.001	1.006	0.942	1.670	1.140	1.048	1.369
	Traveling	0.999	**0.951**	**0.924**	**2.417**	**0.663**	**1.088**	**1.193**
	Relocating	1.000	**0.940**	0.983	**1.945**	**0.555**	**1.127**	0.988

[a]No. of mesic grid cells w/in 2.01 km window.
[b]Degres.
[c]Std. dev. of elevation (m) w/in 810 m window.
[d]Total length (km) w/in 1 km².
[e]No. w/in 1 km².
[f]Percent sagebrush w/in 810 m window.
[g]Std. dev. of percent sagebrush w/in 810 m window.
Bold values indicate estimates where 95% CI does not overlap 1.0. See Table S1 for detailed results.

Table 3. Standardized coefficient estimates of selection for topographic roughness and road density by traveling and relocating greater sage-grouse in central Wyoming, USA, 2008–2010.

Sex	Movement state	Topographic roughness[a]	Road density[a]
Female	Traveling	−0.958	0.035
	Relocating	−0.427	0.034
Male	Traveling	−0.653	0.657
	Relocating	−0.111	0.524

[a]X-Standardized coefficient calculated as: $X\text{-std}(\beta_i) = \beta_i * SD(x_i)$.

wells were developed using data provided by the Wyoming Oil and Gas Conservation Commission. Raster images of all other human modifications of the landscape were developed through heads-up digitizing of 2006 1-m resolution National Agriculture Imagery Program aerial imagery. We used Spatial Analyst in ArcGIS to calculate raster images and to extract values from raster data to location data for all covariates.

Modeling Resource Selection

We used discrete choice models [34] to evaluate resource selection at spatial and temporal scales aligned with the underlying selection process during movement. Discrete choice models are suitable because: 1) they allow availability to be defined uniquely for each point, thus accounting for spatial constraints on availability due to short sampling intervals between successive locations, 2) they allow comparison of true use versus non-use, as an animal cannot simultaneously be at a used location and the paired random locations [35] and 3) they can capture the scale at which individuals perceive the environment and select resources during movement. Individuals likely have made an a priori decision to move in a given direction and thus are choosing locations from nearby combinations of landscape features. This is most important to consider during the traveling and relocating movement states as described above. During the encamped state discrete choice models reflect constrained availability as a function of limited movement behavior.

We matched each used location with a set of 50 non-used (random) locations that were available spatially and temporally but were not chosen. We prevented random locations from occurring within 30 m of each other to minimize pseudoreplication in sampling availability as a function of covariate raster grid cell size (30 m). We used a relatively large number of random locations to estimate selection of relatively uncommon variables, specifically those reflecting human modification of the study area (sensu [36]). Perception of availability among sage-grouse likely varied depending on the underlying movement state, thus we altered the size of the area from which available locations were drawn depending on movement state. For example, when relocating, it is plausible that individuals chose movement paths from a relatively large area, whereas when encamped individuals had already chosen to move short distances and thus had smaller areas that were functionally available to be selected. We sampled random locations within a 250 m radius buffer around encamped locations, a 1 km radius buffer around traveling locations and a 2 km radius buffer around relocating locations.

We used conditional logistic regression [37], [38] to estimate discrete choice models separately for each sex and each movement state. To test for collinearity among predictor variables we assessed pairwise Pearson correlation coefficients. All correlation coefficients were <0.75. We did not perform any model selection or variable reduction procedures and instead relied on the results from our parsimonious model for inference. We used program R for all statistical analyses (R Development Core Team, v. 2.13.2, 2011).

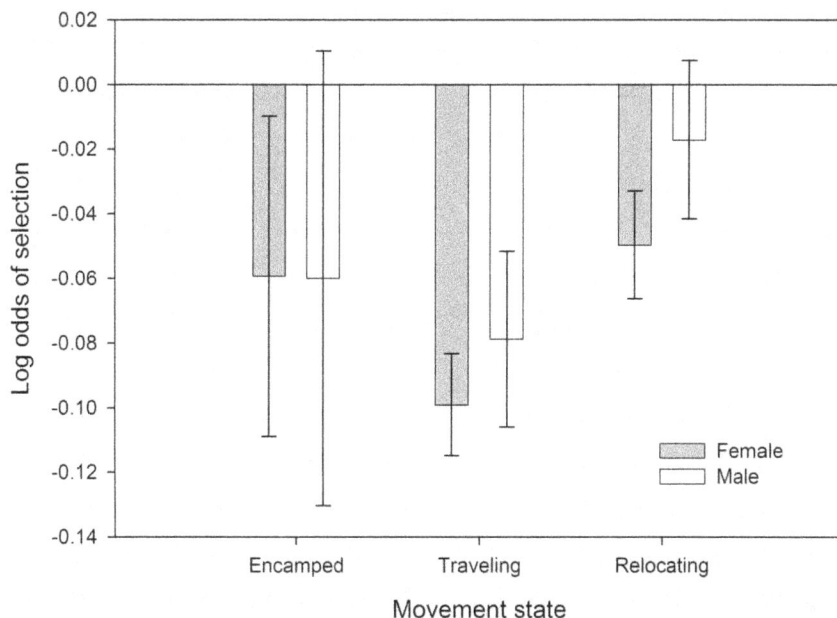

Figure 4. Resource selection by male greater sage-grouse in relation to topographic roughness during different movement states in Wyoming, USA. Topographic roughness was calculated as the standard deviation of elevation within an 800 m window centered on each used or random location. Error bars in column graph are 95% confidence intervals.

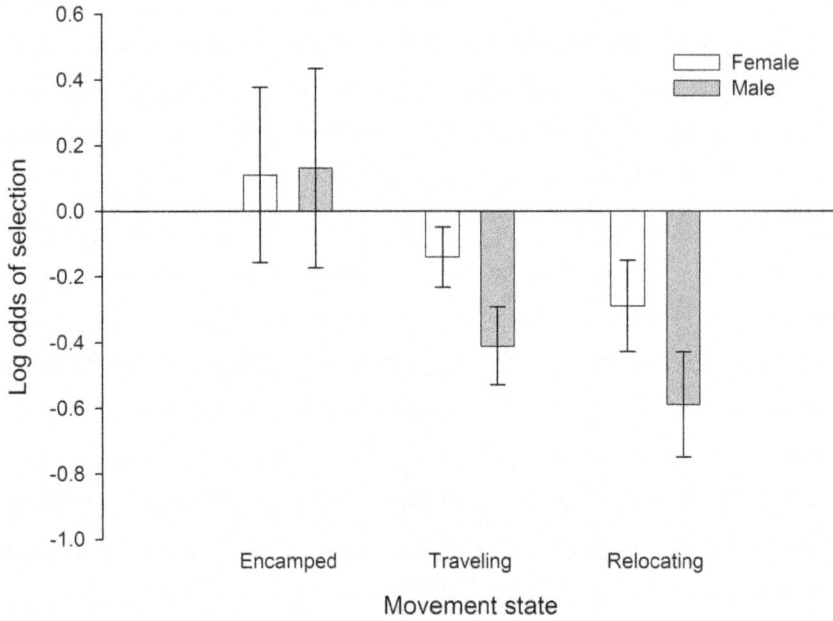

Figure 5. Selection for location in relation to natural gas well density (no. of wells within a 1 km window) by greater sage-grouse during different movement states. Error bars are 95% confidence intervals.

The strengths of this approach include: 1) it relies on known occurrence locations and an index of the movement path (because the true movement path is unknown), 2) it accommodates matrices of varying levels of habitat quality as well as binary habitat types, 3) is based on resource use decisions made by animals within the context of their movement state and perception of the landscape, and 4) allows, but does not assume, fundamentally different selection of resources during different movement states.

Predictive Maps and Validation

We generated predictive probability of use maps by male and female sage-grouse during traveling and relocating movement across the study area. We calculated relative probability of use for each raster grid cell by inputting observed predictor variable values into the final model equation for each sex and movement state:

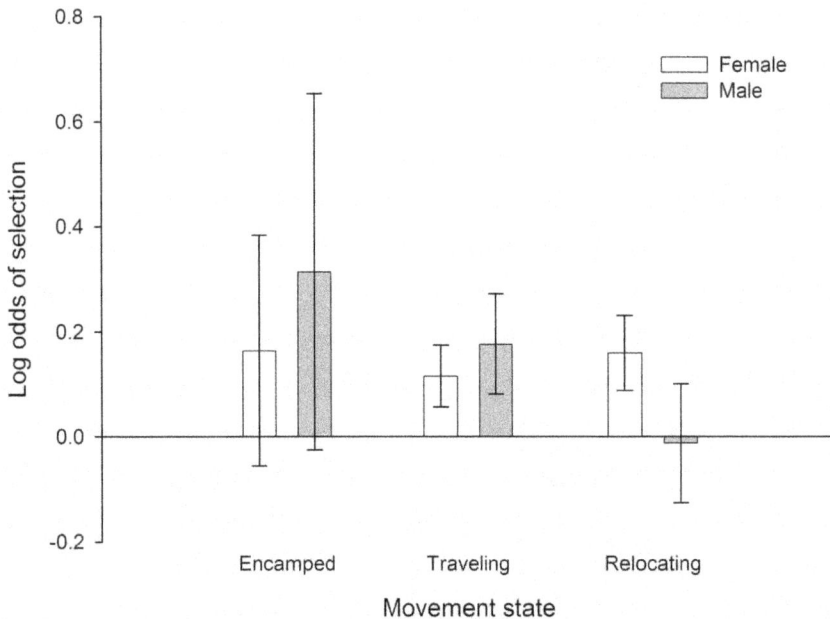

Figure 6. Selection for patchiness of sagebrush by greater sage-grouse in central Wyoming, USA. Patchiness of sagebrush calculated as the standard deviation of the percent coverage of sagebrush within each 30 m grid cell window, based on an 800 m window centered on each used or random location. Error bars are 95% confidence intervals.

A

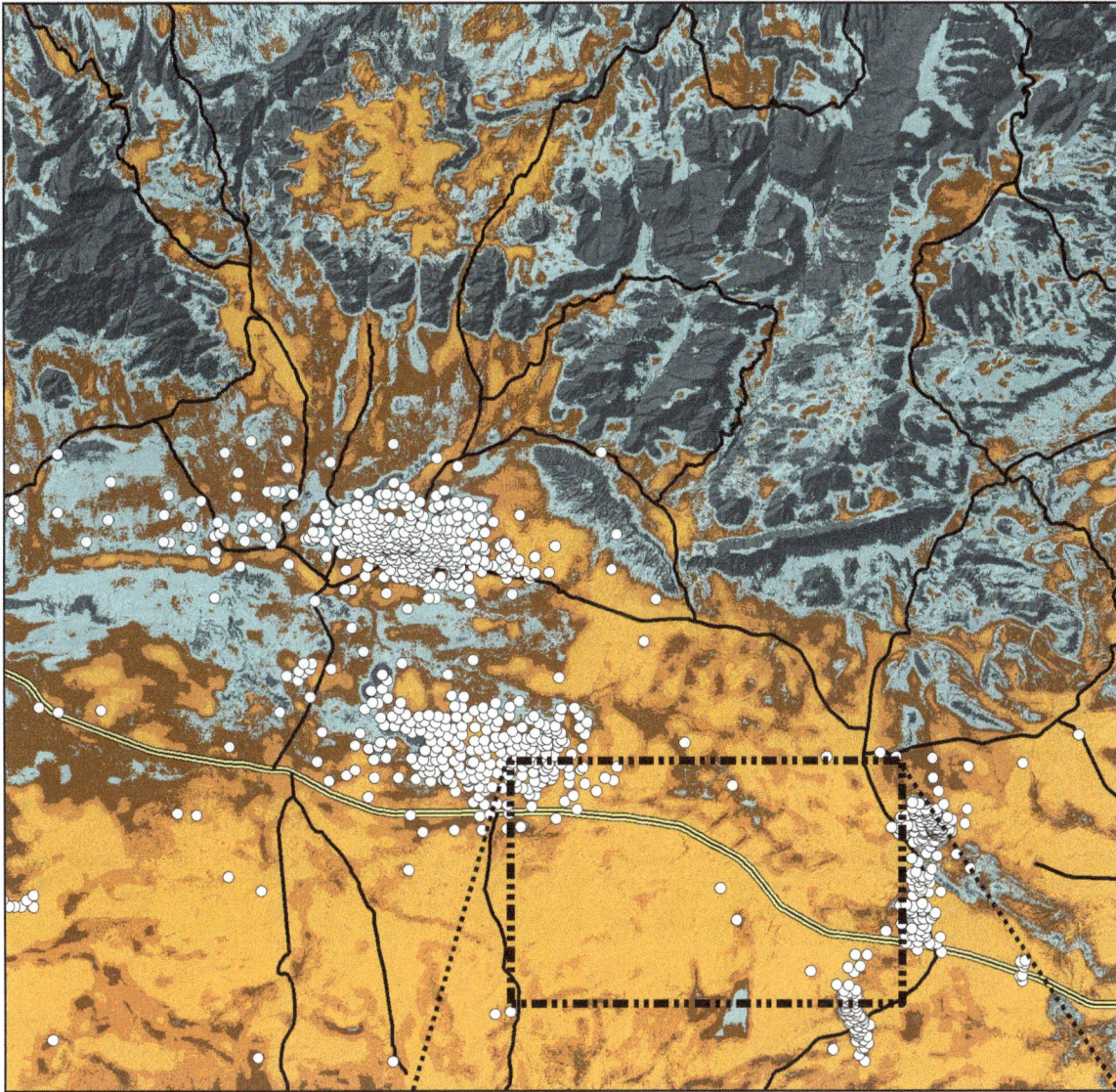

**Predicted quality
of connectivity habitat**

- Lowest
- Low
- Medium
- High
- Highest

**High predicted probability
of seasonal use**

- Nesting
- Winter

- ○ **Natural Gas Well**
- —— **Road**
- ══ **Major Highway**

0   10   20 Kilometers

B

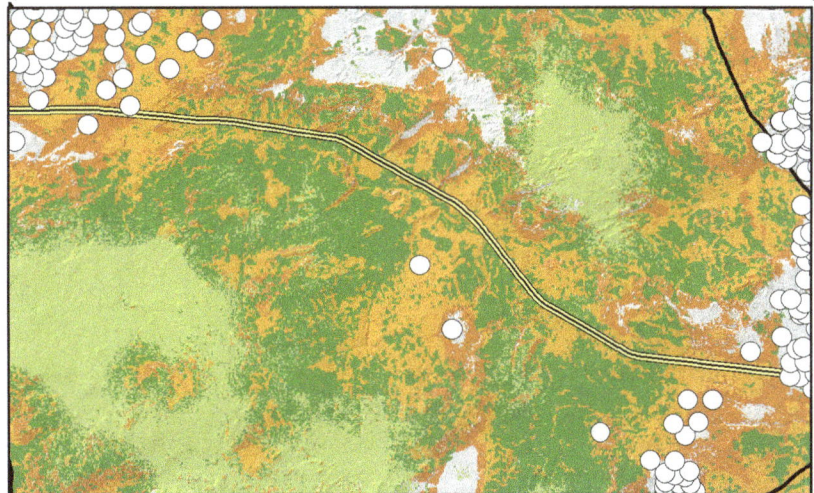

0   5   10 Kilometers

Figure 7. Map of connectivity habitat quality for greater sage-grouse in central Wyoming, USA. Panel 'a' is the entire study area. Panel 'b' illustrates a practical application of the map where critical seasonal use layers are overlaid on top of the two highest connectivity habitat quality layers. The nesting [33] and winter [34] layers are from companion analyses conducted on the same population of sage-grouse during the same time period.

$$Relative\ probability\ of\ selection_j = \exp(\beta_1 x_{1j} + ... + \beta_k x_{kj})$$

for each coefficient estimate (β) and observed value (x) for each predictor variable k at each grid cell j. The RSF value is unit-less, so to compare relative probability of use among the models we binned the RSF values within each model in 5 equal-sized bins (with values from 1 to 5) reflecting relative probability of occurrence in a grid cell ranging from low to high. We then summed the bin values for males and females while traveling and relocating (with values from 4 to 20). This map was then reclassified into 5 equal-sized bins to develop a general connectivity habitat map for all sage-grouse.

To validate this final predictive map we plotted the GPS locations from 20 sage-grouse (16% of all collared sage-grouse) that were withheld from the statistical analysis to evaluate how well the final connectivity habitat map predicted occurrence of independent sage-grouse during traveling and relocating movement states. We then performed Spearman rank correlation Chi-square tests to evaluate whether the difference in the number of observed versus expected locations increased monotonically with increasing probability of use bins.

Results

Total sample sizes across movement states for model-building females and males were 5,125 and 2,251 locations, respectively (Table 1). We used 846 and 365 locations from independent

female and male sage-grouse, respectively, for validation of the predictive connectivity map. The largest contribution of locations from a single sage-grouse to a model-building dataset was 20% (Male encamped). No other model-building dataset had a contribution from a single sage-grouse of more than 10% of the model-building locations (Table 1).

Across sexes and movement states, sage-grouse tended to avoid areas with steep slopes and high topographic roughness (Fig. 4, Table 2). They often occurred in areas of higher road density but lower natural gas well density than available, especially for males that were traveling or relocating (Fig. 5). Both males and females selected for locations with a higher proportion of sagebrush as well as areas with greater patchiness of sagebrush (Fig. 6). One reviewer questioned whether avoidance of topographic roughness was stronger than the apparent selection for higher road density. We calculated x-standardized coefficient estimates (i.e., x-std(β_i) = β_i * SD(x_i)) to compare the strength of selection/avoidance for these variables, given that they have different units of measurement. The x-standardized coefficient estimates indicated that selection for higher road density was as strong as or stronger than avoidance of topographic roughness for males, whereas female avoidance of high topographic roughness was an order of magnitude stronger than their mild selection for higher road density (Table 3). The pairwise correlation coefficient between road density and topographic roughness was −0.26 for all non-used locations (non-used locations are an objective assessment of underlying spatial correlation of landscape feature values). See Table S1 for detailed

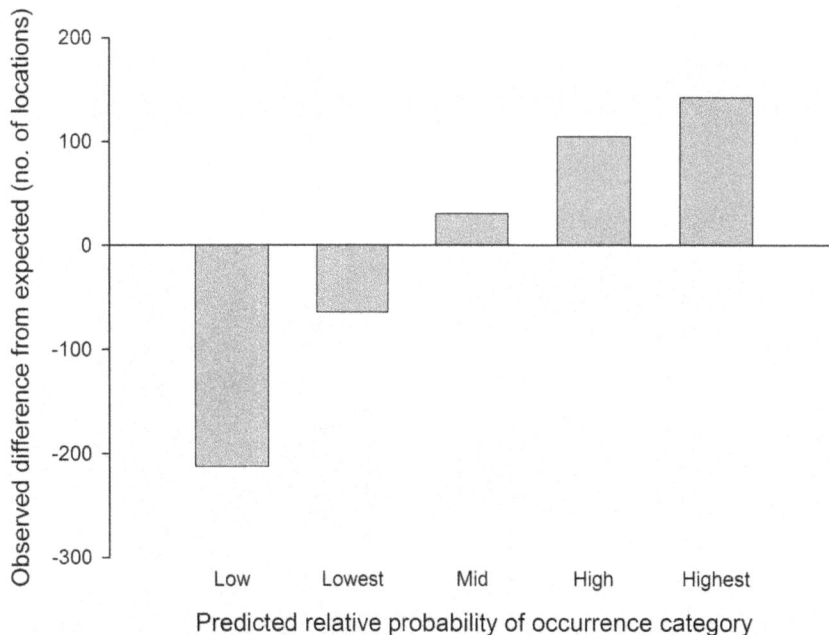

Figure 8. Difference in the number of observed versus expected independent greater sage-grouse GPS locations central Wyoming, USA. Fewer observations than expected (negative values) and more observations than expected (positive values) in lower and higher use categories, respectively, indicate the connectivity model performed well at predicting occurrence and resource selection of independent greater sage-grouse. Only locations from traveling and relocating movement states of independent birds were used for validation.

results and Table S2 for summary statistics of predictor variables at 'non-used' locations for each movement state.

Discrete connectivity corridors were largely absent from most of the study area, where either high or quality connectivity habitat occurred across broad areas. However, in the northwest quarter of the study area, flat valley bottoms appeared to function as distinct corridors of high quality connectivity habitat (Figure 7a).

The 1,211 validation locations (Table 1) fell more often in high predicted probability of use areas and less often in low predicted probability of use areas than expected at random (Spearman's rho = 1.0, 3 d.f., p<0.001; Fig. 8). This indicates that the combined connectivity map (traveling + relocating, both female and male; Fig. 7a) performed very well at predicting the occurrence of independent male and female sage-grouse during traveling and relocating movement states.

Discussion

Traditionally, connectivity modeling has assumed the presence of discrete corridors as a de facto component of species or community ecology or as a requisite of land use planning in human-modified areas [5], [8], [12]. A general feature of the final connectivity map (Fig. 7a) is that distinct 'connectivity corridors' were largely lacking, although they did appear to occur in some areas. A second feature of the connectivity map is that it conceptually treats connectivity habitat as a spatially variable feature of the landscape. This conceptual treatment is an underpinning of empirically assessing the functional relationships between animals and the landscape [10], [39]. Human assessments of the configuration and value of landscapes are inherently faulty, as they do not fully incorporate perception distances, the animal's weighting of co-occurring landscape features, and the presence of behavioral syndromes among mobile animals [39], [40]. Methods that rely on empirical data overcome this challenge. Approaches to connectivity modeling that do not impose spatial constraints (e.g., least cost path analyses explicitly assume, a priori, that the desired outcome is linear) increase the flexibility of the modeling approach to accurately reflect the functional relationship between animals and the landscape. Flexibility increases the potential for success of connectivity management actions and allows for connectivity management to be integrated with other management priorities.

The utility of this approach for maintaining connectivity of animal populations is strengthened by integration with GIS [33], [41], [42], [43]. Here, we used the final resource selection models for traveling and relocating male and female sage-grouse to develop a comprehensive 'connectivity' map, essentially identifying areas likely to be selected or avoided by sage-grouse when they are moving moderate to long distances (Fig. 7a). This map can be used to maximize the return on conservation and management dollars by identifying areas where enhancements or restrictions would most benefit sage-grouse, minimize unnecessary restrictions on development by identifying areas where restrictions would have minimal benefit to sage-grouse connectivity and ultimately increase the positive benefit to sage-grouse populations under multiple land use scenarios by simultaneously increasing stakeholder buy in to management plans, maximizing the effectiveness of management actions, and increasing the spatial extent of areas to be conserved. In our case study, the spatially-explicit connectivity map could be used to ensure landscape connectivity between other areas important to sage-grouse, such as critical nesting, brood-rearing or winter habitat (Fig. 7b; [33], [44]), possibly by delineating areas of high density of high quality connectivity habitat [8]. Connectivity maps could also be combined with spatially-explicit models of mortality risk to identify

attractive sinks where managers may want to discourage connectivity [32], [33], [45]. An alternative application would be modeling connectivity of undesirable animal populations (e.g. invasive species) to identify areas where management actions would be most effective at reducing connectivity of populations..

The approach we used overcomes many shortcomings of alternative methods of modeling connectivity for non-corridor species, although it has its own limitations. We set different scales of availability for each movement state, which has been noted to make comparisons among analyses difficult [46]. We did this because it seems reasonable that an individual animal's perception of availability is related to its underlying movement state and maintaining a single scale of availability across movement states could generate results inconsistent with the underlying biological process of resource selection [47]. Thus we accepted potential sampling errors in order to gain potential ecological reality. There are two tests of these potential limitations. The first is to compare the landscape to determine if what we characterize as 'available' resources is systematically influenced by our scale of defining what is available [46]. We found that the characterization of the landscape was similar for each movement state, especially given wide variation in observed values for each of our predictor variables (Table S2). The second test is in the validation of the predictive RSF models. The final sage-grouse connectivity map performed well at predicting occurrence of independent sage-grouse during traveling and relocating movement states, suggesting that any methodological limitations did not extensively hamper inference and application.

These results also offer general inference and specific application to managing connectivity for the focal species greater sage-grouse. Landscape-level connectivity of shrub-steppe habitat has been identified as critical to maintenance of sage-grouse populations [48], [49], [50], [51], [52]. The general assumption has been that managers need to delineate corridors (e.g., [51], [52]). In this study, there was general lack of distinct corridors due to the spatial configuration of sage-grouse connectivity habitat within the study area. What characterized high quality connectivity habitat mirrors the factors that numerous studies have identified as high quality sage-grouse habitat during critical times of year, such as nesting, brood-rearing, and wintering (e.g., [22], [33], [53]). Specifically, we also documented avoidance of natural gas and oil wells, steep slopes, and areas of rough topography and a preference for higher coverage (and patchiness) of big sagebrush. Perhaps surprisingly, we also documented a preference for portions of the landscape with higher density of roads. This is the opposite pattern observed in a separate analysis we performed on winter habitat selection by this same population of sage-grouse, where we documented avoidance of roads during both day and night [44]. It is important to clarify that sage-grouse (particularly males) selected for locations that had higher road density than was available to them at that time and place rather than what was available at the landscape level. A low landscape-level correlation between road density and topographic roughness, combined with an equal strength of selection for these variables by male sage-grouse (Table 3), indicates that male selection of higher road density when traveling and relocating was a real pattern in the study area. We therefore hypothesize that perhaps road density does not act as a barrier to connectivity even though roads may be negatively associated with occurrence during particular seasons. This finding highlights the value of an empirical approach to modeling connectivity, as an expert-based assessment of connectivity habitat would have likely considered high road density as a barrier to movement. Using this generalized conceptual and analytical approach to modeling landscape connectivity allows for empirical-based models that do not constrain the spatial pattern of connectivity into corridors, unless corridors are a natural part of how animals interact with the landscape.

Acknowledgments

We thank BJ Lukins for providing valuable contributions during the early years of the project. We also thank Clay Bowers, Trennan Dorval, Craig Okraska, Brett Basler, and numerous other field technicians for assistance with fieldwork. C. Hedley, S.L. Webb and 4 anonymous reviewers provided insightful comments. Sue Oberlie with the U.S. Bureau of Land Management and Greg Anderson with the Wyoming Game and Fish Department provided invaluable support for this project.

Author Contributions

Conceived and designed the experiments: SMH CVO MRD. Performed the experiments: CVO. Analyzed the data: SMH. Contributed reagents/materials/analysis tools: JPM JBW. Wrote the paper: SMH CVO MRD.

References

1. Fischer J, Lindenmayer DB (2007) Landscape modification and habitat fragmentation: a synthesis. Glob Ecol Biogeogr 16: 265–280.
2. Sawyer H, Kauffman MJ, Nielson RM, Horne JS (2009) Identifying and prioritizing ungulate migration routes for landscape-level conservation. Ecol Appl 19: 2016–2025.
3. Ricketts TH (2001) The matrix matters: effective isolation in fragmented landscapes. Am Nat 158: 87–99.
4. Dzialak MR, Lacki MJ, Larkin JL, Carter KM, Vorisek S (2005) Corridors affect dispersal initiation in reintroduced peregrine falcons. Anim Conserv 8: 421–430.
5. Chetkiewicz CB, St. Clair CC, Boyce MS (2006) Corridors for conservation: integrating pattern and process. Annu Rev Ecol Evol Syst 37: 317–342.
6. Horne JS, Garton EO, Krone SM, Lewis JS (2007) Analyzing animal movements using brownian bridges. Ecol 88: 2354–2363.
7. Barraquand F, Benhamou S (2008) Animal movements in heterogeneous landscapes: identifying profitable places and homogeneous movement bouts. Ecol 89: 3336–3348.
8. Beier P, Majka DR, Spencer WD (2008) Forks in the road: choices in procedures for designing wildland linkages. Conserv Biol 22: 836–851.
9. Richard Y, Armstrong DP (2010) Cost distance modelling of landscape connectivity and gap-crossing ability using radio-tracking data. J Appl Ecol 47: 603–610.
10. Chetkiewicz CB, Boyce M (2009) Use of resource selection functions to identify conservation corridors. J Appl Ecol 46: 1036–1047.
11. Simberloff D, Cox J (1987) Consequences and costs of conservation corridors. Conserv Biol 1: 63–71.
12. Rayfield B, Fortin MJ, Fall A (2011) Connectivity for conservation: a framework to classify network measures. Ecol 92: 847–858.
13. Lendrum PE, Anderson Jr CR, Long RA, Kie JG, Bowyer RT (2012) Habitat selection by mule deer during migration: effects of landscape structure and natural-gas development. Ecosphere 3: 82. http://dx.doi.org/10/1890/ES12-00165.1.
14. Blackwood J, Hastings A, Costello C (2010) Cost-effective management of invasive species using linear-quadratic control. Ecol Econ 69: 519–527.
15. Connelly JW, Braun CE (1997) Long-term changes in sage grouse Centrocercus urophasianus populations in western North America. Wildl Biol 3: 229–234.
16. Schroeder MA, Aldridge CL, Apa AD, Bohne JR, Braun CE, et al. (2004) Distribution of sage-grouse in North America. Condor 106: 363–376.
17. United States Fish and Wildlife Service (2010) Endangered and threatened wildlife and plants; 12-month finding for petitions to list the greater sage-grouse (Centrocercus urophasianus) as threatened or endangered. Fed Regist 75: 13910–14014.
18. State of Wyoming (2011) Greater sage-grouse core area protection. State of Wyoming Executive Department Executive Order 2011–5. Office of the Governor, Cheyenne, WY, USA.
19. United States Bureau of Land Management (2012) Greater sage-grouse habitat management policy on Wyoming Bureau of Land Management (BLM) administered lands including the federal mineral estate. Instruction Memorandum No. WY-2012–019. Bureau of Land Management, Cheyenne, WY, USA.
20. Rowland MM, Wisdom MJ, Suring LH, Meinke CW (2006) Greater sage-grouse as an umbrella species for sagebrush-associated vertebrates. Biol Conserv 129: 323–335.
21. Connelly JW, Schroeder MA, Sands AR, Braun CE (2000) Guidelines to manage sage grouse populations and their habitats. Wildl Soc Bull 28: 967–985.
22. Doherty KE, Naugle DE, Walker BL, Graham JM (2008) Greater sage-grouse winter habitat selection and energy development. J Wildl Manage 72: 187–195.
23. Harju SM, Dzialak MR, Taylor RC, Hayden-Wing LD, Winstead JB (2010) Thresholds and time lags in effects of energy development on greater sage-grouse populations. J Wildl Manage 74: 437–448.
24. Holloran MJ, Kaiser RC, Hubert WA (2010) Yearling greater sage-grouse response to energy development in Wyoming. J Wildl Manage 74: 65–72.
25. Wakkinen WL, Reese KP, Connelly JW, Fischer RA (1992) An improved spotlighting technique for capturing sage-grouse. Wildl Soc Bull 20: 425–426.
26. Bedrosian B, Craighead D (2007) Evaluation of techniques for attaching transmitters to common raven nestlings. Northwest Nat 88: 1–6.
27. Johnson CJ, Parker KL, Heard DC, Gillingham MP (2002) Movement parameters of ungulates and scale-specific responses to the environment. J Anim Ecol 71: 225–235.
28. Schick RS, Loarie SR, Colchero F, Best BD, Boustany A, et al. (2008) Understanding movement data and movement processes: current and emerging directions. Ecol Lett 11: 1338–1350.
29. Van Moorter B, Visscher DR, Jerde CL, Frair JL, Merrill EH (2010) Identifying movement states from location data using cluster analysis. J Wildl Manage 74: 588–594.
30. Connelly JW, Browers HW, Gates RJ (1988) Seasonal movements of sage grouse in southeastern Idaho. J Wildl Manage 52: 116–122.
31. Nams VO (2006) Animal movement rates as behavioural bouts. J Anim Ecol 75: 298–302.
32. Aldridge CL, Boyce MS (2007) Linking occurrence and fitness to persistence: habitat-based approach for endangered greater sage-grouse. Ecol Appl 17: 508–526.
33. Dzialak MR, Olson CV, Harju SM, Webb SL, Mudd JP, et al. (2011) Identifying and prioritizing greater sage-grouse nesting and brood rearing habitat for conservation in human-modified landscapes. PLoS ONE 6: e26273.
34. Cooper AB, Millspaugh JJ (1999) The application of discrete choice models to wildlife resource selection studies. Ecol 80: 566–575.
35. Duchesne T, Fortin D, Courbin N (2010) Mixed conditional logistic regression for habitat studies. J Anim Ecol 79: 548–555.
36. Fortin D, Beyer HL, Boyce MS, Smith DW, Duchesne T, et al. (2005) Wolves influence elk movements: behavior shapes a trophic cascade in Yellowstone National Park. Ecol 86: 1320–1330.
37. Arthur SM, Manly BFJ, McDonald LL, Garner GW (1996) Assessing habitat selection when availability changes. Ecol 77: 215–227.
38. Compton BW, Rhymer JM, McCollough M (2002) Habitat selection by wood turtles (Clemmys insculpta): an application of paired logistic regression. Ecol 83: 833–843.
39. Sawyer SC, Epps CW, Brashares JS (2011) Placing linkages among fragmented habitats: do least-cost models reflect how animals use landscapes? J Appl Ecol 48: 668–678.
40. Baguette M, Van Dyck H (2007) Landscape connectivity and animal behavior: functional grain as a key determinant for dispersal. Landsc Ecol 88: 1117–1129.
41. Manly BF, McDonald L, Thomas DL, McDonald TL, Erickson WP (2002) Resource selection by animals. Kluwer Academic Publishers, Netherlands. 240 p.
42. Johnson CJ, Seip DR, Boyce MS (2004) A quantitative approach to conservation planning: using resource selection functions to map the distribution of mountain caribou at multiple spatial scales. J Appl Ecol 41: 238–251.
43. Harju SM, Dzialak MR, Osborn RG, Hayden-Wing LD, Winstead JB (2011) Conservation planning using resource selection models: altered selection in the presence of human activity changes spatial prediction of resource use. Anim Conserv 14: 502–511. doi: 10.1111/j.1469-1795.2011.00456.x.
44. Dzialak MR, Olson CV, Harju SM, Webb SL, Winstead JB (2012) Temporal and hierarchical spatial components of animal occurrence: conserving seasonal habitat for greater sage-grouse. Ecosphere 3: 30. http://dx.doi.org/10.1890/ES11-00315.1.
45. Schlaepfer MA, Runge MC, Sherman PW (2002) Ecological and evolutionary traps. Trends Ecol Evol 17: 474–480.
46. Beyer HL, Haydon DT, Morales JM, Frair JL, Hebblewhite M, et al. (2010) The interpretation of habitat preference metrics under use-availability designs. Philos Trans R Soc Lond B Biol Sci 365: 2245–2254.
47. Wilson RR, Gilbert-Norton L, Gese EM (2012) Beyond use versus availability: behaviour-explicit resource selection. Wildl Biol 18: 424–430.
48. Aldridge CL, Nielsen SE, Beyer HL, Boyce MS, Connelly JW, et al. (2008) Range-wide patterns of greater sage-grouse persistence. Divers Distrib 14: 983–994.
49. Bush KL, Aldridge CL, Carpenter JE, Paszkowski CA, Boyce MS, et al. (2010) Birds of a feather do not always lek together: genetic diversity and kinship

structure of greater sage-grouse (*Centrocercus urophasianus*) in Alberta. Auk 127 343–353.

50. Knick ST, Hanser SE (2011) Connecting pattern and process in greater sage-grouse populations and sagebrush landscapes. In: Knick ST, Connelly JW, editors. Greater sage-grouse: ecology and conservation of a landscape species and its habitats. Berkeley, University of California Press. 383–405.

51. Fedy BC, Aldridge CL, Doherty KE, Beck JL, Bedrosian B, et al. (2012) Interseasonal movements of greater sage-grouse, migratory behavior, and an assessment of the core regions concept in Wyoming. J Wildl Manage 76: 1062–1071.

52. Tack JD, Naugle DE, Carlson JC, Fargey PJ (2012) Greater sage-grouse Centrocercus urophasianus migration links the USA and Canada: a biological basis for international prairie conservation. Oryx 46: 64–68.

53. Carpenter J, Aldridge C, Boyce MS (2010) Sage-grouse habitat selection during winter in Alberta. J Wildl Manage 74: 1806–1814.

Seasonal Dynamics of Atlantic Herring (*Clupea harengus* L.) Populations Spawning in the Vicinity of Marginal Habitats

Florian Eggers[1,2]*, Aril Slotte[1], Lísa Anne Libungan[3], Arne Johannessen[2], Cecilie Kvamme[1], Even Moland[4], Esben M. Olsen[4,5,6], Richard D. M. Nash[1]

1 Institute of Marine Research, Bergen, Norway, 2 Department of Biology, University of Bergen, Bergen, Norway, 3 Department of Life and Environmental Sciences, University of Iceland, Reykjavík, Iceland, 4 Institute of Marine Research, Flødevigen, Norway, 5 Centre for Ecological and Evolutionary Synthesis (CEES), Department of Biosciences, University of Oslo, Oslo, Norway, 6 Department of Natural Sciences, Faculty of Science and Engineering, University of Agder, Kristiansand, Norway

Abstract

Gillnet sampling and analyses of otolith shape, vertebral count and growth indicated the presence of three putative Atlantic herring (*Clupea harengus* L.) populations mixing together over the spawning season February–June inside and outside an inland brackish water lake (Landvikvannet) in southern Norway. Peak spawning of oceanic Norwegian spring spawners and coastal Skagerrak spring spawners occurred in March–April with small proportions of spawners entering the lake. In comparison, spawning of Landvik herring peaked in May–June with high proportions found inside the lake, which could be explained by local adaptations to the environmental conditions and seasonal changes of this marginal habitat. The 1.85 km^2 lake was characterized by oxygen depletion occurring between 2.5 and 5 m depth between March and June. This was followed by changes in salinity from 1–7‰ in the 0–1 m surface layer to levels of 20–25‰ deeper than 10 m. In comparison, outside the 3 km long narrow channel connecting the lake with the neighboring fjord, no anoxic conditions were found. Here salinity in the surface layer increased over the season from 10 to 25‰, whereas deeper than 5 m it was stable at around 35‰. Temperature at 0–5 m depth increased significantly over the season in both habitats, from 7 to 14°C outside and 5 to 17°C inside the lake. Despite differences in peak spawning and utilization of the lake habitat between the three putative populations, there was an apparent temporal and spatial overlap in spawning stages suggesting potential interbreeding in accordance with the metapopulation concept.

Editor: Brian R. MacKenzie, Technical University of Denmark, Denmark

Funding: These authors have no support or funding to report.

Competing Interests: The authors have declared that no competing interests exist.

* Email: eggersf@t-online.de

Introduction

Typically, fish species may be split into populations based on their degree of reproductive isolation from each other in space and/or time, which could be reflected in genetic or phenotypic differences driven by diverging environmental conditions [1–3]. Under such circumstances exploitation on one population should have little effect on the population dynamics of a neighboring population, and therefore it is also common to assess and manage such populations separately [4,5]. On the other hand, there are also examples where populations are recognized to be separate with diverging spawning season and/or spawning area, but due to mixing in other seasons a separate management of the populations may be difficult [6,7]. The need to identify the different populations, especially where exploitation occurs on mixtures of populations is important for successful management [8,9]. Fisheries biologists therefore often use the term stock instead of population in their fisheries advice; i.e. sometimes a population is harvested and therefore managed as one stock and at other times several separate populations are harvested and managed as one

stock. In Begg et al. [10] the concept of a fish stock was simply defined as characteristics of semi-discrete groups of fish with some definable attributes, which are of interest to fishery managers. The definition of ICES [11] for a stock as a part of a fish population usually with a particular migration pattern, specific spawning grounds, and subject to a distinct fishery, will be used hereby. In theory, all individual fish in an area, being part of the same reproductive process, are comprised as a stock. When referring to fisheries management, the term "stock" is used, otherwise the term "population" is preferred.

Atlantic herring (*Clupea harengus* L.) is characterized by highly complex population structure and migration patterns [12]. It is an iteroparous clupeid, becoming sexually mature at two or three years of age, and a total spawner that aggregates at spawning, laying benthic eggs on shells, gravel, coarse sand and small stones at depths down to 250 m [13]. The larvae hatch after 2–4 weeks depending on temperature [14,15]. They drift with currents until metamorphosis [16–18], with vertical migration increasing throughout ontogeny [19,20] and affecting the dispersal trajectories of larvae. The different herring populations are generally classified according to their spawning grounds, which, due to the specific spawning substratum requirements, are fixed geographically and used at a predictable time of the year. Due to physical and geographical barriers, such as prevailing currents and general location of nursery areas, there is often little mixing of larvae, thus tending to isolate the different populations. However, there are occasions where larvae and juveniles may co-occur. Under these circumstances identification of individuals or groups of individuals is undertaken using otolith or meristic characters [1,21–24] as well as genetic markers [25–28]. In the 1950–60s experimental studies [29–31] demonstrated that myotome counts in herring were influenced by both temperature (negatively) and salinity (positively) experienced during the incubation period. The consequence is that mean vertebral count of adult herring is an indicator of spawning ground and spawning times and in some cases also population.

In Norwegian waters some herring populations occupy marginal habitats along the coastline and deep inside fjords, most of which are thought to be stationary with adaptations to local conditions. Hence, they are often phenotypically and, in some occasions, genotypically different from the nearby oceanic population. Examples of such local herring populations are Trondheimsfjord herring [32,33], Borge Poll herring [34], Lusterfjord herring [35], Lindåspollene herring [36], Balsfjord herring [37], Lake Rossfjord herring [38] and the summer/autumn spawners in northern Norway [39]. Despite the discovery of these local populations, the overall research effort targeting marginal areas along the Norwegian coast has been rather low, and it is therefore expected that a number of additional local populations may exist.

Migratory coastal or oceanic populations may occasionally enter the marginal habitats along the Norwegian coast and mix with local herring. This is in accordance with the metapopulation concept, where two or more distinguished subpopulations have variable but moderate interbreeding and significant gene flow [40]. Temporal and spatial overlap during spawning may allow genetic exchange between subpopulations, which is a prerequisite for the existence of metapopulations. An example of such an overlap was demonstrated by Johannessen et al. [41],[42] in the local Lindåspollene herring, where significant changes in life history traits over a 50 year period were linked to genetic exchange with the oceanic population according to the metapopulation concept.

An important mixing area for herring is the northeastern North Sea and Skagerrak, where three different stocks may occur, Norwegian Spring Spawners (NSS), North Sea Autumn Spawners (NSAS) and Western Baltic Spring Spawners (WBSS). Some of these stocks comprise different herring populations, such as coastal Skagerrak spring spawners or more local herring populations, which are not directly subjected to a distinct fishery. The different populations (stocks) can be distinguished by spawning site, spawning season, meristic characters such as the number of vertebrae (VS) and otolith characteristics [23,41].

Of particular interest in the Skagerrak area is a brackish water environment inside Landvikvannet, an inland lake in southern Norway connected to the open sea through an artificial channel. The Institute of Marine Research (IMR) has been sampling herring in Landvikvannet on regular basis since 1984, mainly in May. Data from these investigations demonstrate that herring inside the lake are normally ripe or with running gonads, with a low mean vertebral number (<56.0), slow growth and high fecundity [43,44]. This has led to the hypothesis that the lake is visited on an annual basis by a herring population with specific adaptations to spawning in these brackish water environments.

However, in the coastal areas outside the lake, ripe and spawning herring with higher growth and mean vertebral numbers (56.0–57.5) have occurred in samples over the period February–June [43]. This indicates that there may be a mixture of several populations in the area with some temporal and spatial overlap in spawning, which could be linked to spatial seasonal differences in environmental conditions. Such metapopulation dynamics may be revealed by a more detailed seasonal sampling outside the May period normally focused on in IMR's investigations in Landvikvannet. Hence, the principal objective of the present study was to explore the overlap in time, space and maturation stages of phenotypically different herring appearing in Landvikvannet and neighboring fjord areas and their dependence on seasonal changes in environmental conditions.

Material and Methods

Study area

Landvikvannet is a 1.85 km^2 lake located on the Norwegian Skagerrak coast (Figure 1). In 1877 a 3 km long channel (Reddal channel, Figure 1) was constructed, connecting the lake to the open sea. This narrow 1–4 m deep channel transformed Landvikvannet into a brackish system and in addition lowered the water level in the lake by 3 m. At the entrance of the lake there is a small 25 m deep basin. Further into the lake the bottom depth decreases rapidly to 7–10 m. Most of the shoreline is covered by reeds; otherwise the shore is rocky and steep. There is inflow of saltwater over the tidal cycle, whereas freshwater empties into the lake from streams, resulting in a halocline. Oxygen is depleted in the lower layers whereas the surface layer is oxygen rich. In Landvikvannet, herring have been caught by floating gillnets together with trout (*Salmo trutta*) and other freshwater fish since shortly after the channel was opened.

The Reddal channel drains into Strandfjorden (Figure 1), where conditions are estuarine. The outer Strandfjorden is narrow and shallow (1–7 m), whereas the inner part is deeper (10–13 m). Most herring samples were collected in the inner part, close to the mouth of the Reddal channel. The shore is rocky and steep with sparse macroalgae in the upper few meters. At depths >5–6 m the bottom consists of sand and mud. The outermost fjord (Bufjorden, Figure 1) is small with direct connection to Skagerrak. Strandfjorden is connected to the open ocean via Bufjorden (Figure 1). The entrance of Bufjorden is characterized by a 54 m deep basin. The physical environment is similar to Strandfjorden, only less influenced by fresh water runoff. Access to Bufjorden is from the south or east.

Environmental data

To explore whether potential differences in habitat utilization and timing of peak spawning among herring populations were dependent on seasonal changes in environmental conditions, sampling of environmental data was undertaken between March and June 2012 both inside and outside the lake habitat. Note, that no stations could be sampled in February due to ice cover. Water samples were collected at the site where gillnets were moored in the inner part of Strandfjorden and at the entrance of Landvikvannet in the first basin (Figure 1). We measured temperature and salinity at depth with a CTD (STD/CTD – model SD204, SAIV Ltd. Environmental sensors and Systems, Bergen, Norway), while oxygen and hydrogensulfide concentrations were analyzed in the laboratory at the Institute of Marine Research (IMR). In the lake, water samples were collected each 0.5 meter down to the depth of oxygen depletion (hypoxic depth), which was found using the Winkler test [45], thereafter water samples were taken at 5 m

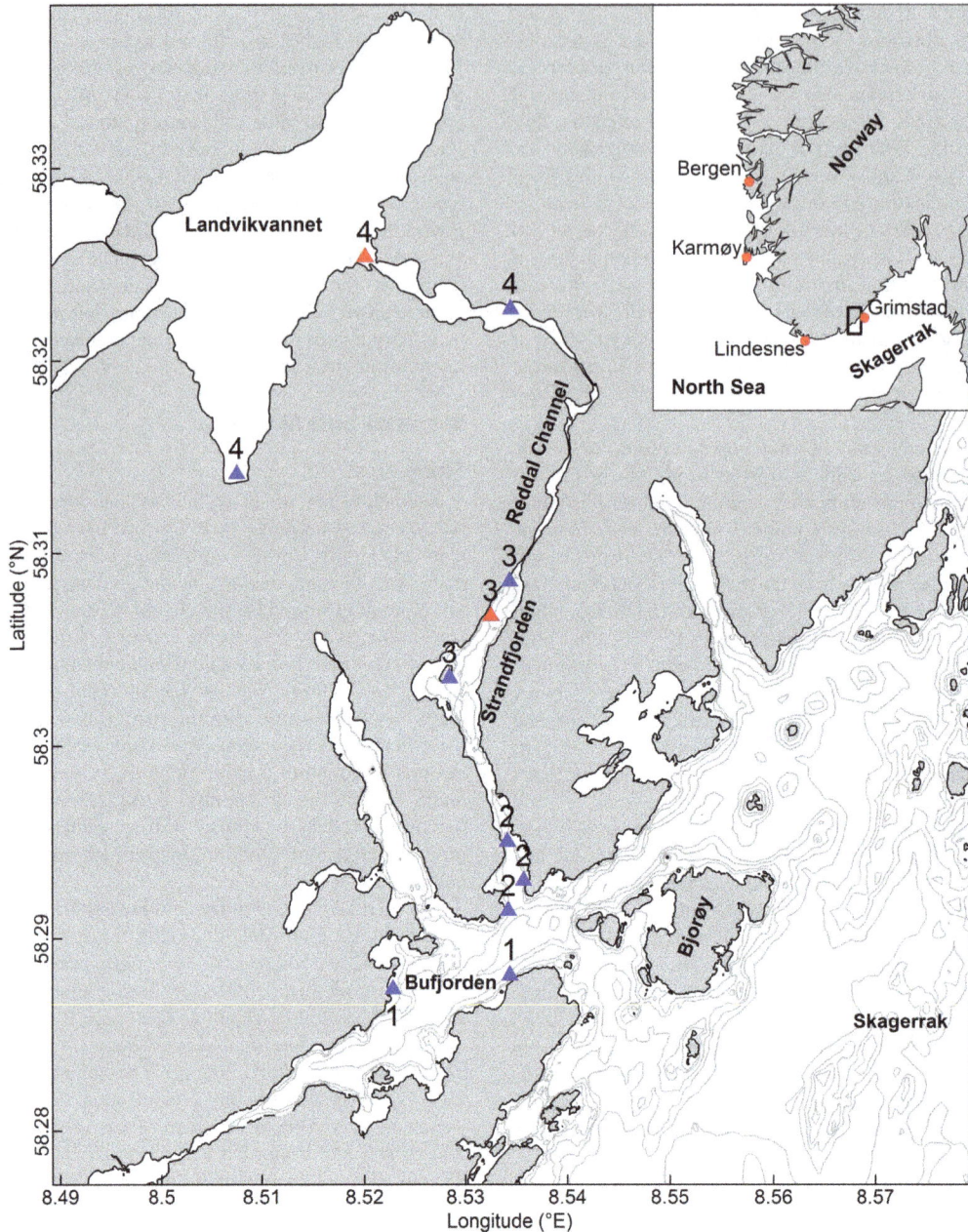

Figure 1. Map of the study area. The map shows CTD-stations (red) and gillnet stations (blue) in 1 = Bufjorden, 2 = Outer part of Strandfjorden, 3 = Inner part of Strandfjorden, 4 = Landvikvannet.

depth intervals. The choice of position for sampling environmental data inside the lake is based on the depth contours of the area. The lake itself is rather shallow, and the bottom depth at most gillnet stations is 2–4 m. However, at the entrance the lake is at its deepest (25 m), which is why this position has been used since investigations started in the area in the 1980s. The environmental conditions at this site between 0 and 10 m have been examined thoroughly over a number of years and are comparable to conditions elsewhere in the lake and as such can be used to characterize the whole lake. These data are therefore representative of all gill net sampling sites.

Biological data

To explore the potential overlap in time, space and maturation stages of phenotypically different herring appearing inside and outside the lake habitat, herring were sampled with gillnet over the full spawning season in 2012 (February–June) concurrently in both habitats (Figure 1, Table 1). In February, due to ice cover both in the lake and inner fjord habitats of Strandfjorden, samples were only taken further out in Bufjorden. The floating gillnets with a mesh size of 26 mm and 29 mm, a depth of 8 m and a length of approximately 10 m were used randomly in all areas. Soak time was 24 hours. This experiment was approved by the Norwegian committee for the use of animals in scientific experiments (FDU). Special permission to fish with floating gillnet inside

Landvikvannet and in the connected fjord system in 2012 was given by the County Governor of Aust-Agder, Department of Climate and Environment, Ragnvald Blakstadsv. 1, Postbox 788 Stoa, 4809 Arendal, Norway. The permission was given to the Institute of Marine Research under the prerequisite that details on the catch were reported when the investigations were finished. The report was delivered to the authorities according to the plan. Our study did not involve endangered or protected species.

Biological samples were analyzed according to IMR standard protocols [46]. The maximum sample size was 100 herring. Biological parameters included in the present study were total length (nearest 0.5 cm below), weight (nearest gram below), sex, stage of maturity, age (otolith readings) and vertebral count (VS). Maturity stages were determined by visual inspection of gonads according to the following scale: immature = 1–2, maturing = 3–4, ripe = 5, spawning/running = 6, spent = 7 and recovering = 8 [46].

Image and shape analyses

Individuals of NSS herring were identified from otoliths, based on a sharper distinction between winter and summer rings compared to local spring spawners (Figure 2). This distinction was also independently tested using image and shape analyses of the otoliths. The rest of the individuals were divided into two populations based on sampling location: local Landvikvannet herring (LV) sampled inside Landvikvannet and coastal Skagerrak spring spawners (CSS) sampled outside Landvikvannet (Table 2). We expected that LV herring would mainly consist of individuals with similar biological characteristics as normally found in May, whereas the CSS herring would mainly consist of spring spawners with characteristics normally found along the Skagerrak coast during February–June. However, some mixture of the two populations would be expected, and this would be evident from results of the biological analyses. To investigate changes in the mixture of NSS, CSS and LV herring in the two habitats, selected biological characters (otolith shape, vertebral count, growth and maturation stage) were analyzed over the full season. The numbers analyzed by month and population are given in Table 2.

Otolith shape was analyzed using the programming language R [47]. Outlines of otoliths were collected from digital images using the package pixmap [48], and applying the conte function [49] to record a matrix of X and Y coordinates (Figure 2a). Mean shape of otoliths differed among the populations, where the modifications in the shape of otoliths mainly were found at the excisura major and antirostrum areas (Figure 2b).

To remove size-induced bias, otolith sizes were standardized to equal area by dividing the coordinates of each otolith with the square root of the otolith area. Equally spaced radiis were drawn

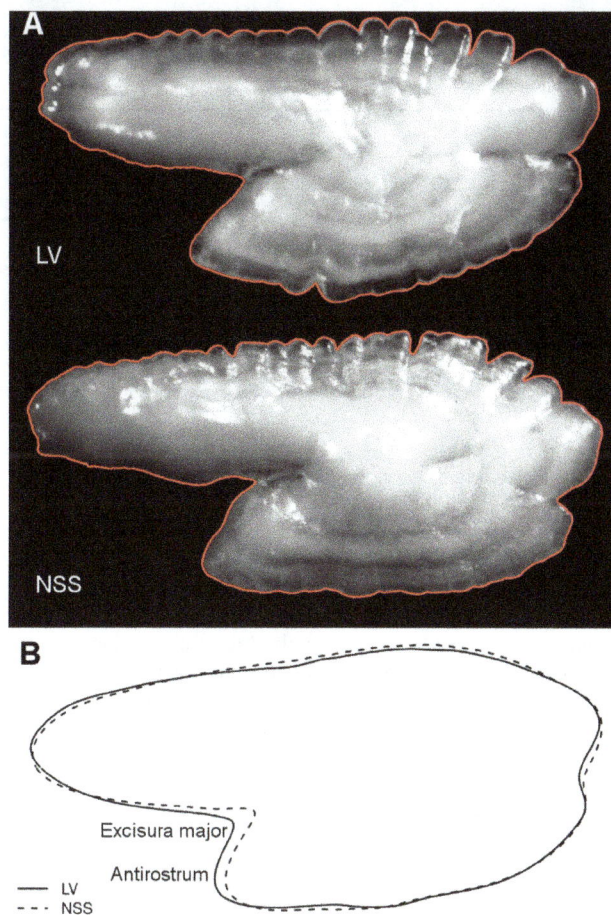

Figure 2. Example of otolith characteristics from two herring populations. A) Example of otoliths used for the shape analysis from Landvikvannet herring (LV) and Norwegian spring-spawning herring (NSS), both at the age of 3 years. Individuals of NSS herring were subjectively identified based on a sharper distinction between winter (dark areas) and summer rings (white areas). Red outline marks the shape of the otolith which was used to compare among populations. B) shows the mean shape of otoliths for the two populations, where the excisura major and antirostrum areas are the most variable areas.

from the otolith centroid to the otolith outline, using the regular radius function [49]. Independent Wavelet shape coefficients were obtained by conducting a Discrete Wavelet transform on the

Table 1. Total number of herring caught in the local area for 2012, in brackets number of gillnets; ice = no sampling possible because the area was covered by ice.

Date	Landvikvannet	Inner Strandfjorden	Outer Strandfjorden	Bufjorden
15/2	Ice cover	Ice cover	28 (1)	11 (1)
6/3	4 (3)	129 (1)	119 (1)	
20/3	47 (3)	542 (1)		
26/3	115 (3)	486 (1)		100 (1)
11/4	290 (2)	663 (1)		
14/5	177 (1)	69 (1)		
21/6	82 (1)	66 (1)		
Total	715	1955	147	111

Table 2. Total number of herring analyzed in 2012 by month for the three putative herring populations, Norwegian spring spawners (NSS), Coastal Skagerrak spring spawners (CSS) and Landvik herring (LV), in brackets number of NSS inside Landvikvannet.

Month	NSS	CSS	LV
2	7 (0)	32	0
3	108 (38)	440	113
4	32 (14)	68	86
5	8 (5)	61	95
6	0 (0)	66	77
Total	155 (57)	667	371

equally spaced radiuses using the wavethresh package [50]. To determine the number of Wavelet coefficients needed for the analysis, the deviation of the reconstructed Wavelet otolith outline from the original outline was evaluated. To correct for fish length, an ANCOVA was performed on the wavelet coefficients taking fish length as a covariate. Coefficients which could not be adjusted by linear relationships on fish length, due to interaction between the origin and length were excluded from the analysis [51–53]. To adjust the Wavelet coefficients for allometric growth, a normalization technique based on regression was applied to scale the Wavelet coefficients [54].

Data analyses

The number of gillnets varied between Landvikvannet and the neighboring fjord area. Therefore, to estimate the proportions of the LV, CSS and NSS herring, the total catches landed were standardized by catch per unit effort (CPUE), i.e. catch per gillnet.

All statistical analyses were conducted in R (version 3.0.1; [47]). A significance level of $\alpha = 0.05$ was used for all statistical tests. For the plots, mean and standard error (1 SE) are shown. Some samples had very few or no data, and samples with N<5 were excluded.

Analysis of Covariance (ANCOVA) was used to test for sex differences in the biological characters (length, age, VS and stage of maturity). Differences in VS among different herring populations were assessed using Analysis of Variance (ANOVA), and a Kruskal-Wallis test for length and age variables as these were not normally distributed. For pairwise comparisons of VS a paired T-test was used, and the Mann-Whitney test for length and age comparisons.

Length-at-age data, used as a proxy for growth of individual herring, were fitted to the von Bertalanffy growth model (VBGM) [55]:

$$L_t = L_\infty (1 - e^{-K(t - t_0)})$$

where L_t is the average length at age t, L_∞ is the asymptotic maximum length, K is the von Bertalanffy growth rate coefficient, i.e. the rate at which length approaches the maximum length asymptote and t_0 is the intercept on the time axis. Growth was compared between the different groups using ANOVA.

Variation in otolith shape, as reflected by the scaled Wavelet coefficients, was analyzed with Canonical Analysis of Principal coordinates (CAP) [56] using the capscale function in the vegan package in R [57]. Using multivariate data to represent otolith shape, an ANOVA like permutation test (vegan package) was used to assess the significance of constraints using 5000 permutations.

Variation in otolith shape was analyzed with CAP, while length and VS were compared with ANOVA with respect to herring group: NSS, LV and CSS, the month in which they were caught over the sampling period (Feb–June) and age in years (3–12) using the following models: shape~herring population*month*age, length~herring population*month*age and VS~herring population*month*age. Non-significant interaction terms (p>0.05) were excluded from the models. P-values for all posteriori comparisons were corrected with the Bonferroni correction [58]. Possible trends of length and VS within herring populations were tested for significance using linear regression, while the stage of maturity was tested with the Spearman's rank correlation coefficient. For the comparisons of environmental data at time of spawning with the VS of herring, measurements from 3 m were used for Landvikvannet due to the depth of oxygen depletion in combination with previous (2010) acoustic observations of school depth [43]. In Strandfjorden, measurements from 5 m were used, based on acoustic observations of herring school depth during tagging experiments and the gillnet sampling [43].

Results

Environmental conditions

The environmental conditions differed considerably between Landvikvannet and the neighboring fjord, and changed over the spawning season in both locations (Figure 3). Anoxic conditions were found in Landvikvannet at increasing depths from 2.5 m in March to 5 m in June. Salinity ILV at 0–1 m increased over the season from 1‰ in March to 7‰ in June, but was stable around 20–25‰ deeper than 10 m. In comparison, there were no anoxic conditions in Strandfjorden, the salinity at 0–1 m increased from 10‰ in March to 25‰ in June and was stable at 35‰ deeper than 5 m. The temperature at 0–5 m depth increased from March to June from 5 to 17°C in Landvikvannet, and from 7 to 14°C in Strandfjorden.

Population structure

A total of 1260 herring were analyzed during the 2012 spawning season. Total length ranged from 22.0–34.5 cm (mean: 28.3 cm) and age from 2–12 years (mean: 4.2 years). None of the biological characters varied between sexes (p>0.05). Hence, all further analyzes were carried out with sexes combined.

Mean length, age and vertebral count (VS) differed significantly among the three herring populations (p<0.001, Figure 4). For age and length, pairwise comparisons were also significant (p<0.001), with the exception of CSS versus LV for age (p>0.05). The vertebral count differed significantly (p<0.001) for all pairwise comparisons. The main tendency was a significant increase in

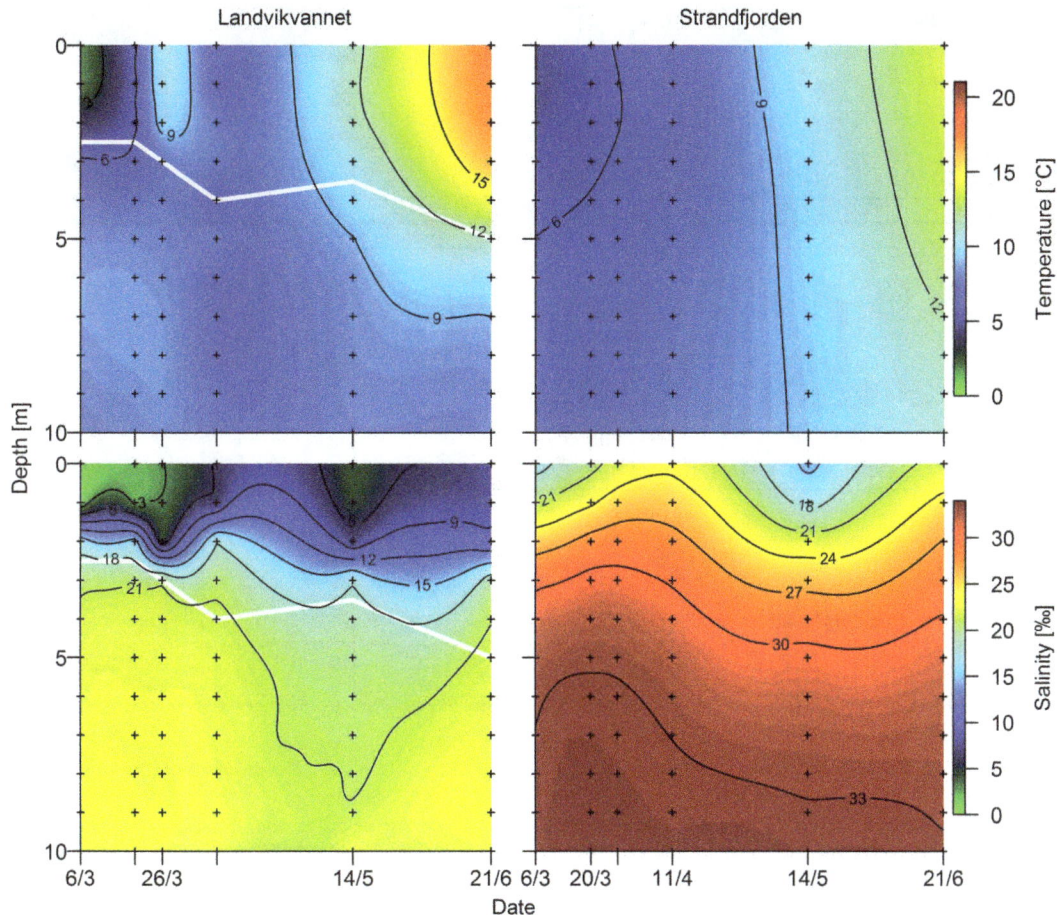

Figure 3. Seasonal change in temperature and salinity by depth. Temperature (upper) and salinity (lower) in Landvikvannet and in Strandfjorden over the study period from March to June. White line indicates the depth of oxygen depletion.

mean body length and VS when moving from LV to CSS to NSS, whereas men age decreased. The most common age was 3 years for NSS, CSS and LV herring. The 4 year olds were also abundant in CSS and LV herring, but hardly present among NSS herring.

Length-at-age data indicated the highest growth for NSS herring, and lowest for LV herring ($p<0.01$) (Figure 5). The von Bertalanffy growth model supported these growth differences (Table 3). Consequently, there were three categories: 'high growth rate' (NSS herring), 'moderate growth rate' (CSS herring) and 'low growth rate' (LV herring).

Between February and June there was a change in the abundance of the different populations (Figure 6). During February–April CPUE was highest for CSS and NSS herring with a low proportion of LV herring ($<20\%$). Also the proportion of NSS herring entering Landvikvannet was insignificant ($<10\%$). The proportion of spawning and spent herring during this period was highest in NSS herring and a little lower for CSS herring, but still indicating peak spawning of two different populations in the fjord habitat during this period. Among the LV herring analyzed in March–April an even lower proportion were in spawning and spent stages than for CSS herring, indicating a later spawning peak for LV herring. This was further demonstrated in the May–June sampling showing a spatial shift in CPUE towards higher abundance of LV than CSS and NSS herring.

Otolith shape differed among the three herring populations ($p<0.001$, Table 4, Figure 7) and also varied though the spawning season ($p<0.001$, Figure 8A). Vertebral count and length differed between the populations ($p<0.001$) and between months ($p<0.001$, Figure 8B, C). Age was a significant factor for all characters ($p<0.001$) and therefore incorporated in the model for all comparisons. Posteriori comparisons showed that LV and CSS differed in otolith shape, VS and length ($p<0.04$, Figure 8, Table 4). NSS and LV ($p<0.001$) as well as NSS and CSS ($p<0.02$) also differed, while no differences were detected for NSS caught inside or outside the lake ($p>0.05$). There was a signifiant ($p<0.001$) negative trend in the mean Canonical scores (CAN1) derrived from the CAP analysis of otolith shape, vertebral count and length for LV and CSS herring at standardized ages over the spawning season, but not for NSS (Figure 8). This indicates that LV herring, characterized by slow growth and low vertebral count, were arriving and mixing with CSS herring.

Maturation and spawning time

Herring in spawning condition were present and overlapped in time for LV, CSS and NSS herring, however, maturation and timing of spawning was delayed in LV compared to NSS and CSS herring (Figure 6). This indicates an adaptation to the environmental conditions and seasonal change in Landvikvannet. Since differences in vertebral count are linked to environmental conditions, the temperature and salinity at depth and time of

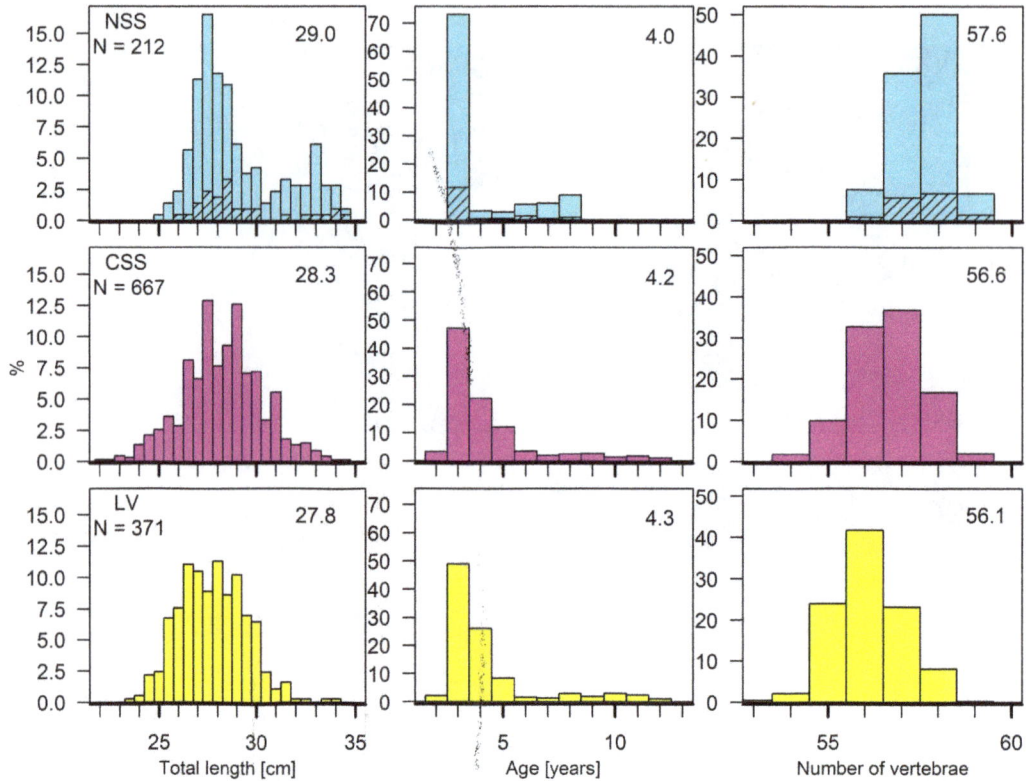

Figure 4. Distribution of length, age and vertebral counts of different herring populations. Comparison between Norwegian spring spawning (NSS), Coastal Skagerrak spring spawning (CSS) and Landvik (LV) herring. Shaded areas are NSS herring inside Landvikvannet. The mean values are included.

spawning affects the vertebral count. The salinity at expected spawning depth in Landvikvannet was distinctly lower (10–15‰) than in the adjacent fjord (>30‰), which could explain the low vertebral count observed in Landvikvannet. The vertebral count was not significantly related to change in salinity over season within habitats; there was negligible change at assumed spawning depth. However, there were significant changes in temperature over season in both habitats, coinciding with a significant decrease in vertebral count at spawning time for both CSS and LV herring ($p < 0.05$).

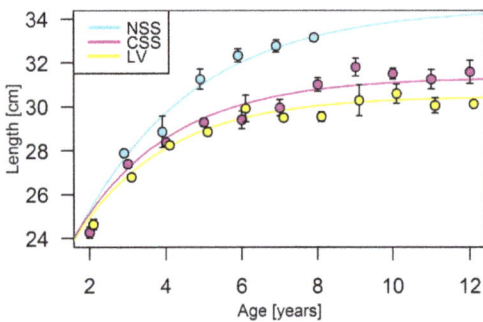

Figure 5. Growth curves of different herring populations. Length-at-age for Norwegian spring spawning (NSS, N = 212), Coastal Skagerrak spring spawning (CSS, N = 667) and Landvik (LV, N = 371) herring in samples pooled over the 2012 spawning season. Means and standard error (1 SE) are given, lines show van Bertalanffy growth models fitted to data.

Discussion

This study reveals strong seasonal dynamics involving three populations of a pelagic migratory fish, the Atlantic herring, in the vicinity of a marginal inland brackish water lake habitat (Landvikvannet) on the Norwegian Skagerrak coast. Gillnet sampling was standardized, implying that the observed differences between herring populations and over season dynamics were not affected by the selectivity normally experienced with gillnet sampling [59]. Three putative herring populations were identified; Norwegian spring spawners (NSS), Landvik herring (LV) and Coastal Skagerrak spring spawners (CSS). Individual NSS herring were identified subjectively based on otolith growth characteristics, and statistically based on otolith shape and mean vertebral count (57.5). NSS herring also had higher growth than the other populations, which is typical for this stock [13,43]. Identification of individual CSS and Landvik herring was not possible. Individuals sampled inside the lake were all classified as LV herring, whereas those sampled outside the channel connecting the lake to the sea were assigned as CSS herring. However, there was a significant decrease in vertebral count over the sampling season in both LV and CSS herring, from levels known as typical for CSS herring (56.5–56.9) in March–April to levels typical for Landvik herring (<56.0) in May–June, again based on historic data [43]. This trend in vertebral count was followed by a decrease in size and change in otolith shape, and a marked change in the relative proportions of the two populations.

The observed seasonal dynamics in biological characters clearly indicate that the assignment of individual fish into CSS and LV herring simply based on sampling location was uncertain, and that

Table 3. Von Bertalanffy growth parameters (L_∞, k, and t_0) of herring populations Norwegian spring spawners (NSS), Coastal Skagerrak spring spawners (CSS) and Landvik herring (LV).

	L_∞	K	t_0
NSS	34.51	0.33	−1.98
CSS	31.31	0.41	−1.98
LV	30.33	0.43	−1.98

the two populations were mixing both inside and outside the lake habitat together with NSS herring showing a different peak occurrence. Early in the season in February–April the biological characteristics indicated that NSS and CSS herring predominated, with only small numbers entering the lake. There was a clear temporal and spatial overlap in spawning individuals from these two populations, although proportions spawning in CSS were comparatively lower than in NSS herring. In May–June there was a significant change with the appearance of a new spawning wave of LV herring, with the highest proportion found inside the lake. Still, the immigration of this population was evident throughout both habitats, where many of the herring found in the fjord would be expected to enter the lake. The data on otolith shape, vertebral count and growth in May tended to differ from the observations in June in both locations, which indicated a spatial and temporal overlap in May between minor proportions of NSS and CSS herring completing their spawning season at the same time as the LV herring was peaking.

All three putative populations were caught at the same location, in the same gillnets, at the same time with running gonads, suggesting that the populations together form a metapopulation [40]. However, there is doubt as to whether interbreeding between distinct populations is occurring despite their proximity in spawning condition. Since breeding was not observed directly,

one cannot exclude the possibility that the populations separate for spawning events. Such a full separation seems unlikely for NSS and CSS herring because of the high temporal and spatial overlap; whereas it seems more likely for LV herring considering the limited temporal and spatial overlap with the other populations.

The idea that LV herring is reproductively isolated from other populations may be supported by the low vertebral count and concept of natal homing. Differences in vertebral count stem from the incubation phase and thus reflect the origin of the fish at spawning [60]. In general, there is a positive correlation with salinity [31] and negative with temperature [21,29,61] experienced prior to hatching. Hence, the warmer and less saline ambient environment for herring occurring inside Landvikvannet in May–June compared with that experienced by CSS in March–April in the fjord habitat, could result in the observed differences in vertebral count. The low vertebral count of LV herring and the late timing of spawning is an indication of spawning and adaptations to the environmental conditions of the lake habitat. However, this also implies that natal homing [62,63] of Landvik herring occurs on an annual basis. The vertebral number for LV herring in May has been remarkably stable (55.5–55.8) since 1984 [43], supporting natal homing. The principle of natal homing is central to the discrete population concept [12]. Moreover, recent genetic studies support the occurrence of natal homing of herring in the North and Baltic Seas [6,64]. Likewise, Brophy et al. [65] suggested that spawning season and location of Atlantic herring could be predetermined and not learnt from repeated spawning [66]. Support for natal homing and adaptations of Landvik herring to environmental conditions of its marginal habitat also originates from a recent genetic study using 20 microsatellite markers, where Landvikvannet differed from other local herring in Lindåspollene, Lusterfjord and Trondheimsfjord as well as from other herring populations surrounding the Norwegian Sea [67]. Unpublished results on the microsatellite locus Cpa112, which is non-neutral to salinity variability with allele frequencies varying from 45% in the Baltic to 2–4% in the North Sea [27], have shown that Landvik herring is obvious with a frequency of 15% (Carl André, pers. Comm., Department of Biology and Environmental Sciences - Tjärnö, University of Gothenburg, Strömstad, Sweden).

It seems clear from this study that we can refute the hypothesis of a resident local population inside the lake; LV herring definitely migrates into the lake habitat from coastal areas. In this sense the Landvik herring differs from other local herring populations, such as the Trondheimsfjord or Lindås herring, which can be observed throughout the year in their local areas [32,33,36,41]. This may simply be because of the unsuitability of this location as a nursery area for juveniles and feeding grounds for adults. Both CSS and LV herring may still represent more stationary coastal populations not undertaking large scale oceanic migrations. The observed relatively low investment costs in reproduction (low GSI) of NSS compared with that of LV herring supports the assumption that

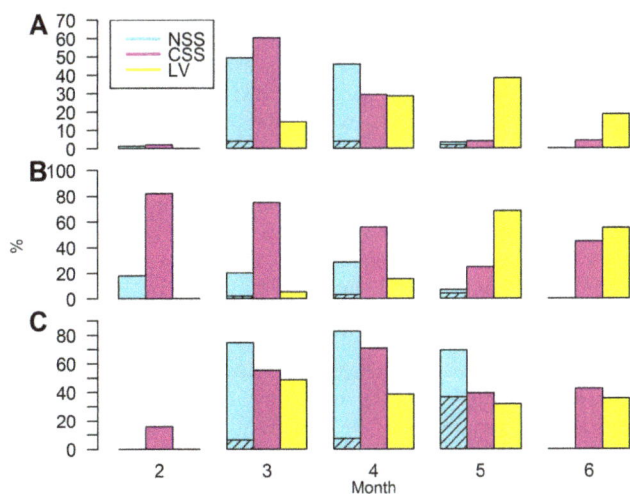

Figure 6. Seasonal change in proportion of different herring populations. Proportion (%), standardized to one gillnet per sample and area, by month of Norwegian spring spawning (NSS), Coastal Skagerrak spring spawning (CSS) and Landvik (LV) herring relative to a) total number analyzed over entire study period (see Table 1 for N), b) total number at month and c) spawning and spent herring (stage of maturity> = 6) relative to total number at month (see Table 2 for N). Shaded areas are NSS herring inside Landvikvannet.

Table 4. Comparing otolith shape, vertebral count (VS) and length among herring populations Norwegian spring spawners (NSS), Coastal Skagerrak spring spawners (CSS) and Landvik herring (LV).

Comparison		N	Otolith shape				Vertebral count				Fish length			
			df	Var	F	P	df	Mean Sq	F	P	df	Mean Sq	F	P
Overall	NSS vs LV vs CSS	897	2	3.28	5.36	<0.001	2	109.95	136.44	<0.001	2	129.80	102.58	<0.001
	Month		1	1.20	3.91	<0.001	1	71.49	88.71	<0.001	1	690.00	545.44	<0.001
	Age		10	4.49	1.47	0.001	10	3.87	4.80	<0.001	10	178.20	140.90	<0.001
	Residuals		883	270.41			867	0.81			867	1.30		
Posteriori	LV vs CSS	745	1	0.69	2.22	0.04	1	32.10	36.69	<0.001	1	13.10	10.08	0.006
	NSS vs LV	500	1	1.45	4.76	<0.001	1	219.80	276.99	<0.001	1	250.45	196.30	<0.001
	NSS vs CSS	549	1	0.84	2.72	0.02	1	115.53	149.39	<0.001	1	178.20	114.88	<0.001
	NSS-ILV vs NSS-OLV	152	1	0.20	0.65	>0.05	1	0.23	0.47	>0.05	1	1.85	1.65	>0.05

NSS herring were also compared between sampling locations, inside (NSS-ILV) and outside (NSS-OLV) Landvikvannet. ANOVA like permutation tests were used to assess the difference in otolith shape and ANOVA for the vertebral count and fish length comparisons. For otolith shape: df: degrees of freedom, Var: Variance among populations, F: pseudo F-value, P: proportion of permutations which gave as large or larger F-value than the observed one. For the vertebral count and fish length: df: degrees of freedom, Mean Sq: Mean Square, F: F-value, P: P-value. P-values for posteriori comparisons have been corrected with a Bonferroni correction. P<0.05 indicates a significant effect.

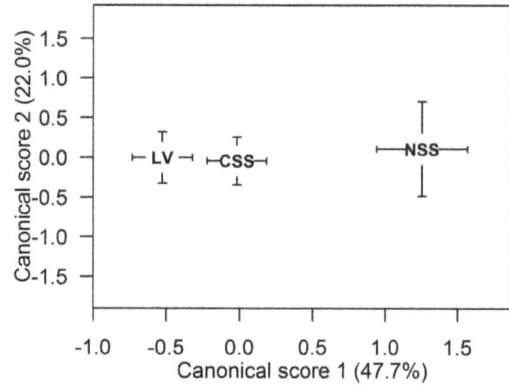

Figure 7. Otolith shape compared for different herring populations. Canonical scores for Norwegian spring spawning (NSS, N = 152), Coastal Skagerrak spring spawning (CSS, N = 397) and Landvik (LV, N = 348) herring are shown on discriminating axes 1 and 2. Black letters represent the mean canonical value for each group with standard error of the mean (1 SE).

NSS is more migratory [44]. The fact that growth of CSS was higher than in LV herring, further suggest that these two populations may not overlap much during the nursery period or at adult feeding grounds. In fact, there is probably little or no spatial overlap for most of the year, with overlap only occurring during the spawning season.

The movements of herring between the fjord and Landvikvannet habitats have also been studied with acoustic telemetry [43,68]. The telemetry study showed that some fish moved in and out of the lake habitat, whereas others stayed inside the lake for more than two weeks. Those fish that arrived and only stayed for a short period of time were interpreted as being NSS or CSS, whereas the ones remaining in the area for extended periods of time were thought to be local LV herring. It is likely that some NSS and CSS herring have short visits to the lake as exploratory migrations searching for good habitats cued by the current from the Reddal channel, but migrate out again to spawn in areas which are more characteristic of their normal spawning habitat. Conversely, fish that stay for two weeks inside the lake before leaving is a reasonably good indication of an established adaptation to the lake and to potential spawning within the lake.

The appearance of NSS herring in the habitats within Landvikvannet and adjacent fjords probably does not represent natal homing. The predominance of 3-year-olds among the NSS stock as well as the high stability of growth and meristic characters over the season, suggest independent selection of spawning grounds, as supported by Slotte and Fiksen [69]. In NSS herring specifically, the use of spawning grounds other than their natal ground is common. NSS herring have a tendency to change their spawning ground as they grow older with larger fish tending to migrate further, in this case southward, and thus potentially increase their life time fitness [69–71]. Such straying from natal spawning grounds results in considerable gene flow [72,73]. The predominance of 3-year-old NSS mixing with CSS and Landvik herring in 2012 may be explained by the relatively unusual spawning migrations of NSS herring in 2009–2010. During these two years a significant proportion of the adult NSS migrated from wintering grounds in the northern Norwegian Sea to areas south of 60°N, resulting in the largest fishery in the fjords (e.g. Boknafjorden) east of the traditional spawning grounds off Karmøy since the 1950s [74]. Based on vertebral count and growth data, it was apparent that the fishery was targeting NSS

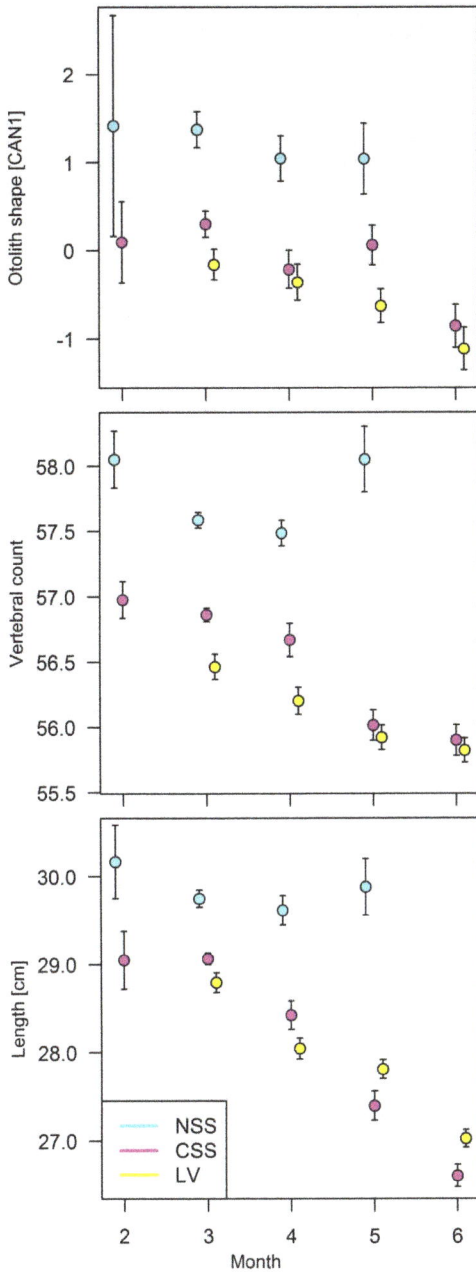

Figure 8. Seasonal changes of otolith shape, vertebral counts and length for different herring populations. For standardized ages. Comparison between Norwegian spring spawning (NSS), Coastal Skagerrak spring spawning (CSS) and Landvik (LV) herring (see Table 2 for N). Values given are means and standard errors (1 SE).

herring [75] and the abundance was high as evaluated by catch levels (Table 5). One hypothesis is that the 3 year old NSS mixing with CSS and Landvik herring in 2012 was a result of this significant spawning at the southern grounds in 2009. Generally, if first time spawners of NSS do not meet older conspecifics and learn to follow their migration towards the spawning grounds then the location of the spawning ground is a chance event [70,71,76,77]. In addition, NSS herring tend to migrate upstream to spawn [69]. Therefore it is not unlikely that NSS from Boknafjorden or further south may have spawned close to their nursery areas or even migrated further south-eastwards against the

Table 5. Commercial catches of herring off Karmøy 2005–2012.

Month	Year of catch							
	2005	2006	2007	2008	2009	2010	2011	2012
1					0.1			
2	21.2	32.6			172.0	3302.9	609.1	897.3
3	24.5		16.5		19052.0	14877.0	6528.4	6283.2
4	129.2	0.7	1.0	4.8	2301.2	1000.3	52.0	13.4
8	1.0							
9			0.1		0.9			
10								
11		72.8				0.5		
12	0.2							
Total	176.1	106.1	17.6	4.8	21526.2	19180.7	7189.5	7193.9

Live weight (tons) calculated from landed weight to live weight equivalent for Norwegian spring spawning herring in the Norwegian statistical area 08 (SW coastal Norway) by month and year as registered in the Directorate of Fisheries database.

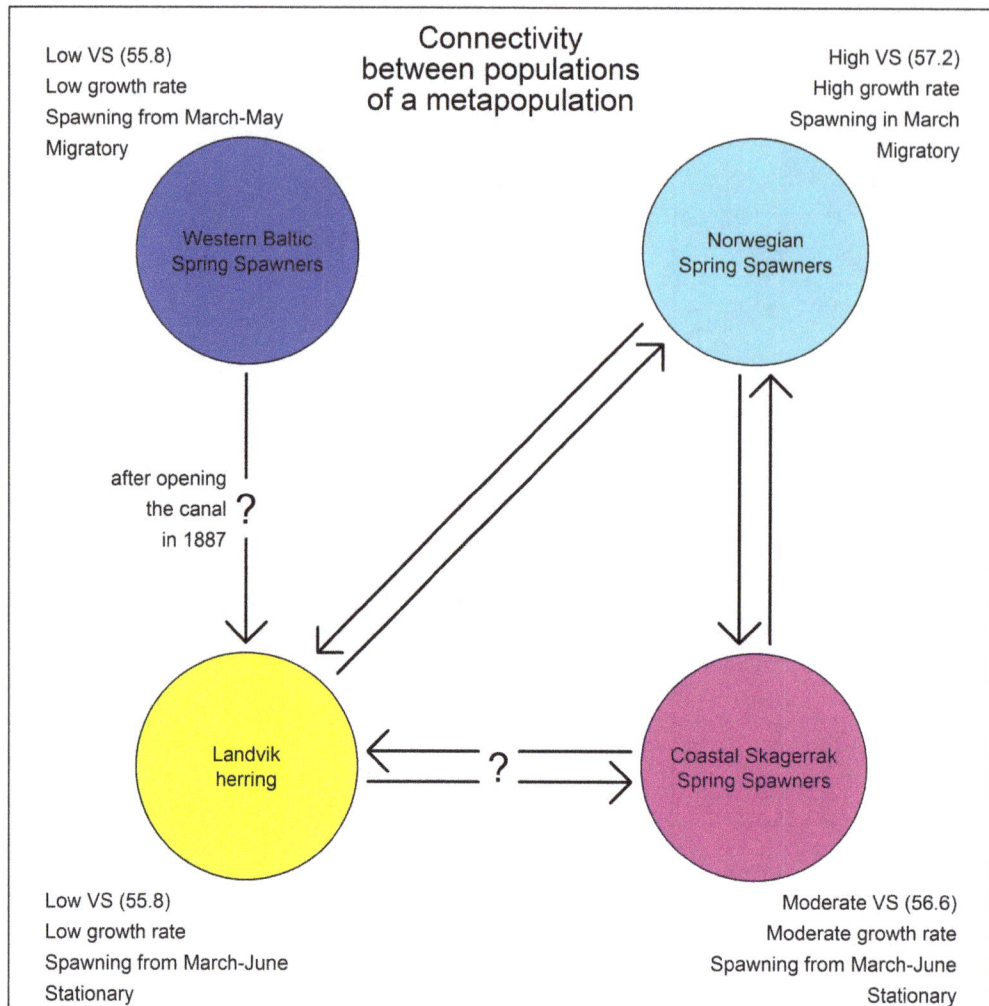

Figure 9. A schematic model of potential metapopulation dynamics in the study area. Potential connectivity between populations of a metapopulation in the study area of Landvikvannet and the connected fjords as hypothesized based on the results of the present study. The biological characteristics (VS = vertebral counts) of the different populations are given.

coastal current to spawn. In addition, school composition tends to involve size-matching among individuals [78], in this case younger, smaller NSS. Three year old NSS (mostly first-time spawners), may have adopted the behavior of the joint local populations with whom they mix during the nursery period as postulated in the adopted-migrant hypothesis [40,79].

From an evolutionary perspective, the Landvikvannet habitat has only been available for marine species for a relatively short period of time. This raises the question of the origin of the herring first colonizing the lake after the opening of the Reddal channel (Figure 9). One possibility is that CSS herring entered the lake sometime after the opening of the channel and successfully spawned there. Due to lower salinity and higher temperature in the lake the offspring developed significantly divergent characters over the years. A strong natal homing effect of herring would lead to the development of a new local population inside Landvikvannet. Hendry and Kinnison [80] concluded that a time span less than 100 years can be sufficient for significant microevolution to develop in response to local agents of selection. Also, Neb [81] demonstrates that such a time interval and differences in salinity are sufficient for herring to diverge in meristic characters. This explanation assumes reproductive isolation during spawning

between the original CSS herring and the "new" Landvik herring. A second possibility is that the origin of Landvik herring could be Western Baltic Spring Spawners (WBSS) herring. First time, or even repeated, spawners could have established a new spawning ground in Landvikvannet. The reason for not conducting an annual migration to the original spawning grounds off the island Rügen may be a trade-off between survival of progeny and physiological migration constraints, as shown for NSS by Slotte [70]. WBSS close to their feeding grounds in the Skagerrak could have "discovered" Landvikvannet, cued by similar environmental conditions as those of their original spawning grounds. The continued link to Landvikvannet may have been a result of a fidelity to this site rather than for joining conspecifics in a migration back in to the Baltic region. Huse et al. [76] demonstrate that a high ratio of first-time spawners could lead to the establishment of new wintering grounds. In the case of Landvik herring, it may have led to a new spawning ground.

In conclusion, the present study provides evidence for a distinct small local population of herring associated with Landvikvannet, partly mixing with NSS and CSS herring. This population of LV herring resides, during part of the year in brackish water with many morphometric characteristics indicative of spawning in warm and low salinity environments. Whilst ripe and spent fish have been found in the area, there is no direct evidence of spawning in the lake. If spawning does occur there are no data to indicate likely survival rates or even the residence time of offspring in the lake. There has been one attempt to find eggs with a diver for 1 hour at one of the many bays in the lake, without success. Also, limited plankton net sampling in selected parts of the lake have failed to capture any larvae. The only evidence of potential spawning in the lake, is from two eels with stomachs full of fertilized herring eggs. There is also no clear evidence of the origin of this population, however, they could have arisen from either WBSS or other local CSS. The presence of mixtures of these and other stocks and populations in the Skagerrak area have been shown previously [6,82]. Recent genetic studies using microsatellite DNA [83] have demonstrated differences between Landvik herring and many other stocks, in addition, unpublished results on one microsatellite locus (Carl André, pers. Comm., Department of Biology and Environmental Sciences - Tjärnö, University of Gothenburg, Strömstad, Sweden) suggesting that Landvikvannet herring has not recently immigrated from the Baltic.

The results of the present study may also have some implications for the official ICES stock assessment of herring in the North Sea and Skagerrak area. The present work demonstrates that there can be a fairly complex population structure in the areas with more than one 'stock' which can be mixed. Whilst this may not be a significant problem for the assessment of NSAS or WBSS due to the relatively small abundances of CSS and LV herring, there is a possibility that these smaller populations could be very vulnerable to overfishing [9]. This is probably not unique for coastal areas as there are a number of relatively small populations bordering the North Sea and Skagerrak area [84].

From management point of view, probably the most striking result of the present study is the conclusive evidence of NSS herring as far southeast as in the Skagerrak. This is the first time that individuals from this historically large herring stock have been studied in the Skagerrak area. By definition this stock is not exploited south of 62°N, with exception of the spawning period when they previously have been found as far south as to Lindesnes (Figure 1). This signifies that migration dynamics and population connectivity among herring in the Northeastern Atlantic may be more dynamic than previously assumed, and this must be taken into account in the future development and implementation of new management strategies.

Acknowledgments

Knut Hansen is thanked for his very valuable contribution with running the sampling program throughout the 2012 spawning season and being in charge of biological analyses. The following are also thanked for their contributions to the research: Åse Husebø undertook the initial work on photographing and outlining otoliths for the shape analyses, Jostein Røttingen and Inger Henriksen contributed to sampling and biological analyses, Terje Jåvold analyzed the water samples and Øystein Paulsen undertook photographing and assisted with the sampling. The two anonymous reviewers are thanked for their very valuable suggestions for improvements to the manuscript.

Author Contributions

Conceived and designed the experiments: FE AS LAL. Performed the experiments: FE AS LAL. Analyzed the data: FE AS LAL. Contributed reagents/materials/analysis tools: FE AS LAL AJ EMO EM. Wrote the paper: FE AS LAL RDMN. Reviewing the manuscript: FE AS LAL AJ EMO EM CK RDMN.

References

1. Heincke F (1898) Naturgeschichte des Herings. Berlin: Otto Salle Verlag.
2. Sinclair M, Iles DT (1988) Population richness of marine fish species. Aquat Living Resour 1: 71–83.
3. McPherson AA, Stephenson RL, O'Reilly PT, Jones MW, Taggart CT (2001) Genetic diversity of coastal Northwest Atlantic herring populations: implications for management. J Fish Biol 59: 356–370.
4. Wallace RK, Fletcher KM (1997) Understanding Fisheries Managment: A Manual for understanding the Federal Fisheries Management Process, Including Analysis of the 1996 Sustainable Fisheries Act: Mississippi-Alabama Sea Grant Consortium.
5. Cochrane KL (2002) A fishery manager's guidebook. Management measures and their application. Rome: FAO.
6. Ruzzante DE, Mariani S, Bekkevold D, André C, Mosegaard H, et al. (2006) Biocomplexity in a highly migratory pelagic marine fish, Atlantic herring. Proc R Soc B Biol Sci 273: 1459–1464.
7. Stephenson RL, Melvin GD, Power MJ (2009) Population integrity and connectivity in Northwest Atlantic herring: a review of assumptions and evidence. ICES J Mar Sci 66: 1733–1739.
8. Kell LT, Dickey-Collas M, Hintzen NT, Nash RDM, Pilling GM, et al. (2009) Lumpers or splitters? Evaluating recovery and management plans for metapopulations of herring. ICES J Mar Sci 66: 1776–1783.
9. Hintzen NT, Roel B, Benden D, Clarke M, Egan A, et al. (2014) Managing a complex population structure: exploring the importance of information from fisheries-independent sources. ICES J Mar Sci In press.
10. Begg GA, Friedland KD, Pearce JB (1999) Stock identification and its role in stock assessment and fisheries management: an overview. Fish Res 43: 1–8.
11. ICES (2012) Report of the ICES Advisory Committee 2012.
12. Iles TD, Sinclair M (1982) Atlantic herring: Stock discreteness and abundance. Science 215: 627–633.
13. Runnström S (1941) Quantitative investigations on herring spawning and its yearly fluctuations at the West coast of Norway FiskDir Skr Ser HavUnders 6.
14. Meyer HA (1878) Beobachtungen über das Wachstum des Herings im westlichen Theilen der Ostsee. Wiss Meer 2: 227.
15. Soleim PA (1942) Årsaker til rike og fattige årganger av sild. FiskDir Skr Ser HavUnders 7.
16. Corten A (1986) On the causes of the recruitment failure of herring in the central and northern North Sea in the years 1972–1978. J Cons int Explor Mer 42: 281–294.
17. Dragesund O, Hamre J, Ulltang Ø (1980) Biology and population dynamics of the Norwegian spring- spawning herring. Rapp P-v Réun Cons Int Explor Mer 177: 43–71.
18. Russell FS (1976) The eggs and planktonic stages of British marine fishes. London: Academic Press.
19. Blaxter JHS, Parrish BB (1965) The importance of light in shoaling, avoidance of nets and vertical migration by herring. J Cons int Explor Mer 30: 40–57.
20. Woodhead PMJ, Woodhead AD (1955) Reactions of herring larvae to light: a mechanism of vertical migration. Nature 176: 349–350.
21. Hulme TJ (1995) The use of vertebral counts to discriminate between North Sea herring stocks. ICES J Mar Sci 52: 775–779.
22. Bekkevold D, Clausen LAW, Mariani S, André C, Christensen TB, et al. (2007) Divergent origins of sympatric herring population components determined using genetic mixture analysis. Mar Ecol Prog Ser 337: 187–196.
23. Clausen LAW, Bekkevold D, Hatfield EMC, Mosegaard H (2007) Application and validation of otolith microstructure as a stock identification method in mixed Atlantic herring (Clupea harengus) stocks in the North Sea and western Baltic. ICES J Mar Sci 64: 377–385.
24. Geffen AJ, Nash RDM, Dickey-Collas M (2011) Characterisation of herring populations to the west of the British Isles: an investigation of mixing between populations based on otolith microchemistry. ICES J Mar Sci 68: 1447–1458.
25. Bekkevold D, André C, Dahlgren TG, Clausen LAW, Torstensen E, et al. (2005) Environmental correlates of population differentiation in Atlantic herring. Evolution 59: 2656–2668.
26. Jørgensen HBH, Hansen MM, Bekkevold D, Ruzzante DE, Loeschcke V (2005) Marine landscapes and population genetic structure of herring (Clupea harengus L.) in the Baltic Sea. Mol Ecol 14: 3219–3234.
27. André C, Larsson LC, Laikre L, Bekkevold D, Brigham J, et al. (2011) Detecting population structure in a high gene-flow species, Atlantic herring (Clupea harengus): direct, simultaneous evaluation of neutral vs putatively selected loci. Heredity 106: 270–280.
28. Limborg MT, Helyar SJ, De Bruyn M, Taylor MI, Nielsen EE, et al. (2012) Environmental selection on transcriptome-derived SNPs in a high gene flow marine fish, the Atlantic herring (Clupea harengus). Mol Ecol 21: 3686–3703.
29. Hempel G (1953) Die Temperaturabhängigkeit der Myomerenzahl beim Hering (Clupea harengus L.). Naturwissenschaften 17: 467–468.
30. Blaxter JHS (1957) Herring rearing III - The effect of temperature and other factors on myotome counts. Mar Res Scot 1: 1–16.
31. Hempel G, Blaxter JHS (1961) The experimental modification of meristic characters in herring (Clupea harengus L.). J Cons int Explor Mer 26: 336–346.

32. Broch H (1908) Norwegische Heringsuntersuchungen während der Jahre 1904–1906. Bergen Museums Årbok 1.

33. Runnstrøm S (1941) Racial analysis of the herring in Norwegian waters. FiskDir Skr Ser HavUnders 6.

34. Rasmussen T (1942) The Borge Poll Herring. FiskDir Skr Ser HavUnders 7: 63–73.

35. Aasen O (1952) The Lusterfjord herring and its environment. FiskDir Skr Ser HavUnders 10.

36. Lie U, Dahl O, Østvedt OJ (1978) Aspects of the life history of the local herring stock in Lindåspollene, western Norway. FiskDir Skr Ser HavUnders 16: 369–404.

37. Jørstad KE, Pedersen SA (1986) Discrimnation of herring populations in a northern Norwegian fjord: genetic and biological aspects. ICES CM 1986/H:63.

38. Hognestad PT (1994) The Lake Rossfjord herring (Clupea harengus L.) and its environment. ICES J Mar Sci 51: 281–292.

39. Husebø Å, Slotte A, Clausen LAW, Mosegaard H (2005) Mixing of populations or year class twinning in Norwegian spring spawning herring? Mar Freshw Res 56: 763–772.

40. McQuinn I (1997) Metapopulations and the Atlantic herring. Rev Fish Biol Fish 7: 297–329.

41. Johannessen A, Nøttestad L, Fernö A, Langård L, Skaret G (2009) Two components of Northeast Atlantic herring within the same school during spawning: support for the existence of a metapopulation? ICES J Mar Sci 66: 1740–1748.

42. Johannessen A, Skaret G, Langård L, Slotte A, Husebø Å, et al. (2014) The Dynamics of a Metapopulation: Changes in Life-History Traits in Resident Herring that Co-Occur with Oceanic Herring during Spawning. PLoS ONE 9: e102462.

43. Eggers F (2013) Metapopulation dynamics in Atlantic herring (Clupea harengus L.) along the coast of southern Norway and in the local area of Landvikvannet. Bergen: University of Bergen. 100 p.

44. Silva FFG, Slotte A, Johannessen A, Kennedy J, Kjesbu OS (2013) Strategies for partition between body growth and reproductive investment in migratory and stationary populations of spring-spawning Atlantic herring (Clupea harengus L.). Fish Res 138: 71–79.

45. Winkler LW (1888) Die Bestimmung des im Wasser gelösten Sauerstoffes. Ber Dtsch Chem Ges 21: 2843–2854.

46. Mjanger H, Hestenes K, Svendsen BV, Wenneck TdL (2012) Håndbok for prøvetaking av fisk og krepsdyr. Bergen: Institute of Marine Research.

47. R Core Team (2012) R: A language and environment for statistical computing. R Foundation for Statistical Computing, Vienna, Austria. ISBN 3-900051-07-0, URL http://www.R-project.org.

48. Bivand R, Leisch F, Maechler M (2011) pixmap: Bitmap Images ("Pixel Maps"). R package version 0.4-11. http://CRANR-projectorg/package=pixmap.

49. Claude J (2008) Morphometrics with R. Springer, New York, USA. 316 pp.

50. Nason G (2012) wavethresh: Wavelets statistics and transforms. R package version 4.5. http://CRAN.R-project.org/package=wavethresh.

51. Agüera A, Brophy D (2011) Use of saggital otolith shape analysis to discriminate Northeast Atlantic and Western Mediterranean stocks of Atlantic saury, Scomberesox saurus saurus (Walbaum). Fish Res 110: 465–471.

52. Begg GA, Overholtz WJ, Munroe NJ (2001) The use of internal otolith morphometrics for identification of haddock (Melanogrammus aeglefinus) stocks on Georges Bank. Fish Bull 99: 1–14.

53. Longmore C, Fogarty K, Neat F, Brophy D, Trueman C, et al. (2010) A comparison of otolith microchemistry and otolith shape analysis for the study of spatial variation in a deep-sea teleost, Coryphaenoides rupestris. Environ Biol Fish 89: 591–605.

54. Lleonart J, Salat J, Torres GJ (2000) Removing Allometric Effects of Body Size in Morphological Analysis. J Theor Biol 205: 85–93.

55. Bertalanffy Lv (1934) Untersuchungen über die Gesetzlichkeit des Wachstums. Wilhelm Roux Arch Entwickl Mech Org 131: 613–652.

56. Anderson MJ, Willis TJ (2003) Canonical Analysis of Principal Coordinates: A useful method of constrained ordination for Ecology. Ecology 84: 511–525.

57. Oksanen J, Blanchet FG, Kindt R, Legendre P, Minchin PR, et al. (2013) vegan: Community Ecology Package. R package version 2.0-7. http://CRAN.R-project.org/package=vegan.

58. Sokal RR, Rohlf FJ (1995) Biometry. 3rd edn. New York, NY: W. H. Freeman and Company.

59. Hamley JM (1975) Review of gillnet selectivity. J Fish Res Bd Can 32: 1943–1969.

60. Pavlov DA, Shadrin AM (1998) Development of variation in the number of myomeres and vertebrae in the White Sea herring, Clupea pallasi marisalbi, under the influence of temperature. J Ichthyol 38: 251–261.

61. Johnston II, Cole N, Vieira VV, Davidson II (1997) Temperature and developmental plasticity of muscle phenotype in herring larvae. J Exp Biol 200: 849–868.

62. MacLean JA, Evans DO (1981) The stock concept, discreteness of fish stocks, and fisheries management. Can J Fish Aquat Sci 38: 1889–1898.

63. Horrall RM (1981) Behavioral stock-isolating mechanisms in great lakes fishes with special reference to homing and site imprinting. Can J Fish Aquat Sci 38: 1481–1496.

64. Gaggiotti OE, Bekkevold D, Jørgensen HBH, Foll M, Carvalho GR, et al. (2009) Disentangling the effects of evolutionary, demographic, and environmental factors influencing genetic structure of natural populations: Atlantic herring as a case study. Evolution 63: 2939–2951.

65. Brophy D, Danilowicz BS, King PA (2006) Spawning season fidelity in sympatric populations of Atlantic herring (Clupea harengus). Can J Fish Aquat Sci 63: 607–616.

66. Haegele CW, Schweigert JF (1985) Distribution and characteristics of herring spawning grounds and description of spawning behavior. Can J Fish Aquat Sci 42: 39–55.

67. Pampoulie C, Slotte A, Óskarsson GJ, Helyar SJ, Jónsson Á, et al. (2014) Stock structure of Atlantic herring (Clupea harengus L.) in the Norwegian Sea and adjacent waters: Concordant genetic patterns between neutral and selective microsatellite loci? Submitted to Mar Ecol Prog Ser.

68. Eggers F, Olsen EM, Moland E, Slotte A (2014) Individual habitat transitions of Atlantic herring (Clupea harengus L.) in a human-modified coastal system. Submitted to Mar Ecol Prog Ser.

69. Slotte A, Fiksen Ø (2000) State-dependent spawning migration in Norwegian spring-spawning herring. J Fish Biol 56: 138–162.

70. Slotte A (1999) Effects of fish length and condition on spawning migration in Norwegian spring spawning herring (Clupea harengus L). Sarsia 84: 111–127.

71. Slotte A (2001) Factors influencing location and time of spawning in Norwegian spring spawning herring: An evaluation of different hypotheses. In: Funk F, Blackburn J, Hay D, Paul AJ, Stephenson R et al., editors. Herring: Expectations for a New Millennium: University of Alaska Sea Grant. pp. 255–278.

72. Hourston AS (1959) The relationship of the juvenile herring stocks in Barkley sound to the major adult herring populations in British Columbia. J Fish Res Bd Can 16: 309–320.

73. Smith PJ, Jamieson A (1986) Stock discreteness in herrings: A conceptual revolution. Fish Res 4: 223–234.

74. Directorate of Fisheries (2013) Landing- and sales documents (Landings- and sluttsedler) from Norwegian vessels landed in Norway and abroad. Statistics Department, Bergen, Norway.

75. Slotte A, Stenevik EK, Kvamme C (2009) A note on NSS herring fishery south of 62°N in 2009. Pelagic Fish Research Group, Institute of Marine Research, Bergen. 3 p.

76. Huse G, Fernö A, Holst JC (2010) Establishment of new wintering areas in herring co-occurs with peaks in the first time/repeat spawner ratio. Mar Ecol Prog Ser 409: 189–198.

77. Petitgas P, Secor DH, McQuinn I, Huse G, Lo N (2010) Stock collapses and their recovery: mechanisms that establish and maintain life-cycle closure in space and time. ICES J Mar Sci 67: 1841–1848.

78. Pitcher TJ, Magurran AE, Edwards JI (1985) Schooling mackerel and herring choose neighbours of similar size. Mar Biol 86: 319–322.

79. Corten A (2002) The role of "conservatism" in herring migrations. Rev Fish Biol Fish 11: 339–361.

80. Hendry AP, Kinnison MT (1999) Perspective: the pace of modern life: measuring rates of contemporary microevolution. Evolution 53: 1637–1653.

81. Neb K-E (1970) Über die Heringe des Wendebyer Noors. Ber Dtsch Wiss Komm Meeresforsch 21: 265–270.

82. Bekkevold D, Clausen LAW, Mariani S, Carl A, Hatfield EMC, et al. (2011) Genetic mixed-stock analysis of Atlantic herring populations in a mixed feeding area. Mar Ecol Prog Ser 442: 187–199.

83. Skírnisdóttir S, Ólafsdóttir G, Helyar S, Pampoulie C, Óskarsson GJ, et al. (2012) A Nordic network for the stock identification and increased value of Northeast Atlantic herring (HerMix). Matís ohf., Reykjavík, Iceland. 50 p.

84. Dickey-Collas M, Nash RDM, Brunel T, van Damme CJG, Marshall CT, et al. (2010) Lessons learned from stock collapse and recovery of North Sea herring: a review. ICES J Mar Sci 67: 1875–1886.

4

Spatio-Temporal Patterns of Major Bacterial Groups in Alpine Waters

Remo Freimann[1,2]*, Helmut Bürgmann[3], Stuart E. G. Findlay[4], Christopher T. Robinson[2]

1 Institute of Molecular Health Sciences, Professorship of Genetics, ETH Zurich, Zurich, Switzerland, 2 Department of Aquatic Ecology, Swiss Federal Institute of Aquatic Science and Technology, Eawag, Dübendorf, Switzerland and Institute of Integrative Biology, ETH-Zurich, Zurich, Switzerland, 3 Department of Surface Waters – Research and Management, Swiss Federal Institute of Aquatic Science and Technology, Eawag, Kastanienbaum, Switzerland, 4 Cary Institute of Ecosystem Studies, Millbrook, New York, United States of America

Abstract

Glacial alpine landscapes are undergoing rapid transformation due to changes in climate. The loss of glacial ice mass has directly influenced hydrologic characteristics of alpine floodplains. Consequently, hyporheic sediment conditions are likely to change in the future as surface waters fed by glacial water (kryal) become groundwater dominated (krenal). Such environmental shifts may subsequently change bacterial community structure and thus potential ecosystem functioning. We quantitatively investigated the structure of major bacterial groups in glacial and groundwater-fed streams in three alpine floodplains during different hydrologic periods. Our results show the importance of several physico-chemical variables that reflect local geological characteristics as well as water source in structuring bacterial groups. For instance, *Alpha-*, *Betaproteobacteria* and *Cytophaga-Flavobacteria* were influenced by pH, conductivity and temperature as well as by inorganic and organic carbon compounds, whereas phosphorous compounds and nitrate showed specific influence on single bacterial groups. These results can be used to predict future bacterial group shifts, and potential ecosystem functioning, in alpine landscapes under environmental transformation.

Editor: Jack Anthony Gilbert, Argonne National Laboratory, United States of America

Funding: Funding provided by Swiss National Science Foundation (No. 31003A-119735) to CR HB SF RF. The funders had no role in study design, data collection and analysis, decision to publish, or preparation of the manuscript.

Competing Interests: The authors have declared that no competing interests exist.

* Email: remofreimann@gmail.com

Introduction

Heterotrophic bacteria are key players in the functional ecology of aquatic ecosystems, alpine waters in particular [1]. Their high metabolic capacity, phylogenetic variation and abundance enable bacterial assemblages to process and retain nutrients and chemical compounds under varying environmental conditions. Bacteria also represent an integral component within trophic food webs and global carbon cycling [2]. Glaciated alpine floodplains are an important source of fresh water as they are regions of high precipitation that is stored as snow/ice in winter and released during warm periods. Alpine environments not only modulate flow patterns and affect water chemistry but also provide microbe-mediated ecosystem services to waters used intensely by humans at lower elevations [3,4]. Furthermore, alpine headwaters can have a high variability in species diversity within relative small geographical distance and thus contribute to the maintenance of microbial and functional diversity in the fluvial network [5,6].

Members of bacterial groups are associated with metabolic traits and occupy habitat niches depending on the physico-chemical environment and apparent bacterial (single cell) functional plasticity [7]. Glaciated alpine floodplains comprise a mosaic of groundwater-fed (krenal) and glacial (kryal) streams differing in physico-chemical characteristics and thus different habitat niches [8]. The hyporheic zone of streams, in particular, is known as a biological hotspot of microbial abundance and functioning within these landscapes [9]. Environmentally-induced recession of glaciers and current shifts in precipitation patterns will affect alpine waters and associated habitats by reducing kryal systems at the landscape scale [10,11]. This reduction includes shifts in quality, quantity and timing of glacially-released organic matter and nutrients as well as changes in flow regimes that can alter hyporheic sediment characteristics [12]. Thus, ecological shifts in bacterial structure in conjunction with potential functioning are likely to occur. The underlying mechanics are dependent on the magnitude and rate of environmental change and are at least partially determined by the properties of the contemporary bacterial assemblage [13]. The magnitude of ecological change will ultimately depend on the degree of functional redundancy/ plasticity and is likely manifest in a combination of changes involving cell abundances, single cell metabolic activities and shifts in bacterial composition [7].

In previous studies, we documented a strong linkage between bacterial structure and function in hyporheic sediments [6]. We showed how ecological strategies of communities (generalists vs. specialists) influence the trajectory of community changes due to altered physico-chemical properties [13]. However, these earlier results only focused on the structure of bacterial assemblages without taking into account phylogenetic identities. In the present study, we quantitatively examined the phylogeny of major freshwater bacterial groups present in hyporheic sediments via catalyzed reporter deposition fluorescence in-situ hybridization (CARD-FISH) within three alpine glacial catchments. The catchments differed in the degree of deglaciation and thus covered a wide range of habitat types within krenal and kryal systems. These different physic-chemical conditions were associated with differences in bacterial groups and the broad range in habitat and bacterial attributes probably encompass future scenarios for alpine landscapes under transformation.

Material and Methods

We collected hyporheic sediments (~10 to 20 cm depth) from 45 stream sites in three alpine floodplains, Val Roseg (VR, 9°53′53″E, 46°29′24″N), Loetschental (L, 07°49′03″E, 46°25′08″N) and Macun (M, 10°07′31″E, 46°43′51″N), that differed in their degree of deglaciation (% of the catchment area covered by ice, data provided by the Federal Office for Environment and the Swiss National Park) and general landscape features such as the presence of interconnected lakes (Table 1, Figure 1). Sampling sites along streams were categorized into krenal and kryal systems, depending on the connectivity to glaciers/glacial meltwater and based on water chemistry patterns (Figure 1) [6]. Macun had a perennially-reduced glacial water input, and more specific characterizations of these catchments can be found elsewhere [6]. Sites were sampled during three distinct hydrological periods: glacial ablation in summer (August: A), winter stagnation (October: O) and snowmelt input in spring (June: J) in Val Roseg and Loetschental and for A and O in Macun. No specific permission was required for the sampling in VR and L. Permission for the sampling campaign in M was issued by the Swiss National Park. The study did not involve endangered or protected species.

A 0.5 ml aliquot of collected sediment was suspended in 1.11 ml paraformaldehyde (2%, final concentration) in an Eppendorf tube and fixed for 24 h at 4°C followed by three washing steps with 1× PBS and 5 min centrifugation at 10,000 g between washing steps. Samples were then stored at −20°C in a 1:1 mix of PBS/ethanol until further processing [14]. Cell detachment was done by sonication (Branson Digital Sonifier 250, Danbury, USA, 5-mm tapered microtip, actual output of 20 W, 30 s). Catalyzed reporter deposition fluorescence in-situ hybridization (CARD-FISH) was performed following the protocol of Pernthaler et al. [15] paired with a high throughput imaging system [16]. Horseradish labeled FISH probes (Biomers Inc, Ulm) EUB I-III targeting the domain

Bacteria (EUBI-III) [17], Alf968 and Bet42a affiliated with classes of Alphaproteobacteria (Alph) and Betaproteobacteria (Bet) of the phylum Proteobacteria, respectively [18,19], and CF319a assigned to the class of Cytophaga-Flavobacteria (CF) within the phylum Bacteroidetes [20] were used to quantify microbial groups within the stream sediments. The domains and classes are expressed as percentage of total bacterial abundance as assessed by counter-staining bacterial cells with 4′,6-diamidino-2-phenylindole (DAPI) (Sigma-Aldrich Co) [21].

Specific conductance (μS cm^{-1} at 20°C) and temperature were measured in the field with a conductivity meter (LF323, WTW, Weilheim, Germany). Surface water samples were analyzed for dissolved organic carbon (DOC), particulate organic carbon (POC), total inorganic carbon (TIC), ammonium (NH_4-N), nitrite (NO_2-N), nitrate (NO_3-N), dissolved organic nitrogen (DON), particulate organic nitrogen (PN), phosphate (PO_4-P), dissolved phosphorus (DP) and particulate phosphorus (PP) according to standard protocols detailed in Tockner et al. [22]. Sediments were analyzed for pH [23] and organic matter content (OM) as ash free dry mass. Grain size distribution of sediment was assessed by sieving with mesh sizes of 6.3, 2.0, 1.0, 0.5, 0.25, 0.125 and 0.063 mm. The D90/D10 gradation index was then calculated using GRADISTAT software [24]. The raw physico-chemical data have been published elsewhere [6].

To assess the explained variance of physico-chemical variables on CARD-FISH based bacterial community composition, we performed a redundancy analysis (RDA) based on forward selected environmental variables [25]. Significance of physico-chemical constraining variables and constrained RDA axes were tested by permutation tests (999 permutations) [26]. The relative contribution of physico-chemical variables to the constraint variation of single RDA axis was assessed via canonical correlation coefficients. Factor (catchment, water source and season) and vector (i.e. physico-chemical variables) fitting was performed on the first two RDA axes to assess their significance and relationship (r^2) to the phylogenetic patterns shown on these two axes. We tested all levels of interactions for the factor fitting, i.e. the influence of a single factor and all double and triple interactions of the respective factors. Pearson's product moment correlation was used to correlate physico-chemical variables with single bacterial groups and correlations were tested using Algorithm AS 89 [27]. Lastly, comparison of hybridization rates between water source, seasons and catchments were done using ANOVAs followed by Tukey's honest significance text (Tukey's HSD). Analyses were based on $\arcsin(\sqrt{x})$ transformed percentage values of CARD-FISH results

Table 1. Basic characteristics of the three catchments.

Catchment	Val Roseg	Loetschental	Macun
Coordinates	9°53′53″E, 46°29′24″N	07°49′03″E, 46°25′08″N	10°07′31″E, 46°43′51″N
Altitude [m a.s.l.]	1766–4049	1375–3200	2616–3046
Catchment area [km², (% glaciated)	66.5 (30.1)	77.8 (36.5)	3.6 (18.8)*
Annual precipitation [m]	1.6	1.1	0.9
Mean discharge [m³ s⁻¹]	28.5	37.2	ND
Mean water temperature of main channel [°C] (range)	3.6 (1–12)	4 (0.1–10.9)	2.9 (0.1–19.2)
Interconnected lakes	No	No	Yes
Geology, dominating minerals	Crystalline bedrock, diorite, granite	Crystalline bedrock, amphibolite, gneiss	Crystalline bedrock, ortho-gneiss

Abbreviation: ND, no data; *rock glaciers.

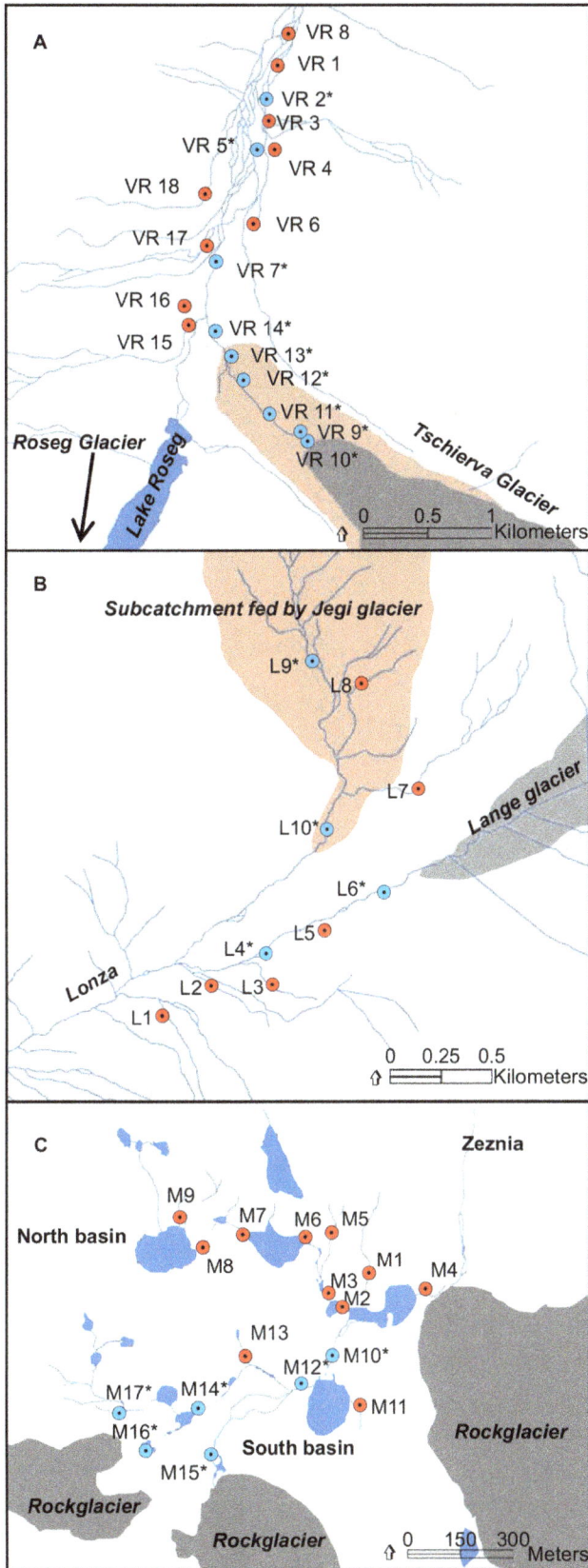

Figure 1. Map of the study sites in the three catchments, A: Val Roseg (VR), B: Loetschental (L) and C: Macun (M). Kryal sites are depicted in blue and annotated with an asterisk. Glaciers and the moraine area in VR and the sub-catchment in L are depicted in light grey and orange, respectively.

and log +1 transformed physico-chemical variables. All analysis were done using the R statistical environment [28].

Results

Sediments from L and VR had a higher mean *EUBI-III* hybridization rate than M (Figure 2, Table 2, Tukey's HSD: P< 0.001). *EUBI-III hybridization rate* in Macun showed proportionally the smallest overlap with the further defined lower taxonomic level (classes, Figure 2). We refer to the not further defined *Bacteria* group (i.e., %EUBI-III − (%Alph + %Bet + %CF)) as $EUBI\text{-}III_{(undef)}$ (Figure 3). There was a higher *EUBI-III*

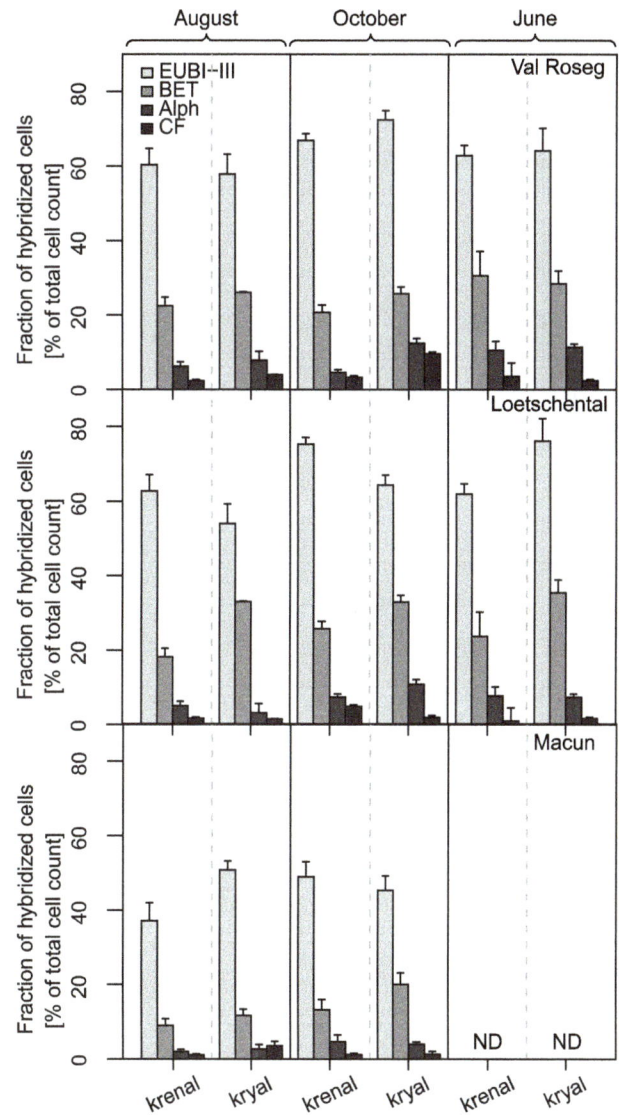

Figure 2. Bar plots (+1SE) of the relative abundance of major bacterial groups in the Val Roseg (A), Loetschental (B) and Macun (C) catchment. Bar groups are split within a panel by season and water source. ND: No data.

Table 2. ANOVA results (F values) for bacterial group abundance.

F-statistic parameter	Catchment (C)	Sampling date (S)	Water source (W)	C×S	C×W	S×W	C×S×W
EUBI-III	7.49***	6.68***	4.55*	1.19	3.26*	3.89*	3.1*
VR		4.68*	0.4	-	-	0.85	-
L		4.14*	0.11	-	-	5.08*	-
M		0.16	0.49	-	-	1.64	-
BET	20.22***	1.54	5.78*	1.39	4.18*	0.79	3.1
VR		3.49*	1.02	-	-	1.06	-
L		0.67	12.84**	-	-	0.48	-
M		5.44*	3.14	-	-	0.59	-
Alph	5.57**	5.98***	4.25*	1.06	1.23	1.51	1.05
VR		2.73	5.79*	-	-	2.3	-
L		6.48**	0.11	-	-	1.65	-
M		3.35	0	-	-	0.37	-
CF	8.26***	3.18*	5.37*	4.18**	4.68*	0.02	7.38***
VR		10.31***	10.89***	-	-	9.61***	-
L		1.87	0.76	-	-	1.4	-
M		3.13	4.04	-	-	3.45	-

Catchment (C), Sampling date (S) and Water source (W) were used as independent variables.

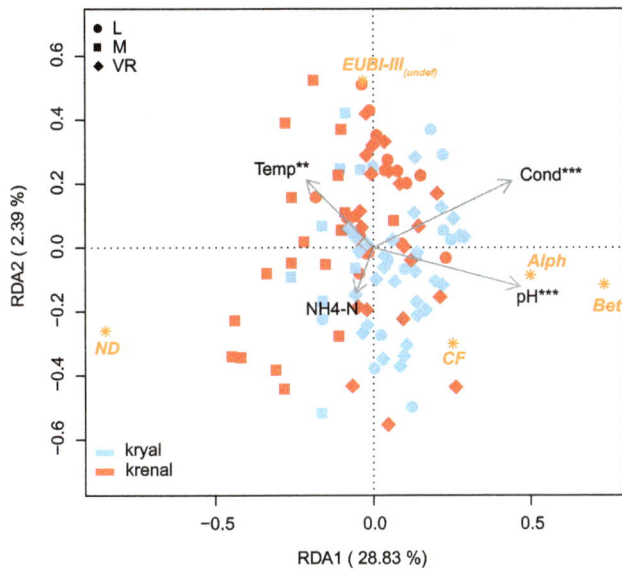

Figure 3. Biplot of the redundancy analysis based on the CARD-FISH data. Red symbols correspond to krenal sampling sites, whereas blue symbols depict kryal sites. Centroids of the respective probes are given: beta-*proteobacteria* (*Bet*), alpha-*proteobacteria* (*Alph*), Cytophaga-Flavobacteria (*CF*), Eubacteria excluding *Alph*, *Bet* and *CF* (*EUBI-III*(undef)) and not-defined DAPI positive cells (ND). Arrows depict the forward selected physico-chemical variables. Asterisks depict significantly tested variables (* p<0.05, ** p<0.01). Explained variation for the first two constraint axes is given.

hybridization rate in winter compared to summer (Tukey's HSD: P<0.01).

Bet was the most abundant group according to CARD-FISH and was lowest in M compared to the other two catchments (Tukey's HSD: p<0.001). *Bet* had on average highest abundance in kryal sediments in L (Tukey's HSD: P<0.001). *Alph* showed highest abundance in kryal sediments in VR (Tukey's HSD: p<0.05) and higher abundance in winter compared to spring in L (Tukey's HSD: p<0.01). *CF* was least abundant and VR had higher *CF* abundance compared to the other two catchments (Tukey's HSD: p<0.001) and showed a peak in kryal sediments in winter (Tukey's HSD: p<0.001). See also table 2 for detailed results of ANOVAs.

RDA based on forward selected environmental variables explained 31.9% of the total variation and revealed a differentiation of the three catchments concerning their phylogenetic group structuring (Figure 3). Catchment and water source were significantly (p<0.05) fitted on the RDA biplot ($r^2 = 0.09$ and 0.05, respectively). This was also true for the interaction term of catchment, sampling date and water source (P<0.001, $r^2 = 0.30$). The ordination ellipses in figure 4 depict the standard error of weighted average scores (confidence limit = 0.95) of the interaction sampling date × water source split by catchments. This figure is based on the RDA biplot of figure 3. Significant differences between water sources within a specific season can be expected when the respective ellipses do not overlap. The pH, conductivity, temperature and NH4-N contributed 23.2%, 2.1%, 1.7% and 1.8% to the total 28.8% explained variation on the first RDA axis (Figure 3). The pH, conductivity and temperature could also be fitted a-posteriori as independent variables on the biplot ($r^2 = 0.53$, 0.46 and 0.13, respectively, p<0.01). NH4-N could not be significantly fitted on the RDA (p = 0.73) (Figure 3). Additionally,

we fitted all non-forward selected environmental variables on the first two RDA axes of the forward selected model to assess their relative importance in structuring the bacterial communities (Table S1). OM, TIC and PP showed significant a-posteriori fitting on the first two RDA axes (p<0.01). The abundances of the different bacterial groups could be linked to several physico-chemical parameters (Figure 5). All bacterial groups were negatively correlated with OM and temperature, and *Bet* also was negatively correlated with POC and PN. TIC, Cond and pH were positively correlated with *Alph*, *Bet* and *CF*. PP was positively correlated with *Bet* and *CF*, whereas *Alph* was correlated with DN. PO4-P and NO2-N showed a positive Pearson correlation with *CF*, whereas *Bet* was correlated with NO3-N.

Discussion

Our results showed that bacterial group composition in hyporheic sediments of streams in glaciated alpine floodplains have a strong spatio-temporal dynamic. Physico-chemical differences between catchments, such as pH, conductivity and temperature, were related to geological and geographical characteristics, and dictate the coarse-scale boundary conditions on bacterial group composition. These factors have been described previously to drive assemblage composition and diversity in soils as well as in stream sediments, although not at the group level assessed here [6,29,30]. Importantly, these previously documented variables affecting bacterial structural patterns at a higher phylogenetic level (i.e. automated ribosomal intergenic spacer analysis) within these catchments were largely congruent to the variables structuring the lower phylogenetic level structures (i.e. CARD-FISH) in this study [13]. This congruence was also true for drivers (i.e. OM, POC, PP, TIC and NO2-N) linked more specifically to single bacterial groups (see Figure 5). Furthermore, enzymatic activities measured in the above mentioned study were used as constraints variables in an RDA analysis based on the CARD-FISH data and revealed significant correlations to the phylogenetic group structuring (29.6% total explained variation, see Figure S1). Results from the automated ribosomal intergenic spacer analysis and the enzyme activities were also fitted a-posteriori on the RDA biplot produced here (Figure S2). Taken together, these results underpin the separation of the different catchments based on the CARD-FISH data and indicate a substantial coupling of bacterial groups and their potential metabolic capabilities.

Most of the examined variables differed between the two water sources (kryal vs krenal) [6] and partly induced the observed seasonal turnover in assemblage structure (Figure 4). Kryal systems are rich in PP during summer ablation and favored *Bet* and *CF* within kryal habitats in VR and L (Figure 5) [22]. Catchment M showed reduced abundance of *Bet*, which was correlated to reduced PP input into kryal streams here [6]. Also, NH4-N and NO2-N levels are high in summer in kryal waters with the latter being positively correlated with the abundance of *Alph*, *Bet* and *CF* [13]. During winter, kryal systems become physicochemically more equal to krenal systems as glacial water input diminishes. Nevertheless, there was no distinct shift in kryal group composition towards a krenal one at the floodplain scale within the three catchments, suggesting relatively resistant local bacterial assemblages despite the aforementioned coupling of structure and function (see dispersion ellipses in figure 4). Experiments where kryal and krenal bacterial communities were cross-transplanted between their natural habitats revealed a high structural resistance along with a pronounced functional plasticity in response to physico-chemical disturbance [13]. Nevertheless, at this higher

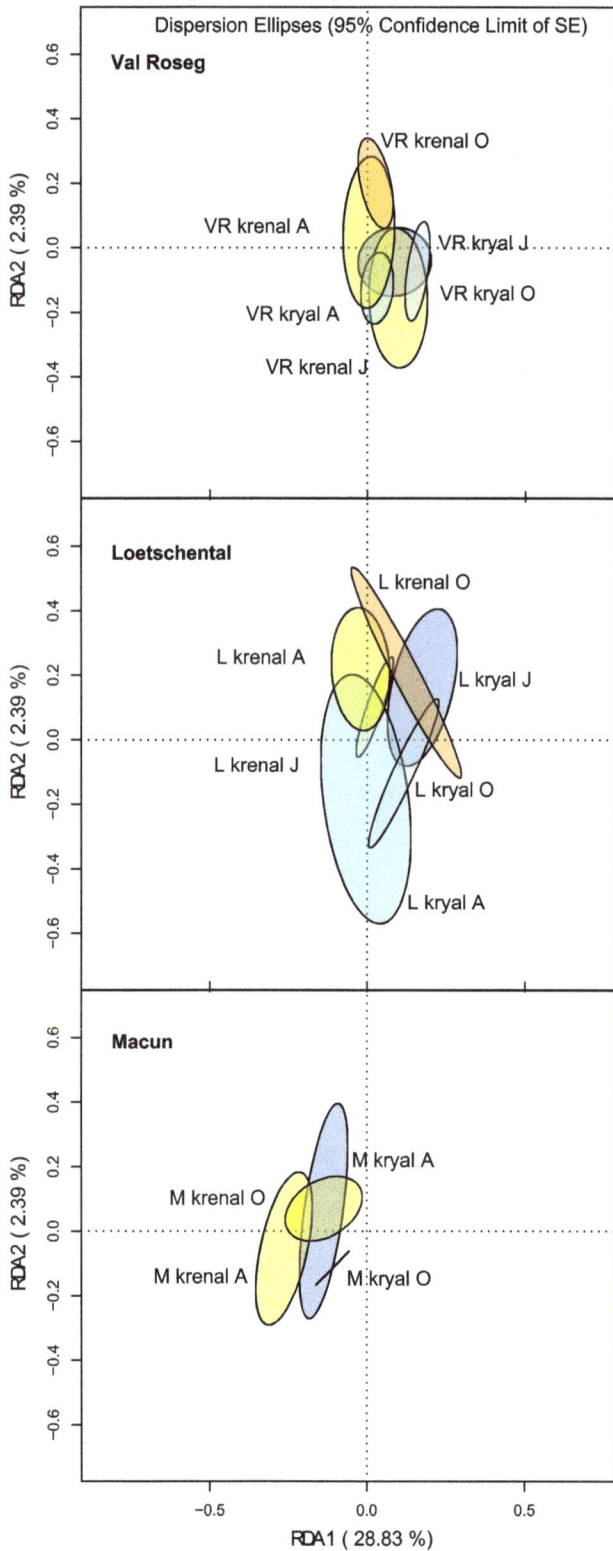

Figure 4. Dispersion ellipses fitted on the biplot of the redundancy analysis based on the CARD-FISH data constraint by physico-chemical variables (Figure 3). Dispersion ellipses split by catchments for different water sources and seasons are shown and depict the standard error of weighted average scores (confidence limit = 0.95).

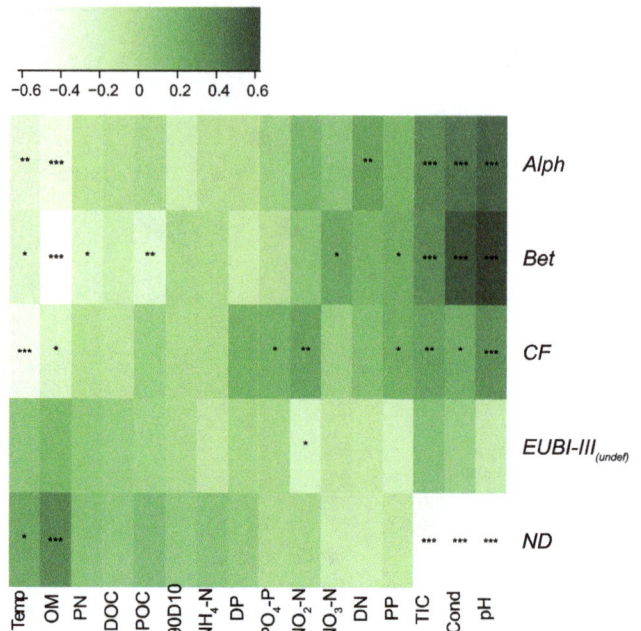

Figure 5. Heat-map of the Pearson correlations of beta-proteobacteria (*Bet*), alpha-*proteobacteria* (*Alph*), Cytophaga-Flavobacteria (*CF*), Eubacteria excluding *Alph*, *Bet* and *CF* (*EUBI-III(undef)*) and not-defined DAPI positive cells (ND) to physico-chemical variables. Asterisk indicate the p-value (*: p< 0.05; **: p<0.01; ***: p<0.001) of correlations between paired samples.

phylogenetic level and at the landscape scale, there was a difference in coupling of bacterial structure and functions within krenal and kryal systems with the latter having higher congruency [6]. At the phylogenetic level studied here, it seems that functional shifts, which depend on apparent redundancy/plasticity, are linked to structural changes of similar extent in both water systems. This discrepancy is likely due to non-detected seasonal changes of bacterial species within bacterial groups. That is, kryal systems can be seen as dominated by specialists whereas krenal systems tend to harbor generalists (on the species level) [6]. Thus, kryal community composition changes in response to new functional requirements, since the level of plasticity/redundancy in the community is low. Krenal communities, on the other hand, don't shift as strongly, since the communities plasticity/redundancy is high. Such differences in community shifts between the systems cannot be detected at the level of bacterial groups, if the community shift occurs mainly within and not across groups. Regardless, the coupling of bacterial groups to functions becomes evident when comparing different catchments.

A dominance of *Proteobacteria* has been shown before in snowpacks as well as in different glacial habitats [31,32]. Thus, the low total abundance of *Proteobacteria* in catchment M may be due to an interactive effect of decreased glacial water input and a generally harsh environment due to the high altitude (Table 1). *Bet* and partly *Alph* are often predominant in sediments in lower elevation streams [33,34]. *Bet* has been described as a diverse group dominating freshwater systems of different oligotrophic states [35,36] and are highly competitive at the initial state of biofilm development [37]. This trait may be the reason that they can compete well in kryal sediments in VR and L, as these habitats experience mechanic abrasion induced by flow as well as low OM. *Bet* also have been shown to be involved in the degradation of pollutants, and thus may provide beneficial ecosystem functions within alpine floodplains. Indeed, precipitation driven pollutant inputs may favor *Bet* as proposed by Brümmer et al. [35]. Biofilm bacterial assemblages in urban rivers have been shown to be dominated by *Bet* and *CF*, which may be linked to their pollution load [38]. Similar to our findings in VR, Araya *et al.* also found a peak of *CF* during the winter season in urban rivers which can be linked to lower temperatures (Figure 5) [8,38].

In summary, ongoing (and rapid) glacial recession and shift in water source (i.e. physico-chemical habitat template) will likely influence bacterial group composition in glaciated alpine floodplains. Differences between kryal and krenal systems were not as distinct at this taxonomic resolution (Figure 3, Figure S2) such that a dramatic change in bacterial group diversity at the landscape scale is expected. At a longer time scale, there will be other environmental changes concomitant to the glacial mass loss induced physico-chemical habitat shifts, such as changes in terrestrial vegetation, increasing OM inputs and increased water temperatures. These changes will likely induce subtle shifts in major bacterial groups towards a krenal assemblage composition and thus a homogenization of bacterial assemblages. Ultimately, alpine bacterial assemblage structure may become more similar to present lower elevation stream bacterial assemblages, thereby affecting ecosystem functioning and services.

Supporting Information

Figure S1 Biplot of the redundancy analysis based on the CARD-FISH data constrained by enzymatic activities measured in a previous study. Dark grey dots correspond to krenal sampling sites, whereas light grey dots depict kryal sites. Centroids of the respective probes are given: Beta*proteobacteria* (*Bet*), Alpha*proteobacteria* (*Alph*), Cytophaga-Flavobacteria (*CF*), Eubacteria excluding *Alph*, *Bet* and *CF* (*EUBI-III(undef)*) and not-defined DAPI positive cells (ND). Arrows depict the forward selected Hellinger transformed enzyme activities (expressed as nmol m^2 h^{-1}). Asterisks depict significantly tested enzymes (* p< 0.05, ** p<0.01). Dispersion ellipses for the catchments and

different water sources are shown and depict the standard error of weighted average scores (confidence limit = 0.95). Explained variation for the first two constraint axes is given.

Figure S2 Biplot of the redundancy analysis based on the CARD-FISH data constrained by physico-chemical variables (not shown but equal to Figure 3). Dark grey dots correspond to krenal sampling sites, whereas light grey dots depict kryal sites. Centroids of the respective probes are given: Beta*proteobacteria* (*Bet*), Alpha*proteobacteria* (*Alph*), Cytophaga-Flavobacteria (*CF*), Eubacteria excluding *Alph*, *Bet* and *CF* (*EUBI-III(undef)*) and not-defined DAPI positive cells (ND). Arrows depict the a posteriori fitted enzymatic activities (p< 0.05) and operational taxonomic units (p<0.01) from a previous study. Dispersion ellipses for the catchments and different water sources are shown and depict the standard error of weighted average scores (confidence limit = 0.95). Explained variation for the first two constraint axes is given.

Acknowledgments

We thank the Swiss National Park for access to the Macun catchment, Prof. Dr. Jakob Pernthaler and Dr. Michael Zeder for providing the infrastructure and in-house software for conducting CARD-FISH analysis in the most rapid way, and Simone Blaser and Christa Jolidon for field and laboratory assistance.

Author Contributions

Conceived and designed the experiments: RF HB SF CR. Performed the experiments: RF HB SF CR. Analyzed the data: RF. Contributed reagents/materials/analysis tools: RF SF SF CR. Wrote the paper: RF HB SF CR.

References

1. Battin TJ, Kaplan LA, Findlay S, Hopkinson CS, Marti E, et al. (2008) Biophysical controls on organic carbon fluxes in fluvial networks. Nature Geoscience 1: 95–100.
2. Pollarda PC, Ducklow H (2011) Ultrahigh bacterial production in a eutrophic subtropical Australian river: Does viral lysis short-circuit the microbial loop? Limnol Oceanogr 56: 1115–1129.
3. Battin TJ (1999) Hydrologic flow paths control dissolved organic carbon fluxes and metabolism in an alpine stream hyporheic zone. Water Resour Res 35: 3159–3169.
4. Hood E, McKnight DM, Williams MW (2003) Sources and chemical character of dissolved organic carbon across an alpine/subalpine ecotone, Green Lakes Valley, Colorado Front Range, United States. Water Resour Res 39: 1188–1199.
5. Besemer K, Singer G, Quince C, Bertuzzo E, Sloan W, et al. (2013) Headwaters are critical reservoirs of microbial diversity for fluvial networks. Proc R Soc B Biol Sci 280.
6. Freimann R, Bürgmann H, Findlay SEG, Robinson CT (2013) Bacterial structures and ecosystem functions in glaciated floodplains: Contemporary states and potential future shifts. ISME J 7: 2361–2373.
7. Comte J, del Giorgio PA (2011) Composition influences the pathway but not the outcome of the metabolic response of bacterioplankton to resource shifts. PLoS ONE 6: e25266.
8. Tockner K, Malard F, Uehlinger U, Ward JV (2002) Nutrients and organic matter in a glacial river-floodplain system (Val Roseg, Switzerland). Limnol Oceanogr 47: 266–277.
9. Hendricks SP (1993) Microbial ecology of the hyporheic zone: A perspective integrating hydrology and biology. J N Am Benthol Soc 12: 70–78.
10. IPCC (2007) Climate change 2007: The physical science basis. New York.
11. Horton R, Schaefli B, Mezghani A, Hingray B, Musy A (2006) Assessment of climate-change impacts on alpine discharge regimes with climate model uncertainty. Hydrol Process 20: 2091–2109.

12. Ward JV (1994) Ecology of alpine streams. Freshwat Biol 32: 277–294.
13. Freimann R, Bürgmann H, Findlay SEG, Robinson CT (2013) Response of lotic microbial communities to altered water source and nutritional state in a glaciated alpine floodplain. Limnol Oceanogr 58: 951–965.
14. Pernthaler J, Glöckner FO, Schönhuber W, Amann R (2001) Fluorescence in situ hybridization (FISH) with rRNA-targeted oligonucleotide probes. In: Paul JH, editor. Methods in Microbiology: Academic Press. pp. 207–226.
15. Pernthaler A, Pernthaler J, amann R (2004) Sensitive multicolour fluorescence in situ hybridization for the identification of environmental organisms. Molecular Microbial Ecology Manual 2: 711–726.
16. Zeder M, Pernthaler J (2009) Multispot live-image autofocusing for high-throughput microscopy of fluorescently stained bacteria. Cytometry Part A 75: 781–788.
17. Daims H, Brühl A, Amann R, Schleifer KH, Wagner M (1999) The domain-specific probe EUB338 is insufficient for the detection of all bacteria: Development and evaluation of a more comprehensive probe set. Syst Appl Microbiol 22: 434–444.
18. Neef A, Amann R, Schleifer KH (1997) Spezifischer und schneller in situ - Nachweis von Mikroorganismen aus Aerosolen mit Gensonden. Zentralbl Hyg Umweltmed 199: 410.
19. Manz W, Amann R, Ludwig W, Wagner M, Schleifer KH (1992) Phylogenetic oligodeoxynucleotide probes for the major subclasses of proteobacteria: Problems and solutions. Syst Appl Microbiol 15: 593–600.
20. Manz W, Amann R, Ludwig W, Vancanneyt M, Schleifer KH (1996) Application of a suite of 16S rRNA-specific oligonucleotide probes designed to investigate bacteria of the phylum cytophaga-flavobacter-bacteroides in the natural environment. Microbiology 142: 1097–1106.
21. Porter KG, Feig YS (1980) The use of DAPI for identifying and counting aquatic microflora. Limnol Oceanogr 25: 943–948.

22. Tockner K, Malard F, Burgherr P, Robinson CT, Uehlinger U, et al. (1997) Physico-chemical characterization of channel types in a glacial floodplain ecosystem (Val Roseg, Switzerland). Archiv für Hydrobiologie 140: 433–463.

23. Schofield RK, Taylor AW (1955) The measurement of soil pH. Soil Sci Soc Am J 19: 164–167.

24. Simon J. Blott KP (2001) GRADISTAT: A grain size distribution and statistics package for the analysis of unconsolidated sediments. ESPL 26: 1237–1248.

25. Blanchet FG, Legendre P, Borcard D (2008) Forward selection of explanatory variables. Ecology 89: 2623–2632.

26. Legendre P, Oksanen J, ter Braak CJF (2011) Testing the significance of canonical axes in redundancy analysis. Methods in Ecology and Evolution 2: 269–277.

27. Best DJ, Roberts DE (1975) Algorithm AS 89: The Upper Tail Probabilities of Spearman's Rho. Journal of the Royal Statistical Society Series C (Applied Statistics) 24: 377–379.

28. R (2014) R Development Core Team (2014). R: A language and environment for statistical computing. 3.1.0 ed. Vienna, Austria: R core team.

29. Fierer N, Jackson RB (2006) The diversity and biogeography of soil bacterial communities. Proc Natl Acad Sci U S A 103: 626–631.

30. Fierer N, Morse JL, Berthrong ST, Bernhardt ES, Jackson RB (2007) Environmental controls on the landscape-scale biogeography of stream bacterial communities. Ecology 88: 2162–2173.

31. Amato P, Hennebelle R, Magand O, Sancelme M, Delort AM, et al. (2007) Bacterial characterization of the snow cover at Spitzberg, Svalbard. FEMS Microbiol Ecol 59: 255–264.

32. Xiang SR, Shang TC, Chen Y, Jing ZF, Yao T (2009) Dominant bacteria and biomass in the Kuytun 51 Glacier. Appl Environ Microbiol 75: 7287–7290.

33. Kloep F, Manz W, Röske I (2006) Multivariate analysis of microbial communities in the River Elbe (Germany) on different phylogenetic and spatial levels of resolution. FEMS Microbiol Ecol 56: 79–94.

34. Brablcová L, Buriánková I, Badurová P, Rulík M (2013) The phylogenetic structure of microbial biofilms and free-living bacteria in a small stream. Folia Microbiol (Praha) 58: 235–243.

35. Brümmer I, Fehr W, Wagner-Döbler I (2000) Biofilm community structure in polluted rivers: Abundance of dominant phylogenetic groups over a complete annual cycle. Appl Environ Microbiol 66: 3078–3082.

36. Manz W, Szewzyk U, Ericsson P, Amann R, Schleifer K, et al. (1993) In situ identification of bacteria in drinking water and adjoining biofilms by hybridization with 16S and 23S rRNA-directed fluorescent oligonucleotide probes. Appl Environ Microbiol 59: 2293–2298.

37. Manz W, Wendt-Potthoff K, Neu TR, Szewzyk U, Lawrence JR (1999) Phylogenetic composition, spatial structure, and dynamics of lotic bacterial biofilms investigated by fluorescent in situ hybridization and confocal laser scanning microscopy. Microb Ecol 37: 225–237.

38. Araya R, Tani K, Takagi T, Yamaguchi N, Nasu M (2003) Bacterial activity and community composition in stream water and biofilm from an urban river determined by fluorescent in situ hybridization and DGGE analysis. FEMS Microbiol Ecol 43: 111–119.

5

Patterns and Variability of Projected Bioclimatic Habitat for *Pinus albicaulis* in the Greater Yellowstone Area

Tony Chang*, Andrew J. Hansen, Nathan Piekielek

Department of Ecology, Montana State University, Bozeman, Montana, United States of America

Abstract

Projected climate change at a regional level is expected to shift vegetation habitat distributions over the next century. For the sub-alpine species whitebark pine (*Pinus albicaulis*), warming temperatures may indirectly result in loss of suitable bioclimatic habitat, reducing its distribution within its historic range. This research focuses on understanding the patterns of spatiotemporal variability for future projected *P.albicaulis* suitable habitat in the Greater Yellowstone Area (GYA) through a bioclimatic envelope approach. Since intermodel variability from General Circulation Models (GCMs) lead to differing predictions regarding the magnitude and direction of modeled suitable habitat area, nine bias-corrected statistically down-scaled GCMs were utilized to understand the uncertainty associated with modeled projections. *P.albicaulis* was modeled using a Random Forests algorithm for the 1980–2010 climate period and showed strong presence/absence separations by summer maximum temperatures and springtime snowpack. Patterns of projected habitat change by the end of the century suggested a constant decrease in suitable climate area from the 2010 baseline for both Representative Concentration Pathways (RCPs) 8.5 and 4.5 climate forcing scenarios. Percent suitable climate area estimates ranged from 2–29% and 0.04–10% by 2099 for RCP 8.5 and 4.5 respectively. Habitat projections between GCMs displayed a decrease of variability over the 2010–2099 time period related to consistent warming above the 1910–2010 temperature normal after 2070 for all GCMs. A decreasing pattern of projected *P.albicaulis* suitable habitat area change was consistent across GCMs, despite strong differences in magnitude. Future ecological research in species distribution modeling should consider a full suite of GCM projections in the analysis to reduce extreme range contractions/expansions predictions. The results suggest that restoration stragies such as planting of seedlings and controlling competing vegetation may be necessary to maintain *P.albicaulis* in the GYA under the more extreme future climate scenarios.

Editor: Ben Bond-Lamberty, DOE Pacific Northwest National Laboratory, United States of America

Funding: This work was supported by the National Aeronautics and Space Administration Applied Sciences Program (Grant 10-BIOCLIM10-0034); Funder URL: http://www.nasa.gov (AJH TC). It also received support from the National Science Foundation Experimental Program to Stimulate Competitive Research (EPSCoR) Track-I EPS-1101342 (INSTEP 3); Funder URL: http://www.nsf.gov/div/index.jsp?div = EPSC (NBP TC); and the North Central Climate Science Center (G13AC00392-G-8829-1); Funder URL: http://www.doi.gov/csc/northcentral/index.cfm (AJH NBP). The funders had no role in study design, data collection and analysis, decision to publish, or preparation of the manuscript.

Competing Interests: The authors have declared that no competing interests exist.

* Email: tony.chang@msu.montana.edu

Introduction

Over the next century, it is expected that most of North America will experience climate changes related to increased concentrations of anthropogenic greenhouse gas emissions and natural variability [1]. At regional scales these changes are highly variable and can result in areas of increased mesic, xeric, or even hydric habitat conditions relative to present day. These shifting climates in turn also transform the suitable habitat for individual species that may result in changes in species composition and dominant vegetation types.

Whitebark pine (*Pinus albicaulis*) is a native conifer of the Western U.S. that is considered a keystone species in the sub-alpine environment. It provides a food source for animals such as the grizzly bear (*Ursus arctos*), red squirrel (*Tamiasciurus hudsonicus*), and Clark's nutcracker (*Nucifraga columbiana*) [2]. It also serves the ecosystem functions of stabilizing soil, moderating snow melt and runoff, and facilitating establishment for other species [2,3]. Whitebark pine has experienced a notable decline in

the past two decades within the U.S. Northern Rockies due to high rates of infestation from the mountain pine beetle (*Dendroctonus ponderosae*) and infections from white pine blister rust (*Cronartium ribicola*), resulting in an 80% mortality rate within the adult population [4–7]. Given the potential loss of important ecosystem functions that whitebark pine contribute to the landscape under this mortality event, there is an emphasis to understand the climate characteristics of its habitat to identify the restoration strategies and locations that may aid the persistence of the species under future climates.

One method of understanding species response to climate change is through bioclimate niche modeling, which has become a common practice for assessing potential vegetation shifts under new environmental conditions [8–13]. Ecological niche theory proposes there exists some range of bioclimatic conditions within which a species can persist [14]. In bioclimatic niche modeling, the realized niche is modeled by empirical relationships between the presence or absence of a species and the associated abiotic, and

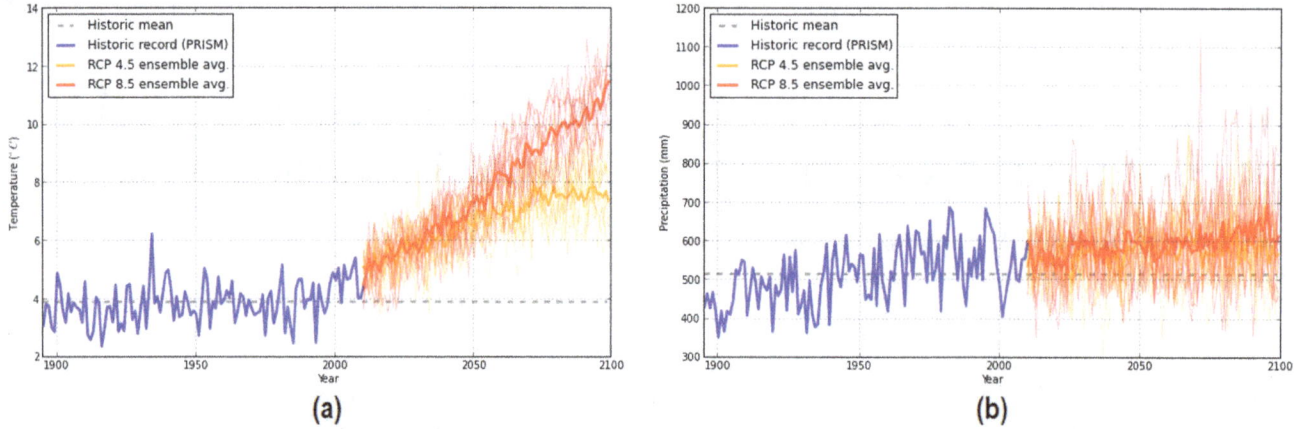

Figure 1. Historic and projected climates variables for the GYA from 1895–2099 under RCP 4.5 and 8.5 scenarios. Light shaded orange and red lines represent individual GCMs for RCP 4.5 and 8.5 respectively. Bold lines represent GCM ensemble average. (a) Mean annual temperature (b) Mean annual precipitation.

sometimes biotic, variables that describe the niche space. Bioclimatic models assume that species are in equilibrium with their environment and that the current abiotic relationships reflect a species environmental preferences which may be retained into the future [15,16]. At macro scales, bioclimatic approaches have demonstrated success at predicting current distributions of species [17,18]. Most bioclimatic models do not explicitly consider the many additional ecological factors that ultimately influence a

species distribution such as dispersal, disturbance, or biotic interaction. Thus the approach does not predict where a species will actually occur in the future, but rather it predicts locations where climatic conditions will be suitable for the species.

Bioclimatic niche methodology has demonstrated utility in modeling historic ranges of species for conservation and management applications. By modeling the present day suitable habitat and then projecting those habitats into the future, bioclimatic

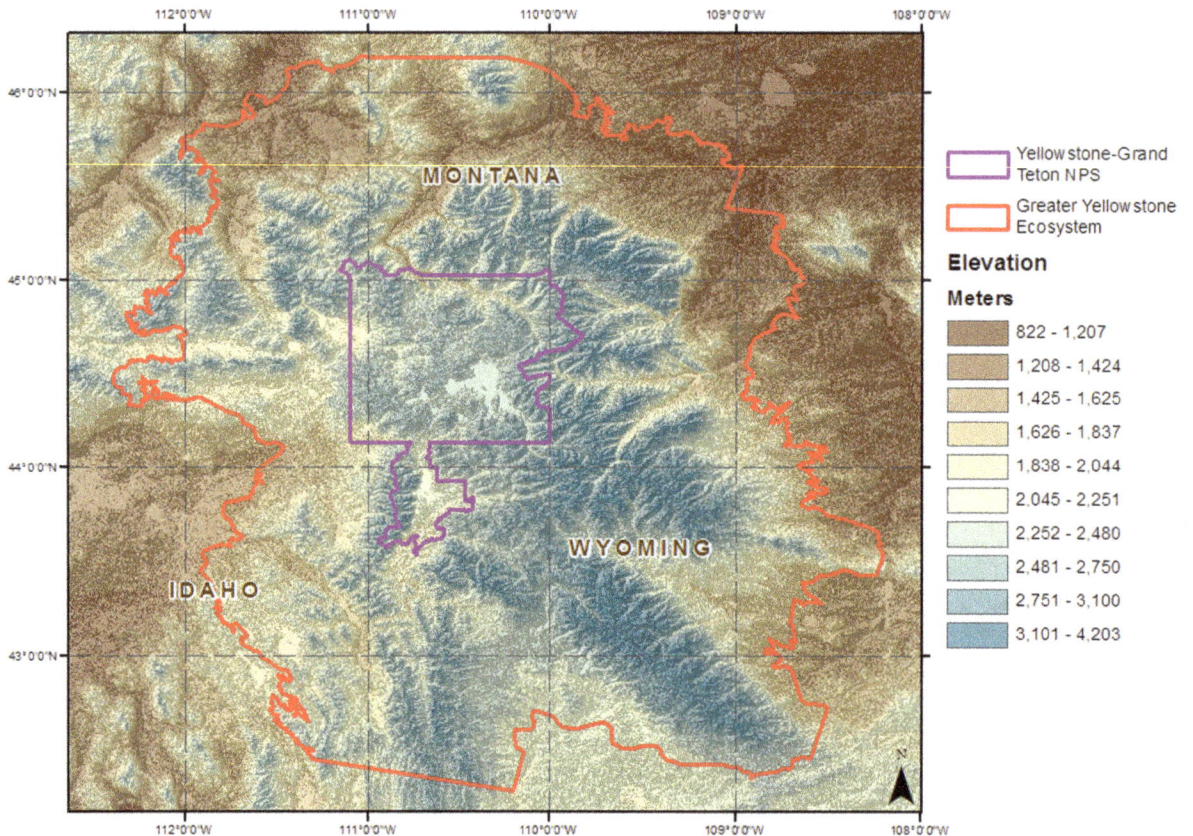

Figure 2. The Greater Yellowstone Area, representing an area of 150,700 km² with an elevational gradient from 522–4,206 m.

Table 1. General Circulation Models for analysis.

Name	Institute	Country
CESM1-CAM5	National Center for Atmospheric Research	US
CCSM4	National Center for Atmospheric Research	US
CESM1-BGC	National Center for Atmospheric Research	US
CNRM-CM5	Centre National de Recherche Meteorologiques	FR
HadGEM2-AO	Met Office Hadley Centre Climate Programme	UK
HadGEM2-ES	Met Office Hadley Centre Climate Programme	UK
HadGEM2-CC	Met Office Hadley Centre Climate Programme	UK
CMCC-CM	Centro Euro-Mediterraneo per Cambiamenti Climatici	ITA
CanESM2	Canadian Centre for Climate Modelling and Analysis	CAN

Selection of AR5 GCMs that represent historic climate in the U.S. Pacific Northwest region for future bioclimate habitat modeling.

niche models can serve as the first step filter for conservation action plans, such as mapping suitable species reintroduction sites or habitat reserve selection [19–21]. For *P.albicaulis*, McLane and Aitken [22] utilized bioclimate niche models to successfully implement experimental assisted migration on persisting climate habitat in British Columbia. Additionally, models of *Pinus flexis*, a closely related species of five needle pine, have been used to evaluate management options in Rocky Mountain National Park [23]. Given these examples, an effort to model and projected suitable climate habitat for *P.albicaulis* within a regional domain can provide valuable insight to land resource managers.

In this study, we present a bioclimatic habitat model for *P.albicaulis* within the Greater Yellowstone Area (GYA). Although *P. albicaulis* has a range-wide distribution that is split into two broad sections, one along Western North America: the British Columbia Coast Range, the Cascade Range, and the Sierra Nevada; and the other section in the Intermountain West that covers the Rocky Mountains from Wyoming to Alberta [2,24]; the GYA was selected as the primary geographic modeling domain for three reasons: 1) evidence that the *P. albicaulis* sub-population in the GYA is genetically distinct from other regional populations with different climate tolerances [25]; 2) the high regional investment in *P. albicaulis* conservation in the area [6]; 3) the high density of climate stations within the region. Climate within the GYA is highly heterogenous due to complex topography, and sharp elevational gradients. Current knowledge of the region expects climate to shift towards increased mean annual temperatures and earlier spring snowmelt [26,27]. This shift is expected to have an impact on the total suitable habitat area for *P. albicaulis*. Modeling at a regional scale can provide a finer resolution spatially explicit description of the bioclimatic envelope of *P. albicaulis* in the GYA.

Here we also present an opportunity to investigate the effect of future climate variability on projected species distributions. In 2013, the World Climate Research Programme Coupled Model released the new generation General Circulation Model (GCM) projections through the Coupled Model Intercomparison Project Phase 5 (CMIP5) [28]. These new GCM projections also include four possible climate futures are modeled with each GCM under the Representative Concentration Pathways (RCP) of greenhouse gas/aerosol. These RCP scenarios designate four different levels of radiative forcing (2.6, 4.5, 6.0 and 8.5 W/m^2) that may occur by the year 2099 [29]. In practice, research of future species suitable climate generally use a small suite of GCM/RCP combinations to project future climate [8,11,30]. However, internal variability in

these GCMs that arise from modeled coupled interactions among the atmosphere, oceans, land, and cryosphere can result in atmospheric circulation fluctuations that are characteristic of a stochastic process [31]. Such intrinsic atmospheric circulation variations from model structure induce regional changes in air temperature and precipitation on the multi-decadal time scale [31]. For the GYA specifically, this GCM variability has be observed with mean annual temperatures projected to increase by $2-9°C$ and mean annual precipitation to change by -50 to $+225$ mm (Fig. 1). This suggests that magnitude and direction of projected species distributions at a regional scale can vary depending on the GCM selected and the modeled species response to more xeric or mesic future climate conditions [32].

To summarize, this study presents a bioclimate niche model for *P. albicaulis* based on historic climate observations and field sampling of *P. albicaulis* presence and absence. Using this modeled bioclimate envelope, projections of future total climate suitable habitat area under nine GCMs and two RCP scenarios will be measured. Since different GCMs may project a diverging spectrum of climates, it is expected that measures of total suitable habitat will reduce with varying degrees of area loss. It is also expect that number and size of continous patches of *P. albicaulis* habitat will reduce due to the limited available number of subalpine areas distributed within the landscape. This research provides an analysis of the variability of biotic response under a large suite of GCMs to provide managers/researchers with a measure of the uncertainty associated with future species distribution models. Furthermore, this analysis explicitly describes the spatial patterns of bioclimatic niches for *P. albicaulis* to gain a better understanding of topographic characteristics, such as elevation, on suitable habitat. Changes in these spatial patterns are examined through quantifying landscape patch dynamic that may result from GCM projections to understand the species trends for persistence on the landscape.

Methods

Study area

The GYA, which includes Yellowstone National Park, Grand Teton National Park, and a number of state and federally managed forests, is a mid- to high-latitude region in the Northern Rocky Mountains of western North America. Conifers are dominant in the range, with forest types composed of *Pinus contorta*, *Abies lasiocarpa*, *Pseudotsuga menziesii*, *Pinus albicaulis*, *Juniperus scopulorum*, *Pinus flexis* and *Picea engelmannii*, although the deciduous hardwood *Populus tremuloides*, is also wide spread. Plateaus and lowlands are dominated by species of *Artemisia tridentata* and open grasslands of mixed composition. The GYA study area encompasses 150,700 km^2 with an elevational gradient from 522–4,206 m that represents 14 surrounding mountain ranges (Fig. 2).

Data

Biological data. Field observations of *P. albicaulis* adult presences and absences were compiled from three data sources. First, 2,545 observations from the Forest Inventory and Analysis (FIA) program were assembled. FIA plots are located on a regular gridded sampling design with one plot at approximately every 2,500 forested hectares, with swapped and fuzzed exact plot locations within 1.6 km to protect privacy [33]. Gibson et al. [34] found that model accuracy to not be dramatically affected by data fuzzing, but to provide the most spatial accuracy, this study culled FIA field points where measured elevation were >300 m different from a 30 m USGS DEM [35]. To capitalize on additional field observations of *P. albicaulis* within the study area, and because false absences are one of the most problematic data issues in constructing bioclimatic niche models [36]; supplementary points were drawn from the Whitebark/Limber Pine Information System (WLIS) [37], and long-term monitoring plots established by the

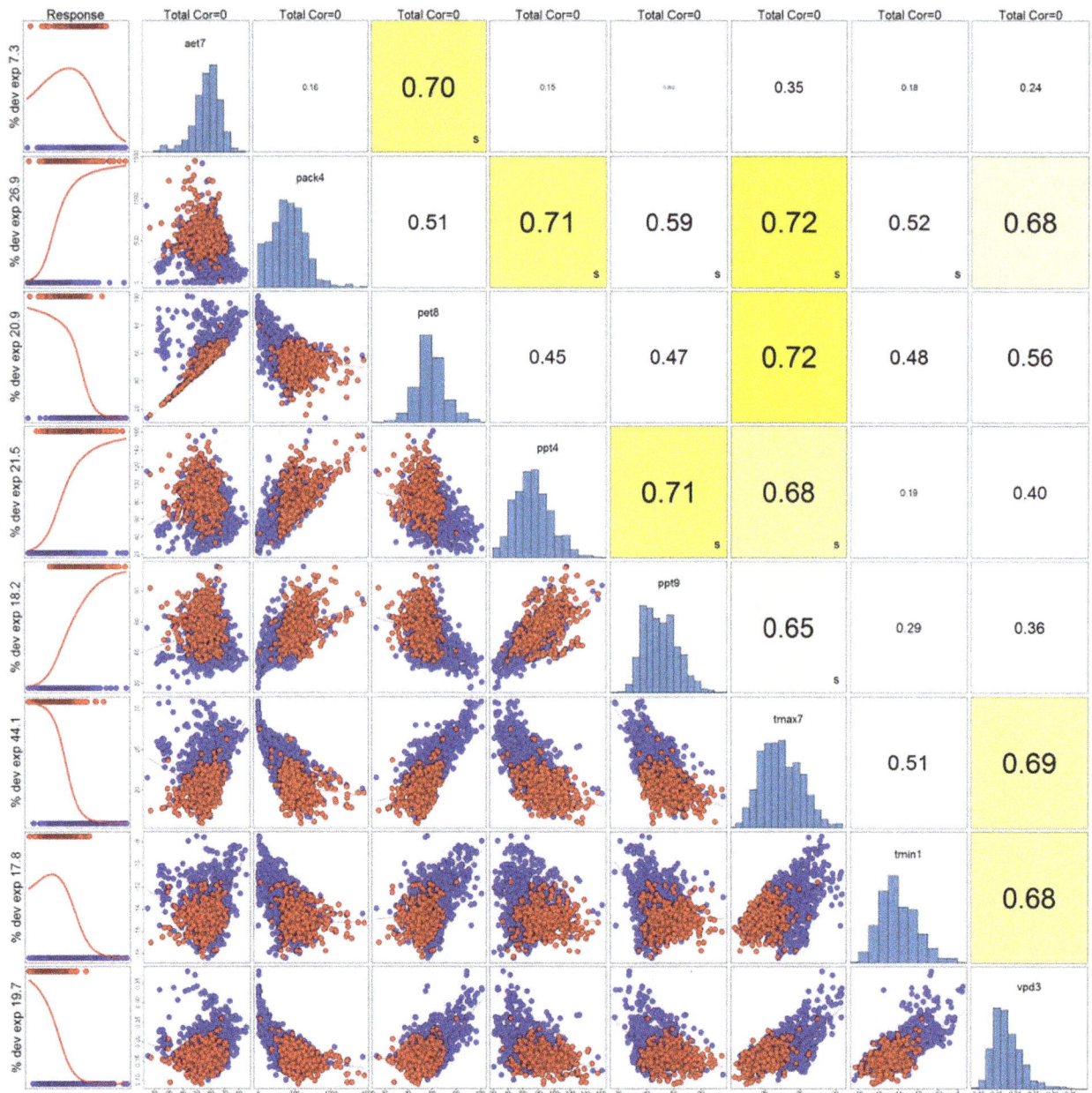

Figure 3. Selected predictor variables based on Principal Component Analysis and a maximum correlation filter of ≤0.75. Scatter plots represent one-to-one covariate plots where red points represent *P. albicaulis* presence, and blue points represent absence from field data. Far-left columns display logistic-regression of covariates from Generalized Additive Modeling using the Software for Assisted Habitat Modeling (SAHM [59]).

Table 2. Bioclimatic predictor variable list.

Code	Predictor Variable
tmin1	Minimum Temperature January
vpd3	Vapor Pressure Deficit March
ppt4	Precipitation April
pack4	Snow Water Equivalent April
tmax7	Maximum Temperature July
aet7	Actual Evapotranspiration July
pet8	Potential Evapotranspiration August
ppt9	Precipitation September

Final predictor variable set for Random Forest modeling. All variables were calculated as a 30-year climate mean from 1950–1980.

National Park Service Greater Yellowstone Inventory and Monitoring Network (GYRN) [38]. The presences in these two additional datasets were collocated within predictor pixels of FIA absence to correct for false absences. In doing so, only one *P. albicaulis* presence or absence record was associated per predictor pixel, thereby avoiding issues associated with sampling bias that are common when building bioclimate niche models with data from targeted surveys [39]. This compilation of data represents an effort for "completeness" as described by Kadmon et al. [40] and Franklin [36], to capture all climate conditions where a species does exist. New data sources added 119 *P. albicaulis* presences that would have been missed by using FIA data alone, for a total of 938 presences and 1,633 absences.

"Adult" class *P. albicaulis* were selected for modeling based on a recorded diameter at breast height (DBH) > 20 cm. *P. albicaulis* within the Central Montana are reported to reach 100 years of age at approximately 8–12 m in height with DBHs between 15–20 cm

[41]. Given previous silvicultural studies, it was assumed that 20 cm DBH *P. albicaulis* represent adult class individuals for the GYA, with potential to reproduce [24]. Furthermore, this study focused on adult size class due to difficulties distinguishing younger age class *P. albicilus* from *P. flexis*.

Historic climate data. Climate inputs for modeling were acquired from the 30-arc-second (~800 m) monthly Parameter-elevation Regressions on Independent Slopes Model (PRISM), a derived product that interpolates local station measurements across a continuous grid [42]. PRISM data includes monthly average minimum temperatures (T_{min}), maximum temperature (T_{max}), mean temperature (T_{mean}), and mean precipitation (Ppt). All monthly data were averaged for the temporal extent of 1950–1980 for bioclimatic niche model fitting. The 1950–1980 temporal extent was selected for modeling since: 1) a sufficient density of weather stations were operating by 1950 to provide a reasonable network; 2) evidence of anthropogenic warming that begins in the

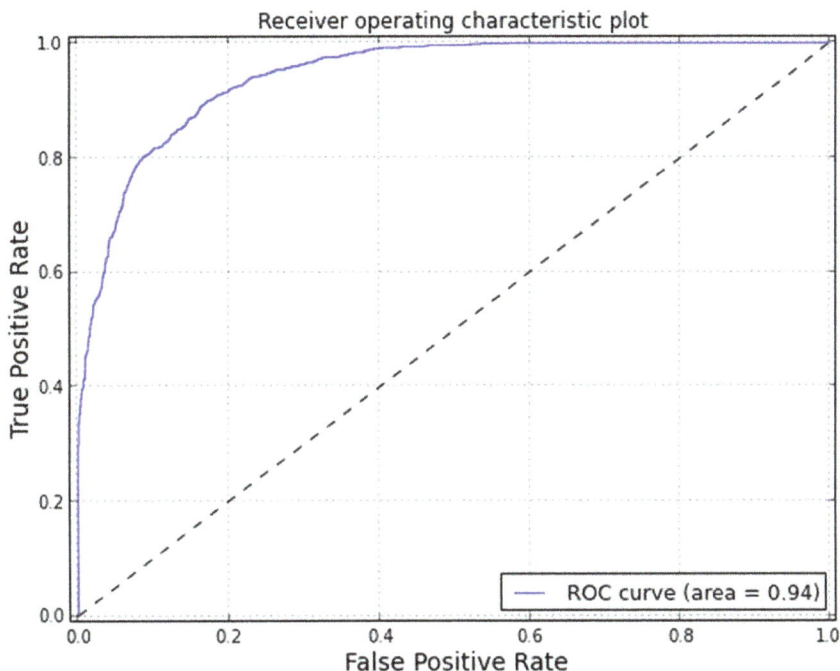

Figure 4. Area under curve for the receiver operating characteristic plot suggests adequate performance from the Random Forest modeling.

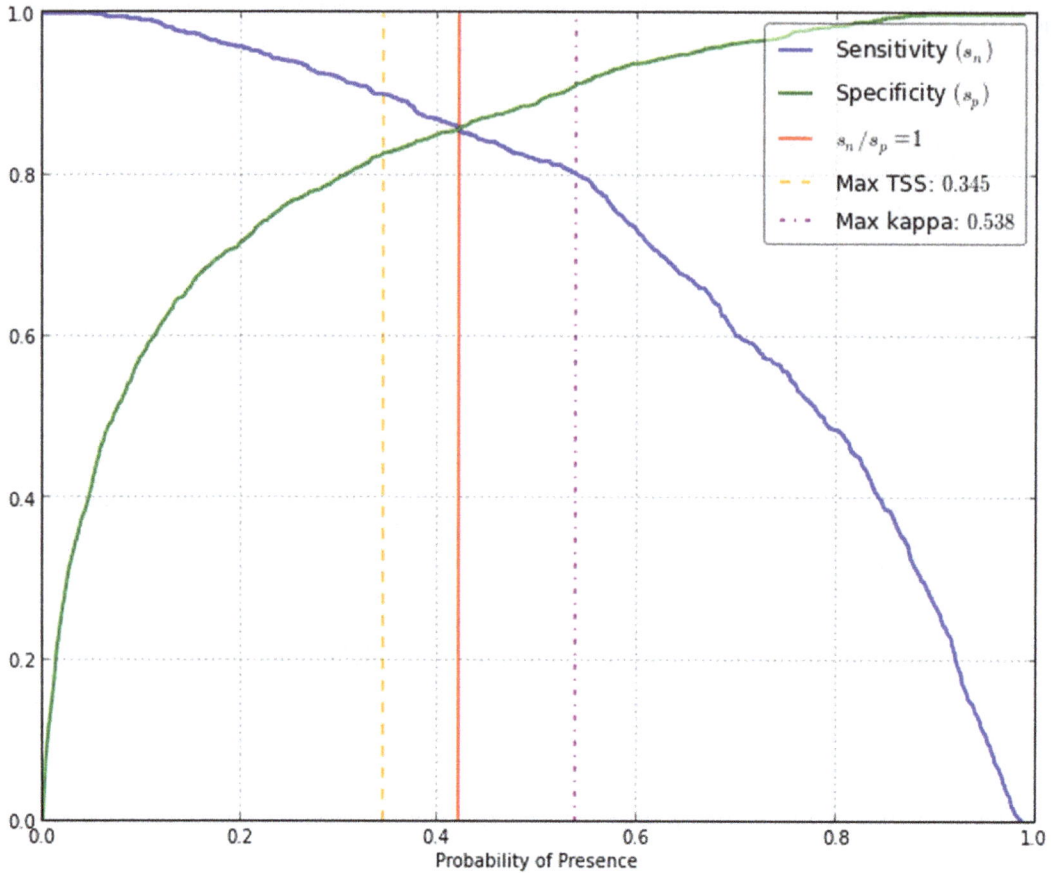

Figure 5. Threshold for probability of presence of 0.421 determined at the intersection of true positive rate (TPR) and true negative rate (TNR). Equivalent TPR and TNR, displayed a compromise between the maximum true skill statistic (TSS : 0.345) and maximum Kappa statistic (κ : 0.538).

late 1980s; 3) trees old enough to bear seeds today likely established under a similar climates to the 1950–1980 period.

Water balance. A Thornthwaite-based dynamic water balance model was used to estimate a number of variables that include actual evapotranspiration (AET) and potential evapotranspiration (PET) [43–45]. The model required only monthly mean temperatures, dew point temperatures, and precipitation (see Text S1). Water was stored as soil moisture or in surface snowpack, with the excess taking the form of evaporated vapor or loss through seepage/runoff. In addition to the climatic variables, latitude and physical characteristics of the soil were required to define water holding capacity. Soil attributes assigned by the Soil Survey Geographic (STATSGO) datasets were allocated from the Natural Resource Conservation Service at a 30-arc-second

resolution to determine soil water holding capacity and estimates for soil depth [46]. All water balance variables, which include PET, AET, soil moisture, vapor pressure deficit (vpd), and snow water equivalent (pack), were averaged by month over 1950–1980 to match with historic climate data for bioclimate model fitting.

GCM data. The general circulation model (GCM) experiments conducted under CMIP5 for the Intergovernmental Panel on Climate Change Fifth Assessment Report provided future projected climate data sets for assessing the effects of global climate change. Using a Bias-Correction Spatial Disaggregation (BCSD) approach, an archive of statistically down-scaled CMIP5 climate projections for the conterminous United States at 30-arc-second spatial resolution was assembled by the NASA Center for Climate Simulation NEX-DCP30 [47]. For this analysis, a subset of the

Table 3. Confusion matrix from out-of-bag analysis.

		Validation data set	
		Presence	Absence
Model	Presence	763 (81.9%)	169 (13.1%)
	Absence	176 (10.9%)	1437 (89.1%)

Random Forest tree estimators displays higher OOB specificity than sensitivity. Area Under Curve (AUC) value of 0.94 suggests model has high predictive capacity for projecting future suitable bioclimate habitat.

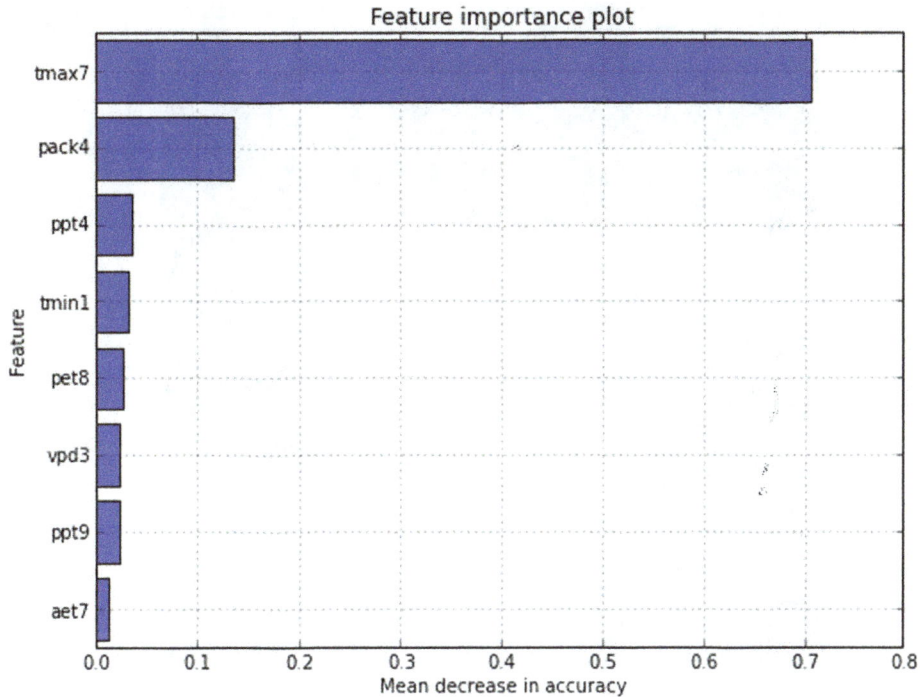

Figure 6. Random Forest out-of-bag variable importance plots find removal of maximum temperatures for July and April snow water equivalent to create the greatest reducing in model accuracy.

total GCM models available from NASA were selected that best represent the Northwestern US. Rupp et al. [48] recently presented an analysis of GCM performance versus the observed historic climate in the U.S. Pacific Northwest under 18 specified climate metrics. In their analysis, Rupp et al. ranked GCMs for accuracy using an empirical orthogonal function (EOF) analysis of the total normalized error compared to reference data. This analysis selected models with a normalized error score <0.5 as a threshold to cull the full suite of GCMs to the top nine models. These GCMs were used to project modeled *P. albicaulis*

distributions into the future (Table 1). Two RCP scenarios were selected to understand effects of differing carbon futures under climate change from 2010 to 2099. RCP 4.5 was the first, representing increased radiative forcing until stabilization of greenhouse emissions between 2040–2050 and total radiative forcing of 4.5 W/m^2 by 2099. RCP 8.5 was the second, representing the "business as usual" scenario, with uncontrolled radiative forcing increasing with stabilization of 8.5 W/m^2 by 2099 [49,50].

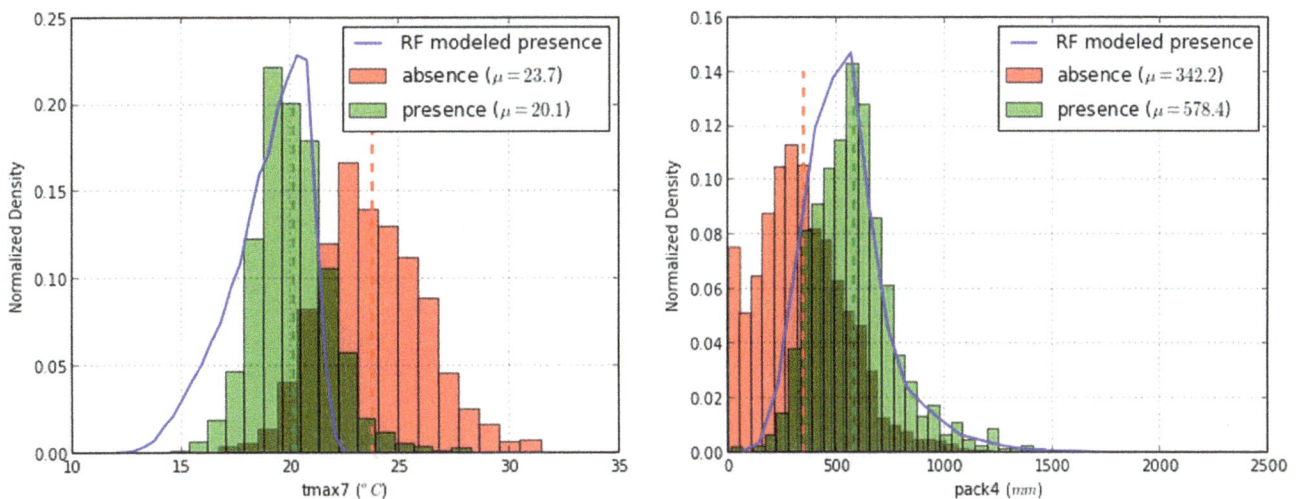

Figure 7. Modeled binary presence for *P.albicaulis* under 1980–2010 mean July maximum temperatures and mean April snow water equivalent bioclimate variables shows agreement with field presence data. Dotted lines designate climate means for corresponding *P. albicaulis* field points. Blue lines represent the distribution of Random Forest modeled presence within the GYA.

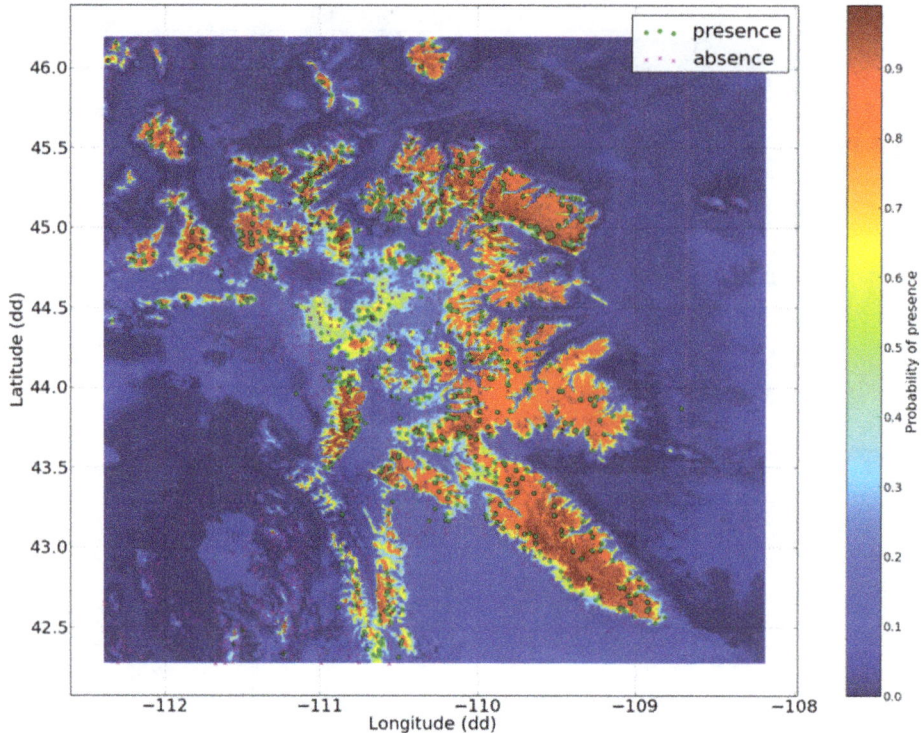

Figure 8. Probabiliy presence for *P.albicaulis* $\geq$ 20 cm DBH within the GYA for the 2010 climate period.

Modeling methods

A random forest (RF) [51] algorithm was used to create a bioclimate niche model of *P. albicaulis* in the GYA. Random forest is an ensemble learning technique that generates independent random classification trees using a subset of the total predictor variables and classifies a bootstrap random subsample of the data. These trees are aggregated and a majority vote over all trees in the random forest defines the resulting response class. This method of random trees with subsampling ensures a robust ensemble classification reducing overfitting and collinearity issues, especially with a large number of trees [9,51–53]. The python programming language (Python 3.3) and the Scikit-Learn library was used to fit the random forest model and predict current habitat niche, with parameters for number of trees ($n_{estimators} = 1000$), number of variables ($max_{features} = 4$), and node size ($min_{samplesleaf} = 20$) [54].

First pass filtering of environmental covariates was performed using Principal Component Analysis (PCA) to generate proxy sets [55–57]. After initial list was constructed, an additional filter was imposed on the variables with a 0.75 maximum correlation threshold to avoid collinearity issues (Fig. 3) [55]. Physiologically relevant variables to *P. albicaulis* presence were given precedence in final culling in cases of correlation above the specified maximum threshold. The final variable list selected were tmin1, vpd3, ppt4, pack4, tmax7, aet7, pet8, ppt9 (Table 2). The Software for Assisted Habitat Modeling (SAHM) was used to visualize correlations with the pairs function embeded in the VisTrails scientific workflow management system [58,59].

Model evaluation was performed under a variety of methods. An out-of-bag (OOB) error estimate was calculated by comparing the modeled probability of presence using approximately two-thirds of the field data, while withholding a subset of the remainder. Accuracy was evaluated by calculating: 1) the sensitivity, representing the true positive rate (TPR), 2) the specificity, representing the true negative rate (TNR), 3) the receiver operator characteristic curve (AUC). Importance of a specific predictor variable was calculated by examination of the increase in prediction error within the OOB sample when the predictor variable was permuted while others were held constant [54,60]. The rate of prediction error with permutation of a specified variable can be interpreted as the level of dependence of presence or absence response to that variable [61].

Projections for *P. albicaulis* were computed using 30 year moving climate averages for the period from 2010–2099 for both RCP 4.5 and 8.5 climate scenarios. Changes of suitable habitat area were determined using a binary classification of expected presence and absence. Binary class assignment was made under a probability of presence threshold where the ratio of sensitivity and specificity equalled 1. This method ensured an equal ability of the model to detect presence and absence. The Kappa and True Skill Statistic (TSS) were also calculated to observe how sensitivity and specificity responded under differing probability thresholds [62]. Survey plots predicted as suitable under climatic conditions in 2010 served as a reference for projections. The presence classifications were evaluated as the amount of suitable habitat changed over time, confined within specified elevational limits. To account for the need for a minimum patch size, total number of patches and median sizes using the an eight-neighbor rule (see Text S1) for patch identification were tracked over time [63].

Results

Model evaluation

The random forest model displayed an out-of-bag (OOB) error rate of 16.1% with greater errors of commission (13.1%) than omission (10.9%) (Table 3). The AUC was 0.94, displaying high

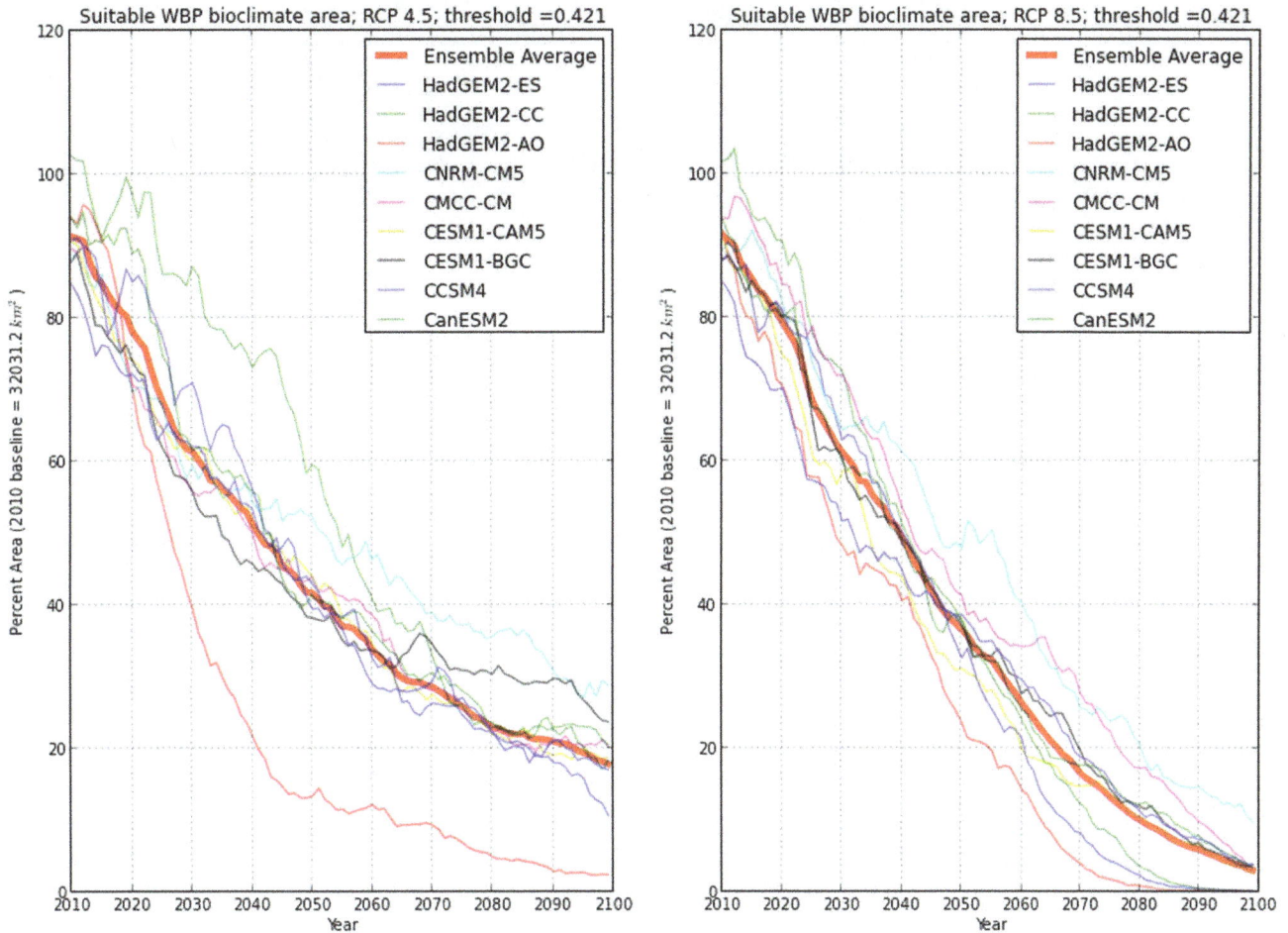

Figure 9. Bioclimate projections for *P.albicaulis* for 2010 to 2099 under 30-year moving averaged climates.

specificity and sensitivity (Fig. 4). Threshold probability of presence for a binary classification was selected at 0.421 (i.e where sensitivity = specificity). A probability threshold where TPR and TNR were equal was compared to the maximum Kappa statistic (0.538) and the maximum True Skill Statistic (TSS) (0.345) and found to be a compromise between the diagnostics (Fig. 5).

Estimates of variable importance plots revealed that permutation of maximum temperatures of summer months from all random trees resulted in a large drop in mean accuracy for distinguishing presence and absence of *P. albicaulis* (0.706 decrease in mean accuracy). This was followed by spring time snowpack (0.137 decrease in mean accuracy) (Fig. 6). Histogram plots of July maximum temperatures and April snowpack provided evidence of discrimination for presence and absence that are consistent with the modeled probability of presence for the year of calibration (Fig. 7).

Spatially explicit probability plots for the 2010 climate displayed highest probability of presence values within the ≥ 2500 m mountain ranges of the GYA in agreement with studies employing aerial imagery and remote sensing [4,5] (Fig. 8). Assuming that the modeled suitable bioclimate for *P. albicaulis* remains similar in the next century, the model demonstrated capacity to predict probable future *P. albicaulis* suitable habitat under projected climate conditions.

Model projections

Under both RCP 4.5 and 8.5, there was a predicted steady reduction of suitable bioclimate habitat for *P. albicaulis* over the course of this century, with RCP 8.5 displaying steeper declines than RCP 4.5 (Fig. 9). Under the RCP 4.5 and 8.5 scenarios, suitable habitat shifts from 100–85% to 2–29% by 2099, and 100–85% to 0.04–10% by 2099 respectively (Table 4).

CNRM-CM5, CMCC-CM, and CESM1-BGC projections showed the highest probabilities for suitable habitat area at the end of the century, while HadGEM2-AO, HadGEM2-ES, and HadGEM2-CC indicated the lowest probabilities. The standard deviations per year for both RCPs progressively decreased over time (Fig. 10). Among climate scenarios, standard deviations for both RCPs display low variability for the first five projection years and a rapid increase of variability peaking at 2043. For RCP 4.5, high variability existed primarily due to differing climate projections by models HadGEM2-AO and HadGEM2-CC, resulting in uncertainties in probabilities of presence fluctuating between 8 and 15% until 2068, after which variability was between 6–8%. Under RCP 8.5, standard deviations between GCMs were consistently lower than RCP 4.5. Regardless of the GCM, by 2079 the areas of suitable habitat converged to similar values.

Spatially explicit mapping of probability surfaces presented similar contractions of *P. albicaulis* habitat suitability toward the

Table 4. Projected binary *P. albicaulis* presence area within GYA to 2099.

Ensemble Average RCP 4.5	2010	2040	2070	2099
Area (km^2)	29250.9	16381.2	9151.1	5685.9
	(27134–32858)	(6918–23359)	(2962–12477)	(763–9194)
% Total Threshold Area*	91.3	51.1	28.6	17.8
	(85–103)	(22–73)	(9–39)	(2–29)
Mean Elevation (m)	2875.7	3020.2	3128.0	3217.9
	(2842–2895)	(2938–3182)	(3055–3297)	(3114–3471)
2.5 Percentile Elevation (m)	2356.3	2494.2	2595.0	2691.5
	(2320–2376)	(2433–2656)	(2506–2758)	(2571–3041)
97.5 Percentile Elevation (m)	3521.9	3603.5	3677.8	3734.6
	(3507–3530)	(3551–3701)	(3636–3783)	(3673–3905)
Ensemble Average RCP 8.5	**2010**	**2040**	**2070**	**2099**
Area (km^2)	29259.3	15746.0	5271.5	960.0
	(27188–32604)	(12985–19581)	(1247–8850)	(13–3105)
% Total Threshold Area*	91.3	49.2	16.5	3.0
	(85–102)	(40–61)	(4–28)	(0–10)
Mean Elevation (m)	2874.7	3022.5	3225.5	3470.5
	(2845–2893)	(2974–3061)	(3116–3412)	(3255–3749)
2.5 Percentile Elevation (m)	2353.1	2492.2	2691.3	3001.5
	(2322–2369)	(2436–2547)	(2553–2934)	(2622–3401)
97.5 Percentile Elevation (m)	3522.1	3605.7	3739.5	3908.7
	(3508–3530)	(3576–3631)	(3677–3866)	(3775–4063)

Summary of projection outputs under RCP 4.5 and 8.5 climate scenarios displays loss of bioclimate habitat from 2010 to 2099 (low and high probability of presence GCM summaries displayed in parentheses). Projections into 2099 under all 9 GCMs suggest rapid loss of suitable bioclimate habitat to below 70% of the current modeled distribution and shifts towards the limited high elevation zones (>3000 m). *(Percent threshold areas calculated from the 2010 PRISM reference probabilities of presence.)*

upper elevation zones of the GYA that included the Beartooth Plateau and Wind River Ranges (Fig. 11). This implied that rapid warming may lead to conditions outside of the *P. albicaulis* niche in lower elevation areas, and limiting the species to the alpine zones. Elevational analysis of cells within threshold presence probabilities over time observed mean elevations of suitable bioclimates shifting from 2,875 to 3,218 m and 2,875 to 3,470 m for RCP 4.5 and 8.5 respectively. By 2099, ensemble averaged GCM projections displayed over 70% loss of habitat under both scenarios.

P. albicaulis patches from the 2010 baseline observed 202 patches with median patch size of ~180 km^2. Projected patch dynamics analysis denoted a quadratic relationship of patch size over time. Patch dynamics displayed a slow increase in number of *P. albicaulis* patches to a maximum at 2074 and 2057 for RCP 4.5 and 8.5 respectively, followed by a decreasing trend. RCP 4.5 patch numbers were more sporadic, displaying fluctuations across the time period compared to RCP 8.5 associated with the greater interannual climate variability amongst GCM models. Median patch size saw a steady decrease from 72–65 km^2 to 21–8 km^2 for RCP 4.5 and 8.5 respectively, for the projection period, suggesting habitat loss through fragmentation (Fig. 12).

Discussion

In this analysis, the spatiotemporal patterns for *P. albicaulis* distributions were assessed under nine climate models and two emissions scenarios. Bioclimate modeling of *P. albicaulis* illustrat-

ed that presence and absence were strongly separated by summer temperatures and spring snowpack. This was in agreement with empirical findings of *P. albicaulis* presence in cool summertime environments where July temperatures range between 4–18°C [64]. Concordantly, these cool summer regions were synonymous with late snow melt, supporting snowpack as an important feature in distinguishing presence and absence.

Future projections by all nine GCMs suggested a contraction in suitable *P. albicaulis* climate area by the end of the century to <30% of current conditions. This was consistent with the results from various other research using either niche models or hybrid process models, predicting similar amounts of *P. albicaulis* contraction [8,9,65]. Variability among projected suitable habitat areas under differing GCMs decreased as all projected maximum temperatures increased above 1°C from the 100 year historic mean. This pattern of warming convergence occurred earlier for the GCMs under the RCP 8.5 scenario than those under RCP 4.5, resulting in the observed low variability of *P. albicaulis* suitable habitat area under RCP 8.5. Despite temperature variability remaining relatively constant amongst GCMs within a RCP, once mean annual temperatures increased beyond *sim* 1°C from the historic average, all bioclimatic habitat models exhibited a pattern of contracting total area and variability. These results lead to the conclusion that explicit selection of a GCM to model under may not necessarily matter for *P. albicaulis* bioclimatic niche modeling studies, especially if the direction of change is solely of concern. However, if investigation of the magnitude of change is relevant,

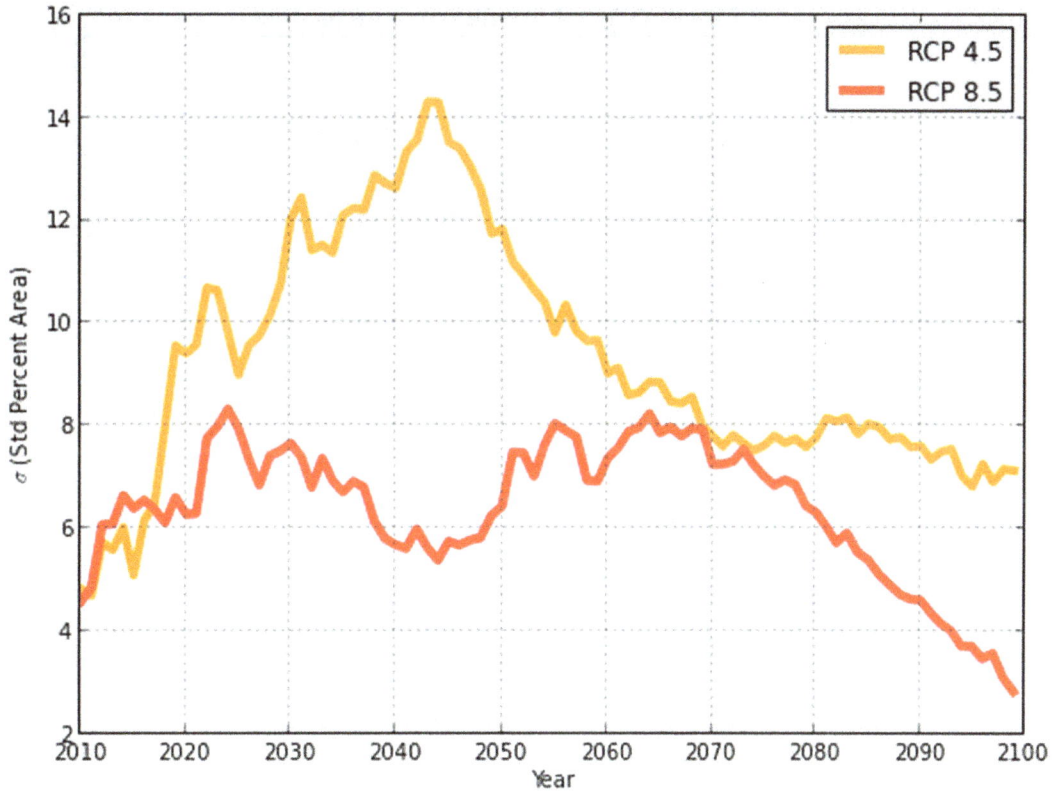

Figure 10. Evaluation of the standard deviation σ **for percent suitable habitat area by RCP scenario.**

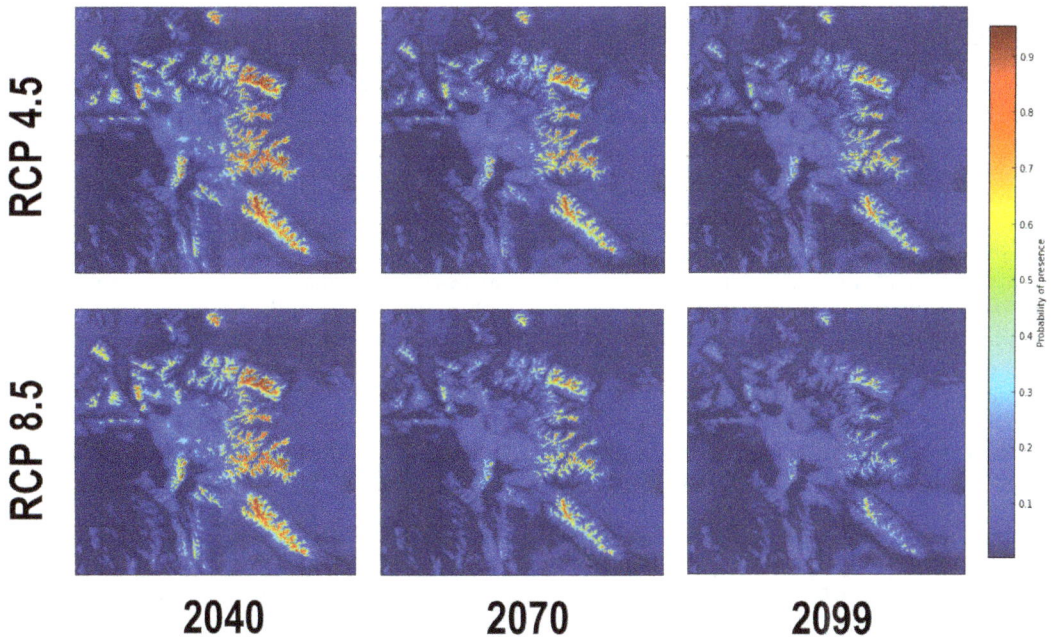

Figure 11. Spatially explicit probabilty surfaces for 2040 to 2099 suggest contraction of suitable bioclimatic habiatat for *P.albicaulis* **into the** $\geq$ **2500 m elevation zones.**

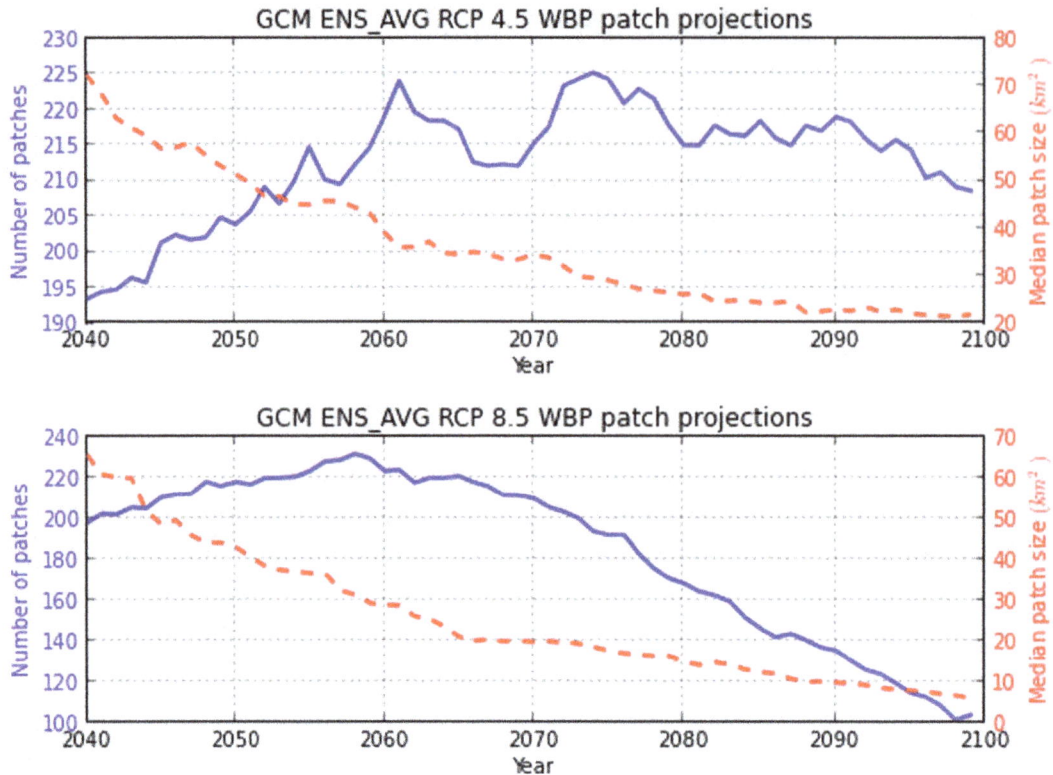

Figure 12. Patch dynamics of modeled *P.albicaulis*. Time series of *P.albicaulis* patch projections for number of patches and median patch size to 2099.

then GCM selection may directly influence the projected total suitable habitat area. This can be observed with RCP 4.5 habitat projection models differing by as much as 27% total suitable habitat area by the year 2099. Therefore arbitrary selection of a GCM for future projection modeling is likely inappropriate since it could lead to overly optimistic/pessimistic results for the species of concern.

Temporal patch dynamic analysis present an increase in fragmentation of the larger *P. albicaulis* suitable habitats over the next five decades, suggested through an increase in the total number of continuous patches but decreases in median size. This was followed by a contraction of small patches until they were almost absent from the system. Remaining habitat patches were smaller and less prevalent on the landscape by the end of the century. Reduced habitat patch size and density may reduce the likelihood for *N. columbiana* to disperse successful germinating seed caches, due to the limited size and area of suitable patch space. If changing climate habitats result in mortality within adult patches, genetic diversity may be lost resulting in a population bottleneck, thus reducing the robustness of the species to adapt to future disturbances. Experimental trials of P. albicaulis survival and fecundity under warmer and drier conditions outside the currently known range would provide greater confidence of the species ability to persist under future change. Limited analysis on seedling environmental conditions would also elucidate spatially explicit dispersal ranges and greater understanding of probable ranges for future establishment and survivorship.

Projected distributions of persistent *P. albicaulis* patches displayed a strong trend towards contraction into high elevation zones. Physiologically, there does not appear to be any upper elevation limit for *P. albicaulis* in the GYA. *P. albicaulis* in the

region has been reported to survive in absolute temperatures as low as $-36°C$ [64]. Lab experiments performed on *Pinus cembra*, a related five-needle pine residing in similar climates, were able to endure cold temperature extremes as low at $-70°C$ without cellular tissue damage [66]. Considering the current absolute minimum temperatures the species resides in and cold tolerance of its relatives suggests that *P. albicaulis* treeline in the GYA are not limited by lower temperatures. Controlled laboratory experimentation on *P. albicaulis* tolerances to temperatures would greatly improve this physiological understanding of cold tolerance.

Elevational habitat constriction do not imply that *P. albicaulis* will be completely gone from the region, but merely the loss of suitable climate habitat. Currently pre-established adult age class individuals will likely persist, since projected conditions of increased temperatures and CO_2 concentrations physiologically indicate increased growth rates of *P. albicaulis* [67]. Furthermore, micro-refugia sites may exist in the GYA that support *P. albicaulis* survival into the future, but were failed to have been modeled due to the coarseness of 30-arc-second climate data resolution. Since this bioclimatic envelope modeling approach was parameterized by the realized niche from in-situ data, it was difficult to determine if lower elevation limits are driven by warmer climate conditions or competition for light, water, or nutrients [15,17]. For example, lower treeline limits for *P. albicaulis* maybe driven primarily by competitive exclusion from late seral species *A.lasiocarpa*, *P.contorta*, and *P.engelmannii*. This follows from paleoecological pollen records of competitor migration during the Early Holocene (9000–5000 yr B.P), when climate conditions were warmer and drier. Longer growing seasons allowing competitors to invade likely drove *P. albicaulis* communities +500 m in elevation [68–70]. If future climate conditions become analogous to this Early

Holocene period, invasion of competitor species will likely contract *P. albicaulis* habitat to the limited high elevation zones of the GYA, specifically the Beartooth Plateau region and Wind River range [71].

Conclusion

This analysis examined the future of *P. albicaulis* suitable climate in the GYA and explicitly addressed the question of distribution variability under 9 representative GCMs and 2 emission scenarios. Increases in temperature within the GYA will likely result in a high level of contraction of suitable climate habitat for *P. albicaulis* over the next century. This contraction was consistent for all GCM projections, with approximately 20% uncertainty in total probable area. This analysis recommends that care be taken for species distribution modeling in future studies during the selection of GCMs due to their relevance for magnitudes of change. GCM ensemble averaging may be a solution to this issue, however it should be noted that averaging should take place after an individual GCM is projected in order to maintain interannual variability.

Although other studies have examined *P. albicaulis* species distribution models [8,65,72], this study is a step forward through its focus on relevant regional scale design, expansive local datasets, inclusion of high resolution climate and dynamic water balance variables, and selective projection under the latest AR5 GCMs. It is reiterated that the bioclimate niche model approach has high utility for understanding habitat conditions through correlative relationships with environmental variables, however, it may fail to explicitly model competitive exclusion, disturbance, phenotypic plasticity, and other complex interactions that are vital in determining a species' actual presence as it experiences changes in climate [15,17,73,74]. These unmodeled factors create uncertainties suggesting that this modeling effort does not identify the full potential climatic range of *P. albicaulis* in the future.

Uncertainties also exist regarding new suitable climates that may occur outside the current species range. Despite most rangewide studies confirming our results of total suitable habitat area reduction, there is potential for previously unsuitable habitat to become available under future climate change in the Northern regions [8,22,65]. Caution is therefore advised to individuals interpreting these findings. Changing climate will inevitably result in impacts on biomes and community structures. As such, mitigation and adaptation for potential futures are vital to conservation of climate sensitive species [75]. Future research that combines bioclimatic niche modeling with a mechanistic based disturbance, dispersal, and competition model will likely provide greater insight to the potential range of *P. albicaulis* in a climate changing world [76,77]. It would furthermore provide insight towards informing management options for restoration that may include controlled fire, selected thinning of competitor species, or assisted migration.

Acknowledgments

We are indebted for the insightful reviews from Richard Waring, William Monahan, and Tom Olliff. Many thanks to Marian and Colin Talbert, and Mark Greenwood for the software and statistical consultation.

Author Contributions

Conceived and designed the experiments: TC AJH NBP. Performed the experiments: TC NBP. Analyzed the data: TC. Contributed reagents/materials/analysis tools: TC AJH NBP. Wrote the paper: TC AJH NBP.

References

1. Intergovernmental Panel on Climate Change (2007) Fourth Assessment Report: Climate Change 2007: The AR4 Synthesis Report. Geneva: IPCC.
2. Tomback DF, Arno SF, Keane RE (2001) Whitebark pine communities: ecology and restoration. Island Press.
3. Callaway RM (1998) Competition and facilitation on elevation gradients in subalpine forests of the Northern Rocky Mountains, USA. Oikos 82: pp. 561–573.
4. Macfarlane WW, Logan JA, Kern W (2012) An innovative aerial assessment of greater yellowstone ecosystem mountain pine beetle-caused whitebark pine mortality. Ecological Applications.
5. Jewett JT, Lawrence RL, Marshall LA, Gessler PE, Powell SL, et al. (2011) Spatiotemporal relationships between climate and whitebark pine mortality in the greater yellowstone ecosystem. Forest Science 57: 320–335.
6. Logan JA, Macfarlane WW, Willcox L (2010) Whitebark pine vulnerability to climate-driven mountain pine beetle disturbance in the greater yellowstone ecosystem. Ecological Applications 20: 895–902.
7. Logan JA, Bentz BJ (1999) Model analysis of mountain pine beetle (coleoptera: Scolytidae) seasonality. Environmental Entomology 28: 924–934.
8. Rehfeldt GE, Crookston NL, Sáenz-Romero C, Campbell EM (2012) North American vegetation model for land-use planning in a changing climate: a solution to large classification problems. Ecological Applications 22: 119–141.
9. Rehfeldt GE, Crookston NL, Warwell MV, Evans JS (2006) Empirical analyses of plant-climate relationships for the western United States. International Journal of Plant Sciences 167: 1123–1150.
10. Thuiller W (2004) Patterns and uncertainties of species' range shifts under climate change. Global Change Biology 10: 2020–2027.
11. Iverson LR, Prasad AM, Matthews SN, Peters M (2008) Estimating potential habitat for 134 eastern US tree species under six climate scenarios. Forest Ecology and Management 254: 390–406.
12. Guisan A, Theurillat JP, Kienast F (1998) Predicting the potential distribution of plant species in an alpine environment. Journal of Vegetation Science 9: 65–74.
13. Busby J (1988) Potential impacts of climate change on Australias flora and fauna. Commonwealth Scientific and Industrial Research Organisation, Melbourne, FL, USA.
14. Hutchinson GE (1957) Concluding remarks. Cold Spring Harbor Symposia on Quantitative Biology 22: 415–427.
15. Austin M (2007) Species distribution models and ecological theory: a critical assessment and some possible new approaches. Ecological Modelling 200: 1–19.
16. Austin M (2002) Spatial prediction of species distribution: an interface between ecological theory and statistical modelling. Ecological Modelling 157: 101–118.
17. Pearson RG, Dawson TP (2003) Predicting the impacts of climate change on the distribution of species: are bioclimate envelope models useful? Global Ecology and Biogeography 12.
18. Willis KJ, Whittaker RJ (2002) Species diversity–scale matters. Science 295: 1245–1248.
19. Araújo MB, Cabeza M, Thuiller W, Hannah L, Williams PH (2004) Would climate change drive species out of reserves? an assessment of existing reserve-selection methods. Global Change Biology 10: 1618–1626.
20. Ferrier S (2002) Mapping spatial pattern in biodiversity for regional conservation planning: where to from here? Systematic Biology 51: 331–363.
21. Pearce J, Lindenmayer D (1998) Bioclimatic analysis to enhance reintroduction biology of the endangered helmeted honeyeater (lichenostomus melanops cassidix) in Southeastern Australia. Restoration Ecology 6: 238–243.
22. McLane SC, Aitken SN (2012) Whitebark pine (pinus albicaulis) assisted migration potential: testing establishment north of the species range. Ecological Applications 22: 142–153.
23. Monahan WB, Cook T, Melton F, Connor J, Bobowski B (2013) Forecasting distributional responses of limber pine to climate change at management-relevant scales in Rocky Mountain National Park. PloS ONE 8: e83163.
24. Arno SF, Hoff RJ (1989) Silvics of whitebark pine (pinus albicaulis). Intermountain Research Station GTR-INT-253.
25. Mahalovich MF, Hipkins VD (2011) Molecular genetic variation in whitebark pine (pinus albicaulis engelm.) in the inland west. In: Keane RE, Tomback DF, Murray MP, Smith CM, editors, The future of high-elevation, five-needle white pines in Western North America: Proceedings of the High Five Symposium. 28–30 June 2010; Missoula, MT. Proceedings RMRS.
26. Pederson GT, Gray ST, Ault T, Marsh W, Fagre DB, et al. (2011) Climatic controls on the snowmelt hydrology of the Northern Rocky Mountains. Journal of Climate 24: 1666–1687.

27. Westerling AL, Hidalgo HG, Cayan DR, Swetnam TW (2006) Warming and earlier spring increase western US forest wildfire activity. Science 313: 940–943.
28. Taylor KE, Stouffer RJ, Meehl GA (2012) An overview of CMIP5 and the experiment design. Bulletin of the American Meteorological Society 93.
29. Hibbard KA, van Vuuren DP, Edmonds J (2011) A primer on representative concentration pathways (RCPs) and the coordination between the climate and integrated assessment modeling communities. CLIVAR Exchanges 16: 12–13.
30. Lutz JA, van Wagtendonk JW, Franklin JF (2010) Climatic water deficit, tree species ranges, and climate change in Yosemite National Park. Journal of Biogeography 37: 936–950.
31. Deser C, Phillips AS, Alexander MA, Smoliak BV (2014) Projecting North American climate over the next 50 years: Uncertainty due to internal variability. Journal of Climate 27: 2271–2296.
32. Beaumont LJ, Hughes L, Pitman A (2008) Why is the choice of future climate scenarios for species distribution modelling important? Ecology Letters 11: 1135–1146.
33. Smith WB (2002) Forest inventory and analysis: a national inventory and monitoring program. Environmental Pollution 116: S233–S242.
34. Gibson J, Moisen G, Frescino T, Edwards Jr TC (2014) Using publicly available forest inventory data in climate-based models of tree species distribution: Examining effects of true versus altered location coordinates. Ecosystems 17: 43–53.
35. Gesch D, Oimoen M, Greenlee S, Nelson C, Steuck M, et al. (2002) The national elevation dataset. Photogrammetric engineering and remote sensing 68: 5–32.
36. Franklin J (2009) Mapping species distributions: spatial inference and prediction. Cambridge University Press.
37. Lockman IB, DeNitto GA, Courter A, Koski R (2007) WLIS: The whitebark-limber pine information system and what it can do for you. In: Proceedings of the conference whitebark pine: a Pacific Coast perspective. US Department of Agriculture, Forest Service, Pacific Northwest Region, Ashland, OR. Citeseer, pp. 146–147.
38. Jean C, Shanahan E, Daley R, DeNitto G, Reinhart D, et al. (2010) Monitoring white pine blister rust infection and mortality in whitebark pine in the Greater Yellowstone Ecosystem. Proceedings of the future of high-elevation five-needle white pines in Western North America: 28–30.
39. Edwards Jr TC, Cutler DR, Zimmermann NE, Geiser L, Moisen GG (2006) Effects of sample survey design on the accuracy of classification tree models in species distribution models. Ecological Modelling 199: 132–141.
40. Kadmon R, Farber O, Danin A (2003) A systematic analysis of factors affecting the performance of climatic envelope models. Ecological Applications 13: 853–867.
41. Weaver T, Dale D (1974) Pinus albicaulis in central Montana: environment, vegetation and production. American Midland Naturalist: 222–230.
42. Daly C, Gibson WP, Taylor GH, Johnson GL, Pasteris P (2002) A knowledge-based approach to the statistical mapping of climate. Climate Research 22: 99–113.
43. Thornthwaite C (1948) An approach toward a rational classification of climate. Geographical Review 38: 55–94.
44. Thornthwaite C, Mather J (1955) The water balance. Publication of Climatology 8.
45. Dingman S (2002) Physical hydrology. Prentice Hall.
46. National Resources Conservation Service (2014) Available: http://soildatamart.nrcs.usda.gov. Accessed 2013 Apr 3.
47. Thrasher B, Xiong J, Wang W, Melton F, Michaelis A, et al. (2013) Downscaled climate projections suitable for resource management. Eos, Transactions American Geophysical Union 94: 321–323.
48. Rupp DE, Abatzoglou JT, Hegewisch KC, Mote PW (2013) Evaluation of CMIP5 20th century climate simulations for the Pacific Northwest USA. Journal of Geophysical Research: Atmospheres 118: 10–884.
49. Gent PR, Danabasoglu G, Donner LJ, Holland MM, Hunke EC, et al. (2011) The community climate system model version 4. Journal of Climate 24: 4973–4991.
50. Moss RH, Babiker M, Brinkman S, Calvo E, Carter T, et al. (2008) Towards new scenarios for analysis of emissions, climate change, impacts, and response strategies.
51. Breiman L (2001) Random forests. Machine Learning 45: 5–32.
52. Roberts DR, Hamann A (2012) Method selection for species distribution modelling: are temporally or spatially independent evaluations necessary? Ecography 35: 792–802.
53. Lawrence RL, Wood SD, Sheley RL (2006) Mapping invasive plants using hyperspectral imagery and breiman cutler classifications (randomforest). Remote Sensing of Environment 100: 356–362.
54. Pedregosa F, Varoquaux G, Gramfort A, Michel V, Thirion B, et al. (2011) Scikit-learn: Machine learning in Python. Journal of Machine Learning Research 12: 2825–2830.
55. Dormann CF, Elith J, Bacher S, Buchmann C, Carl G, et al. (2013) Collinearity: a review of methods to deal with it and a simulation study evaluating their performance. Ecography 36: 027–046.
56. Booth GD, Niccolucci MJ, Schuster EG (1994) Identifying proxy sets in multiple linear regression: an aid to better coefficient interpretation. Research paper INT.
57. Tabachnick B, Fidell LS (1989) Using multivariate statistics, 1989. Harper Collins Tuan, PD A comment from the viewpoint of time series analysis Journal of Psychophysiology 3: 46–48.
58. Freire J (2012) Making computations and publications reproducible with vistrails. Computing in Science & Engineering 14: 18–25.
59. Morisette JT, Jarnevich CS, Holcombe TR, Talbert CB, Ignizio D, et al. (2013) Vistrails SAHM: visualization and workflow management for species habitat modeling. Ecography 36: 129–135.
60. Liaw A, Wiener M (2002) Classification and regression by randomforest. R news 2: 18–22.
61. Cutler DR, Edwards Jr TC, Beard KH, Cutler A, Hess KT, et al. (2007) Random forests for classification in ecology. Ecology 88: 2783–2792.
62. Allouche O, Tsoar A, Kadmon R (2006) Assessing the accuracy of species distribution models: prevalence, kappa and the true skill statistic (tss). Journal of Applied Ecology 43: 1223–1232.
63. Turner MG, Gardner RH, O'Neill RV (2001) Landscape ecology in theory and practice: pattern and process. Springer.
64. Weaver T (2001) Whitebark pine and its environment. In: Tomback DF, Arno SF, Keane RE, editors, Whitebark pine communities: ecology and restoration, Washington D.C, USA: Island Press.
65. Waring RH, Coops NC, Running SW (2011) Predicting satellite-derived patterns of large-scale disturbances in forests of the pacific northwest region in response to recent climatic variation. Remote Sensing of Environment 115: 3554–3566.
66. Sakai A, Larcher W (1987) Frost survival of plants. Responses and adaptation to freezing stress. Springer-Verlag.
67. Chapin III FS, Chapin MC, Matson PA, Vitousek P (2011) Principles of terrestrial ecosystem ecology. Springer.
68. Whitlock C, Shafer SL, Marlon J (2003) The role of climate and vegetation change in shaping past and future fire regimes in the Northwestern US and the implications for ecosystem management. Forest Ecology and Management 178: 5–21.
69. Whitlock C (1993) Postglacial vegetation and climate of Grand Teton and southern Yellowstone national parks. Ecological Monographs: 173–198.
70. Bartlein PJ, Whitlock C, Shafer SL (1997) Future climate in the Yellowstone national park region and its potential impact on vegetation. Conservation Biology 11: 782–792.
71. Tausch RJ, Wigand PE, Burkhardt JW (1993) Viewpoint: plant community thresholds, multiple steady states, and multiple successional pathways: legacy of the quaternary? Journal of Range Management: 439–447.
72. Bell DM, Bradford JB, Lauenroth WK (2014) Early indicators of change: divergent climate envelopes between tree life stages imply range shifts in the western united states. Global Ecology and Biogeography 23: 168–180.
73. Keane B, Tomback D, Davy L, Jenkins M, Applegate V (2013) Climate change and whitebark pine: Compelling reasons for restoration. Whitebark Pine Ecosystem Foundation Whitepaper.
74. Guisan A, Thuiller W (2005) Predicting species distribution: offering more than simple habitat models. Ecology Letters 8: 993–1009.
75. Keane RE, Tomback DF, Aubry CA, Bower EM, Campbell CL, et al. (2012) A range-wide restoration strategy for whitebark pine (pinus albicaulis): General technical report. USDA FS, Rocky Mountain Research Station RMRS-GTR-279: 108.
76. Mathys A, Coops NC, Waring RH (2014) Soil water availability effects on the distribution of 20 tree species in Western North America. Forest Ecology and Management 313: 144–152.
77. Morin X, Thuiller W (2009) Comparing niche-and process-based models to reduce prediction uncertainty in species range shifts under climate change. Ecology 90: 1301–1313.

Bathymetric Variation in Recruitment and Relative Importance of Pre- and Post-Settlement Processes in Coral Assemblages at Lyudao (Green Island), Taiwan

Yoko Nozawa*, Che-Hung Lin, Ai-Chi Chung

Biodiversity Research Center, Academia Sinica, Taipei, Taiwan, ROC

Abstract

Studies on coral communities have typically been conducted in shallow waters (~5 m). However, in the face of climate change, and as shallow coral communities become degraded, a greater understanding of deeper coral communities is needed as they become the main reef remnants, playing a central role in the future of coral reefs. To understand the dynamics of deeper coral assemblages, the recruitment and taxonomic composition of different life-stages at 5 and 15 m depths were compared at three locations in Lyudao, southeastern Taiwan in 2010. Coral recruits (<1 cm diameter, <4 months old) were examined using settlement plates. Juvenile corals (1–5 cm, several years old) were examined with quadrats, and adult corals (>5 cm, several years to decades old) were examined using transect lines. Pocilloporid and poritid corals had similar and higher numbers of recruits at 5 m compared to 15 m, whereas acroporid recruits were more abundant at 15 m. The primary cause for the former may be larval behavior, such that they position themselves in shallow waters, while that for the latter may be the dominance of brooding acroporid species (*Isopora* spp.) at 15 m. The taxonomic composition, especially between recruits and juveniles/adults, was more similar at 15 m than at 5 m. These results suggest a change in the relative importance of pre- and post-settlement processes in assemblage determinants with depth; coral assemblages in shallow habitats (more disturbed) are more influenced by post-settlement processes (mortality events), while those in deeper habitats (more protected) are more influenced by pre-settlement processes (larval supply).

Editor: Erik Sotka, College of Charleston, United States of America

Funding: This study was supported by a research grant of Academia Sinica to YN. The funders had no role in study design, data collection and analysis, decision to publish, or preparation of the manuscript.

Competing Interests: The authors have declared that no competing interests exist.

* E-mail: nozaway@gate.sinica.edu.tw

Introduction

Many coral communities have deteriorated as a result of various anthropogenic disturbances, including climate change, and further change is predicted in the future [1,2]. This ongoing trend of deterioration, mainly observed in shallow coral communities, has directed the attention of researchers to deeper coral communities, which are thought to inhabit less disturbed conditions [3–6] and are therefore likely to survive better in an increasingly hostile environment in the future [7–10]. If this prediction is true, deeper coral communities would become the main remnants of future coral reefs, playing a more central role than degraded shallow coral communities, and may also serve as a source of coral recruits for the recovery of shallow coral communities [7,9,10]. Despite a growing need for information concerning deeper coral communities, this knowledge is limited since coral studies have mainly been conducted in shallow waters (~5 m) [10]. Therefore, information on deeper coral communities is valuable, especially information related to their dynamics and interactions between shallow and deep corals.

With recent advances in deep-water survey technologies, new studies of deeper coral communities have begun. However, most of these studies are focusing on "mesophotic coral communities" that occur deeper than 30 m [8,10], and studies of coral communities on reef slopes at 10–30 m depth have not been given equal attention. Given the dramatic change in environmental gradients in the upper 10–20 m of water, especially regarding light intensity and wave action, and associated biotic/abiotic changes with depth (e.g., movement of sand gravels, algal abundance, herbivores and coral growth) [3–5,11,12], the dynamics of deeper coral communities (10–30 m) cannot be assumed to be the same as those of shallow coral communities (~5 m). In fact, coral assemblages at 10–30 m depth possess the highest coral diversity, presumably due to the less disturbed conditions, lower competition for space and sufficient larval supply [4].

The objective of this study was to investigate the recruitment process of deeper coral assemblages to understand their dynamics at Lyudao, southeastern Taiwan. Coral recruitment was examined using settlement plates at 5 m and 15 m depth. Data on the taxonomic composition of juvenile (1–5 cm) and adult corals (>5 cm) were also collected to examine changes in assemblage structure among different life stages (recruits, juveniles and adults). In Southeast Asia, coral recruitment patterns in deeper water (>10 m) have been measured in only a single location (Okinawa, Japan) [13–15], and most studies at similar depths have been undertaken in other regions of the world [6,12,16–23]. The aforementioned studies generally found a higher density of coral recruits at 10–20 m, and argued the importance of the pre-settlement process (larval supply) in relation to the richer coral assemblages at this depth range [4]. In the present study, we observed that coral recruitment varied with depth and that this variation also differed among coral taxa. We also found that similarities in taxonomic composition among the three life-stages varied with depth, and argue that the relative importance of pre- and post-settlement processes in assemblage determinants may vary with depth.

Materials and Methods

Study site

This study was conducted in 2010 at 5 m and 15 m depth on reef slopes at three locations (Chai-kou, Guei-wan, You-zi-hu) around Lyudao (Green Island), southeastern Taiwan (Fig. 1). Permits for this study were granted by Taitung county government. At each depth/site combination, surveys were performed in an area of ~50×50 m. Lyudao is an offshore islet located in the middle of the Kuroshio Current [24] that may transport various marine organisms, including coral larvae, from the up-current coral triangle area. Lyudao is surrounded by clear water, and has well-developed fringing reefs to ~30 m depth containing approximately 250 scleractinian coral species [25]. Several tropical typhoons pass through the vicinity of Lyudao every year, causing significant disturbance of its marine biota [26,27].

Coral recruitment survey

Settlement plates with a "refuge structure" were used to assess coral recruitment. Although settlement plates have been used in coral recruitment studies since the 1970s, most settlement plates have consisted of flat surfaces [except 23], without refuge structures like crevices, pits, and grooves. These substrata often result in no or low recruitment on exposed, upward plate surfaces most likely due to grazing by herbivores [22,28,29]. Previous studies that examined settlement plates with refuges found more coral recruits with a higher taxonomic diversity compared to the traditional settlement plates with plain surfaces [[30,31], Nozawa, Y. unpublished data]. As coral species with small-sized and/or slow-growing recruits are predicted to depend more heavily on refuge structure for post-settlement survival [32], the use of settlement plates with refuges is expected to provide less biased and more artifact-free results in comparison with a plain substratum which is uncommon in coral reef habitats. For this study, we used commercially available unglazed terracotta plates (10 cm×10 cm×2 cm) with two grooved surfaces (14 grooves surface^{-1}, groove size: 5 mm wide, 100 mm long, 2 mm deep). The dimensions of the refuge structure on the settlement plates were determined according to Nozawa [33]. By using settlement plates with refuges in this study, we had more numerous and taxonomically diverse coral recruits than in a previous study using plain settlement plates conducted at the same location [34].

At each depth/site combination, 15–18 settlement plates were deployed haphazardly. Settlement plates were fixed to the sea bottom a few centimeters above the substrata using stainless bolts and nuts. To avoid sediment deposition filling the refuges on plate surfaces and negating their effect [28], settlement plates were fixed at an angle of ~45° to the bottom.

Settlement plates were deployed in early April, approximately 2 to 3 weeks before the main coral spawning period (April–June) [[35], Y. Nozawa unpublished data] in order to biologically condition the plate surfaces. Settlement plates were retrieved 4 months later to cover the main period of coral recruitment predicted for southern Taiwan. The number of settlement plates retrieved at the 5 m and 15 m sites was as follows, respectively: for Guei-wan, 17 and 18; for Chai-kou, 15 and 15; for You-zi-hu, 16 and 16. Environmental data (temperature and light intensity depth^{-1} site^{-1}) were collected during the period in which settlement plates were deployed in 2010, using HOBO pendant temperature/light data loggers (Onset Computer Corp., USA). Retrieved settlement plates were soaked in a dilute chlorine bleach to remove algae and soft-bodied epibenthos. Coral recruits (skeletons) on the top and bottom plate surfaces were counted under a stereomicroscope and taxonomically identified into four family groups (Acroporidae, Pocilloporidae, Poritidae, and others) according to Babcock et al. [36].

Coral juvenile and adult assemblage survey

Coral juvenile and adult assemblage surveys were performed in the same areas used for the deployment of settlement plates at 5 m and 15 m depths at the three sites. Quadrats (25 cm×25 cm) were used to assess juvenile corals. At each depth/site combination, 28–71 quadrats were haphazardly placed on rocky substrata. Juvenile corals (1–5 cm in diameter) that appeared in the quadrats were photographed with a scale and taxonomically identified later. Adult coral (>5 cm in diameter) surveys were performed using a line intercept method with 10-m lines. At each depth, six transects were placed haphazardly along the depth contour. All scleractinian corals below the lines were photographed along with the line and a scale and taxonomically identified later. For each individual, the length of the individual that was intercepted by the line was measured to obtain a cover estimate, and the maximum width of the individual perpendicular to the line was measured for the density estimate [37]. Density was calculated using the following equation:

$$\hat{D} = \left[\frac{1}{L}\right] \sum_{i=1}^{k} \left(\frac{1}{Wi}\right)$$

where $\hat{D}$ = estimate of population density, L = length of all lines combined, Wi = perpendicular width of individuals intersected, k = the total number of individuals intercepted on all lines. We

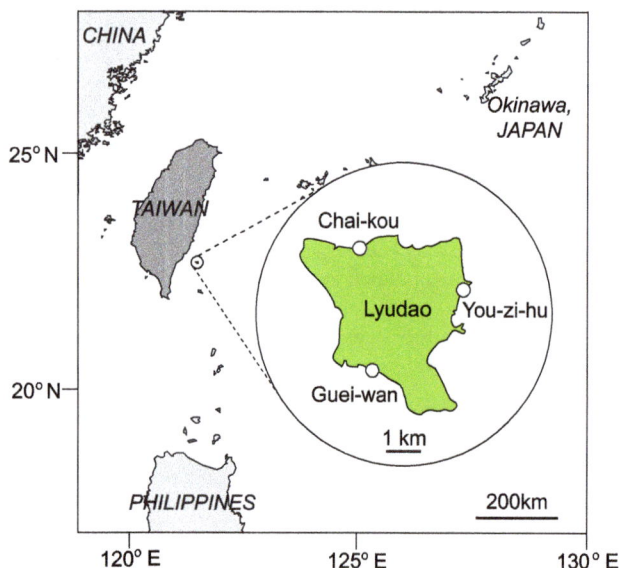

Figure 1. Study location. Coral assemblages at 5 m and 15 m were examined at three locations (Chai-kou, Guei-wan, You-zi-hu) around Lyudao (Green Island), Taiwan in 2010.

separated taxa of juveniles and adult corals into the four family groups (Acroporidae, Pocilloporidae, Poritidae and others), in which the genus *Alveopora* was allocated to Acroporidae following a recent taxonomic revision [38].

Environmental conditions

Seawater temperature and light intensity were measured at 5 m and 15 m depths during the major recruitment season (April–July) in 2010 (Fig. 2). The temperature was higher at 5 m, and the maximum temperature difference between the two depths reached up to 5°C. However, in most cases (>85% of data), the temperature difference was <1°C. The median (25th and 75th percentiles) of the temperature difference was 0.4°C (0.1~0.7) in Chai-kou, 0°C (−0.1~0.2) in Guei-wan and 0.2°C (0.1~0.5) in You-zi-hu. Light intensities at 15 m (max; 4–5×10^4 lx) were about four to five times lower than those at 5 m (<1×10^4 lx).

Statistical analysis

The number of coral recruits per settlement plate (recruits on top and bottom surfaces were pooled) was analyzed using a generalized linear model (GLM) with a Poisson error distribution by the glm function in R (version 3.0.0) [39]. In the GLM, sites and depths were treated as a fixed factor. Pairwise post-hoc comparisons were performed with a Tukey test using the glht function in the package multcomp (version 1.2–17) in R. The same statistical analyses were applied for acroporid, pocilloporid and poritid recruits, respectively.

Similarities in taxonomic composition between coral recruits, juveniles and adult assemblages at 5 m and 15 m were visualized using non-metric multidimensional scaling (MDS) based on Bray–Curtis similarity coefficients with relative density data. A one-way analysis of similarity (ANOSIM) was conducted to determine the significance of any observed differences in taxonomic composition between the three life-stages at each depth [40]. The MDS analysis and ANOSIM test were performed using Primer software, version 6 (Primer-E Ltd, Plymouth, UK).

Results

Coral recruitment

Results on the number of coral recruits per settlement plate are summarized in Figure 3. Of the 97 settlement plates retrieved, most settlement plates (>80%) had more coral recruits on the upward plate surfaces. Among the three sites, You-zi-hu generally had the highest number of recruits, followed by Guei-wan and Chai-kou, in the four recruit categories examined (Fig. 3). Comparison between the two depths (5 m and 15 m) revealed no significant difference in the total number of recruits at Chai-kou and You-zi-hu; however, at Guei-wan more recruits were observed at 15 m (GLM: p<0.001). In the family level analyses, a higher number of acroporid recruits was recorded at 15 m at all three sites (GLM: p<0.001), whereas a higher number of pocilloporid and poritid recruits was recorded at 5 m in Chai-kou and You-zi-hu (GLM: p<0.001); the number of these recruits at Guei-wan was similar at both depths.

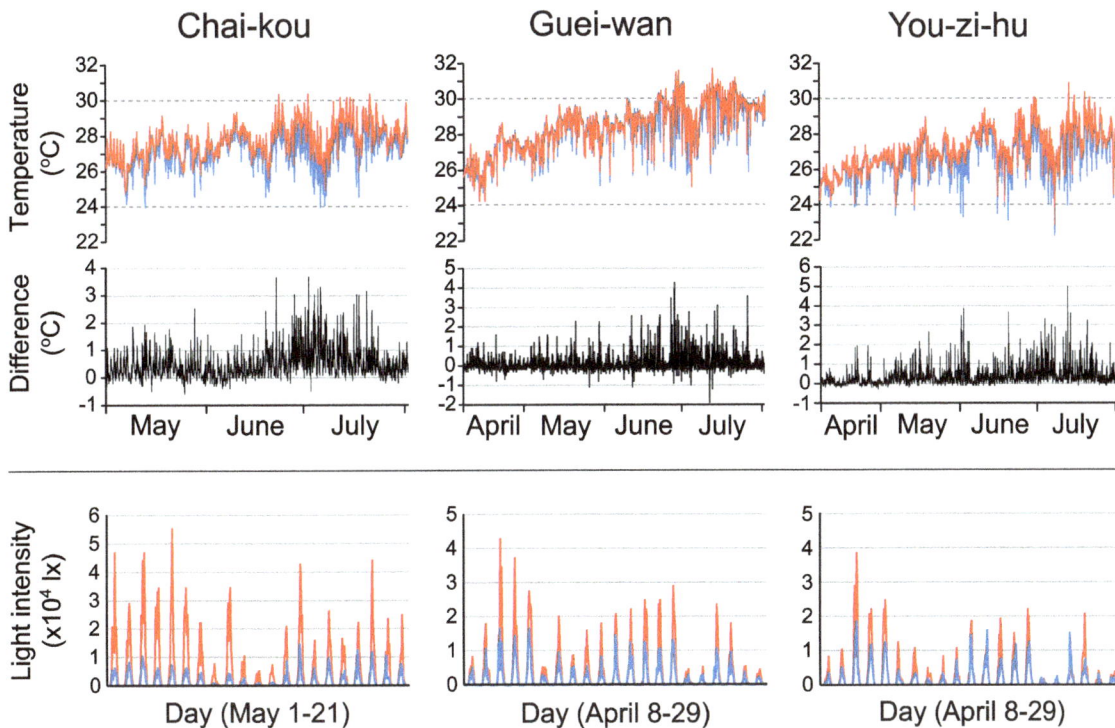

Figure 2. Environmental condition. Seawater temperature and light intensity were monitored in 2010 at the 5 m (red) and the 15 m (blue) sites at three study locations (Chai-kou, Guei-wan, You-zi-hu). For temperature, data were obtained during the 4-month deployment of the settlement plates for the coral recruitment surveys. Absolute differences in temperatures between the 5 m and the 15 m sites are shown below the temperatures. For light intensity, data from the first 3 weeks of the deployment period are shown as the values then gradually declined due to the gradual coverage of the sensor component of the loggers by benthic organisms. Data for Chai-kou started from May due to the loss of initial data-loggers in April.

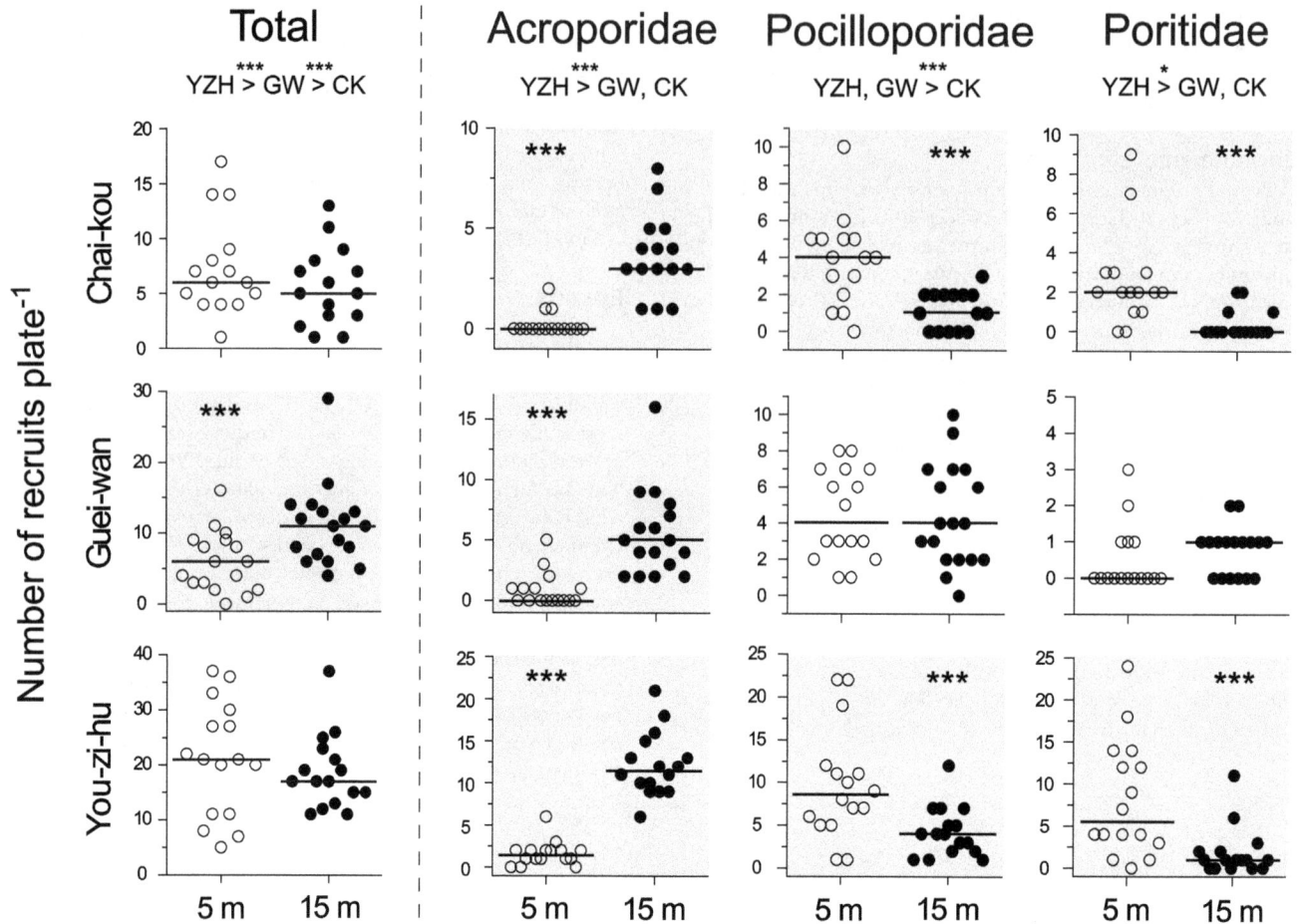

Figure 3. Coral recruits at two depths. The numbers of coral recruits per settlement plate at the 5 m (open circle) and the 15 m sites (closed circle) are shown. The horizontal bar denotes the median. The data were analyzed using a generalized linear model (see the method section for details), and results with a significant difference are highlighted with a grey background (p<0.001). Results on the comparison of recruit numbers between the sites are also presented for each taxonomic group below the heading; CK = Chai-kou, GW = Guei-wan, YZH = You-zi-hu; *** p<0.001, * p<0.05.

Taxonomic composition of recruits and later life stages

Relative abundance data (in density) showed that the dominant recruit groups at 5 m were pocilloporids and poritids (ca. 80% of recruits), whereas those at 15 m were acroporids (48–65%), followed by pocilloporids (12–38%) (Fig. 4A). In comparison with those of juvenile and adult corals, a distinct difference was observed between the recruit and juvenile stages at 5 m, followed by more moderate changes between the juvenile and adult stages. The overall trend at 5 m was that, from the recruit to adult stages, the proportions of pocilloporids and poritids decreased, while those of acroporids and others increased. In contrast, at 15 m, the composition was more similar between the three life-stages, especially at Guei-wan. This observation was supported by the MDS plot of the relative density data, which showed two distinct groups, i.e., coral recruits at 5 m and others (Fig. 4B). Within the group of others, recruits at 15 m were also grouped together, and juvenile and adult assemblages at 15 m were located closer together (i.e., were more similar) than those at 5 m. The ANOSIM results showed a significant difference among the three life-stages at both 5 m and 15 m with a larger R statistic value (i.e., larger difference) at 5 m (5 m: R = 0.745, p<0.01; 15 m: R = 0.407, p<0.05).

Discussion

Recruitment patterns and water depth

Densities of coral recruits were similar at the 5 m and 15 m sites. However, when data were partitioned by family, family-level analyses revealed variations in recruitment patterns between depths. Most acroporid recruits occurred at 15 m depth, while a similar number or more pocilloporid and poritid recruits occurred at 5 m depth.

Isopora species that release planula larvae (i.e., brooders) dominated adult acroporid assemblages at 15 m depth (16–58% of acroporids), and their densities (1–7.6 m^{-2}) were 5–19 times higher than those at 5 m depth (0.2–0.7 m^{-2}). This result would largely account for the higher number of acroporid recruits at the 15 m sites, assuming that many were *Isopora* recruits, because a strong correlation between density of adults and recruits is common in brooding corals [30,41,42].

In contrast, the higher recruitment densities of pocilloporids and poritids at 5 m could not be explained by adult distribution as densities were similar at both depths [0.6–3.2 m^{-2} (5 m) and 0.2–3.5 m^{-2} (15 m) for Pocilloporidae, 1.7–6.8 m^{-2} (5 m) and 1.6–4.2 m^{-2} (15 m) for Poritidae], and the dominant species of the two families were spawners in Lyudao (*Pocillopora verrucosa, P. eydouxi*

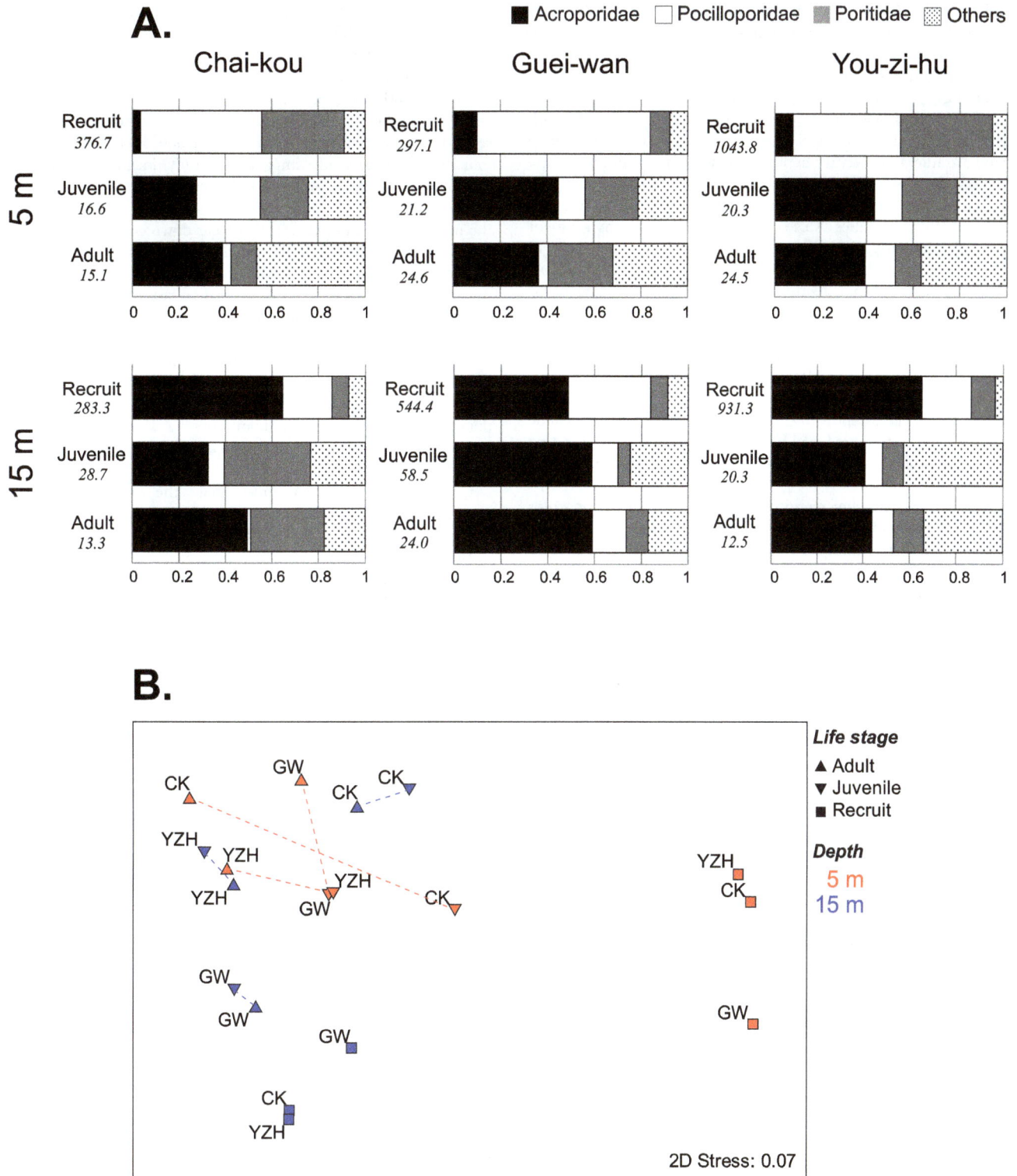

Figure 4. Taxonomic composition of three life-stages at two depths. A) Relative densities of the three dominant families at three life-stages (recruits, juveniles and adults) are shown for the 5 m (above) and 15 m sites (below) at three locations. The number on the left-hand side of each bar denotes the density (m^{-2}). B) Multi-dimensional scaling (MDS) ordination, based on Bray–Curtis similarity coefficients, for the relative density data in Fig. 4A. Juvenile and adult stages of each site are connected by a dashed line. Data for the 5 m sites are shown in red, and data for the 15 m sites are in blue. CK = Chai-kou, GW = Guei-wan, YZH = You-zi-hu.

and massive *Porites* spp.; Y. Nozawa, unpublished data). A potential cause of the recruitment patterns with depth may be larval behavior. Previous studies have reported positive phototaxis and/or negative geotaxis in planula larvae of several coral species, including two pocilloporid species, *P. damicornis* and *Seriatopora hystrix* [43–46]. With a rapid reduction in light intensity with depth, the larval swimming behavior could have created a negative depth gradient in larval supply, enhancing recruitment at the shallow sites [44]. Planula larvae of some coral species are also known to show depth-dependent settlement behavior in response to benthic communities [15,47] and a certain light environment [48–50]. Similarly, it is possible that the larval behavior of acroporids, in addition to the depth gradient of adult *Isopora* density, influenced their recruitment pattern (either enhancing or weakening it).

As an alternative explanation, variation in recruit mortality with depth could have created the same recruitment patterns for the three dominant families. However, this explanation is less likely because previous studies rejected the hypothesis of different mortality between depths (0–11 m) [14,48,51]. A recent genetic study also supported the larval behavior hypothesis, demonstrating some evidence of larval migration from deep to shallow habitats in *S. hystrix* [7].

The pattern of higher recruitment of pocilloporids and poritids at shallow sites (<6 m) has also been reported by several previous studies [21,23] but many found higher recruitment at 10–20 m [16–19,52]. These studies attributed the decline in recruits at shallow depths to higher post-settlement mortality caused by intense grazing of herbivores [cf. 22]. Of these studies, Wallace [23] and the present study are the only studies using settlement substrata with refuges that protect coral recruits from grazers, while others used plain settlement plates. It is therefore likely that the decline in recruits at shallow sites was caused by an absence of refuge structure on the settlement substrata used in previous studies [28,30,31, Y. Nozawa, unpublished data]. Given the complex surface structure seen on natural substrata in coral reef habitats, the recruitment pattern detected by settlement substrata with refuges may better reflect natural patterns.

Assemblage determinants and water depth

Previous studies have demonstrated the importance of pre- and post-settlement processes in determining coral assemblage structure [4,5,13–16,44,47,51,53–56]. In the present study, we found that the relative importance of pre- and post-settlement processes changed with depth in coral assemblages. At the shallow sites, the large change in assemblage structures among the three life-stages, especially when comparing recruits and juveniles, suggested that post-settlement processes (mortality events) had a strong influence, whereas at the deep sites, the less prominent difference among the

life-stages suggested a prevalence of pre-settlement processes (larval supply and behavior).

Higher disturbance frequencies and competition in shallower habitats are common on most reefs [5,12,16,44,55], and are attributed to the fact that the richest coral species-diversity occurs at 10–30 m [4]. In Taiwan, typhoons are generally the most serious natural disturbance affecting shallow reefs, and three to four typhoons typically impact the study locations each year [26,27]. Strong wave action created by typhoons causes serious damage to shallower coral assemblages [27,53,57]. In particular, the movement of sand gravels that wear down coral tissues during typhoons and smother coral recruits and juveniles after typhoons has been implicated in the high coral mortality of early life-stages in shallow reef habitats [5,53,58]. In Lyudao, the 5 m sites were dominated by encrusting acroporid corals (mainly *Montipora* spp.) and domed faviid corals, while various other corals inhabited the 15 m sites, including tabular and branching corals. As the disappearance of non-encrusting corals from shallow habitats is commonly observed after typhoons [[59], Nozawa, Y. unpublished data], the success of the dominant coral taxa at the 5 m sites in Lyudao may largely be attributable to the frequent typhoon disturbance typical in Taiwanese waters.

Connell et al. [53] concluded that "the dynamics of coral assemblages can be understood through the variation in types and scales of disturbances and other ecological processes where disturbances are rare". This may be applicable to the dynamics of coral assemblages at different depths, which are exposed to a negative depth gradient of disturbance, and may explain the change with depth in assemblage determinants. Although this observed change in assemblage determinants with depth may be a generality at most reef sites, as the negative depth gradient of disturbance is a general pattern [4], the conclusion of Connell et al. [53] also suggests that the relative importance of pre- and post-settlement processes at each depth may vary (i.e., are not fixed), depending on the variation in type and scale of disturbance. When (and where) disturbances reach deeper habitats, the influence of post-settlement mortality may prevail over depth, and vice versa.

Acknowledgments

The authors thank T.-Y. Huang, C.-M. Hsu, C.-Y. Kuo, S.-L. Chen, and the Chiu Fu diving shop for their valuable assistance and P. Edmunds, D. Vianney, C.-M. Hsu, and three anonymous reviewers for comments that improved an earlier version of this paper.

Author Contributions

Conceived and designed the experiments: YN. Performed the experiments: YN CHL ACC. Analyzed the data: YN. Contributed reagents/materials/analysis tools: YN. Wrote the paper: YN.

References

1. Wilkinson C (2008) Status of the coral reefs of the world: 2008. Townsville Australia: Global Coral Reef Monitoring Network and Reef and Rainforest Research Centre. 296 p.

2. Hoegh-Guldberg O, Mumby PJ, Hooten AJ, Steneck RS, Greenfield P, et al. (2007) Coral reefs under rapid climate change and ocean acidification. Science 318: 1737–1742.

3. Vermeij MJA, Bak RPM (2003) Species-specific population structure of closely related coral morphospecies along a depth gradient (5–60 M) over a Caribbean reef slope. Bull Mare Sci 73: 725–744.

4. Huston MA (1985) Patterns of species diversity on coral reefs. Annu Rev Ecol Syst 16: 149–177.

5. Bak RPM, Engel MS (1979) Distribution, abundance and survival of juvenile hermatypic corals (Scleractinia) and the importance of life history strategies in the parent coral community. Mar Biol 54: 341–352.

6. Bak RPM, Nieuwland G (1995) Long-term change in coral communities along depth gradients over leeward reefs in the Netherlands Antilles. Bull Mar Sci 56: 609–619.

7. van Oppen MJ, Bongaerts P, Underwood JN, Peplow LM, Cooper TF (2011) The role of deep reefs in shallow reef recovery: an assessment of vertical connectivity in a brooding coral from west and east Australia. Mol ecol 20: 1647–1660.

8. Lesser MP, Slattery M, Leichter JJ (2009) Ecology of mesophotic coral reefs. J Exp Mar Biol Ecol 375: 1–8.

9. Sinniger F, Morita M, Harii S (2012) "Locally extinct" coral species *Seriatopora hystrix* found at upper mesophotic depths in Okinawa. Coral Reefs 32: 153.

10. Bongaerts P, Ridgway T, Sampayo EM, Hoegh-Guldberg O (2010) Assessing the 'deep reef refugia' hypothesis: focus on Caribbean reefs. Coral Reefs 29: 309–327.

11. Huston M (1985) Variation in coral growth rates with depth at Discovery Bay, Jamaica. Coral Reefs 4: 19–25.

12. Birkeland C (1977) The importance of rate of biomass accumulation in early successional stages of benthic communities to the survival of coral recruits. Proc 3rd Int Coral Reef Symp 1: 15–21.

13. Suzuki G, Arakaki S, Kai S, Hayashibara T (2012) Habitat differentiation in the early life stages of simultaneously mass-spawning corals. Coral Reefs 31: 535–545.

14. Suzuki G, Hayashibara T, Toyohara H (2009) Role of post-settlement mortality in the establishment of *Acropora* reef slope zonation in Ishigaki Island, Japan. Galaxea 11: 13–20.

15. Suzuki G, Hayashibara T (2011) Do epibenthic algae induce species-specific settlement of coral larvae? J Mar Biol Assoc U.K. 91: 677–683.

16. Rogers CS, III HCF, Gilnack M, Beets J, Hardin J (1984) Scleractinian coral recruitment patterns at Salt River Submarine Canyon, St. Croix, U.S. Virgin Islands. Coral Reefs 3: 69–76.

17. Birkeland C, Rowley D, Randall RH (1981) Coral recruitment patterns at Guam. Proc 4th Int Coral Reef Symp 2: 339–344.

18. Smith SR (1997) Patterns of coral settlement, recruitment and juvenile mortality with depth at Conch reef, Florida. Proc 8th Int Coral Reef Sym 2: 1197–1202.

19. Adjeroud M, Penin L, Carroll A (2007) Spatio-temporal heterogeneity in coral recruitment around Moorea, French Polynesia: Implications for population maintenance. J Exp Mar Biol Ecol 341: 204–218.

20. Carlon DB (2001) Depth-related patterns of coral recruitment and cryptic suspension-feeding invertebrates on Guana Island, British Virgin Islands. Bull Mar Sci 68: 525–541.

21. Harriott VJ (1985) Recruitment patterns of scleractinian corals at Lizard island, Great Barrier Reef. Proc the 5th Int Coral Reef Cong 4: 367–372.

22. Penin L, Michonneau F, Baird AH, Connolly SR, Pratchett MS, et al. (2010) Early post-settlement mortality and the structure of coral assemblages. Mar Ecol Prog Ser 408: 55–64.

23. Wallace CC (1985) Seasonal peaks and annual fluctuations in recruitment of juvenile scleractinian corals. Mar Ecol Prog Ser 21: 289–298.

24. Liang WD, Tang TY, Yang YJ, Ko MT, Chuang WS (2003) Upper-ocean currents around Taiwan. Deep Sea Res Part 2 Top Stud Oceanogr 50: 1085–1105.

25. Dai CF, Horng S (2009) Scleractinia fauna of Taiwan I. The complex group. Taipei: National Taiwan University. 172 p.

26. Wu CC, Kuo YH (1999) Typhoons affecting Taiwan: current understanding and future challenges. Bull Am Meteorol Soc 80: 67–80.

27. Kuo CY, Meng PJ, Ho PH, Wang JT, Chen JP, et al. (2010) Damage to the reefs of Siangjiao Bay Marine Protected Area of Kenting National Park, southern Taiwan during typhoon Morakot. Zool Stud 50: 85.

28. Nozawa Y (2008) Micro-crevice structure enhances coral spat survivorship. J Exp Mar Biol Ecol 367: 127–130.

29. Christiansen NA, Ward S, Harii S, Tibbetts IR (2009) Grazing by a small fish affects the early stages of a post-settlement stony coral. Coral Reefs 28: 47–51.

30. Nozawa Y, Tanaka K, Reimer JD (2011) Reconsideration of the surface structure of settlement plates used in coral recruitment studies. Zool Stud 50: 53–60.

31. Suzuki G, Kai S, Yamashita H, Suzuki K, Iehisa Y, et al. (2011) Narrower grid structure of artificial reef enhances initial survival of *in situ* settled coral. Mar Poll Bull 62: 2803–2812.

32. Nozawa Y (2010) Survivorship of fast-growing coral spats depend less on refuge structure: the case of *Acropora solitaryensis*. Galaxea 12: 31–36.

33. Nozawa Y (2012) Effective size of refugia for coral spat survival. J Exp Mar Biol Ecol 413: 145–149.

34. Soong K, Chen MH, Chen CL, Dai CF, Fan TY, et al. (2003) Spatial and temporal variation of coral recruitment in Taiwan. Coral Reefs 22: 224–228.

35. Dai CF, Soong K, Fan TY (1992) Sexual reproduction of corals in northern and southern Taiwan. Proc 7th Int Coral Reef Symp 1: 448–455.

36. Babcock RC, Baird AH, Piromvaragorn S, Thomson DP, Willis BL (2003) Identification of scleractinian coral recruits from Indo-Pacific reefs. Zool Stud 42: 211–226.

37. Krebs CJ (1999) Ecological methodology (2nd ed.). Menlo Park, CA.: Addison-Wesley Educational Publishers Inc. 620 p.

38. Fukami H, Chen CA, Budd AF, Collins A, Wallace C, et al. (2008) Mitochondrial and nuclear genes suggest that stony corals are monophyletic but most families of stony corals are not (Order Scleractinia, Class Anthozoa, Phylum Cnidaria). PLoS ONE 3: e3222.

39. R Core Team (2013) R: A language and environment for statistical computing. Vienna, Austria: R Foundation for Statistical Computing.

40. Clarke KR (1993) Non-parametric multivariate analyses of changes in community structure. Aust J Ecol 18: 117–143.

41. Tioho H, Tokeshi M, Nojima S (2001) Experimental analysis of recruitment in a scleractinian coral at high latitude. Mar Ecol Prog Ser 213: 79–86.

42. Underwood JN, Smith LD, Van Oppen MJH, Gilmour JP (2006) Multiple scales of genetic connectivity in a brooding coral on isolated reefs following catastrophic bleaching. Mol Ecol 16: 771–784.

43. Kawaguti S (1941) On the physiology of reef corals. V. Tropisms of coral planulae, considered as a factor of distribution of the reefs. Palao Trop Biol Stat Stud 2: 319–328.

44. Raimondi PT, Morse ANC (2000) The consequences of complex larval behavior in a coral. Ecology 81: 3193–3211.

45. Szmant AM, Meadows MG (2010) Developmental changes in coral larval buoyancy and vertical swimming behavior: implications for dispersal and connectivity. Proc 10th Int Coral Reef Symp 1: 431–437.

46. Bassim KM, Sammarco PW (2003) Effects of temperature and ammonium on larval development and survivorship in a scleractinian coral (*Diploria strigosa*). Mar Biol 142: 241–252.

47. Baird AH, Babcock RC, Mundy CP (2003) Habitat selection by larvae influences the depth distribution of six common coral species. Mar Ecol Prog Ser 252: 289–293.

48. Babcock R, Mundy C (1996) Coral recruitment: Consequences of settlement choice for early growth and survivorship in two scleractinians. J Exp Mar Biol Ecol 206: 179–201.

49. Mundy CN, Babcock RC (1998) Role of light intensity and spectral quality in coral settlement: Implications for depth-dependent settlement? J Exp Mar Biol Ecol 223: 235–255.

50. Mason B, Beard M, Miller MW (2011) Coral larvae settle at a higher frequency on red surfaces. Coral Reefs 30: 667–676.

51. Mundy CN, Babcock RC (2000) Are vertical distribution patterns of scleractinian corals maintained by pre-or post-settlement processes? A case study of three contrasting species. Mar Ecol Prog Ser 198: 109–119.

52. Sammarco PW (1991) Geographically specific recruitment and postsettlement mortality as influences on coral communities: The cross-continental shelf transplant experiment. Limnol Oceanogr 36: 496–514.

53. Connell JH, Hughes TP, Wallace CC (1997) A 30-year study of coral abundance, recruitment, and disturbance at several scales in space and time. Ecol Monogr 67: 461–488.

54. Edmunds PJ, Bruno JF, Carlon DB (2004) Effects of depth and microhabitat on growth and survivorship of juvenile corals in the Florida Keys. Mar Ecol Prog Ser 278.

55. Miller MW, Hay ME (1996) Coral-seaweed-grazer-nutrient interactions on temperate reefs. Ecol Monogr 66: 323–344.

56. Suzuki G, Hayashibara T, Shirayama Y, Fukami H (2008) Evidence of species-specific habitat selectivity of *Acropora* corals based on identification of new recruits by two molecular markers. Mar Ecol Prog Ser 355: 149–159.

57. Hughes TP (1994) Catastrophes, phase shifts, and large-scale degradation of a Caribbean coral reef. Science 265: 1547–1551.

58. Mumby PJ (1999) Bleaching and hurricane disturbances to populations of coral recruits in Belize. Mar Ecol Prog Ser 190: 27–35.

59. Nozawa Y, Tokeshi M, Nojima S (2008) Structure and dynamics of a high-latitude scleractinian coral community in Amakusa, southwestern Japan. Mar Ecol Prog Ser 358: 151–160.

Catchment-Scale Conservation Units Identified for the Threatened Yarra Pygmy Perch (*Nannoperca obscura*) in Highly Modified River Systems

Chris J. Brauer[1], Peter J. Unmack[2], Michael P. Hammer[3,4,5], Mark Adams[3,5], Luciano B. Beheregaray[1]*

1 Molecular Ecology Laboratory, School of Biological Sciences, Flinders University, Adelaide, South Australia, Australia, 2 Institute for Applied Ecology and Collaborative Research Network for Murray-Darling Basin Futures, University of Canberra, Canberra, Australian Capital Territory, Australia, 3 School of Earth and Environmental Sciences, University of Adelaide, South Australia, Australia, 4 Curator of Fishes, Museum and Art Gallery of the Northern Territory, Darwin, Northern Territory, Australia, 5 Evolutionary Biology Unit, South Australian Museum, Adelaide, South Australia, Australia

Abstract

Habitat fragmentation caused by human activities alters metapopulation dynamics and decreases biological connectivity through reduced migration and gene flow, leading to lowered levels of population genetic diversity and to local extinctions. The threatened Yarra pygmy perch, *Nannoperca obscura*, is a poor disperser found in small, isolated populations in wetlands and streams of southeastern Australia. Modifications to natural flow regimes in anthropogenically-impacted river systems have recently reduced the amount of habitat for this species and likely further limited its opportunity to disperse. We employed highly resolving microsatellite DNA markers to assess genetic variation, population structure and the spatial scale that dispersal takes place across the distribution of this freshwater fish and used this information to identify conservation units for management. The levels of genetic variation found for *N. obscura* are amongst the lowest reported for a fish species (mean heterozygosity of 0.318 and mean allelic richness of 1.92). We identified very strong population genetic structure, nil to little evidence of recent migration among demes and a minimum of 11 units for conservation management, hierarchically nested within four major genetic lineages. A combination of spatial analytical methods revealed hierarchical genetic structure corresponding with catchment boundaries and also demonstrated significant isolation by riverine distance. Our findings have implications for the national recovery plan of this species by demonstrating that *N. obscura* populations should be managed at a catchment level and highlighting the need to restore habitat and avoid further alteration of the natural hydrology.

Editor: Stefano Mariani, School of Environment & Life Sciences, United Kingdom

Funding: Financial support for this study was provided by the Australian Research Council (LP100200409 to LB Beheregaray, J Harris & M Adams). Support was provided to CJB by the AJ & IM Naylon Honours scholarship. PJU was supported as a Murray-Darling Basin futures research fellow through the Australian Government's Collaborative Research Networks (CRN) Program. MPH was provided support by the Cooperative Research Centre for Freshwater Ecology and an Australian Postgraduate Award. The funders had no role in study design, data collection and analysis, decision to publish, or preparation of the manuscript.

Competing Interests: The authors have declared that no competing interests exist.

* E-mail: luciano.beheregaray@flinders.edu.au

Introduction

Human activities such as land development, agriculture, and exploitation of natural resources have long been acknowledged as driving processes behind habitat fragmentation and degradation [1,2]. This can decrease population connectivity through reduced migration and gene flow, leading to higher genetic differentiation among populations and lowered levels of genetic diversity within [3]. When populations become isolated they become vulnerable to extirpation due to environmental [1], demographic [4] and genetic [5] processes that increase the chances of local extinction. If habitat fragmentation is widespread on a regional scale then there is the potential for loss of biodiversity and species extinctions [6]. It is therefore important for conservation and natural resource managers to consider patterns and processes related to population connectivity and gene flow at both local and regional scales [7].

Conservation genetics is the application of evolutionary principles and molecular methods to species and biodiversity conservation [8]. Riverscapes have long been recognised for their ecological complexity and sensitivity to human impacts [9,10], and have been the focus of many recent conservation genetics studies [11–15]. Understanding the spatial scale of patterns of genetic diversity is also important for species conservation in order to identify the evolutionary processes shaping these patterns and to detect when populations become demographically and genetically independent [16,17]. This information can be used to estimate the geographical extent of conservation units defined by genetic criteria [18,19], such as the popularly used Evolutionarily Significant Units (ESUs) and Management Units (MUs) (see Moritz [20] and Crandall [21] for definitions). These conservation units can inform conservation management strategies by recognising the historical isolation of evolutionary lineages (i.e. ESU), and the functional and demographic independence of populations (i.e. MU) [22].

In this study we identify units for conservation in a threatened freshwater fish, *Nannoperca obscura* (Yarra pygmy perch), found across two biogeographic provinces [23] in a series of adjacent, but highly fragmented Australian riverine ecosystems. One is the

Murray-Darling Basin (MDB), arguably Australia's most important agricultural region, given it contributes 50% of the water used for agricultural irrigation in the country [24]. Modifications to the natural flow regime, water abstraction, drainage of wetlands, and the introduction of exotic species have all contributed to the decline of native fishes across the MDB, and this trend is expected to continue in the face of future climate change [25]. In recent years drought has led to extremely low inflows and reduced water levels throughout the system and this is especially evident in the Lower Lakes region of the MDB [26,27]. Bass Province is the second biogeographically distinct area which contains many smaller separate catchments encompassing most of the range of *N. obscura* (Fig. 1). Most drainages have experienced considerable alterations due to major agricultural activity and as a result the region contains some of the most highly disturbed waterways in Australia [28]. The extensive alterations to the natural surface water hydrology here have had a major impact, resulting in wetland drainage and reduction in freshwater habitat [29].

Nannoperca obscura (family Percichthyidae) is a small (up to 75 mm total length) freshwater fish endemic to coastal drainages of southeastern Australia [30] (Fig. 1). This species prefers slow flowing habitat along the edge of streams and rivers, with abundant submerged vegetation [31]. They are largely sedentary, with large demersal larvae suggesting limited dispersal potential at all life stages [32]. *Nannoperca obscura* has declined since European

settlement, and is currently listed as vulnerable by the IUCN [33], protected and critically endangered in South Australia [34], vulnerable in Victoria [35], and has a National Species Recovery Plan [36]. Habitat fragmentation has been exacerbated by recent drought and is considered a major threat along with competition and predation by introduced species and climate change [30]. The decline of *N. obscura* in the MDB and southeastern South Australia was recently exacerbated by extreme drought, which resulted in the extirpation of all populations in the lower Murray River [37]. However, remnants of these populations were rescued before their extirpation and form the basis for a conservation breeding program, with offspring from this effort being released back into former habitats since drought conditions eased in 2011 [27]. In order to guide conservation efforts of *N. obscura*, Hammer et al. [32] conducted a phylogeographic study using mtDNA and allozymes. They identified four evolutionary lineages as ESUs, which they designated Eastern, Merri, Central, and MDB, and tentatively suggested that most catchments contained separate MUs.

Here we employed recently developed microsatellite DNA markers to provide finer resolution of recent genetic divergences. This, combined with greatly increased sampling, allows a more detailed assessment of population structure and genetic diversity over multiple spatial scales (i.e. within and between catchments) across the entire distribution of *N. obscura*. In order to address

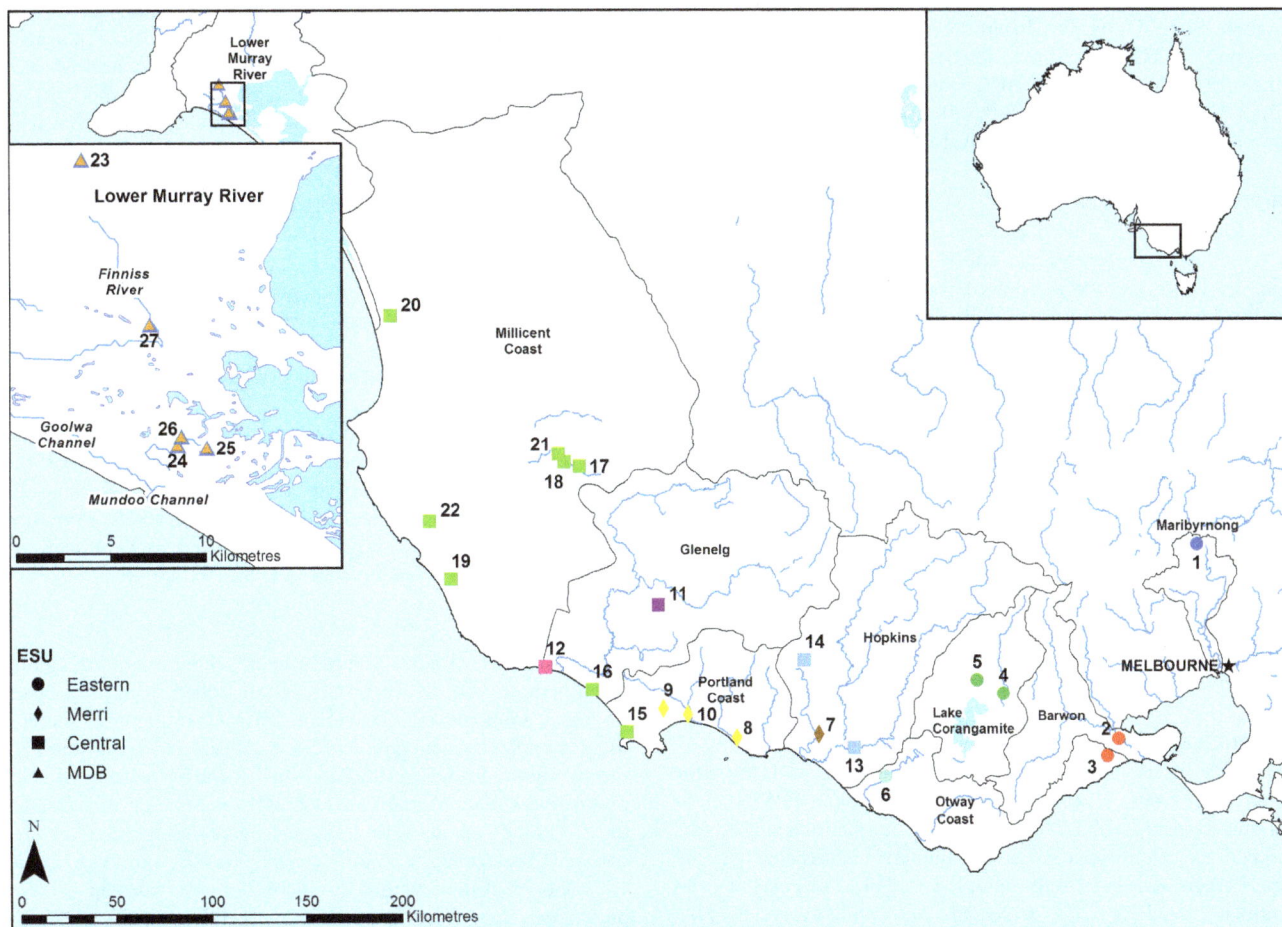

Figure 1. *Nannoperca obscura* **sampling locations and proposed Evolutionarily Significant Units (ESUs).** Inset shows close-up of the MDB sites. Colours denote genetic clusters (MUs) described in Figure 2.

issues of relatedness between isolated river systems, we treat each river with an independent connection to the ocean as a separate unit, herein defined as a catchment. Firstly, we used these data to test the proposed ESUs of Hammer et al. [32], which were defined with more conservative molecular markers using a relatively small sample of individuals (n = 156) and localities (18). Secondly, we tested the hypothesis derived from life history and ecological requirements that *N. obscura* are poor dispersers and should therefore display very low connectivity between catchments and low connectivity between sites within catchments, with many populations representing different MUs within highly structured ESUs. For this hypothesis, we use a combination of frequency-based and genotypic-based statistical methods to assess the number and spatial distribution of MUs within inferred ESUs. Finally, we explore factors that have shaped the spatial distribution of genetic diversity in *N. obscura* and highlight the direct implications of our findings for conservation management.

Materials and Methods

Ethics Statement

Permission to undertake field work and collect specimens was obtained under the following permits: Victorian Fisheries research permits RP 581 and RP 945, Victorian Flora and Fauna permits 10002072 and 10004939, Victorian National Parks permit 10004939, South Australian Primary Industries and Resources - Section 59 and 115 Exemptions. Specimens were obtained under Arizona State University Institutional Animal Care and Use Committee (IACUC) approval 09-1018R, Brigham Young University IACUC approval 070403, University of Adelaide Animal Ethics Committee approval S-32-2002 and Flinders University Animal Welfare Committee approval E313.

Sampling and genotyping

A total of 541 individuals were sampled from 27 locations (n = 5–40 per site) across the entire extant range of *N. obscura* (Fig. 1). DNA was extracted from caudal fin clips following a modified salting out process [38]. Fourteen microsatellite loci designed specifically for *N. obscura* were amplified in two polymerase chain reaction (PCR) multiplexes of six and eight loci, respectively [39] (Table S1). The PCR conditions were based on a modified touchdown procedure [40]. The PCR product was diluted 1:5 with H_2O, sized with GS500LIZ size standard and analysed using an automated ABI 3130 capillary electrophoresis system (Applied Biosystems) with one run per multiplex PCR. Genotypes were binned and scored visually with GeneMapper 4.0 (Applied Biosystems). Genotypes were checked for scoring errors putatively related to null alleles, stuttering, and large allele drop-out using Micro-Checker 2.2.3 [41]. To ensure all loci were scored consistently, we repeated amplification and genotyping procedures for 86 individuals.

Genetic variation

Fisher's exact test of linkage disequilibrium and tests for departures from Hardy-Weinberg equilibrium (HWE) were conducted using GENEPOP 4.1.4 [42], and GenoDive 2.0 [43], respectively. Significance levels were Bonferroni-corrected to avoid type I errors associated with multiple tests [44]. For each site, the number of alleles (N_A), expected (H_E) and observed (H_O) heterozygosity, and inbreeding coefficient (F_{IS}) were calculated in GenoDive 2.0 [43]. Allelic richness (A_R) was estimated using the rarefaction procedure in HP-RARE [45], and the percentage of polymorphic loci was calculated in GenAlEx 6.5 [46].

Population structure

Population genetic structure was assessed at multiple spatial scales, both across the entire distribution and within each of the proposed ESUs, using a combination of several frequency-based and genotype-based statistical methods. Pairwise F_{ST} [47] and R_{ST} [48] tests were performed in Arlequin 3.5.1.2 [49] to evaluate between-site differentiation. Given the potential for temporal variation in population structure, an assessment of pairwise F_{ST} and R_{ST} was also performed between years at 12 sites where samples were collected on multiple occasions. In order to determine if either F_{ST} or R_{ST} was more appropriate for this study, the relative contribution of genetic drift and mutation to population differentiation was assessed [50]. SPAGeDi 1.3 [51] was used to permutate global allele sizes for each locus and to compare observed R_{ST} with permutated R_{ST} (pR_{ST}) values. Arlequin was used to perform an analysis of molecular variance (AMOVA) with 1000 permutations based on F_{ST} [47]. Hierarchical structure was assessed using AMOVA among the major genetic lineages, among sites within lineages, and among individuals within sites. Separate AMOVAs were also performed for each primary lineage.

A Bayesian clustering analysis of individual genotypes using STRUCTURE 2.3.4 [52] was initially performed using all samples to identify primary population structure across the entire distribution, before repeating the analysis within each of the primary clusters to assess hierarchical population structure at smaller spatial scales [53]. Twenty independent runs for each K value (1–27) were completed to ensure reproducibility [54], using a burn in of 100 000 followed by 1 million Monte-Carlo Markov chain (MCMC) iterations. We used the admixture model, with independent allele frequencies among populations and no prior information on sampling location. The most likely K value was inferred using the Evanno et al. method [53] implemented in STRUCTURE HARVESTER [55]. Results of the 20 replications were then combined using the software CLUMPP 1.1.2 [56], and visualised using Distruct 1.1 [57]. A different analytical approach based on assignment of individual genotypes was performed using GeneClass2 [58]. This was conducted using the Bayesian approach of Rannala and Mountain [59] to calculate the probability that each individual originates from its sampling locality or from other sites.

Principal coordinates analysis (PCA) was also employed to allow visual examination of the genetic affinities of individuals across the entire distribution and to clusters identified within each lineage. Pairwise genetic distances [60] between individuals were first calculated before those results were subjected to PCA analysis. Both procedures were completed in GenAlEx 6.5 [46].

Gene flow

We used BayesAss 3.0 [61] to estimate recent migration among lineages and also between clusters identified within lineages. BayesAss implements a Bayesian MCMC resampling method using multilocus genotypes to estimate asymmetrical rates of recent migration, where migration (*m*) is the proportion of each population having migrant ancestry. First generation migrants, or the offspring of at least one first generation migrant, are considered as having migrant ancestry. The software was run for 10 million iterations with a 1 million iteration burn in. Mixing parameters for allele frequencies, inbreeding coefficients and migration rate were adjusted to achieve optimum acceptance rates of 20–40% [61]. Convergence was confirmed by plotting the cumulative log likelihoods of the iterations using the program Tracer 1.5 [62]. Each run was also repeated five times using different seeds and the posterior estimates compared for consistency [61].

Spatial analyses

Spatial genetic structure was assessed at both the population and individual level. At the population level isolation by distance (IBD) was assessed using Mantel tests. These tests were applied to each of the three lineages that displayed evidence of population structure to determine the association between pairwise population F_{ST} and geographic distance. Euclidean distance was calculated because most sites occur in isolated catchments and are not connected by continuous stream lengths.

Results of a test for IBD can be difficult to interpret when population structure is strongly influenced by sharply divided spatial groups [63]. This can occur when hierarchical structure due to the presence of strong barriers for dispersal (physical or ecological) creates clusters of populations that are not necessarily better explained by spatial distances between demes. To assess this possibility partial Mantel tests were performed to assess for hierarchical population structure, such as that identified by the AMOVAs and STRUCTURE. These tests assessed the association between F_{ST} and geographic distance while controlling for hierarchical population structure by using a binary model matrix describing whether comparisons were made either between or within the identified clusters [64]. Finally a simple Mantel test was performed for C4, the largest genetic cluster identified within a lineage (no other clusters contained a sufficient number of sites), in order to test for IBD at a local scale. Performing this test separately within predefined clusters removes any potential bias of hierarchical structure [63]. All Mantel and partial Mantel tests were performed in GenoDive 2.0 using 1000 permutations.

To further evaluate spatial dimensions of genetic structure at an individual level within each lineage we used spatial autocorrelation [65] in two ways. Firstly, correlograms were constructed using the method of Smouse and Peakall [60] implemented in GenAlEx 6.5. The autocorrelation coefficient (r) was plotted as a function of discrete distance classes, partitioned so as to achieve a similar number of pairwise comparisons for each class [18]. A positive r value indicates the presence of IBD and the x intercept can provide an estimate of the extent of IBD for each lineage [17–19]. Peakall et al. [17] also suggest a second autocorrelation method to accurately identify the scale at which population genetic structure is detectable. In this case, r was calculated using multiple distance class analysis, also in GenAlEx 6.5. This method plots r as a function of increasing distance class sizes [17]. The first class is based on the minimum distance between sites (0–10 km for all lineages) and each successive class adds individuals from more distant groups (i.e. 0–10 km, 0–20 km, 0–30 km, etc.). When significant IBD exists, the value of r is expected to decrease with the increasing size of each distance class. The last distance class for which r is significant is considered the limit of detectable IBD [17]. Significance was assessed for both tests using 95% confidence intervals for the null hypothesis of no spatial structure using 999 random permutations, and for estimates of r by bootstrapping 1000 pairwise comparisons for each distance class [46].

Results

Data quality and genetic variation

There was no consistent evidence for stuttering or large allele drop-out for any locus, or for linkage disequilibrium between any pairs of loci. Null alleles were detected for Nob26 at sites 14 and 26, Nob30 at sites 2, 5 and 17 and Nob35 at site 27. However as these findings were not consistent across populations, and when analyses were run without these loci similar results were obtained, all loci were retained. After Bonferroni correction, only one sampled site (#2, Waurn Ponds Creek) was found to deviate significantly from expectations of Hardy-Weinberg equilibrium (Table 1). The 14 microsatellite loci contained between 4 and 26 alleles, with a mean of 11.9 alleles per locus. Despite the highly polymorphic nature of the markers used, the overall levels of genetic variation were very low for *N. obscura*, with mean observed heterozygosity of 0.318 and mean allelic richness of 1.92 (Table 1). No major differences in genetic variation were apparent between sampled sites or among putatively different lineages. All repeated individuals generated the same genotype.

Population structure

Clustering analysis in STRUCTURE demonstrated high levels of differentiation, identifying four major clusters (Fig. 2A). These generally correspond with the proposed ESUs of Hammer et al. [32], although our microsatellite data displayed evidence of significant admixture between the Eastern and Merri ESUs at site 6 (Curdies River, Eastern ESU; Hammer et al. [32]), to the extent that site 6 was assigned to the Merri cluster. Since contemporary genetic and demographic processes are the focus of the present study, Curdies River was included as part of the Merri genetic lineage for all analyses. Apart from the lack of differentiation within the MDB, strong population structure was evident within the other lineages. Three genetic clusters were identified within the 'pure Eastern' lineage, three within Merri/Curdies, and four within the Central lineage (Fig. 2B). Population assignment results from GeneClass2 strongly support the population structure identified by STRUCTURE (Table S2). Most individuals (66.9%) were correctly assigned to their sampling location, with very few assigned to sites outside of their proposed lineage for probabilities greater than 5%. Within the four primary genetic lineages, sites that shared a high probability of assignment closely correspond to the genetic clusters identified by STRUCTURE.

As predicted by the STRUCTURE analyses, significant and very high levels of genetic structure were evident between most demes of *N. obscura* (341 pairwise comparisons were significant out of 351 tests), with F_{ST} ranging from 0–0.84 (mean F_{ST} = 0.45), and R_{ST} ranging from 0–0.95 (mean R_{ST} = 0.49) (Table S3). Assessment of temporal variation in allele frequencies at sites sampled on multiple years did not reveal any temporal trend in population structure, with only one statistically significant comparison out of 24 pairwise tests (Table S4). The comparison of R_{ST} and pR_{ST} in SPAGeDi revealed that pR_{ST} was significantly greater than R_{ST} for only two loci (Nob2; Nob12). This indicates that, for this dataset, genetic drift contributes more to genetic diversity than mutation. Therefore, F_{ST} was used as the measure of population differentiation [51].

Based on F_{ST}, AMOVA calculated across all sites attributed 34% of the variation to differences among proposed primary lineages (P<0.001), 18% to variation between sites within lineages (P<0.001), and just 2.6% among individuals within sites (P<0.001) (Table 2). When calculated separately, the AMOVA results were similar for each of the pure Eastern, Merri/Curdies, and Central lineages, with ~30% of the variation attributed to among site differences, and among individuals within sites only contributing 0.8% (P = 0.281) for Merri/Curdies, and 5.9% (P<0.001) and 6.3% (P<0.001) for Central and pure Eastern lineages, respectively (Table 2). No significant variation was detected among sites or among individuals within sites in the MDB (Table 2).

Table 1. Information on localities, sample sizes, Evolutionarily Significant Units (ESUs), Management Units (MUs) and summary of genetic diversity for *Nannoperca obscura*.

Site	ESU	MU	Location	Latitude	Longitude	N	N_A	% Poly loci	A_R	H_O	H_E	F_{IS}	P value
1	Eastern	E1	Deep Ck, Lancefield	−37.259	144.713	10	1.9	43%	1.41	0.150	0.149	−0.005	0.504
2		E2	Waurn Ponds Ck, Geelong	−38.189	144.349	33	4.0	93%	2.41	0.468	0.521	0.102	**0.001**
3		E2	Thompson Ck	−38.272	144.290	10	2.9	86%	2.16	0.386	0.424	0.090	0.126
4		E3	Woady Yaloak R, Cressy	−38.024	143.627	7	2.0	64%	1.80	0.347	0.346	−0.002	0.476
5		E3	Gnarkeet Ck, Lismore	−37.972	143.466	29	2.1	57%	1.66	0.241	0.266	0.093	0.064
6*		M1	Curdies R, Curdie	−38.448	142.957	30	2.5	79%	1.77	0.277	0.304	0.090	0.032
7	Merri	M2	Merri R, Grassmere	−38.275	142.542	39	3.7	71%	2.02	0.364	0.377	0.033	0.199
8		M3	Shaw R, Yambuk	−38.315	142.061	40	4.7	79%	2.22	0.378	0.391	0.034	0.147
9		M3	Surry R, Heathmere	−38.200	141.614	30	3.9	93%	2.36	0.452	0.438	−0.032	0.198
10		M3	Fitzroy R, Tyrendarra	−38.221	141.764	8	3.4	79%	2.45	0.390	0.462	0.155	0.014
11	Central	C1	Palmer Ck, Merino	−37.724	141.546	31	2.7	71%	1.94	0.320	0.343	0.066	0.063
12		C2	Crescent Pond, Picks Swamp	−38.040	140.898	20	2.1	86%	1.39	0.136	0.162	0.160	0.030
13		C3	Mount Emu Ck, Panmure	−38.325	142.759	31	2.4	79%	1.56	0.209	0.224	0.066	0.119
14		C3	Mustons Ck	−37.936	142.427	8	2.8	86%	2.10	0.357	0.391	0.086	0.144
15		C4	Bridgewater Lakes, main lake	−38.319	141.405	10	3.0	86%	2.21	0.414	0.425	0.025	0.368
16		C4	Lake Monibeong	−38.133	141.186	10	2.8	93%	2.10	0.407	0.427	0.047	0.308
17		C4	Mosquito Ck, Langkoop	−37.104	141.037	34	2.6	86%	1.80	0.290	0.314	0.077	0.036
18		C4	Mosquito Ck, Wombeena	−37.087	140.945	10	2.3	71%	1.83	0.257	0.333	0.227	0.006
19		C4	Drain 88, Lake Bonney	−37.657	140.316	9	2.3	86%	1.85	0.278	0.330	0.158	0.039
20		C4	Henry Ck, Kingston	−36.450	139.891	16	3.0	86%	1.95	0.366	0.360	−0.018	0.446
21		C4	Mosquito Ck, South Waverley	−37.052	140.908	14	2.2	79%	1.78	0.280	0.319	0.121	0.047
22		C4	Drain M, Elgin Lane	−37.393	140.174	26	2.6	86%	1.91	0.323	0.359	0.100	0.022
23	MDB	MDB	Finniss R, L. Alexandrina	−35.405	138.843	5	2.2	64%	1.90	0.300	0.323	0.072	0.241
24		MDB	Mundoo Channel, Hindmarsh Is.	−35.537	138.905	5	2.0	64%	1.74	0.300	0.273	−0.098	0.238
25		MDB	Mundoo Channel, Hindmarsh Is.	−35.538	138.922	27	3.1	79%	1.85	0.302	0.313	0.037	0.241
26		MDB	Steamer Drain, Hindmarsh Is.	−35.533	138.907	17	2.3	71%	1.82	0.291	0.302	0.037	0.270
27		MDB	Goolwa Ch. Lake Alexandrina	−35.481	138.886	32	2.9	79%	1.82	0.293	0.297	0.015	0.359

N is number of samples, A_R is allelic richness, H_O is observed heterozygosity, H_E is expected heterozygosity, F_{IS} is inbreeding coefficient, P value relates to Hardy-Weinberg equilibrium test (significant value indicated in bold).
*Site 6 (Curdies River) is an Eastern ESU site with high levels of admixture from Merri and as such has been included in Merri for all analyses.

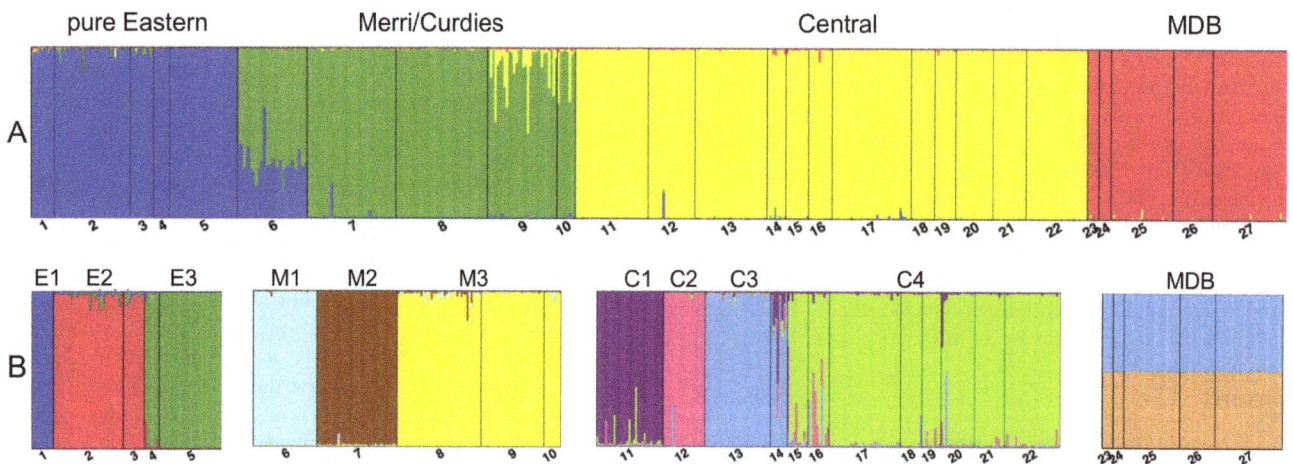

Figure 2. Admixture plots indicating major genetic lineages and Management Units (MUs) for *Nannoperca obscura* produced by the software STRUCTURE. A) K=4 selected as most likely K value by STRUCTURE HARVESTER, and B) hierachical structure indicating proposed MUs. Site numbers correspond to those in Table 1 and Fig. 1.

Table 2. Hierarchical analysis of molecular variance (AMOVA) based on F_{ST} for *Nannoperca obscura*.

Group	Source of variation	d. f.	% of variance	P value
All sites				
	Among lineages	3	34.9%	**<0.001**
	Among sites within lineages	23	18.2%	**<0.001**
	Among individuals within sites	514	2.6%	**<0.001**
pure Eastern				
	Among sites	4	29.7%	**<0.001**
	Among individuals within sites	84	6.3%	**<0.001**
Merri/Curdies				
	Among sites	4	30.6%	**<0.001**
	Among individuals within sites	142	0.8%	0.281
Central				
	Among sites	11	32.2%	**<0.001**
	Among individuals within sites	207	5.9%	**<0.001**
MDB				
	Among sites	4	2.5%	0.131
	Among individuals within sites	81	2.7%	0.200

Significant values indicated in bold.

The PCA results were also generally concordant with the other analyses of geographic population structure. The four primary genetic lineages were well supported in the initial analysis of all individuals (Fig. 3), while the genetic clusters identified within primary lineages (Table 1; Fig. 2B) were mostly supported when each was run separately (Fig. S1). Importantly, as also demonstrated by all other analyses, the PCA plot for MBD provided no evidence for population structure within this lineage (Fig. S1D).

Gene flow

Estimates of recent gene flow in BayesAss demonstrate extremely low exchange of migrants both among lineages (0.2–0.7%) (Table S5) and between clusters within each primary genetic lineage (0.2–2.1%) (Table S6). All pairwise estimates of *m* were within the 95% credible interval and there was no evidence for asymmetric gene flow between populations. These results validate the delineation of genetically and demographically isolated clusters within the four lineages, and are in line with expectations given the high level of population structure indicated by the other analyses.

Spatial analyses

Patterns of IBD and hierarchical population structure were revealed at multiple scales for *N. obscura*. Mantel tests demonstrated a strong, significant association between population F_{ST} and geographic distance (*r* values ranged between 0.411 and 0.833) (Table 3), indicating the presence of IBD between sites within the pure Eastern, Merri/Curdies, and Central lineages (Fig. 4). The results of the partial Mantel tests were not significant for Merri/Curdies (*r* = 0.43, P = 0.192) and Central (*r* = 0.18, P = 0.219) but were significant for the pure Eastern lineage (*r* = 0.0, P = 0.005) (Table 3). Although these tests suggest that hierarchical population structure due to catchment divisions is not evident within two lineages, they suffer from relatively low power due to the general lack of multiple samples representing each catchment. Finally, the strong and highly significant results of the Mantel test within the

C4 genetic cluster (*r* = 0.785, P = 0.001) demonstrate the pattern of IBD also exists at a smaller scale for *N. obscura* (Table 3).

At the individual level, there was significant positive spatial autocorrelation for the first distance class (0–60 km), which intercepted the *x*-axes at 81 km, 84 km, and 115 km for the pure Eastern, Merri/Curdies, and Central correlograms, respectively (Fig. 5). This strongly indicates that on average, individuals from each locality had a higher probability of being born locally, providing support to the IBD signal demonstrated by the Mantel tests. To determine the extent to which this pattern of IBD exists, the autocorrelation coefficient *r* was also calculated for increasing distance class sizes. The positive *r* values became non-significant at 150 km for both pure Eastern and Merri/Curdies lineages (Fig. S2) and at 300 km for the putatively more connected Central lineage (Fig. S3). Significant IBD is therefore confirmed within the pure Eastern and Merri/Curdies lineages for sites up to 140 km apart, beyond which positive spatial autocorrelation is no longer detectable. The results are similar for the Central lineage, where IBD can be detected for sites up to 290 km apart.

Discussion

We employed highly resolving microsatellite markers to assess genetic variation and population structure across the distribution of a threatened freshwater fish, *N. obscura*. Remarkably strong population genetic structure that corresponds with catchment boundaries was detected in a pattern that broadly validates the four ESUs previously proposed for this species [32]. Spatial analysis of genetic variation demonstrated that both significant IBD and hierarchical population structure exist in *N. obscura*. Several MUs were identified within each of the pure Eastern (3 MUs), Merri/Curdies (3 MUs), and Central (4 MUs) lineages (Table 1; Fig. 2B). The MDB however appears to contain a single genetic population which is consistent with the close geographic proximity of sampled sites (Fig. 1). A combination of spatial analytical methods was also implemented to determine the scale and extent of IBD within each ESU. Here we describe the spatial context of the strong subdivision detected, discuss evolutionary processes that might have accounted for these patterns, and consider conservation management implications for this threatened freshwater fish species.

Genetic diversity and population structure

Overall *N. obscura* exhibits very low genetic diversity [66], even when compared to other Australian freshwater fishes (mean $H_O = 0.32$ compared to golden perch $H_O = 0.52$ [67], dwarf galaxias $H_O = 0.40$ [68], southern pygmy perch $H_O = 0.57$ [69], and purple spotted gudgeon $H_O = 0.58$ [70]. In addition to overall low levels of diversity, substantial genetic structure was also observed for *N. obscura*. Natural wet and dry cycles, as well as more recent habitat fragmentation have likely resulted in repeated extirpation and re-colonisation events in this species. This boom-bust cycle is thought to account for the small effective population sizes of other freshwater fishes in Australia [71], and accompanied with habitat specificity may be responsible for the low level of genetic diversity observed here. All results of population structure analyses were in agreement with the delineation of the four ESUs proposed by Hammer et al. [32]. Our study builds substantially on the findings of Hammer et al. [32] by the use of highly resolving microsatellite markers, a greater number of sites, and much larger sample size (n = 156, 18 sites Hammer et al. [32], n = 541, 27 sites this study). The high resolution data generated here provide evidence for recent admixture between Merri and Eastern ESUs at site 6 (Curdies River). Mitochondrial DNA reflects evolutionary

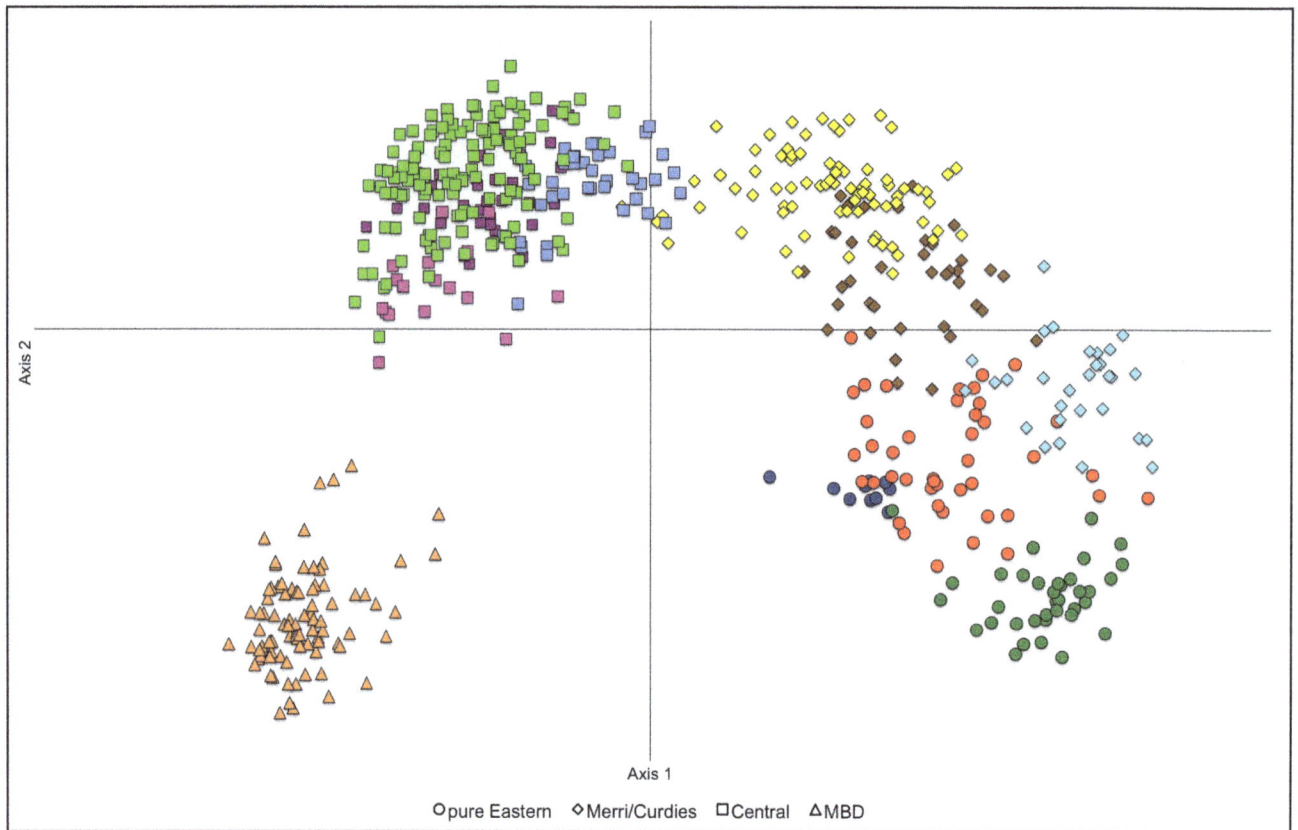

Figure 3. Principal coordinates analysis based on 14 microsatellite loci for *Nannoperca obscura* supporting the delineation of the four genetic lineages. Eigen values for the first and second axes have been plotted, which explain 35% and 25% of the variance, respectively. Colours denote genetic clusters (MUs) described in Figure 2.

history [20] and, in this case, the mtDNA gene tree assigns Curdies River to the Eastern ESU [32], a result also supported by allozyme analysis [32]. However, microsatellites depict more recent and fine-scale structure than either mtDNA or allozymes [72]. Conservation management for this site should therefore consider both the historical and geographical connection with Eastern and also seek to maintain contemporary processes responsible for the more recent association with Merri.

Our fine-scale analysis of genetic structure identified 11 hierarchical clusters nested within the four ESUs (Table 1; Fig. 2B). Notably, the Mt Emu and Mustons Creek sites align genetically with Central (to the west of Merri), however they are tributaries of the Hopkins River, which is east of Merri (Fig. 1). Hammer et al. [32] included only Mt Emu Creek in their study and hypothesised an historical connection between the upper Glenelg and Hopkins rivers to explain the anomaly. The inclusion of Mustons Creek in this study and its clear association with Mt Emu Creek and other Central sites supports this hypothesis. An examination of the local topography here reveals a flat region of swampy wetlands extending between the upper reaches of the Glenelg and Hopkins rivers [32]. A similar pattern of closely related populations between these drainages is also found in the sympatric *N. australis* (southern pygmy perch) [73] and it seems very likely that dispersal could occur between these river basins during wetter periods [23,73]. No genetic structure was detected for the MDB, where the demes appear to be linked by migration. This is probably due to the close proximity of these sites, near

contiguous vegetated littoral habitats, and the likelihood of shared refuges during times of drought.

Across the species range, the pairwise F_{ST} estimates of differentiation were mostly significant and high between sites (Table S4), even within clusters. These results highlight the restricted dispersal potential of this species and the isolation by catchment boundaries at many sites. The genetic clusters identified here are consistent across multiple methods of analysis, and satisfy the requirements for designation as MUs [20]. The negligible level of recent migration detected between MUs further supports and validates the other measures of genetic structure.

Connectivity and the importance of spatial scale

Connectivity is a fundamental ecological and evolutionary process shaping the spatial distribution of genetic variation of species [3]. It is therefore important for conservation management to consider how species biology may alter the way that connectivity is affected by processes such as habitat fragmentation [74,75]. *Nannoperca obscura* is known to possess limited dispersal potential at all life stages. Small, isolated populations occur in permanent wetlands, streams and ponds and these habitat patches are rarely connected, providing even less opportunity for dispersal. Consequently, environmental changes that result in habitat fragmentation, such as wetland drainage and increased salinity, are likely to have a high impact on the already-limited ability of this species to re-colonise demes after local extirpations. Other Australian fishes inhabiting similar environments also exhibit

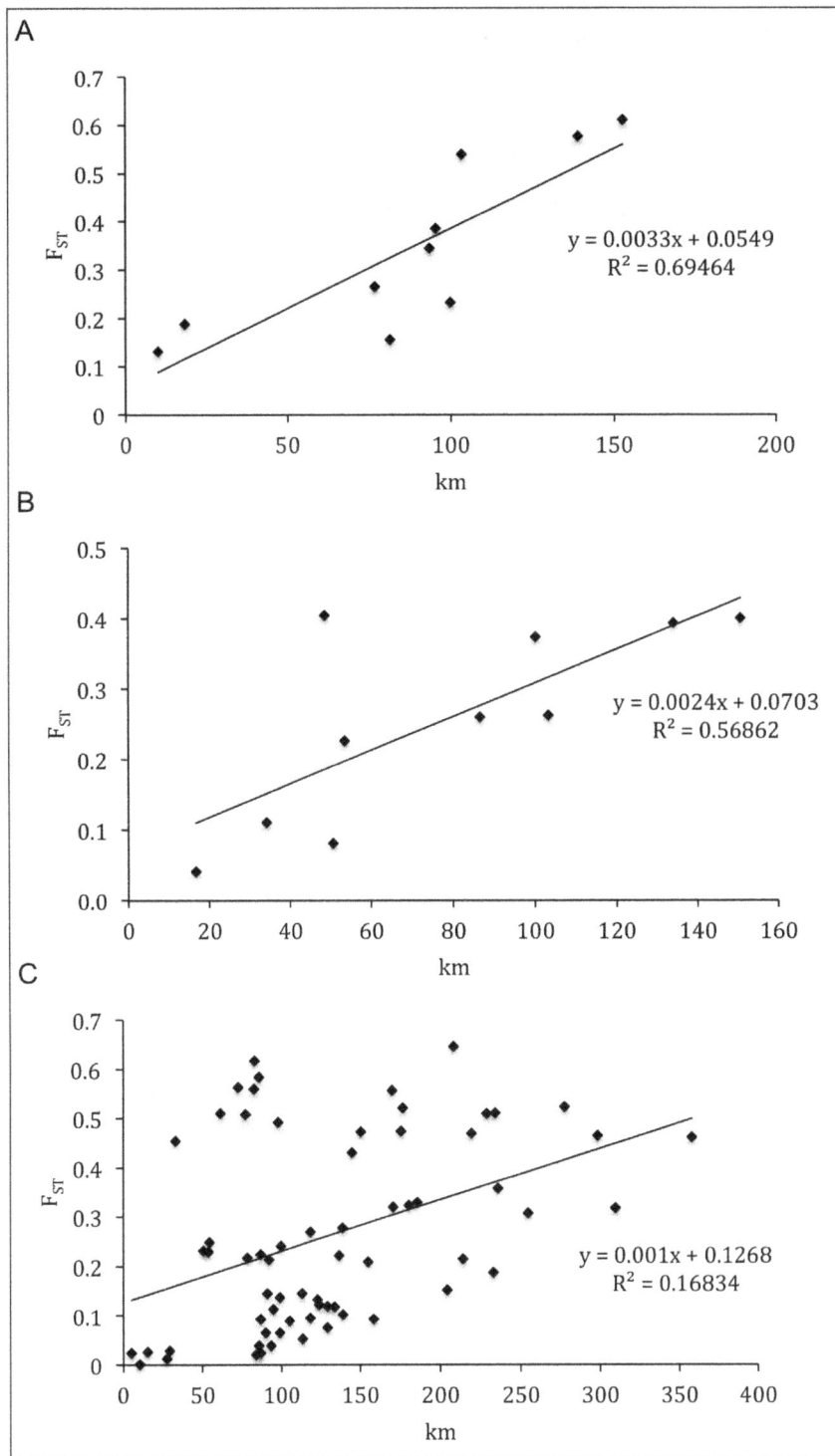

Figure 4. Isolation by distance plots for *Nannoperca obscura* comparing F$_{ST}$ (Weir and Cockerham 1984) with distance between sites within A) pure Eastern, B) Merri/Curdies, and C) Central lineages.

similar patterns of genetic structure and face the same conservation issues [68,69,76]. For example, another native fish *Galaxiella pusilla* (dwarf galaxias) frequently coexists with *N. obscura*, shares many similar life history traits, including restricted dispersal, and is also threatened by habitat fragmentation [68,77]. Overall genetic diversity for populations of *G. pusilla* is low and broadly similar to *N. obscura*, and population structure for both species follow similar patterns of distribution among catchments. *Macquaria australasica* (Macquarie perch) are a larger percichthyid that although disperses more readily than *N. obscura*, today is restricted to

Table 3. Results for Mantel and partial Mantel tests for major genetic lineages, and genetic cluster C4 for *Nannoperca obscura*.

Lineage	Matrix A	Matrix B	Covariate	Mantel's r	P value
pure Eastern	Genetic	Geographic	–	0.833	**0.014**
	Genetic	Geographic	Clusters	0.900	**0.005**
erri/Curdies	Genetic	Geographic	–	0.754	**0.018**
	Genetic	Geographic	Clusters	0.431	0.192
entral	Genetic	Geographic	–	0.411	**0.030**
	Genetic	Geographic	Clusters	0.181	0.219
Cluster					
4	Genetic	Geographic	–	0.785	**0.001**

Genetic distance is F_{ST}, geographic distance is Euclidean distance between sites (km). Significant values indicated in bold.

isolated headwaters and shows reduced connectivity associated with human-induced habitat fragmentation [13]. In contrast, the large-bodied percichthyid *Maccullochella peelii* (Murray cod) inhabits the main river channels and larger tributaries, tends to encounter fewer natural barriers to dispersal and is therefore less affected by habitat fragmentation [78].

Environmental and evolutionary processes affecting populations at a local level may differ from those affecting the same species at a regional level [16,17]. Spatial analyses revealed strong patterns of IBD and population structure at multiple scales for *N. obscura*. At a regional level there is very strong genetic structure with significant divergence between ESUs (Fig. 2A). Within the Eastern ESU a combination of both hierarchical structure (due to population differences between catchments) and IBD exists, while significant IBD was also detectable at a smaller spatial scale within MUs.

Spatial autocorrelation has been used to examine the scale and extent of IBD in several studies [17–19]. By identifying the distance at which samples can be considered genetically and demographically independent, and thus defining the range over which this pattern persists, conservation measures can be designed to ensure maximum genetic diversity is preserved [18,19]. Sampling regimes for genetic monitoring or future ecological studies can also then be designed more efficiently and with confidence that most genetic diversity will be sampled [18]. In the case of *N. obscura*, genetically similar patches are approximately 80 km in diameter for Eastern and Merri and 115 km for Central. This is consistent with the geographic extent of the MUs identified in this study and the proposal that MUs are confined mostly to single catchments.

Evolutionary processes

Identifying the spatial patterns and scale of genetic variation can help to ensure conservation management strategies capture the overall diversity of a species [79]. To maintain species persistence in the long term, it is also important to identify and conserve the evolutionary processes responsible for generating genetic diversity [22]. The complex patterns of genetic structure observed in this study appear to operate at a range of spatial scales. Findings of previous studies of Australian freshwater fishes [11,14,74,80] have often reported a general pattern of genetic structure concordant with the stream hierarchy model proposed by Meffe and Vrijenhoek [81]. In this model, it is expected that genetic structure will be distributed according to hierarchical drainage structure, with gene flow primarily occurring within rather than among

catchments [81]. The strong correspondence between ESU and catchment boundaries across the entire distribution of *N. obscura* suggests that, at a regional scale, the stream hierarchy model is broadly applicable for this species. The significant patterns of IBD evident within three of the ESUs are also predicted under the stream hierarchy model, further supporting this assessment [81].

At a smaller spatial scale however, hydrology has been subjected to significant anthropogenic modification. For example, an extensive network of flood mitigation drains have been constructed throughout the Millicent Coast region [29]. This alteration to the natural hydrology has drained wetlands, and reduced the incidence of flooding by directing surface water to several new coastal outlets [29,82]. The natural path of flood waters parallel to the coast has therefore been disrupted and isolated coastal populations are now less likely to be linked by floods [29]. As a result, re-colonisation following local extirpation events is unlikely to occur. In many ways the modified hydrological regime described above simulates the drier climate and reduction in frequency of flooding predicted for the wider southeast coastal region under future climate change scenarios [83], which may have implications for *N. obscura* conservation. Indeed, since samples were collected for this study, several Central ESU populations have come under threat or become extirpated [32,34]. There is also more recent evidence of an increase in the incidence of hybridisation with the co-distributed *N. australis* at some locations [32] and this may be interpreted as a symptom of populations under environmental stress [84,85]. For instance, Heath et al. [86] showed that a combination of environmental factors was associated with increased levels of hybridisation between sympatric *Oncorhynchus clarkii clarkii* (coastal cutthroat trout) and *O. mykiss* (rainbow trout). The increase in hybridisation observed for *N. obscura* provides further evidence of the negative effects of habitat fragmentation and degradation in this region.

Conservation implications

The findings presented here have several direct implications for conservation management of *N. obscura*. Using a combination of spatial and non-spatial, and both individual and population based analyses, we identified 11 MUs confined mostly to individual catchments, within which individuals are largely genetically and demographically independent. Conservation of these units should therefore be managed separately in order to maintain the genetic integrity of populations. The MDB lineage appears to be one highly connected population, indicating that family groups for the breeding program can be formed using any combination of fish from different MDB sites. Also, offspring from the breeding program can therefore be released into any MDB site regardless of the specific MDB site of their parents' origin. This permits greater flexibility in selecting the best habitat for release sites. However, populations from other ESUs should not be mixed with the MDB ESU as, due to their significant genetic divergence, this could lead to a reduction in overall fitness because of issues related to local adaptation, and perhaps to outbreeding depression [79,87].

In addition to providing information for management of the conservation breeding program, there is potential for the same principles to be used to directly manage fragmented wild populations in other ESUs [79]. Translocation has recently received more attention as a viable tool for managing fragmented populations *in situ* [88,89]. This method has the advantage of maintaining populations in a natural environment, thereby avoiding the potential for adaptation to captivity [89]. The MUs proposed here define the boundaries within which translocations of *N. obscura* might occur should this be considered as a conservation management option in the future. Translocations

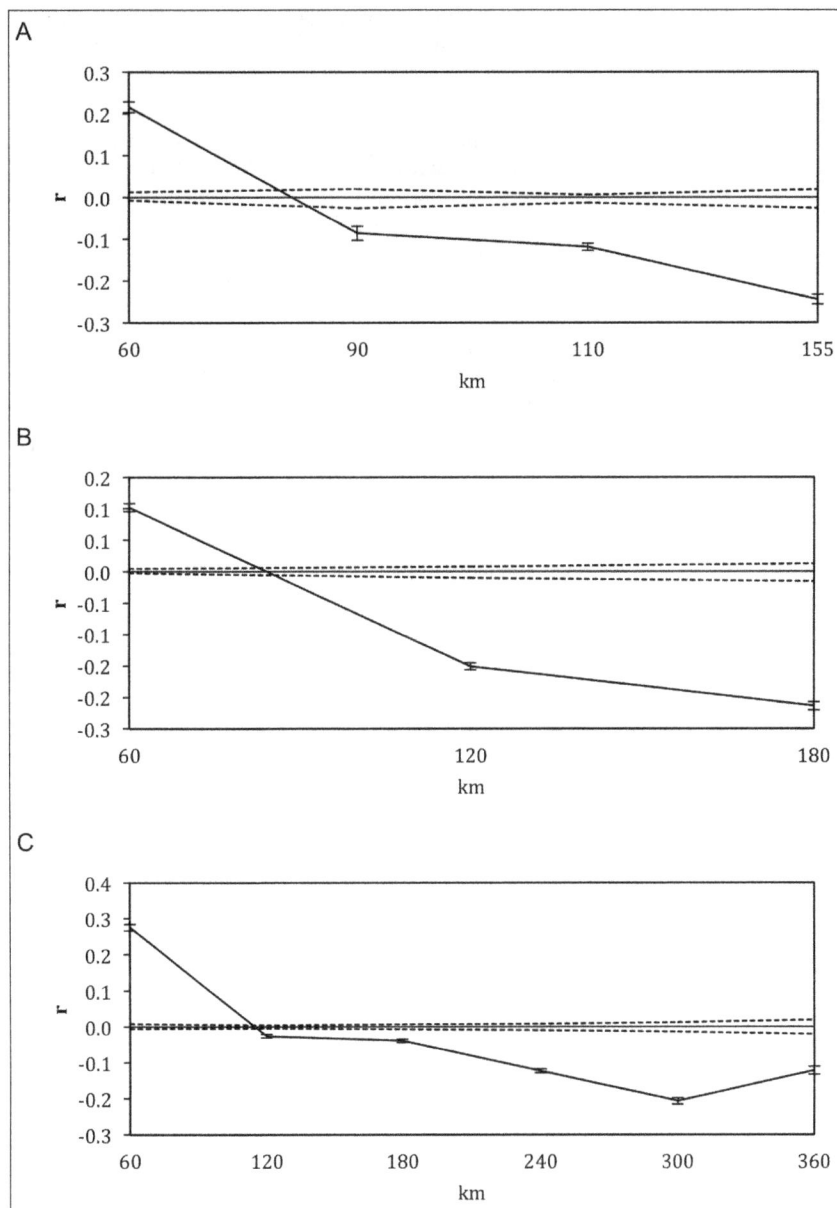

Figure 5. Correlograms showing the autocorrelation coefficient r as a function of distance for A) pure Eastern, B) Merri/Curdies, and C) Central lineages. Distances are the maximum for each class, dashed lines are the 95% CI about the null hypothesis of a random distribution of genotypes, and error bars are 95% confidence of r. Intercept values are 81 km, 84 km, and 115 km for A, B, and C respectively.

have however generally had a low success rate in the past [90]. For instance, first and second generation hybrids of native and mixed source translocated *Cottus cognatus* (slimy sculpin) exhibited reduced fitness in a study of the consequences of freshwater fish translocations in southeast Minnesota [91], highlighting the need for further investigation into the genetic effects of translocations.

Habitat fragmentation has clearly had a major influence on the decline of *N. obscura* across its range. Modifications to natural flow regimes have reduced both the amount of available habitat and population connectivity. Given the patchy distribution and low abundance that characterises *N. obscura*, the low genetic diversity and highly differentiated populations uncovered in this study are not surprising. The limited opportunities for dispersal appear dependent upon intermittent flooding that occasionally connects

isolated habitat patches. It is therefore critical for the conservation of *N. obscura* that no further modifications to the natural hydrology of this region are undertaken. Furthermore, it is vital that habitat is protected wherever extant populations persist and, where possible, connectivity between populations within MUs is restored to allow natural evolutionary processes to continue.

Supporting Information

Table S1 Microsatellite markers (Carvalho et al. 2011) amplified for *Nannoperca obscura*.

Table S2 Geneclass2% probability of individual population assignment of *Nannoperca obscura*.

Table S3 Pairwise population F_{ST} and R_{ST} for *Nannoperca obscura*.

Table S4 Pairwise population F_{ST} and R_{ST} for sites where *Nannoperca obscura* samples were collected in multiple years.

Table S5 Estimated migration rates (m) between Evolutionarily Significant Units (ESUs) and 95% credible intervals (CI) calculated with BayesAss.

Table S6 Estimated migration rates (m) between Management Units (MUs) and 95% credible intervals (CI) calculated with BayesAss.

Figure S1 Principal coordinates analysis based on 14 microsatellite loci for *Nannoperca obscura* individuals from each genetic lineage. A) pure Eastern. Eigen values for the first and second axes have been plotted, which explain 38% and 23% of the variance, respectively. E1, E2 and E3 refer to genetic clusters identified within this lineage; B) Merri/Curdies. Eigen values for the first and second axes explain 40% and 26% of the variance, respectively. M1, M2, and M3 refer to genetic clusters identified within this lineage; C) Central. Eigen values for the first and second axes explain 34% and 26% of the variance, respectively. C1, C2, C3, and C4 refer to genetic clusters identified within this lineage; D) Murray-Darling basin. Eigen values for the first and second axes explain 24% and 22% of the variance, respectively. No genetic structure was apparent within this ESU.

Figure S2 Correlograms showing the autocorrelation coefficient *r* as a function of increasing distance classes for A) pure Eastern, and B) Merri/Curdies ESUs. Distances are the maximum for each class, grey bars indicate 95% CI about the null hypothesis of no genetic structure, and error bars about r indicate 95% CI as determined by bootstrapping.

Figure S3 Correlogram showing the autocorrelation coefficient *r* as a function of increasing distance classes for Central ESU. Distances are the maximum for each class, grey bars indicate 95% CI about the null hypothesis of no genetic structure and error bars about *r* indicate 95% CI as determined by bootstrapping.

Acknowledgments

We thank the many people who helped with field work and other aspects relative to obtaining specimens, especially M. Bachmann, N. Evengelou, C. Kemp, G. Knowles, T. Raadik, T. Ristic, R. Remington, M. Roberts, J. Rowntree, S. Slater, M. Tucker, S. Wedderburn and S. Westergaard. We also thank M. Sasaki for laboratory assistance.

Author Contributions

Conceived and designed the experiments: LBB MA. Performed the experiments: CJB. Analyzed the data: CJB LBB MA. Contributed reagents/materials/analysis tools: LBB PJU MPH MA. Wrote the paper: CJB LBB MA PJU.

References

1. Fischer J, Lindenmayer DB (2007) Landscape modification and habitat fragmentation: a synthesis. Global Ecol Biogeogr 16: 265–280.
2. Lande R (1998) Anthropogenic, ecological and genetic factors in extinction and conservation. Res Popul Ecol 40: 259–269.
3. Lowe WH, Allendorf FW (2010) What can genetics tell us about population connectivity? Mol Ecol 19: 3038–3051.
4. Lande R (1988) Genetics and demography in biological conservation. Science 241: 1455–1460.
5. Frankham R (2005) Genetics and extinction. Biol Conserv 126: 131–140.
6. Hanski I (1998) Metapopulation dynamics. Nature 396: 41–49.
7. Segelbacher G, Cushman S, Epperson B, Fortin M-J, Francois O, et al. (2010) Applications of landscape genetics in conservation biology: concepts and challenges. Conserv Genet 11: 375–385.
8. Frankham R (2010) Where are we in conservation genetics and where do we need to go? Conserv Genet 11: 661–663.
9. Fausch KD, Torgersen CE, Baxter CV, Li HW (2002) Landscapes to riverscapes: bridging the gap between research and conservation of stream fishes. Bioscience 52: 483–498.
10. Palmer MA, Reidy Liermann CA, Nilsson C, Flörke M, Alcamo J, et al. (2008) Climate change and the world's river basins: anticipating management options. Front Ecol Environ 6: 81–89.
11. Hughes JM, Schmidt DJ, Finn DS (2009) Genes in streams: using DNA to understand the movement of freshwater fauna and their riverine habitat. Bioscience 59: 573–583.
12. Cook BD, Kennard MJ, Real K, Pusey BJ, Hughes JM (2011) Landscape genetic analysis of the tropical freshwater fish *Mogurnda mogurnda* (Eleotridae) in a monsoonal river basin: importance of hydrographic factors and population history. Freshwat Biol 56: 812–827.
13. Faulks LK, Gilligan DM, Beheregaray LB (2011) The role of anthropogenic vs. natural in-stream structures in determining connectivity and genetic diversity in an endangered freshwater fish, Macquarie perch (*Macquaria australasica*). Evol Appl 4: 589–601.
14. Huey JA, Baker AM, Hughes JM (2006) Patterns of gene flow in two species of eel-tailed catfish, *Neosilurus hyrtlii* and *Porochilus argenteus* (Siluriformes: Plotosidae), in western Queensland's dryland rivers. Biol J Linn Soc 87: 457–467.
15. Cooke GM, Chao NL, Beheregaray LB (2012) Natural selection in the water: freshwater invasion and adaptation by water colour in the Amazonian pufferfish. J Evol Biol 25: 1305–1320.
16. Anderson CD, Epperson BK, Fortin M-J, Holderegger R, James PMA, et al. (2010) Considering spatial and temporal scale in landscape-genetic studies of gene flow. Mol Ecol 19: 3565–3575.
17. Peakall R, Ruibal M, Lindenmayer DB (2003) Spatial autocorrelation analysis offers new insights into gene flow in the Australian bush rat, *Rattus fuscipes*. Evolution 57: 1182–1195.
18. Diniz-Filho JAF, De Campos Telles MP (2002) Spatial autocorrelation analysis and the identification of operational units for conservation in continuous populations. Conserv Biol 16: 924–935.
19. Primmer CR, Veselov AJ, Zubchenko A, Poututkin A, Bakhmet I, et al. (2006) Isolation by distance within a river system: genetic population structuring of Atlantic salmon, *Salmo salar*, in tributaries of the Varzuga River in northwest Russia. Mol Ecol 15: 653–666.
20. Moritz C (1994) Defining 'Evolutionarily Significant Units' for conservation. Trends Ecol Evol 9: 373–375.
21. Crandall KA, Bininda-Emonds ORP, Mace GM, Wayne RK (2000) Considering evolutionary processes in conservation biology. Trends Ecol Evol 15: 290–295.
22. Moritz C (2002) Strategies to protect biological diversity and the evolutionary processes that sustain it. Syst Biol 51: 238–254.
23. Unmack PJ (2001) Biogeography of Australian freshwater fishes. J Biogeogr 28: 1053–1089.
24. Murray–Darling Basin Authority (2010) Guide to the proposed basin plan: overview. Murray–Darling Basin Authority, Canberra.
25. Lintermans M (2013) Conservation and management. In: Humphries P, Walker K, editors. Ecology of Australian freshwater fishes. Collingwood, VIC: CSIRO Publishing. pp. 283–316.
26. Kingsford RT, Walker KF, Lester RE, Young WJ, Fairweather PG, et al. (2011) A Ramsar wetland in crisis - the Coorong, Lower Lakes and Murray Mouth, Australia. Mar Freshwat Res 62: 255–265.
27. Hammer M, Bice C, Hall A, Frears A, Watt A, et al. (2013) Freshwater fish conservation in the face of critical water shortages in the southern Murray-Darling Basin, Australia. Mar Freshwat Res 64: 807–821.
28. Stein JL, Stein JA, Nix HA (2002) Spatial analysis of anthropogenic river disturbance at regional and continental scales: identifying the wild rivers of Australia. Landscape Urban Plann 60: 1–25.

29. Taffs KH (2001) The role of surface water drainage in environmental change: a case example of the upper south east of South Australia; an historical review. Aust Geogr Stud 39: 279–301.

30. Saddlier S, Koehn J, Hammer M (2013) Lets not forget the small fishes – conservation of two threatned species of pygmy perch in south-eastern Australia. Mar Freshwat Res 64: 874–886.

31. Kuiter RH, Humphries P, Arthington A (1996) Pygmy perches: Family Nannopercidae. In: McDowall RM, editor. Freshwater Fishes of South-Eastern Australia 2nd Edition. Chatswood, NSW: Reed Books.

32. Hammer M, Unmack P, Adams M, Johnson J, Walker K (2010) Phylogeographic structure in the threatened Yarra pygmy perch *Nannoperca obscura* (Teleostei: Percichthyidae) has major implications for declining populations. Conserv Genet 11: 213–223.

33. IUCN (2011) International union for conservation of nature red list of threatned species. Available: http://www.iucnredlist.org/details/39301/0. Accessed 2013 Nov 11.

34. Hammer M, Wedderburn S, van Weenen J (2009) Action plan for South Australian freshwater fishes. Native Fish Australia (SA) Inc. Adelaide.

35. Victorian Government Department of Sustainability and Environment (2013) Advisory list of threatned vertebrate fauna in Victoria. Melbourne.

36. Saddlier S, Hammer M (2010) National recovery plan for the Yarra pygmy perch *Nannoperca obscura*. Department of Sustainability and Environment, Melbourne.

37. Wedderburn S, Hammer M, Bice C (2012) Shifts in small-bodied fish assemblages resulting from drought-induced water level recession in terminating lakes of the Murray-Darling Basin, Australia. Hydrobiologia 691: 35–46.

38. Sunnucks P, Hales DF (1996) Numerous transposed sequences of mitochondrial cytochrome oxidase I–II in aphids of the genus *Sitobion* (Hemiptera: Aphididae). Mol Biol Evol 13: 510–524.

39. Carvalho D, Rodriguez-Zarate C, Hammer M, Beheregaray L (2011) Development of 21 microsatellite markers for the threatened Yarra pygmy perch (*Nannoperca obscura*) through 454 shot-gun pyrosequencing. Conserv Genet Resour 3: 601–604.

40. Beheregaray LB, Möller LM, Schwartz TS, Chao NL, Caccone A (2004) Microsatellite markers for the cardinal tetra *Paracheirodon axelrodi*, a commercially important fish from central Amazonia. Mol Ecol Notes 4: 330–332.

41. Van Oosterhout C, Hutchinson WF, Wills DPM, Shipley P (2004) Micro-checker: software for identifying and correcting genotyping errors in microsatellite data. Mol Ecol Notes 4: 535–538.

42. Rousset F (2008) genepop'007: a complete re-implementation of the genepop software for Windows and Linux. Mol Ecol Resour 8: 103–106.

43. Meirmans PG, Van Tienderen PH (2004) GenoType and GenoDive: two programs for the analysis of genetic diversity of asexual organisms. Mol Ecol Notes 4: 792–794.

44. Rice WR (1989) Analyzing tables of statistical tests. Evolution 43: 223–225.

45. Kalinowski ST (2005) hp-rare 1.0: a computer program for performing rarefaction on measures of allelic richness. Mol Ecol Notes 5: 187–189.

46. Peakall R, Smouse P (2012) GenAlEx 6.5: Genetic analysis in Excel. Population genetic software for teaching and research, an update. Bioinformatics 28: 2537–2539.

47. Weir BS, Cockerham CC (1984) Estimating F-statistics for the analysis of population structure. Evolution 38: 1358–1370.

48. Slatkin M (1995) A measure of population subdivision based on microsatellite allele frequencies. Genetics 139: 457–462.

49. Excoffier L, Lischer HEL (2010) Arlequin suite ver 3.5: a new series of programs to perform population genetics analyses under Linux and Windows. Mol Ecol Resour 10: 564–567.

50. Hardy OJ, Charbonnel N, Freville H, Heuertz M (2003) Microsatellite allele sizes: a simple test to assess their significance on genetic differentiation. Genetics 163: 1467–1482.

51. Hardy OJ, Vekemans X (2002) SPAGeDi: a versatile computer program to analyse spatial genetic structure at the individual or population levels. Mol Ecol Notes 2: 618–620.

52. Pritchard JK, Stephens M, Donnelly P (2000) Inference of population structure using multilocus genotype data. Genetics 155: 945–959.

53. Evanno G, Regnaut S, Goudet J (2005) Detecting the number of clusters of individuals using the software STRUCTURE: a simulation study. Mol Ecol 14: 2611–2620.

54. Gilbert KJ, Andrew RL, Bock DG, Franklin MT, Kane NC, et al. (2012) Recommendations for utilizing and reporting population genetic analyses: the reproducibility of genetic clustering using the program structure. Mol Ecol 21: 4925–4930.

55. Earl D, vonHoldt B (2012) STRUCTURE HARVESTER: a website and program for visualizing STRUCTURE output and implementing the Evanno method. Conserv Genet Resour 4: 359–361.

56. Jakobsson M, Rosenberg NA (2007) CLUMPP: a cluster matching and permutation program for dealing with label switching and multimodality in analysis of population structure. Bioinformatics 23: 1801–1806.

57. Rosenberg NA (2004) distruct: a program for the graphical display of population structure. Mol Ecol Notes 4: 137–138.

58. Piry S, Alapetite A, Cornuet J-M, Paetkau D, Baudouin L, et al. (2004) GENECLASS2: A software for genetic assignment and first-generation migrant detection. J Hered 95: 536–539.

59. Rannala B, Mountain JL (1997) Detecting immigration by using multilocus genotypes. Proceedings of the National Academy of Sciences 94: 9197–9201.

60. Smouse PE, Peakall R (1999) Spatial autocorrelation analysis of individual multiallele and multilocus genetic structure. Heredity 82: 561–573.

61. Wilson GA, Rannala B (2003) Bayesian inference of recent migration rates using multilocus genotypes. Genetics 163: 1177–1191.

62. Rambaut A, Drummond AJ (2009) Tracer v1.5, Available: http://beast.bio.ed.ac.uk/Tracer.Accessed 2013 Nov 11.

63. Meirmans PG (2012) The trouble with isolation by distance. Mol Ecol 21: 2839–2846.

64. Drummond CS, Hamilton MB (2007) Hierarchical components of genetic variation at a species boundary: population structure in two sympatric varieties of *Lupinus microcarpus* (Leguminosae). Mol Ecol 16: 753–769.

65. Legendre P (1993) Spatial autocorrelation: trouble or new paradigm? Ecology 74: 1659–1673.

66. DeWoody JA, Avise JC (2000) Microsatellite variation in marine, freshwater and anadromous fishes compared with other animals. J Fish Biol 56: 461–473.

67. Faulks LK, Gilligan DM, Beheregaray LB (2010) Islands of water in a sea of dry land: hydrological regime predicts genetic diversity and dispersal in a widespread fish from Australia's arid zone, the golden perch (*Macquaria ambigua*). Mol Ecol 19: 4723–4737.

68. Coleman RA, Pettigrove V, Raadik TA, Hoffmann AA, Miller AD, et al. (2010) Microsatellite markers and mtDNA data indicate two distinct groups in dwarf galaxias, *Galaxiella pusilla* (Mack) (Pisces: Galaxiidae), a threatened freshwater fish from south-eastern Australia. Conserv Genet 11: 1911–1928.

69. Cook BD, Bunn SE, Hughes JM (2007) Molecular genetic and stable isotope signatures reveal complementary patterns of population connectivity in the regionally vulnerable southern pygmy perch (*Nannoperca australis*). Biol Conserv 138: 60–72.

70. Hughes JM, Real KM, Marshall JC, Schmidt DJ (2012) Extreme genetic structure in a small-bodied freshwater fish, the purple spotted gudgeon, *Mogurnda adspersa* (Eleotridae). PLoS ONE 7: e40546.

71. Huey JA, Baker AM, Hughes JM (2008) The effect of landscape processes upon gene flow and genetic diversity in an Australian freshwater fish, *Neosilurus hyrtlii*. Freshwat Biol 53: 1393–1408.

72. Sunnucks P (2000) Efficient genetic markers for population biology. Trends Ecol Evol 15: 199–203.

73. Unmack PJ, Hammer MP, Adams M, Johnson JB, Dowling TE (2013) The role of continental shelf width in determining freshwater phylogeographic patterns in south-eastern Australian pygmy perches (Teleostei: Percichthyidae). Mol Ecol 22: 1683–1699.

74. Hughes JM, Huey JA, Schmidt DJ (2013) Is realised connectivity among populations of aquatic fauna predictable from potential connectivity? Freshwat Biol 58: 951–966.

75. Luque S, Saura S, Fortin M-J (2012) Landscape connectivity analysis for conservation: insights from combining new methods with ecological and genetic data. Landscape Ecol 27: 153–157.

76. Hughes J, Ponniah M, Hurwood D, Chenoweth S, Arthington A (1999) Strong genetic structuring in a habitat specialist, the Oxleyan Pygmy Perch *Nannoperca oxleyana*. Heredity 83: 5–14.

77. Unmack PJ, Bagley JC, Adams M, Hammer MP, Johnson JB (2012) Molecular phylogeny and phylogeography of the Australian freshwater fish genus *Galaxiella*, with an emphasis on dwarf galaxias (*G. pusilla*). PLoS ONE 7: e38433.

78. Rourke ML, McPartlan HC, Ingram BA, Taylor AC (2010) Biogeography and life history ameliorate the potentially negative genetic effects of stocking on Murray cod (*Maccullochella peelii peelii*). Mar Freshwat Res 61: 918–927.

79. Frankham R (2010) Challenges and opportunities of genetic approaches to biological conservation. Biol Conserv 143: 1919–1927.

80. McGlashan DJ, Hughes JM (2002) Extensive genetic divergence among populations of the Australian freshwater fish, *Pseudomugil signifer* (Pseudomugilidae), at different hierarchical scales. Mar Freshwat Res 53: 897–907.

81. Meffe GK, Vrijenhoek RC (1988) Conservation genetics in the management of desert fishes. Conserv Biol 2: 157–169.

82. Wear RJ, Eaton A, Tanner JE, Murray-Jones S (2006) The impact of drain discharges on seagrass beds in the south east of South Australia. Final report prepared for the south east natural resource consultative committee and the south east catchment water management board. Adelaide: South Australian Research and Development Institute (Aquatic sciences) and the Department of Environment and Heritage, Coast Protection branch.

83. IPCC (2001) Climate change 2001: working group I: the scientific basis. Contribution of working group I to the third assessment report of the intergovernmental panel on climate change. Eds J. T. Houghton, Y. Ding, D. J. Griggs, M. Noguer, P. J. van der Linden, X. Dai, K. Maskell and C. A. Johnson. Cambridge University Press. Cambridge.

84. Marie AD, Bernatchez L, Garant D (2012) Environmental factors correlate with hybridization in stocked brook charr (*Salvelinus fontinalis*). Can J Fish Aquat Sci 69: 884–893.

85. Seehausen O, van Alphen JJM, Witte F (1997) Cichlid fish diversity threatened by eutrophication that curbs sexual selection. Science 277: 1808–1811.

86. Heath D, Bettles CM, Roff D (2010) Environmental factors associated with reproductive barrier breakdown in sympatric trout populations on Vancouver Island. Evol Appl 3: 77–90.

87. Allendorf FW, Hohenlohe PA, Luikart G (2010) Genomics and the future of conservation genetics. Nat Rev Genet 11: 697–709.

88. Groce MC, Bailey LL, Fausch KD (2012) Evaluating the success of Arkansas darter translocations in Colorado: an occupancy sampling approach. Trans Am Fish Soc 141: 825–840.

89. Weeks AR, Sgro CM, Young AG, Frankham R, Mitchell NJ, et al. (2011) Assessing the benefits and risks of translocations in changing environments: a genetic perspective. Evol Appl 4: 709–725.

90. Fischer J, Lindenmayer DB (2000) An assessment of the published results of animal relocations. Biol Conserv 96: 1–11.

91. Huff DD, Miller LM, Chizinski CJ, Vondracek B (2011) Mixed-source reintroductions lead to outbreeding depression in second-generation descendents of a native North American fish. Mol Ecol 20: 4246–4258.

Projected Polar Bear Sea Ice Habitat in the Canadian Arctic Archipelago

Stephen G. Hamilton[1]*, **Laura Castro de la Guardia**[1,2], **Andrew E. Derocher**[1], **Vicki Sahanatien**[1,3], **Bruno Tremblay**[4], **David Huard**[5]

1 Department of Biological Sciences, University of Alberta, Edmonton, AB, T6G 2E9 Canada, 2 Department of Earth and Atmospheric Sciences, University of Alberta, Edmonton, AB, T6G 2E9 Canada, 3 World Wildlife Fund Canada, PO Box 1750, Iqaluit, NU, X0A 0H0 Canada, 4 Atmospheric and Oceanic Sciences, McGill University, Room 823, Burnside Hall 805 Sherbrooke Street West, Montreal, QC H3A 0B9 Canada, 5 David Huard Solutions, Québec, G1W 4G8 Canada

Abstract

Background: Sea ice across the Arctic is declining and altering physical characteristics of marine ecosystems. Polar bears (*Ursus maritimus*) have been identified as vulnerable to changes in sea ice conditions. We use sea ice projections for the Canadian Arctic Archipelago from 2006 – 2100 to gain insight into the conservation challenges for polar bears with respect to habitat loss using metrics developed from polar bear energetics modeling.

Principal Findings: Shifts away from multiyear ice to annual ice cover throughout the region, as well as lengthening ice-free periods, may become critical for polar bears before the end of the 21st century with projected warming. Each polar bear population in the Archipelago may undergo 2–5 months of ice-free conditions, where no such conditions exist presently. We identify spatially and temporally explicit ice-free periods that extend beyond what polar bears require for nutritional and reproductive demands.

Conclusions/Significance: Under business-as-usual climate projections, polar bears may face starvation and reproductive failure across the entire Archipelago by the year 2100.

Editor: Connie Lovejoy, Laval University, Canada

Funding: Funding was provided by WWF (Canada), ArcticNet, the Canadian Association of Zoos and Aquariums, Canadian Wildlife Federation, Environment Canada, Hauser Bears, Natural Sciences and Engineering Research Council of Canada, Office of Naval Research grant (N000141110977), Pittsburgh Zoo, Polar Continental Shelf Project, Polar Bears International, and Quark Expeditions. The funders had no role in study design, data collection and analysis, decision to publish, or preparation of the manuscript. Co-author David Huard is employed by David Huard Solutions. David Huard Solutions provided support in the form of salary for author David Huard, but did not have any additional role in the study design, data collection and analysis, decision to publish, or preparation of the manuscript. The specific roles of these authors are articulated in the 'author contributions' section.

Competing Interests: The authors have the following interests: Quark Expeditions is a commercial tour company that provided the authors with a donation for research with no conditions. Co-author David Huard is employed by David Huard Solutions. There are no patents, products in development or marketed products to declare.

* Email: stephen.hamilton@ualberta.ca

Introduction

Observed changes in global climate have influenced Arctic sea ice cover more than most models have predicted [1], and ongoing sea ice declines indicate loss of maximum ice cover as well as older, thicker multiyear ice [2,3]. These losses are modifying the Arctic marine ecosystems [4,5], making Arctic and sub-Arctic marine mammals particularly vulnerable to climate change [6,7]. Polar bears (*Ursus maritimus*) are inextricably linked to Arctic sea ice and are sensitive to sea ice loss [6–10]. Polar bears rely on sea ice as a platform for hunting, migrating, and mating, but are forced to move to land in regions where sea ice does not seasonally persist [11–14]. Energetics modeling and population projections indicate that continued sea ice loss with climate warming will negatively affect polar bear survival and reproduction potentially leading to population declines [15–18]. Moreover, of the ice that survives the melt season, insufficient snow cover may limit its viability as habitat for ringed seals (*Pusa hispida*), the primary prey species of polar bears [19].

Optimal polar bear habitat is predicted to decline in the 21st century, with significant losses in the Hudson Bay and peripheral Arctic seas [15,20], though greenhouse gas mitigation and geo-engineering strategies could limit some loss [21,22]. Most sea ice modeling efforts have a crude representation of the geographically complex Canadian Arctic Archipelago (CAA) due to its many narrow channels, which are difficult to resolve. Nevertheless, 7 of the 19 recognized polar bear populations depend on the ice formed within or advected into the CAA (Figure 1). These 7 CAA populations comprise approximately one quarter of the estimated global polar bear population, while covering only 9.1% of the global polar bear range [23].

The CAA and Greenland were thought to have the greatest likelihood of sustaining polar bears to the end of the 21st century [24] although based only on analysis of sea ice conditions in the

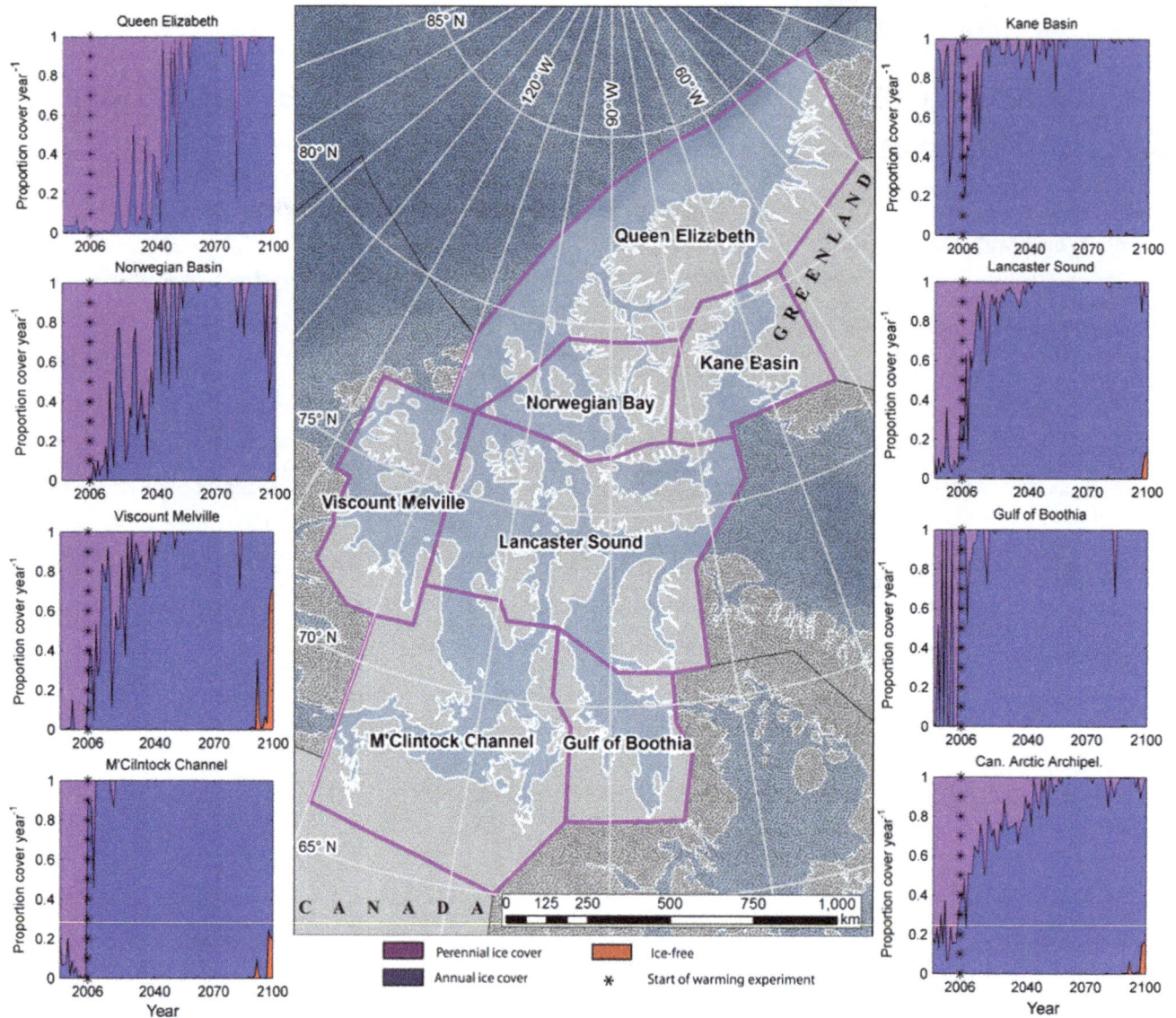

Figure 1. Projected dominance of seasonal sea ice in the polar bear populations of the Arctic Archipelago. The seven populations range from 65–85°N in latitude, with significant variation in the length of ice-free seasons. The proportion of multiyear ice, annual ice, and ice-free waters is given by regional means, and averaged over the total area.

very northern part of the CAA. Here we investigate the impact of projected warming on polar bears within the CAA from projected monthly mean sea ice concentration (SIC), ice thickness, and snow depth between 2006–2100 in comparison to previously established polar bear energetic needs. Polar bears are well-adapted to prolonged periods without food but lose body mass when fasting [24–28]. Body mass is already declining in some polar bear populations with negative consequences on survival and reproduction [16,29,30]. Energy budget models exist in which polar bear survival and reproductive rates can be tied to the availability of access to sea ice [17,18]. Such models are based on the basic energy requirements of animals and are useful when predicting population changes under environmental conditions that have yet to be observed [31]. We examine the seasonality of sea ice and determine when the length of the ice-free period in the CAA may become critically limiting to polar bear foraging and thus negatively affect reproduction and survival.

Materials and Methods

Global climate simulations contributed to the Coupled Model Intercomparison Project Phase 5 (CMIP5) are too coarse to effectively resolve the narrow channels of the CAA. Although numerically challenging, one solution is to dynamically downscale a global simulation onto a finer grid using a regional climate model. A comparison between over 30 different CMIP5 models led us to select one simulation from the Geophysical Fluid Dynamics Laboratory Coupled Physical Model (GFDL-CM3) driven by radiative scenario RCP8.5 to pilot the regional model (available from www.gfdl.noaa.gov/coupled-physical-model-cm3). This simulation includes a realistic spatial distribution of sea ice extent and thickness and simulates the trend in observed minimum sea ice extent during the observational record (1979–2013). This pilot simulation was dynamically downscaled using the ice-ocean Massachusetts Institute of Technology General Circulation Model

(MITgcm) simulation in regional mode over the Arctic at a resolution of 18km (available from http://mitgcm.org). The 3-hourly atmospheric forcing fields from GFDL-CM3 were bias-corrected at the monthly scale using differences for variables (x,y,z) or ratios for variables (u,t,g) between the Japanese 25 year Reanalysis (JRA25) [32] and GFDL-CM3. These biases were calculated over the 2005–2011 period, arguably too short to compute climatological means, to smooth the transition from the JRA25 driven MITgcm simulation to the GFDL-CM3 driven simulation occurring at the start of 2012. We choose a period of 7 years to calculate the biases between the two forcing datasets because of the transitory nature of the climate in the early 21st century with large trends in many of the Arctic climate forcing fields. MITgcm parameters were provided by Nguyen [33] and ocean boundary conditions taken from the Estimating the Circulation and Climate Change of the Ocean Phase 2 (ECCO2) experiment [34]. The MITgcm is run with time steps of 2-hours.

Our model projection is based on the RCP8.5 scenario, which estimates the global average radiative forcing at 8.5 W/m^2 by 2100, and mean global temperature changes of ~3.5°C in 2071–2100 when compared with the historical period of 1961–1990 [35], and represents a worst-case scenario. We compared the seasonal changes in the sea ice cycle between past (1992–2005), near future (2040–2050), and future (2080–2090) by comparing average SIC in each period by month (Figure 2). Population size, survival, and reproduction of polar bears have all been associated with the changes in the seasonal ice cycle, in particular with changes to the ice-free period [10,29,30]. We assume that effects on polar bears within the CAA will be comparable to those observed in other populations.

To study how habitat could change, we classified each pixel within the CAA as multiyear ice, annual ice, or ice-free. The classification was made based on the SIC of the pixel location over a given year [3]. Multiyear ice, which is ice that persists through the height of the melt season (typically March – September), is found when SIC ≥15% year-round. Should SIC dip to <15% before freeze-up begins, but be ≥15% at least once during the year prior, the pixel is classified as annual ice. Ice-free areas are defined as <15% SIC year-round. Polar bears typically avoid or abandon sea ice when concentrations drop below 30–50% although the rate of loss is also important [14,29]. The cutoff of 15% we used is conservative because bears will occupy habitat with as little as 15% SIC [15], but higher concentrations are more closely associated with habitat use and successful predation [36,37].

We defined a critical ice-free period as one in which sea ice was absent in sufficient concentration for ≥180 days or based on energy budget models [17,18]. The ice-free period, with respect to polar bear habitat use, was assessed as the time between break-up (first month in a year with SIC<50%) until freeze-up begins (SIC≥10%). The values of 50% and 10% for break-up and freeze-up, respectively, are correlated with polar bear movements ashore and offshore in regions where there is a seasonal ice cycle [14,29]. If all months had a mean SIC<10%, the ice-free season was twelve months. Conversely, if all months had SIC≥10% the ice-free season was zero months, which may be conservative regarding the impacts of low SIC on polar bears. For example, within the CAA polar bears select for habitat with 90% SIC year-round [38], and in pelagic Arctic regions polar bears tend towards SIC of 75–80% in spring, 65% in summer, 60% in fall, and 95% in winter

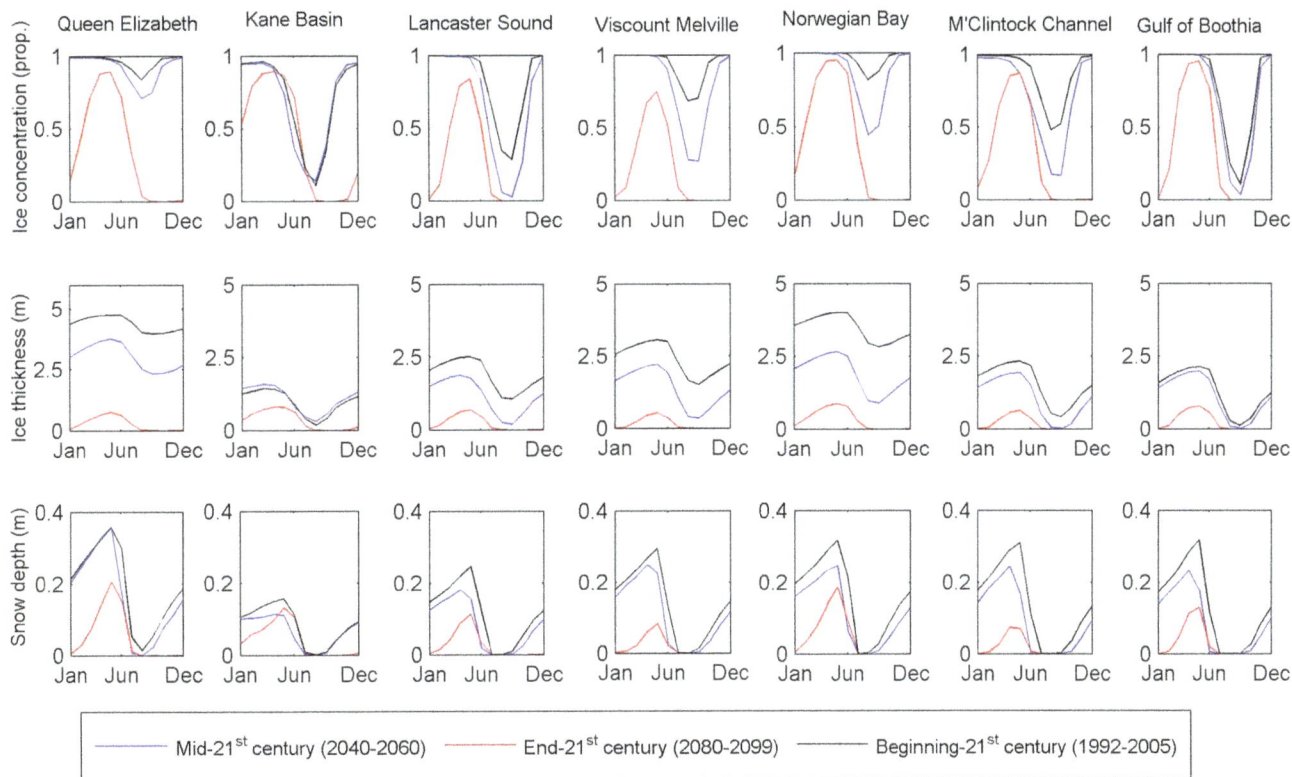

Figure 2. Changes in seasonal sea ice concentration (SIC), thickness, and snow depth over time by region. The mean ice-free season length (in months) for each time period is identifiable by segments of zero SIC or zero ice thickness. All values are monthly means over the respective time periods.

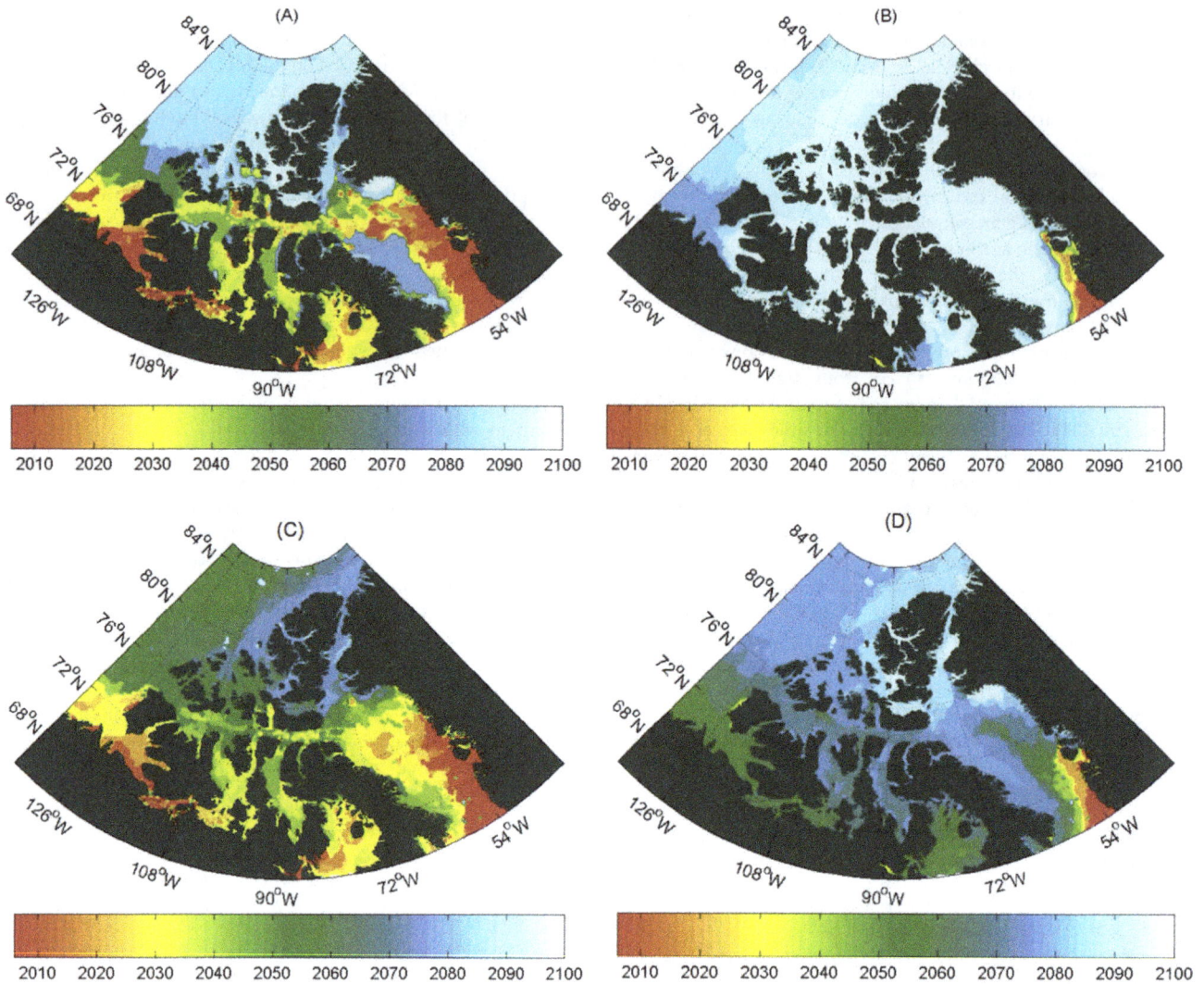

Figure 3. Critical ice-free periods for polar bear survival in the Canadian Arctic. The colors represent the year in which critical habitat loss is reached and never improves in subsequent years. Critical states are reached as starvation sets into adult males at (A) ≥120 days ice-free; (B) ≥180 days ice-free; and reproductive failure occurs in females with (C) break-up in July; and (D) break-up June.

[15,39]. As with the SIC values, we assume that energetic restrictions on polar bears is consistent between populations.

Results and Discussion

All of the CAA exhibits a shift from primarily multiyear ice cover to a primarily seasonally ice-free system by 2100, with the exception of Kane Basin and the Gulf of Boothia, which were largely annually ice-covered regions from the outset (Figure 1). In all cases, the final years of the simulation exhibit some proportion of year-round ice-free areas, where no such areas exist in most of the 21st century.

While multiyear ice is not good hunting habitat due to its low prey abundance [38,40], it provides an alternative habitat for polar bears who otherwise must move onto land during summer, and do not have to wait as long for the new ice to form in the autumn [41]. A shift towards annual ice may seem preferable because it is associated with greater hunting opportunities and ringed seals, the primary prey of polar bears [42,43], may increase in abundance if multiyear ice is replaced by thinner annual ice [7].

However, sea ice must persist long enough for polar bears to take advantage of potential increases in prey density. With the exception of Kane Basin, all polar bear populations in the CAA reached 100% SIC between October and December in the late 20th century, and a non-zero minimum SIC in August or September (Figure 2). By the late 21st century, our simulation projects the southernmost regions (M'Clintock Channel, Gulf of Boothia) and central regions (Viscount Melville, Lancaster Sound) may be entirely ice-free for 5 months, and may no longer reach 100% SIC at maximum ice extent. In the north (Kane Basin, Norwegian Bay, Queen Elizabeth), the simulation estimates a 2–4 month ice-free season by the end of the 21st century, and maximum concentrations <100% in 2080–2090. Ice thickness and snow depth exhibit similar declines throughout the CAA. Ice thickness in the late 20th century was twice to nearly five times the thickness of the projected thickness in the same month of the late 21st century.

Snow depth declines in part due to the reduction in sea ice surface and dates of formation, but also due to a predicted shift in precipitation from snow to rain [19]. Mean snow depth more than

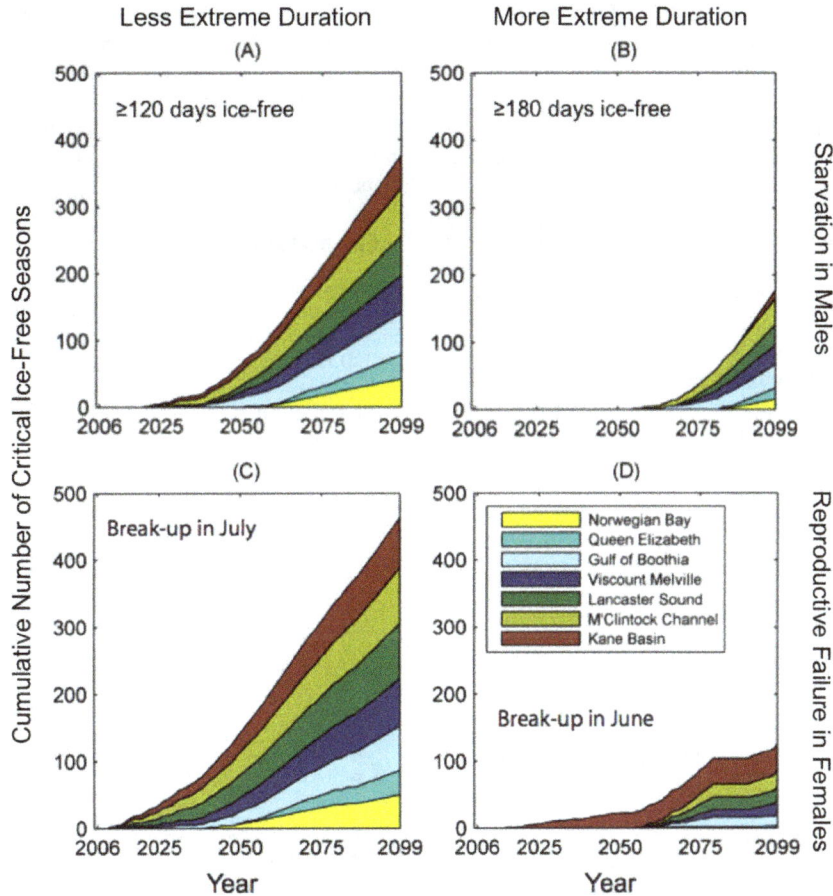

Figure 4. Cumulative number of critical ice-free seasons given by individual polar bear populations in the Canadian Arctic Archipelago. Each color represents the contribution of events in each population to the total number of critical seasons in a given year. Starvation in adult males occurs at (A) ≥120 days ice-free; (B) ≥180 days ice-free. Reproductive failure in females occurs when (C) break-up occurs in July; (D) break-up occurs in June.

halves in the south and central CAA, with the most pronounced changes between the late 20th and late 21st centuries in the western regions (Viscount Melville, M'Clintock Channel). Furthermore, using a conservative estimate of a minimum 20 cm snow depth requirement for seal habitat [19], only the Queen Elizabeth and Norwegian Bay areas may be able to maintain significant ringed seal populations by the end of this century.

Critical Ice-Free Periods

Polar bears fare poorly when sea ice is absent for prolonged periods, losing body mass without the opportunity to hunt [26,44]. Energetics modeling predicts that 2–3% of adult polar bear males could starve when the ice-free period reach 120 days and 9–21% could starve at 180 days of ice-free period with other age and sex classes even more vulnerable [17,37,45]. Similarly, early break-up of sea ice could result in reproductive failure in 55–100% of pregnant females [18]. The frequency of both events would have significant consequences for population trends in abundance. The thresholds established in these energetics models resulted in four types of critical ice-free periods, with the first two being relevant to male starvation rates, and the second two being relevant to female reproductive failure rates (Figure 3): (A) ice-free season >120 days; (B) ice-free season >180 days; (C) break-up occurs in July, and; (D) break-up occurs in June.

We find that sea ice conditions may become unsupportive of polar bear population persistence in the CAA and its surroundings by the late 21st century with ice-free seasons reaching critical duration, and early break-ups occurring in parts of all populations we examined. Similarly, to the east of the CAA, the west coast of Greenland and much of Baffin Bay may no longer be suitable habitat for polar bears before 2050, though ice should persist along the east coast of Baffin Island until much later. Early break-up in the narrow channels of the central CAA may become critical in 2060–70s, whereas the adjacent coastlines of the open Arctic Ocean remain largely non-critical until near 2100.

It is important to consider that what we deem a critical point-of-no-return occurs once the ice-free period crosses our critical threshold and remains critical for the remainder of the modeled period. Nevertheless, it is feasible that single seasons, or clusters of seasons, may become critically ice-free before that point, with subsequent seasons being non-critical. As such, we examined the cumulative number of critical events for of the aforementioned critical periods (A–D) by population (Figure 4). We found that the less extreme critical durations (categories A and C) occur with lower frequency within the first decades of the simulation, and increase in frequency in later decades. When considering more extreme ice-free durations (categories B and D), critical events do not begin to occur until after 2050, with the exception of Kane

Basin, which begins to experience break-up in June before 2020. Nevertheless, the frequency of critical events increases rapidly towards the end of the 21st century.

Implications for Conservation

Without exception, our simulation projects the sea ice habitat in all polar bear populations of the CAA may change from a multiyear to an annual ice system before the end of the century, and the remaining annual ice might not persist sufficiently long each year to allow hunting opportunities for polar bears as we currently understand them. Our model suggests that, by 2070, over 80% of the CAA might experience break-up in July, forcing pregnant females to retreat to land early, with possible negative effects on their reproductive output. Given that our study area comprises approximately one quarter of the world's polar bears, and nearly one-tenth of the total current habitat, our analyses project significant habitat loss and alteration under the business-as-usual model scenario used to estimate sea ice conditions over the coming 21st century.

Conservation efforts to protect polar bear habitat in the Canadian Arctic should focus on regions that are slower to experience change in sea-ice concentration and ice-free period. The Queen Elizabeth and Norwegian Bay populations retain multiyear ice the longest, and their northerly fjords and channels consistently exhibit the fewest critical ice-free events. Nevertheless, by 2100 all regions of the study area may cross the critical point-of-no-return, putting the persistence of the CAA polar bear populations in jeopardy.

Acknowledgments

Computations of the projections were performed on the Guillimin supercomputer at McGill University, under the auspices of Calcul Québec and Compute Canada through an operation grant awarded to B.T. B.T. and D.H. acknowledge Dimitris Menemenlis, Ann Nguyen, and Felix Landerer for assistance with the simulation. The authors acknowledge the comments of Anne Thessen and one anonymous reviewer.

Author Contributions

Conceived and designed the experiments: SGH AED VS LCG. Performed the experiments: SGH LCG BT DH. Analyzed the data: SGH LCG. Wrote the paper: SGH LCG AED VS BT DH.

References

1. Stroeve JC, Kattsov V, Barrett A, Serreze M, Pavlova T, et al. (2012) Trends in Arctic sea ice extent from CMIP5, CMIP3 and observations. Geophysical Research Letters 39.

2. Maslanik JA, Fowler C, Stroeve J, Drobot S, Zwally J, et al. (2007) A younger, thinner Arctic ice cover: Increased potential for rapid, extensive sea-ice loss. Geophysical Research Letters 34.

3. Comiso JC (2012) Large Decadal Decline of the Arctic Multiyear Ice Cover. Journal of Climate 25: 1176–1193.

4. Arrigo KR, van Dijken G, Pabi S (2008) Impact of a shrinking Arctic ice cover on marine primary production. Geophysical Research Letters 35.

5. Bluhm BA, Gradinger R (2008) Regional variability in food availability for arctic marine mammals. Ecological Applications 18: S77–S96.

6. Laidre KL, Stirling I, Lowry LF, Wiig Ø, Heide-Jorgensen MP, et al. (2008) Quantifying the sensitivity of arctic marine mammals to climate-induced habitat change. Ecological Applications 18: S97–S125.

7. Schipper J, Chanson JS, Chiozza F, Cox NA, Hoffmann M, et al. (2008) The status of the world's land and marine mammals: Diversity, threat, and knowledge. Science 322: 225–230.

8. Stirling I, Derocher AE (1993) Possible impacts of climate warming on polar bears. Arctic 46: 240–245.

9. Derocher AE, Lunn NJ, Stirling I (2004) Polar bears in a warming climate. Integrative and Comparative Biology 44: 163–176.

10. Rode KD, Amstrup SC, Regehr EV (2010) Reduced body size and cub recruitment in polar bears associated with sea ice decline. Ecological Applications 20: 768–782.

11. Stirling I (1974) Midsummer observations on behavior of wild polar bears (*Ursus maritimus*). Canadian Journal of Zoology-Revue Canadienne De Zoologie 52: 1191–1198.

12. Ramsay MA, Stirling I (1986) On the mating system of polar bears. Canadian Journal of Zoology-Revue Canadienne De Zoologie 64: 2142–2151.

13. Schliebe S, Rode KD, Gleason JS, Wilder J, Proffitt K, et al. (2008) Effects of sea ice extent and food availability on spatial and temporal distribution of polar bears during the fall open-water period in the Southern Beaufort Sea. Polar Biology 31: 999–1010.

14. Cherry SG, Derocher AE, Thiemann GW, Lunn NJ (2013) Migration phenology and seasonal fidelity of an Arctic marine predator in relation to sea ice dynamics. Journal of Animal Ecology 82: 912–921.

15. Durner GM, Douglas DC, Nielson RM, Amstrup SC, McDonald TL, et al. (2009) Predicting 21st-century polar bear habitat distribution from global climate models. Ecological Monographs 79: 25–58.

16. Hunter CM, Caswell H, Runge MC, Regehr EV, Amstrup SC, et al. (2010) Climate change threatens polar bear populations: a stochastic demographic analysis. Ecology 91: 2883–2897.

17. Molnár PK, Derocher AE, Thiemann GW, Lewis MA (2010) Predicting survival, reproduction and abundance of polar bears under climate change. Biological Conservation 143: 1612–1622.

18. Molnár PK, Derocher AE, Klanjscek T, Lewis MA (2011) Predicting climate change impacts on polar bear litter size. Nature Communications 2.

19. Hezel PJ, Zhang X, Bitz CM, Kelly BP, Massonnet F (2012) Projected decline in spring snow depth on Arctic sea ice caused by progressively later autumn open ocean freeze-up this century. Geophysical Research Letters 39.

20. de la Guardia LC, Derocher AE, Myers PG, van Scheltinga ADT, Lunn NJ (2013) Future sea ice conditions in Western Hudson Bay and consequences for polar bears in the 21st century. Global Change Biology 19: 2675–2687.

21. Amstrup SC, DeWeaver ET, Douglas DC, Marcot BG, Durner GM, et al. (2010) Greenhouse gas mitigation can reduce sea-ice loss and increase polar bear persistence. Nature 468: 955–958.

22. Tilmes S, Jahn A, Kay JE, Holland M, Lamarque J-F (2014) Can regional climate engineering save the summer Arctic sea ice? Geophysical Research Letters 41: 880–885.

23. IUCN/PBSG (2013) Status table for the world's polar bear subpopulations. International Polar Bear Specialist Group, editor. Available online at http://pbsg.npolar.no/en/.

24. Amstrup SC, Marcot BG, Douglas Dc (2008) A Bayesian Network Modeling Approach to Forecasting the 21st Century Worldwide Status of Polar Bears. In: DeWeaver ET, Bitz CM, Tremblay LB, editors. Arctic Sea Ice Decline: Observations, Projections, Mechanisms, And Implications. Washington: Amer Geophysical Union. 213-268.

25. Watts PD, Hansen SE (1987) Cyclic starvation as a reproductive strategy in the polar bear. Symposia of the Zoological Society of London 57: 305–318.

26. Ramsay MA, Stirling I (1988) Reproductive-biology and ecology of female polar bears (*Ursus maritimus*). Journal of Zoology 214: 601–634.

27. Derocher AE, Stirling I (1995) Temporal variation in reproduction and body-mass of polar bears in Western Hudson-Bay. Canadian Journal of Zoology 73: 1657–1665.

28. Atkinson SN, Ramsay MA (1995) The effects of prolonged fasting of the body-composition and reproductive success of female polar bears (*Ursus maritimus*). Functional Ecology 9: 559–567.

29. Robbins CT, Lopez-Alfaro C, Rode KD, Toien O, Nelson OL (2012) Hibernation and seasonal fasting in bears: the energetic costs and consequences for polar bears. Journal of Mammalogy 93: 1493–1503.

30. Stirling I, Lunn NJ, Iacozza J (1999) Long-term trends in the population ecology of polar bears in western Hudson Bay in relation to climatic change. Arctic 52: 294–306.

31. Regehr EV, Lunn NJ, Amstrup SC, Stirling L (2007) Effects of earlier sea ice breakup on survival and population size of polar bears in western Hudson bay. Journal of Wildlife Management 71: 2673–2683.

32. Kooijman SALM (2010) Dynamic energy budget theory for metabolic organisation. Cambridge UK: Cambridge University Press.

33. Onogi K, Tslttsui J, Koide H, Sakamoto M, Kobayashi S, et al. (2007) The JRA-25 reanalysis. Journal of the Meteorological Society of Japan 85: 369–432.

34. Nguyen AT, Menemenlis D, Kwok R (2011) Arctic ice-ocean simulation with optimized model parameters: Approach and assessment. Journal of Geophysical Research-Oceans 116.

35. Menemenlis D, Campin J, Heimbach P, Hill C, Lee T, et al. (2008) ECCO2: High resolution global ocean and sea ice data synthesis. 42 p.

36. Christensen OB, Goodess CM, Harris I, Watkiss P (2011) European and Global Climate Change Projections: Discussion of Climate Change Model Outputs, Scenarios and Uncertainty in the EC RTD ClimateCost Project; Watkiss P, editor. Stockholm, Sweden: Stockholm Environment Institute. 28 p.

37. Rode KD, Regehr EV, Douglas DC, Durner G, Derocher AE, et al. (2013) Variation in the response of an Arctic top predator experiencing habitat loss: feeding and reproductive ecology of two polar bear populations. Global Change

Biology 20: 76–88.

38. Pilfold NW, Derocher AE, Richardson E (2014) Influence of intraspecific competition on the distribution of a wide-ranging, non-territorial carnivore. Global Ecology and Biogeography 23: 425–435.

39. Ferguson SH, Taylor MK, Messier F (2000) Influence of sea ice dynamics on habitat selection by polar bears. Ecology 81: 761–772.

40. Arthur SM, Manly BFJ, McDonald LL, Garner GW (1996) Assessing habitat selection when availability changes. Ecology 77: 215–227.

41. Kingsley MCS, Stirling I, Calvert W (1985) The distribution and abundance of seals in the Canadian High Arctic, 1980–82. Canadian Journal of Fisheries and Aquatic Sciences 42: 1189–1210.

42. Amstrup SC, Durner GM, Stirling I, Lunn NN, Messier F (2000) Movements and distribution of polar bears in the Beaufort Sea. Canadian Journal of Zoology-Revue Canadienne De Zoologie 78: 948–966.

43. Smith TG (1980) Polar bear predation of ringed and bearded seals in the land-fast sea ice habitat. Canadian Journal of Zoology-Revue Canadienne De Zoologie 58: 2201–2209.

44. Thiemann GW, Iverson SJ, Stirling I (2008) Polar bear diets and Arctic marine food webs: insights from fatty acid analysis. Ecological Monographs 78: 591–613.

45. Polischuk SC, Norstrom RJ, Ramsay MA (2002) Body burdens and tissue concentrations of organochlorines in polar bears (*Ursus maritimus*) vary during seasonal fasts. Environmental Pollution 118: 29–39.

46. Molnár PK, Derocher AE, Thiemann GW, Lewis MA (2014) Corrigendum to "Predicting survival, reproduction and abundance of polar bears under climate change" [Biol. Conserv. 143 (2010) 1612–1622]. Biological Conservation 177: 230–231.

9

Conservation Investment for Rare Plants in Urban Environments

Mark W. Schwartz[1,2]*, Lacy M. Smith[2], Zachary L. Steel[3]

1 John Muir Institute of the Environment, University of California Davis, Davis, California, United States of America, 2 Department of Environmental Science & Policy, University of California Davis, Davis, California, United States of America, 3 Graduate Group in Ecology, University of California Davis, Davis, California, United States of America

Abstract

Budgets for species conservation limit actions. Expending resources in areas of high human density is costly and generally considered less likely to succeed. Yet, coastal California contains both a large fraction of narrowly endemic at-risk plant species as well as the state's three largest metropolitan regions. Hence understanding the capacity to protect species along the highly urbanized coast is a conservation priority. We examine at-risk plant populations along California's coastline from San Diego to north of San Francisco to better understand whether there is a relationship between human population density and: i) performance of at-risk plant populations; and ii) conservation spending. Answering these questions can help focus appropriate strategic conservation investment. Rare plant performance was measured using the annualized growth rate estimate between census periods using the California Natural Diversity Database. Human density was estimated using Census Bureau statistics from the year 2000. We found strong evidence for a lack of a relationship between human population density and plant population performance in California's coastal counties. Analyzing US Endangered Species expenditure reports, we found large differences in expenditures among counties, with plants in San Diego County receiving much higher expenditures than other locations. We found a slight positive relationship between expenditures on behalf of endangered species and human density. Together these data support the argument that conservation efforts by protecting habitats within urban environments are not less likely to be successful than in rural areas. Expenditures on behalf of federally listed endangered and threatened plants do not appear to be related to proximity to human populations. Given the evidence of sufficient performance in urban environments, along with a high potential to leverage public support for nature in urban environments, expenditures in these areas appear to be an appropriate use of conservation funds.

Editor: David L. Roberts, University of Kent, United Kingdom

Funding: These authors have no support or funding to report.

Competing Interests: The authors have declared that no competing interests exist.

* E-mail: mwschwartz@ucdavis.edu

Introduction

Determining the degree to which limited resources should be applied to conservation in urban environments remains a critical challenge [1–3]. This challenge is acute in California where a large number of at-risk plant species are restricted to regions of high human density [4–6]. Limited conservation resources require strategic investment in conservation (e.g., [7]). The land cost associated with urban environments is typically far higher than rural areas [8]. Further, conservation opportunities within an urbanized landscape often consist of small, isolated fragments [9–11]. Small, isolated fragments of natural habitats may have lower species richness than larger natural areas [12], and populations within these sites may be at an elevated risk of extinction [13–16]. Rare species, particularly plants and insects, are often spatially associated with urban environments [8,15,17]. Further, the social value of conserving biodiversity that is accessible to urban populations may be high [18–22]. Thus, Lawson et al. [23] raised considerable interest in demonstrating that extant populations of plants in urban areas do not seem to have different population growth rates than those in rural environments.

A constraint of the Lawson et al. [23] study is that it did not distinguish performance of species among habitat types. We might

predict that particular types of habitats (e.g., serpentine outcrops) may be relatively resilient to the kinds of changes in urban environments that put rare plant species at risk. For example, we might expect that the threat of invasive plant species to native populations may be higher in wetland habitats than in edaphically stressful environments (e.g., serpentine). Alternatively, other habitat types (e.g., agricultural landscapes with low human density) may also be vulnerable to similar degradation as in urbanized environments [15]. Protected wetlands, for example, may be threatened by invasive species, nutrient additions, environmental toxins, and other impacts across gradients of human density.

We have two distinct objectives in this paper. First, we further test the hypothesis of Lawson et al. [23] —that urbanization has no detectable effect on performance across a rural to urban gradient. We expand the evaluation of this hypothesis by taking a much closer look at potential confounding factors that may mask a relationship, including asking whether habitat types express differential performance relationships across the rural to urban gradient. We predict that species in some urban habitat types may be more resilient to urbanization than others. If so, then focusing on habitat types that are resilient to urbanization can help increase effectiveness of urban conservation efforts. To assess this hypothesis, we replicate the methods of Lawson et al. [23] to examine population performance of plant species across a human population density gradient. Specifically, we used the California Natural Heritage plant observation data (California Natural Diversity Database, CNDDB [24]) to determine population trends where repeated population size data are available. We then classified species into different habitat types and life forms to examine performance as a function of human density (people/km^2) in more detail.

Our second objective is to assess the degree to which conservation investment in California may be biased by human density. One argument is that the high cost of urban conservation places too high a demand on limited funds relative to the benefits obtained [16]. The results of Lawson et al. [23] challenge the notion that investment in urban conservation is wasted investment. We cannot, unfortunately, relate government spending directly to plant population outcomes because our plant performance data are neither temporally nor spatially linked to expenditures. We can, however, assess patterns in spending on behalf of federally listed endangered plants to determine if it appears inappropriate, given plant performance. We hope that comparing and contrasting these two independent data sets help us to develop better strategies for how expenditures can be effectively applied to plant conservation in California, where there is a high fraction of urban-associated rare species and many of the resources for conservation are generated [25]. Together, these two pieces of information can help guide appropriate investment in conserving plant diversity within the urbanized and urbanizing California landscape.

Methods

We defined the study area to encompass the richest region of rare species in the state of California [6], focusing on at-risk plant species along coastal California from the Mexican border through the San Francisco Bay Area. Our study area was composed of the 17 counties that border the coastline, including San Francisco Bay, from San Diego to Sonoma County (Fig. 1a). Within this region, we aggregated all plant population census data available through the California Natural Diversity Database (CNDDB) [24]. The CNDDB is a spatially explicit database of rare plant and animal occurrences within California. Tracked plants are those defined by the California Native Plant Society [26] as at-risk. Plant occurrence records are based on sightings by both professional and amateur botanists. All records are then vetted by professional biologists before being entered into the database [24]. We use data on at-risk plant populations that contained species, location, and dates of at least two quantitative estimates of population size across a time interval of one year or more. Most CNDDB population locations do not contain repeated estimates of population size, a requirement in order to estimate performance.

We aggregated data on 1,682 population change estimates among 253 species and subspecies across 795 locations. Data included observations from 1897 through 2007. Although data span the 20th century, over 99% of population change estimates are since 1980 and 73% since 1990. Species are also sampled in a skewed distribution with 60% of species appearing in just one or two locations (range = 1 to 21; average 2.86, s.d. = 2.95; median = 2, Appendix S1).

We classified species into three classes of growth form: annual herbs (902 observations, 97 species and subspecies); perennial herbs (662 observations; 121 species or subspecies); and woody shrubs and trees (118 observations; 35 species or sub-species). We also assigned species into one or more of 19 habitat types based on

Jepson Manual species' habitat descriptions and predominant site type descriptions from the CNDDB records. Habitat types are general (e.g., deciduous woodland, grassland, sage scrub) for habitats that are dominant in the landscape, but more specific for edaphically extreme environments in which numerous rare species occur (e.g., vernal pools, alkali sinks, serpentine). Some species were lumped into a non-differentiated "other" category when habitat descriptions were vague (e.g., roadsides) or difficult to place into a particular habitat type (e.g., swales) These difficult to classify habitats are often indicative of stressful environments.

For each population, assessed plant population performance was estimated using an annualized lambda (λ_{ann}) value following Akçakaya et al.[27];

$$\lambda_{ann} = \left[\left(N_{t+y} + 1 \right) / \left(N_t + 1 \right) \right]^{1/y} \quad (1)$$

where N_t is the first population estimate, N_{t+y} is the second population estimate, and y is the number of years between the estimates. Because individual locations may have population estimates from more than two years, each location may have multiple annualized λ estimates. A growing population is indicated by $\lambda_{ann} > 1.0$, while values less than one indicate a declining population. Owing to a strong skew (a small fraction of observations experience very large population change) in the resulting population growth measures, we transformed these measures by taking a natural log of the annual growth measure. We assessed λ_{ann} in three ways; using: i) all observed population changes (n = 1,682), ii) all individual populations (n = 795), and iii) the average of repeated estimates of population changes just for locations with multiple estimates (n = 341).

We categorized plant survey data by the specificity of the population estimate as follows: i) exact numbers less than 200; ii) exact numbers greater than 200 that likely represents a count or a close estimate (e.g., 1,347); or iii) round numbers that likely reflect an order of magnitude estimate (e.g., 1,000's). Exactness of a population estimate is likely strongly correlated with population size (e.g., small populations might be counted, large ones are estimated). Inexact estimates, however, can lead to spurious estimates of population change. For example, the difference between 5 and 10 individuals is likely to be exact and accurate while estimates of a large population may vary substantially among estimators. Hence, there is a trade-off between precision and bias in population data, so we tested for response differences in λ_{ann} depending on estimate specificity and initial population size.

Human density was assessed using census tracts as defined by the US Census Bureau statistics from the 2000 United States population census [28]. Census tracts are long-term county subdivisions containing between 2,500 and 8,800 people. We divided the census tract population by the area of each tract to get a human density (people/km^2) for each plant location. Human density varied from zero (uninhabited islands of the coast) to 12,217 people/km^2. Median density was 40.2 people/km^2. Urban centers are often defined as having population densities of upwards of 400 people/km^2 [29]. Our distribution of human densities was right-skewed, with 24% of observations from urban locations (>400 people/km^2). We used the natural log of human density to partially normalize this distribution from which to predict performance and also used a non-parametric goodness-of-fit test, dividing the data into groups based on human density and λ_{ann}.

We examined the relationship between λ and human density across the entire data set, by growth form (annual, perennial herbaceous plant, and woody plants), by habitat types (n = 19), and individually for the 25 species represented by 20 or more λ_{ann}

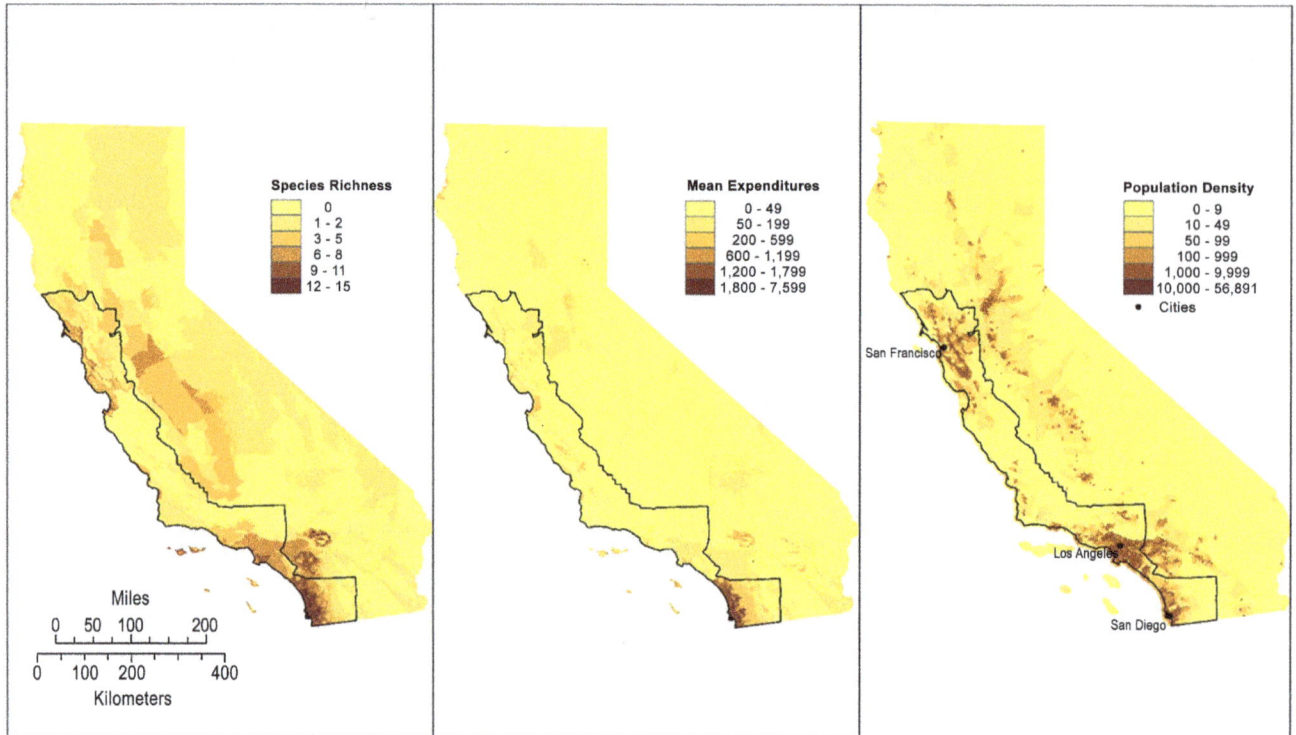

Figure 1. Maps of California highlighting population and expenditure attributes. The outline represents the study area for examining λ_{ann} relative to human density. Color shades represent: (A) species richness of federally listed plant species in California (number of species per square kilometer); and (B) mean annual expenditures for federally listed plant species (dollars per square kilometer per year). Locations of major California cities are included for reference on map (C).

estimates. We used the Human Threat Hypothesis (HTH) of Lawson et al. [23] as a basis for assessing whether λ_{ann} values decline with increasing human density. We expand beyond the work of Lawson et al [23] to test the the relationship between λ_{ann} and human density for specific growth forms (e.g., annual plants) and habitat types (e.g., wetlands). Our alternative hypothesis is that the variability in growth across growth form and habitat type to determine if some growth forms (e.g., annuals) exhibit more variability in population size. In general, greater variability in population size is expected in annuals than in perennial herbs or woody plants. In addition, we assessed whether mean growth rate is lower for particular growth forms or habitat types that are particularly prone to environmental degradation, (e.g., wetlands). This test serves to assess the sensitivity of the data to detect change. Finally, we combined the HTH with a null hypothesis of no difference among habitat types to test the hypothesis that relationship of λ_{ann} to human density does not differ among habitat types.

Urban populations change through time and densities estimated by the 2000 census may reflect impacts at the time of the CNDDB plant surveys with variable accuracy. Therefore, we tested for a relationship between λ_{ann} and human density using the data for the time period across which plant population change was estimated as a covariate. Our analysis consists of linear models of log transformed λ_{ann} as a function of growth form and habitat type, and log transformed human density and interactions among these variables. All analyses were done using JMP 9.0 (SAS institute).

Separately, we sought to understand conservation support across the gradient of high to low population density. We

aggregated all recovery expenditures on behalf of 175 listed plant species in California using three years (2006–2008) of reported expenditures [30–32]. Eight species were not included as a consequence of recent listing action or taxonomic change. These 175 species represent over 95% of the total three-year endangered species expenditures. These expenditures are spent largely on managing populations and not on habitat acquisition, which is tracked separately. A separate treatment by Underwood et al. [33] reported on spending on behalf of acquisition, and is discussed in the context of our analysis. With a far smaller number of listed (n = 183) [34] rather than tracked rare species (n>2,000) [26], the geographic distribution of federally listed plant species (Fig. 1a) is less coastal than the distribution of California endemic plants [6].

We used CALJEP [35], a geospatial database of plant species distributions in California, to identify the distribution of these 175 federally listed species. We placed occurrences within a 1 km^2 grid over the entire state. We counted polygons identified as "present" and "probably present" as occurrences, and "possible" and "not recorded" as absences [35]. We averaged the per 1 km^2 expenditures across the three years based on an assumption that expenditures were evenly distributed across the range of a species. Most listed species are narrow endemics found in a very small portion of the state [4,6], hence we find this an acceptable abstraction of the geographic distribution of spending on behalf of listed species. We summed the total expenditures within grid cells based on all listed species occurring in each grid cell when more than one species were found in a cell. We then mapped human density, based on census tract numbers, at the same spatial scale across the entire state, and estimated the relationship between expenditures and density at this 1 km^2 grid cell scale for our coastal study region. In addition, we summed expenditures within census tracts and compared expenditures by human density at the census tract scale for the coastal counties for which we assessed λ_{ann} .

Finally, we also analyzed these data treating species as replicates. Here, we compared the average human density in all tracts in which a species occurs to the total three years of expenditures on behalf of the species. Results were similar using either species or census tract as the unit of observation. We focused on the data that depict spending by census tract as it relates more directly to our geographic depiction of the distribution of spending. All variables were log transformed to better fit the expectation of normality.

Results

Plant population growth by human density

Our results agree with those of Lawson et al. [17] in finding no relationship between human population density and λ_{ann} (n = 1,682, F = 0.038, p = 0.85, coefficient = −0.003) (Table 1, Fig. 2). This strongly non-significant result persists irrespective of how we assessed the data including across parametric versus nonparametric tests and all subsets of data based on sampling, sampling date, or sample specificity (Table 1). We found no bias in the examination of residuals for any of these tests. The effect size of human density on λ_{ann}, as estimated by 1,000 replicates of a randomized bootstrap sample of 1,000 observations, was nearly zero (mean correlation coefficient = −0.006; s.d. = 0.0235).

We observed no effect of year on λ_{ann} (n = 1,682; F = 0.42; p = 0.84). There was a slight positive relationship between initial plant population size and human density (n = 1,682; F = 10.3; p = 0.001; coefficient = 0.095) and a negative correlation between λ_{ann} and initial population size (n = 1,682; F = 141.7; p<0.001; coefficient = −0.154). This combination of results should make it more likely to find an adverse impact of human population on λ_{ann}, yet we do not find this to be the case.

We also conducted a goodness-of-fit test on data classified into categories based on human density and λ_{ann}. Human density

Table 1. Plant population growth response as predicted by human density using 13 different tests of the relationship performed to assure consistency of results depending on how we treated (A) repeat sampling within a plant population; (B) violations of normal distributions driving non-parametric considerations; (C) initial population size and the ease of gaining an accurate population assessment; (D) when the plant population was assessed; and (E) growth form.

Criteria	N	P	Coefficient
A. Unit of observations			
All population change estimates	1682	0.845	−0.0032
Average of observations from each population	795	0.519	−0.0083
Average of observations with multiple estimates	341	0.596	−0.008
B. Nonparametric correlation*			
All population change estimates, Kendall's tau	1682	0.4	−0.0014
All population change estimates, Spearmans p	1682	0.406	−0.0203
C. Population estimate specificity			
Population size less than 200	654	0.613	−0.011
Exact estimate, population larger than 200	504	0.51	0.232
Rounded number estimate, population >200	521	0/296	−0.032
D. Population count end date			
Population change end date prior to 1990	446	0.221	−0.0435
Population change end date 1990 or later	1234	0.527	0.0116
E. Life Form			
Annual herbaceous plants	902	0.814	0.0067
Perennial herbaceous plants	660	0.857	0.0029
Woody trees and shrubs	118	**0.046**	−0.0498

*-nonparametric versions of most of correlations by population for varying units of observations (A) and estimate specificity (B) were similarly, not significant and had correlations close to 0.

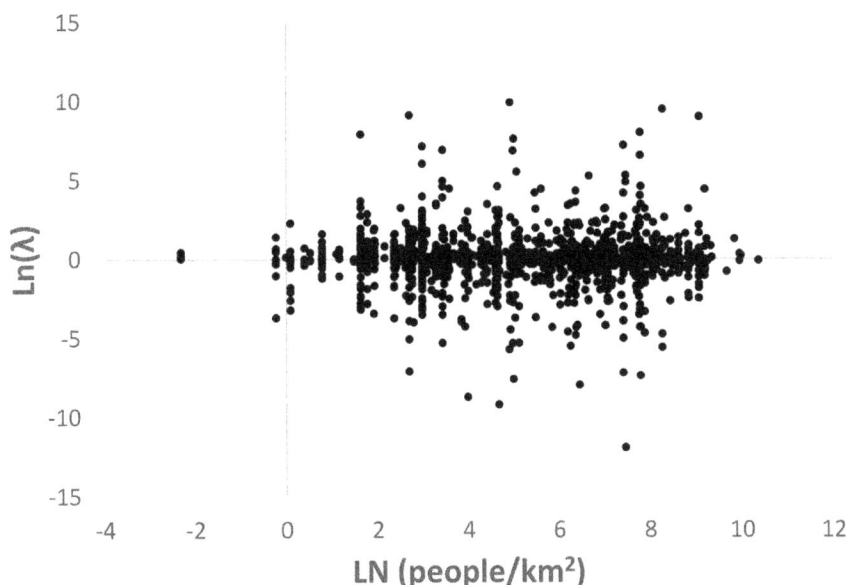

Figure 2. A scatterplot of the relationship between the natural log of annualized plant population growth (λ_{ann}) and human density (people/km². The plot shows no relationship between human density and λ_{ann}.

classes consisted of rural (<40 people/km^2, n = 785), peri-urban (40–400 people/km^2, n = 488), and urban (>400 people/km^2, n = 409). We classified λ_{ann} into shrinking (ln(λ_{ann})< −0.2; n = 560), stable (−0.2<ln(λ_{ann})<0.2, n = 595) and growing (ln(λ_{ann})>0.2, n = 520) populations approximately by equal frequency categories. Goodness-of-fit test results were significant (chi square = 10.79; df = 4; p = 0.029; Table 2). The pattern of this significance suggests an increase in variability of plant growth responses in urban environments. We found more observations of both λ_{ann} growth and retraction in urban environments and fewer than expected stable transitions (−0.2>λ_{ann}>0.2). Conversely, the lowest human density sites exhibited fewer population declines and more stable transitions than expected (Table 2). We further examined this relation bycreating a more stringent criteria for growing and shrinking populations, where values between −1 and +1 were considered stable, and larger values were characterized as collapse (ln(λ_{ann})< −1.0; n = 218) or growth (ln(λ_{ann})>1; n = 226). Although the general pattern observed above remained, the test result was not significant (chi-square = 4.46; p = 0.35)

Given that the chi-square tests suggest that urban populations may be more variable than rural ones, and that variability can lead to decreased persistence likelihoods, we further assessed variability in λ_{ann}. We restricted our assessment to observations with at least five repeat population change estimates (six survey dates). We found no relationship between the coefficient of variation (standard deviation/mean) as a function of human density (n = 80; F = 0.43; p = 0.51). An alternative explanation for increased variability of λ_{ann} in urban environments could be a differential distribution of life forms, with annual herbaceous plants found, on average, in higher human densities. This, in fact, is the case (one-way ANOVA, n = 1,682, df = 2, F = 25.7; p<0.001). Half (50.4%) of all observations were annual plants, yet 58.7% of urban λ_{ann} were from annual plants. These effects, however, are small. A linear model predicting λ_{ann} by growth form and human density yields no significant effects or interactions.

Assessing λ_{ann} by growth form indicated no differences among life forms in their population responses to human density for annual or perennial herbaceous plants (Table 1). Woody trees and shrubs, however, exhibited a significant decrease in λ_{ann} with increasing human density (Table1). This is the single significant result we observed relating λ_{ann} to human density. Examining the distribution of life form across human density showed a significant difference (ANOVA; n = 1,682, F = 25.70; p<0.001) with woody plants being found at significantly lower human densities.

Finally, we assessed λ_{ann} by species and by habitat type. We found no significant (p<0.1) patterns with human density, and observed an even balance of those with positive (n = 11) and negative (n = 11) coefficients relative to human density for the 22 species with 20 or more λ_{ann} estimates (Table 3). We constructed

independent tests of λ_{ann} versus human density for 19 habitat types. Again, we found no significant (p = 0.1) correlations in either direction (Table 4). Further, we achieved the same results when we reduced the number of data points by removing data of lower overall quality and restricting estimates to those data with specific estimates of the number of plants (Table 4). We observed no effect of human density, nor an interaction of habitat type by population. We did not correct for family error rates in either case as not a single test was significant even at a 0.10 level. There were considerable differences in the distribution of habitat types with respect to human density, but no relationship between the distribution of human densities sampled within a habitat type and the overall mean performance for that habitat type (Table 4).

In summary, we found no support for the hypothesis that increasing human density results in decreasing plant performance other than for the case of woody plants. Among the 19 habitat types assessed, mean λ_{ann} varied with 12 habitat types exhibiting a positive net growth rate and seven habitat types with an average negative growth rate. No individual habitat type showed significant variation in λ_{ann} across a gradient of human density within habitats (Table 4).

Plant Endangered Species Expenditures by Human Population Density

Our assessment of endangered species expenditures shows that there is a weak, but positive, relationship between the natural log of human density and the natural log of endangered species expenditures using both census tract (coefficient = 0.186;

Table 3. Correlations of λ_{ann} with human density for the 22 species with 20 or more individual observations.

Species	n	F	P	Coefficient
Acanthomintha ilicifolia	38	0.37	0.55	0.074
Amsinckia grandiflora	21	0.848	0.37	−0.167
Astragalus brauntonii	23	1.003	0.33	0.142
Blennosperma bakeri	27	0.027	0.87	−0.039
Camissonia benitensis	35	0.001	0.97	0.009
Clarkia franciscana	66	0.89	0.35	0.347
Cordylanthus maritimus ssp. maritimus	20	0.045	0.84	−0.063
Cordylanthus maritimus ssp. palustris	57	2.71	0.11	−0.101
Cordylanthus mollis ssp. mollis	26	0.032	0.86	0.045
Dichanthelium lanuginosum var. thermal	30	0.182	0.67	−0.049
Dudleya multicaulis	25	0.03	0.86	0.01
Dudleya setchellii	11	1.82	0.4	0.063
Fritillaria liliacea	54	0.126	0.72	−0.02
Gilia tenuiflora ssp. arenaria	22	0.297	0.59	−0.103
Helianthella castanea	20	1.77	0.2	0.219
Hesperolinon congestum	26	0.013	0.91	0.012
Holocarpha macradenia	101	0.552	0.46	0.083
Lasthenia conjugens	24	0.1	0.76	−0.184
Lupinus tidestromii	21	0.047	0.83	−0.055
Pentachaeta lyonii	34	0.566	0.46	0.083
Phacelia insularis var. continentis	26	0.001	0.97	−0.009
Triphysaria floribunda	35	0.301	0.59	−0.049

No significant (p<0.1) correlations and equal numbers of positive and negative correlation coefficients were observed.

Table 2. A contingency table of λ_{ann} relative to human density showing observed numbers of populations within each class and the expected numbers parenthetically; goodness-of-fit likelihood ratio = 10.8 (p = 0.03).

	Plant population change		
Human Population	Shrink	Stable	Grow
0–40 people/km^2	245 (261.4)	298 (277.7)	242 246.0)
40–400 people/km^2	110 (109.2)	125 (116.0)	93 (102.8)
>400 people/km^2	205 (189.4)	172 (201.3)	192 (178.3)

Table 4. Summary statistics for correlation of population performance (natural log of annualized mean λ) with human density.

Habitat	Population Growth Mean ln(ë)	All Observations			Highly specific estimates			Distribution of occurrences by human population density		
		n	coefficient	p	n	coefficient	p	Median People/km2	F	p
Conifer Forest	0.272	60	−0.014	0.81	49	−0.087	0.25	14.1	6.57	**0.011**
Alkali Sink	0.191	35	−0.08	0.68	18	−0.116	0.84	13.8	3.32	0.069
Desert	0.179	29	−0.008	0.9	22	−0.009	0.92	5.7	13.3	**0.003**
Deciduous Woodland	0.111	10	−0.001	0.97	8	0.026	0.63	35.3	0.902	0.342
Vernal Pool	0.103	162	0.04	0.64	54	−0.013	0.21	61	1.51	0.219
Sand Dunes	0.075	194	0.003	0.94	162	0.012	0.78	40	0.316	0.574
Freshwater Wetland	0.074	208	−0.024	0.56	147	−0.009	0.86	7.8	26.23	**0.0001**
Ocean Bluffs	0.069	64	−0.021	0.66	54	0	0.99	6.2	19.51	**0.0001**
Serpentine	0.04	300	−0.01	0.76	249	0.004	0.83	59	9.09	**0.003**
Sagescrub	0.012	249	0.032	0.48	215	0.034	0.47	38.6	7.04	**0.008**
Chaparral	0.011	288	0.016	0.68	227	0.046	0.92	59.3	2.04	0.154
Woodland	−0.015	161	−0.037	0.3	114	−0.026	0.5	20.4	34.63	**0.0001**
Riparian	−0.02	48	−0.03	0.57	16	0.052	0.56	7.8	2.69	0.101
Grassland	−0.03	752	0.005	0.83	511	0.033	0.26	59.3	6.71	**0.01**
Rocky Slopes	−0.035	104	0.007	0.78	77	0.017	0.57	40	0.351	0.553
Other (stress)	−0.104	60	0.047	0.31	51	0.047	0.4	7.4	0.455	0.5
SaltMarsh	−0.184	154	−0.094	0.16	91	−0.138	0.14	7.2	3.31	0.07

Each regression was conducted on two subsets of data: all observations; and data that meet our data quality criteria of having high precision. The key point to note is that there are no significant relationships. Also of note are the differences in mean λ_{ann} by habitat. Habitats are ranked from the highest mean λ_{ann} to the lowest (leftmost data column). Many individual habitats were found in significantly more urban or rural environments (right columns), but the overall habitat performance was not predicted by either more urban or rural distributions. Values <0.05 are in bold face.

$r^2 = 0.029$; p<0.001 n = 5,064) and species (coefficient = 0.112; $r^2 = 0.025$; p<0.039; n = 171) as the replicate. We tried several additional transformations but the results did not substantively vary by transformation. Investigating the effect of range size on expenditures suggest that species with larger distributions have slightly more expenditures and that controlling for this effect reverses the effect of human density on expenditures (coefficient = −0.17; $r^2 = 0.037$; p<0.01 n = 171).

This coarse evaluation masks important detail, as San Diego County receives far more endangered species recovery expenditures than any other county in our study region (or likely, in the US) [26–28] (Table 5, Fig. 1). This county contains both regions of high and low human density. Excluding San Diego County actually increases both the coefficient and the fit of the positive relationship between human density and spending among census tracts (coefficient = 0.234; $r^2 = 0.081$; p<0.001 n = 4,459).

Among the 17 counties we included in this study, the relationship between human density and spending was significant in 10 counties, of which seven (Contra Costa, Los Angeles, Monterey, San Diego, San Francisco, Santa Cruz, Sonoma) were positive and three (Alameda, Orange, Santa Clara) were negative (Table 5). At a coarser scale, counties with very high average human densities tended to be on the highest end of expenditures (San Diego, Los Angeles Counties) or toward the low end of expenditures (Orange, Santa Clara Counties) with no apparent relationship between spending and human density.

Discussion

Our results both support and strengthen the conclusions of Lawson et al. [23] that plant populations perform equally well across the gradient from rural to urban locations. Using a larger dataset over a more extensive coastal region, and numerous additional analyses, we find no evidence that at-risk plant populations perform poorly in areas with high human density. Out of 65 tests of λ versus human population density, only a single result was significant. That result suggests that woody plants have lower average growth rates in areas of high human density than they do in more rural areas. However, this result is brought into question by the observation that only 16 of our 118 observations were from urban (>400 people/km) locations, and these represented just 6 of 35 woody species. Among the four woody species found in both urban (>400 people/km^2) and non-urban (<400 people/km^2) areas, three actually had higher average growth rates in urban populations than in their rural ones. In addition, this negative correlation is not driven by low λ_{ann} values in urban populations, but high λ_{ann} values in rural populations (Fig. 3).

The results of our study strongly suggest a potential for successful plant conservation efforts within the urban environments of California. We do not mean to imply, however, that populations of at-risk plants are not threatened by urban environments. Habitat loss, and the extirpation of populations, remains a critical issue. Our data suggest that protecting these populations from habitat loss may provide opportunities for protection of these at-risk biological resources. By focusing on those populations that remained extant across survey periods, we

Table 5. County level statistics of human population for coastal California from the San Francisco Bay metropolitan region southward and spending on behalf of federally listed endangered and threatened plant species from 2006–2008.

COUNTY	POPULATION	Mean $/km²	Coefficient	P
Alameda	1,443,741	57.47	−0.003	0.04
Contra Costa	948,816	58.87	0.008	0.0003
Los Angeles	9,453,140	85.69	0.002	<0.0001
Marin	246,104	90	−0.001	0.518
Monterey	407,907	372.35	0.035	0.0001
Napa	119,901	8.36	−0.003	0.133
Orange	2,852,710	64.61	−0.003	0.0019
San Benito	40,838	4.8	−0.003	0.157
San Diego	3,056,509	1591.99	0.012	0.0003
San Francisco	790,240	128.24	0.001	0.0004
San Luis Obispo	207,490	108.03	0.022	0.18
San Mateo	708,709	131.97	0.037	0.013
Santa Barbara	495,933	16.49	−0.001	0.435
Santa Clara	1,675,965	19.66	−0.003	<0.001
Santa Cruz	250,245	80.95	0.004	0.008
Solano	394,542	67.99	0.003	0.565
Sonoma	458,614	51.02	0.012	<0.0001
Ventura	806,420	50.99	0.002	0.412

Human density (people/km²) estimates are from the 2000 US census, summarized in the California Department of Finance (www.dof.ca.gov). Expenditure estimates are from the US Fish and Wildlife endangered and threatened species expenditure reports for 2006 through 2008. Expenditures are reported by species and we used species distribution maps to identify species within counties and averaged species expenditures within census tracts across distributions. Regression coefficients and p-values are for human density versus the expenditures by census tract.

found that λ_{ann} was no different across the spectrum from rural to urban environments.

A goal of plant conservation should be to protect viable populations over long time periods. The CNDDB data do not provide the capacity to conduct viability assessments using these data. Our observation of slightly greater rates of inter-annual variability for plants in areas of high human density, and high inter-annual variability generally, may result in higher long-term vulnerability of urban populations. Analyzing magnitude of this effect is beyond the scope of this project and the capacity of these data.

Our results expand on prior results. We consider variability in response in a variety of attributes, including habitat, growth form and time period, analyzing the CNDDB data in far more detail than previous studies. Second, we distinguish among growth forms and habitat types to assess performance across the gradient from rural to urban habitats. Our results suggest that specialized habitat types, such as alkali sinks, vernal pools and desert, may be more resilient to threats associated with urbanization as demonstrated by larger average λ values than other habitat types (Table 4). If homogenization of urban floras is driven by invasive species [15,16], then it stands to reason that habitat types that are less prone to invasion may be more resilient in urban systems. This does not help explain the relatively large net positive growth rates among species found in coniferous or deciduous forests, but may help explain the generally negative mean growth rates for riparian, grassland and salt marsh species (Table 4).

An alternative possible interpretation of the Lawson et al. [23] and these results is that the CNDDB data are not sufficiently specific and detailed to assess differences in λ_{ann}. The CNDDB data are haphazardly collected by a broad and diverse suite of individuals across a long timeframe using different methods. Our finding—that habitat types differ in performance in ways that we would predict based on habitat resilience—is encouraging because it supports the assertion that these data provide a signal of performance that can be assessed. If there were no habitat type

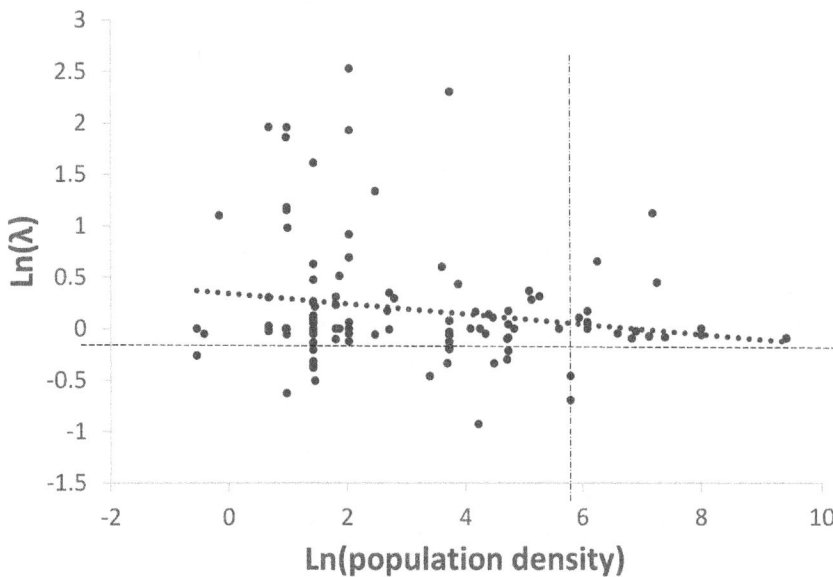

Figure 3. The correlation between human population density and plant population growth (λ) for woody plants. The only significant correlation observed between human density and plant performance (λ_{ann}) was for woody plants (n = 118, F = 4.06; p = 0.046; coefficient = −0.050). However this relationship is driven by high λ_{ann} in rural areas as opposed to strong negative λ_{ann} in urban settings. The horizontal dashed line represents no population change, the vertical dashed line separates urban (right) from non-urban (left) populations, with the dotted line representing the best fit correlation.

differences, then we might conclude that the data are simply too coarse to distinguish performance. However, since λ_{ann} values differ across habitat types, and those habitat types that are less prone to degradation through invasive species appear more resilient, this provides evidence that our data are, in fact, informative.

Finding no relationship between λ_{ann} and human density, we reinforce the assertion by Lawson et al. [23] that plant performance within protected habitats is not diminished simply by virtue of having a high local human density. This is encouraging given the significant conservation investment in and around major California urban areas [33]. Our study was not designed to determine why spending is allocated differentially across urban areas, with San Diego County receiving far more recovery funding than other California urban centers of high rare plant density. Nevertheless, we find it likely that some of the differential spending in California is driven by systematic conservation planning through tools such as the state's Natural Community Conservation Planning (NCCP) [36] and the Federal Habitat Conservation Planning (HCP) [37] .

Our geographic treatment of endangered species recovery expenditures is, necessarily, an abstraction of real expenditures. We do not know exactly where money was applied or to which species. However, our intent is to present an overview of the geography of federal expenditures in support of plant conservation, and not to assess the effectiveness of those reported expenditures.

Examining this result geographically, there is a complex relationship between human density (Fig. 1c), listed species density (Fig. 1a), and expenditures (Fig. 1b). Rural and medium density regions appear to uniformly receive little support for their endangered species (Fig. 1b, 1c). Similarly, the San Francisco and Los Angeles metropolitan regions have high human densities, high rare species richness, and modest expenditures (Fig. 1). In contrast, the San Diego region, with high human density, is rich in listed species and these species garner high levels of financial support (Fig. 1). This region appears unique in that respect. We speculate that the plants of San Diego County garner more funding because of the extensive multi-species conservation plans in place within the region [38].

The US Endangered Species expenditure data suggest that there are significant resources being applied to urban conservation in San Diego County, but not generally across the region. Underwood et al. [33], analyzing land protection expenditures in California, also found a large focus on spending along the urbanized coast of California. Schwartz et al. [4] made the case that providing for protection in urban environments can have collateral positive impacts on conservation through social engagement in the process of conservation. Whether by chance or design, evidence that λ_{ann} does not diminish with human density provides an endorsement of conservation expenditures in California. This is simply because such a large percentage of the unique flora of California is isolated to coastal, and often urban, regions of the state.

If plants are generally in need of protection, and human dominated landscapes are both rich in populations of rare species and under the most severe immediate threat, then it is sensible to skew investment toward these urban areas (San Francisco, Los Angeles, San Diego). However, this would be a poor investment of conservation resources if there were evidence that protected populations in urban environments were less likely to persist. Our study helps to assure conservation managers that these populations, once protected, do not appear to be at undue risk simply by virtue of urban proximity.

The case for investing in urban plant conservation [4] is further strengthened by examining extinction processes in urban environments [11,14,39]. The California Native Plant Society (CNPS) maintains a database on rare and endangered plants. This list includes 27 species considered extirpated in California [26]. Among these 27 extirpated species, 12 were formerly found within our study region. Among these 12 regional extirpated species, only two (*Ribes divericatum* var *parishii*, 1980, Los Angeles and San Bernardino counties; *Castilleja uliginosa*, 1987, Sonoma county) have been observed since 1954. In other words, just two of 12 documented extirpations have occurred since the onset of modern conservation measures such as the Endangered Species Act. These data do not indicate how many populations of plants may have been lost through time, nor do they report on individual county extirpations, but they do give an overall indication that California, with a high number of narrow endemics (defined as occurring in 1–2 counties [23]), has lost very few species through 30 years of conservation management. Given these arguments, we maintain that conservation investment in California urban centers can successfully protect rare plants over the medium term and provide strong incentives for local conservation value. Given the parallel conservation need to engage the urban population in protecting nature [21], investment of limited resources in urban plant conservation appears as likely as any to succeed.

Acknowledgments

We thank D. Lawson for her assistance and advice. We thank K. White and R. Bittman from the CA Department of Fish and Wildlife for their assistance with the CNDDB.

Author Contributions

Conceived and designed the experiments: MWS LMS ZLS. Performed the experiments: LMS ZLS. Analyzed the data: LMS ZLS MWS. Wrote the paper: MWS LMS ZLS.

References

1. Moilanen A, Anderson BJ, Eigenbrod F, Heinemeyer A, Roy DB, et al. (2011) Balancing alternative land uses in conservation prioritization. Ecological Applications 21: 1419–1426.

2. von der Dunk A, Gret-Regamey A, Dalang T, Hersperger AM (2011) Defining a typology of peri-urban land-use conflicts - A case study from Switzerland. Landscape and Urban Planning 101: 149–156.

3. Seabloom EW, Dobson AP, Stoms DM (2002) Extinction rates under nonrandom patterns of habitat loss. Proceedings of the National Academy of Sciences of the United States of America 99: 11229–11234.

4. Schwartz MW, Jurjavcic NL, O'Brien JM (2002) Conservation's disenfranchised urban poor. Bioscience 52: 601–606.

5. CDFG (2003) Atlas of the Biodiversity of California. Sacramento: California Department of Fish and Game.

6. Loarie SR, Carter BE, Hayhoe K, McMahon S, Moe R, et al. (2008) Climate Change and the Future of California's Endemic Flora. Plos One 3.

7. Ando A, Camm J, Polasky S, Solow A (1998) Species distributions, land values, and efficient conservation. Science 279: 2126–2128.

8. Schwartz MW (1999) Choosing the appropriate scale of reserves for conservation. Annual Review of Ecology and Systematics 30: 83–108.

9. Gibb H, Hochuli DF (2002) Habitat fragmentation in an urban environment: large and small fragments support different arthropod assemblages. Biological Conservation 106: 91–100.

10. Radeloff VC, Hammer RB, Stewart SI, Fried JS, Holcomb SS, et al. (2005) The wildland-urban interface in the United States. Ecological Applications 15: 799–805.

11. Williams NSG, Schwartz MW, Vesk PA, McCarthy MA, Hahs AK, et al. (2009) A conceptual framework for predicting the effects of urban environments on floras. Journal of Ecology 97: 4–9.

12. McKinney ML (2002) Urbanization, biodiversity, and conservation. Bioscience 52: 883–890.

13. Bastin L, Thomas CD (1999) The distribution of plant species in urban vegetation fragments. Landscape Ecology 14: 493–507.

14. Hahs AK, McDonnell MJ, McCarthy MA, Vesk PA, Corlett RT, et al. (2009) A global synthesis of plant extinction rates in urban areas. Ecology Letters 12: 1165–1173.

15. McKinney ML (2006) Urbanization as a major cause of biotic homogenization. Biological Conservation 127: 247–260.

16. Schwartz MW, Thorne JH, Viers JH (2006) Biotic homogenization of the California flora in urban and urbanizing regions. Biological Conservation 127: 282–291.

17. Dobson AP, Rodriguez JP, Roberts WM, Wilcove DS (1997) Geographic distribution of endangered species in the United States. Science 275: 550–553.

18. Dallimer M, Irvine KN, Skinner AMJ, Davies ZG, Rouquette JR, et al. (2012) Biodiversity and the Feel-Good Factor: Understanding Associations between Self-Reported Human Well-being and Species Richness. Bioscience 62: 47–55.

19. Fuller RA, Irvine KN, Devine-Wright P, Warren PH, Gaston KJ (2007) Psychological benefits of greenspace increase with biodiversity. Biology Letters 3: 390–394.

20. Miller JR, Hobbs RJ (2002) Conservation where people live and work. Conservation Biology 16: 330–337.

21. Schwartz MW (2006) How conservation scientists can help develop social capital for biodiversity. Conservation Biology 20: 1550–1552.

22. West P, Igoe J, Brockington D (2006) Parks and peoples: The social impact of protected areas. Annual Review of Anthropology. pp. 251–277.

23. Lawson DM, Lamar CK, Schwartz MW (2008) Quantifying plant population persistence in human-dominated landscapes. Conservation Biology 22: 922–928.

24. CNDDB (2009) Californa Natural Diversity Database. In: Game BDBDoFa, editor.

25. Press DM (2002) Saving Open Space: The Politics of Local Land Preservation In California. Berkeley: University of California Press.

26. CNPS (2011) Onlne Inventory of Rare and Endangered Plants. In: Society CNP, editor. 8 ed. Sacramento: California Native Plant Society.

27. Akcakaya HR, Burgman MA, Ginzberg LR (1999) Applied population ecology: principles and computer exercises. Sunderland Massachusetts: Sinauer Associates.

28. Bureau" USC Population Statistics. Washington DC: US Census Bureau.

29. Nations" U (2005) Demographic Yearbook 2005. In: Division P, editor. The Hague: United Nations.

30. USFWS (2006) Federal ad State Endangered and Threatened Species Expenditures; Fiscal years 2005 and 2006. In: Interior UDf, editor. Washigton DC: US Fish and Wildlife Service.

31. USFWS (2007) Federal and State Endangered and Threatened Species Expenditures; Fiscal year 2007. In: Interior Do, editor. Washington DC: Department of Interior.

32. USFWS (2008) Federal and State Endangered and Threatened Species Expenditures. In: Interior Do, editor. Washington DC: Department of Interior.

33. Underwood EC, Klausmeyer KR, Morrison SA, Bode M, Shaw MR (2009) Evaluating conservation spending for species return: A retrospective analysis in California. Conservation Letters 2: 130–137.

34. USFWS (2009) Environmental Conservation Online System. Washington DC: United States Fish and Wildlife Service.

35. Viers JH, Thorne JH, Quinn JF (2006) CalJep: a spatial distribution database of CalFlora and Jepson plant species. San Francisco Estuary and Watershed Science. Davis, CA: eSchlolarship, University of California.

36. Feldman TD, Jonas AEG (2000) Sage scrub revolution? Property rights, political fragmentation, and conservation planning in Southern California under the federal Endangered Species Act. Annals of the Association of American Geographers 90: 256–292.

37. Thomas CW (2001) Habitat conservation planning: Certainly empowered, somewhat deliberative, questionably democratic. Politics & Society 29: 105–130.

38. Franklin J, Regan HM, Hierl LA, Deutschman DH, Johnson BS, et al. (2011) Planning, implementing, and montoring multi-species habitat conservation plans. American Journal of Botany 98: 559–571.

39. Duncan RP, Clemants SE, Corlett RT, Hahs AK, McCarthy MA, et al. (2011) Plant traits and extinction in urban areas: a meta-analysis of 11 cities. Global Ecology and Biogeography 20: 509–519.

10

Defining Landscape Resistance Values in Least-Cost Connectivity Models for the Invasive Grey Squirrel: A Comparison of Approaches Using Expert-Opinion and Habitat Suitability Modelling

Claire D. Stevenson-Holt[1]*, Kevin Watts[2], Chloe C. Bellamy[3], Owen T. Nevin[4], Andrew D. Ramsey[5]

1 Centre for Wildlife Conservation, University of Cumbria, Ambleside, Cumbria, United Kingdom, **2** Centre for Ecosystems, Society and Biosecurity, Forest Research, Farnham, Surrey, United Kingdom, **3** Centre for Ecosystems, Society and Biosecurity, Forest Research, Roslin, Midlothian, United Kingdom, **4** School of Medical and Applied Sciences, Central Queensland University, Gladstone, Queensland, Australia, **5** School of Biological and Forensic Sciences, University of Derby, Derby, Derbyshire, United Kingdom

Abstract

Least-cost models are widely used to study the functional connectivity of habitat within a varied landscape matrix. A critical step in the process is identifying resistance values for each land cover based upon the facilitating or impeding impact on species movement. Ideally resistance values would be parameterised with empirical data, but due to a shortage of such information, expert-opinion is often used. However, the use of expert-opinion is seen as subjective, human-centric and unreliable. This study derived resistance values from grey squirrel habitat suitability models (HSM) in order to compare the utility and validity of this approach with more traditional, expert-led methods. Models were built and tested with MaxEnt, using squirrel presence records and a categorical land cover map for Cumbria, UK. Predictions on the likelihood of squirrel occurrence within each land cover type were inverted, providing resistance values which were used to parameterise a least-cost model. The resulting habitat networks were measured and compared to those derived from a least-cost model built with previously collated information from experts. The expert-derived and HSM-inferred least-cost networks differ in precision. The HSM-informed networks were smaller and more fragmented because of the higher resistance values attributed to most habitats. These results are discussed in relation to the applicability of both approaches for conservation and management objectives, providing guidance to researchers and practitioners attempting to apply and interpret a least-cost approach to mapping ecological networks.

Editor: Benjamin Lee Allen, University of Queensland, Australia

Funding: This project was funded by the Forestry Commission GB and the National School of Forestry at the University of Cumbria. The funders had no role in study design, data collection and analysis, decision to publish, or preparation of the manuscript.

Competing Interests: The authors have the following competing interest: This work was funded by the Forestry Commission GB and National School of Forestry at the University of Cumbria.

* Email: claire.stevenson@cumbria.ac.uk

Introduction

Effective biodiversity conservation within fragmented landscapes often requires the modelling of connectivity to define the extent of the problem, target conservation activities and to evaluate the impacts of landscape change [1]. Connectivity is defined as the degree to which the landscape facilitates or impedes species movement among resource patches [2]. A landscape consists of a complex, often dynamic, heterogeneous mixture of habitats and land uses which may impact on important ecological processes, such as species movement, habitat selection and survival, and influence behavioural and physiological responses [2–5]. The study of the impacts of the matrix on species movement, known as functional connectivity [6], is now the subject of much research within modified and fragmented landscapes [7]. Assessing functional connectivity is commonly used to aid conservation strategies by identifying potential movement pathways across fragmented landscapes for species of conservation concern [8–10]. It has also been used to help predict the potential dispersal and movement of invasive species to aid species management by identifying areas to target resources [11,12].

Geographical Information System (GIS), raster-based least-cost analysis techniques are often used to assess functional connectivity by modelling the impact of permeability of the surrounding landscape matrix on species movement [10]. It has been used in conservation [8–10] and invasive species management contexts [11,12]. For example, the population expansion of the grey squirrel (*Sciurus carolinensis*) in Britain, following its first introduction in 1876 [13], has had negative effects upon the forestry industry and native biodiversity [14–16]. In particular, it has occurred simultaneously with the decline and replacement of native red squirrel (*Sciurus vulgaris*) populations through resource competition and disease [14–16]. Therefore, an understanding of how grey squirrels utilise and move through the landscape is essential for effective red squirrel conservation and grey squirrel management. By using least-cost modelling it is possible to identify the potential dispersal areas, in addition to the most probable dispersal corridors, to assess the extent of spread [11]. Developing these models involves defining a species' 'core' or 'source' habitat and assigning resistance values to the surrounding landscape features, based on the actual or perceived impact to species movement at a particular resolution [17]. A cell with a high resistance value is used to represent an area that an individual is unlikely to traverse under typical conditions because of high energy, mortality, or other ecological costs [18]. Using information on a species' maximum dispersal distance, the area around a core habitat patch that is accessible to a species can be mapped with a simple Euclidean buffer. The permeability buffer zone is then taken into account so that the buffer is compressed or stretched according to the cumulative resistance scores assigned to the underlying landscape features. Overlapping buffers therefore signify connections where the species is assumed to be able to move between core habitat patches, forming a functionally connected habitat network.

It is widely acknowledged [4,18,19] that a critical step in least-cost modelling is defining resistance values for each type of landscape feature. Beier et al. [19] highlighted three ranked choices for estimating landscape resistance values with the first being the most highly ranked option: 1. empirical animal movement data, genetic distance or rates of inter-patch movements; 2. animal occurrence, density or fitness; 3. literature review and expert opinion. Ideally, resistance values should be informed and parameterised with independent field data, such as extensive mark release recapture studies, actual movement data from radio-telemetry or Global Positioning System (GPS) studies [11,20], data from experimental studies to record movement through different land cover types [21], or inferred movement data from landscape genetics [9]. However, as these resistance values are species and landscape specific, there is an understandable shortage of such empirical data [22]. Zeller et al. [23] reviewed the different types of data used to parameterize least-cost models and concluded that expert-opinion and occurrence data are most often used. However, they also suggest that comparative studies on the data used to derive resistance values are needed.

Although the use of expert-opinion to parameterise least-cost models is seen as subjective and out performed by values informed by empirical data [24], many studies utilise this type of information to parameterise models [3,12,25]. The use of expert-opinion may be appropriate in some cases, such as where there is a particular shortage of empirical data, an urgency to act, or a focus on general principles, focal species or particular species traits. However, in an attempt to make the setting of landscape resistance values less biased and more data-driven, some researchers [26–31] are starting to utilise species distribution models, such as MaxEnt [32], to parameterise least-cost connectivity models (defined as option 2 by Beier [19]). This study uses MaxEnt, a species distribution model which utilises maximum entropy principles to predict a species' use of a landscape based upon occurrence data and a selected set of environmental predictors [32]. The habitat suitability indices provided by the models can then be used in calculations [26–31] to create least-cost connectivity models. Given that resistance values informed by empirical data are ranked higher [19] and seen to outperform expert-opinion values [24], it is hypothesised that the HSM-informed values will produce a more accurate least-cost network

than expert-opinion data. The aim of this study is to investigate how expert-derived resistance values compare against values informed by habitat suitability modelling (HSM). The results of this study provide guidance to researchers and practitioners on the suitability of these approaches for informing management and research objectives relating to both species of conservation concern and invasive species spread.

Materials and Methods

Ethical statement

Ethical clearance for this study was approved by the University of Cumbria Ethics Committee, ref 09/17. This was a desk based study with no field work required. Therefore, research permits and licences were not required.

Study site

To compare expert-derived resistance values against HSM-informed values, grey squirrel within the county of Cumbria UK (Figure 1), are used as the study species. Whilst six large woodlands in Cumbria are designated red squirrel refuge reserves (Figure 1), the grey squirrel remains throughout the county. A number of previous studies have used expert-derived least-cost models to define habitat connectivity for Britain's native red squirrels and invasive grey squirrels [33–36], providing expert-opinion on land cover resistance. In addition, Cumbria has an extensive collection of grey squirrel distribution records available with which to create HSM-informed data for comparison. Cumbria covers an area of 6,768 km² and has a sparse population of 490,000 people. The Lake District National Park is located in the centre of Cumbria and has legislation and planning restrictions to conserve the landscape. The National Park Authority are responsible for implementing legislation and planning decisions aimed at conserving the landscape and its species, which means that little has changed regarding land use during the time frame that the species presence data used within this study were recorded (2000–2009). The topography is varied with the Cumbrian Mountain range (≤978 m a.s.l.) that runs approximately west to east across the middle of the county. The majority of land at these higher elevations is used for grazing with little woodland habitat. However, at lower elevations there are numerous woodlands, and other semi-natural habitats, scattered within an agricultural matrix which may provide greater potential for squirrel movement.

Identifying least-cost networks

Land cover types from a highly accurate and up to date vector land cover map (Ordnance Survey Master Map) were reclassified into 21 broad land cover categories for Cumbria (Table 1). The map was rasterised at 10 m resolution to ensure accurate representation of narrow linear features, such as strips of woodland. All woodland patches were classed as core habitat as squirrels use these areas for nesting and breeding [37,38]. This map was then parameterised with five alternative expert-derived resistance sets from previous studies (Table 1). The resistance values given in the different studies varied substantially. An additional set of values was developed by the authors by refining Stevenson's [35] scores (referred to as new expert-derived), following a review of the literature and the ecological underpinning of the values that had been applied previously, as described below.

Coniferous, mixed and broadleaved woodland were all assigned the lowest resistance value of 1, as core habitat. Scrub, coppice, orchard, and garden were given relatively low resistance values because they often contain tree species and are commonly used by

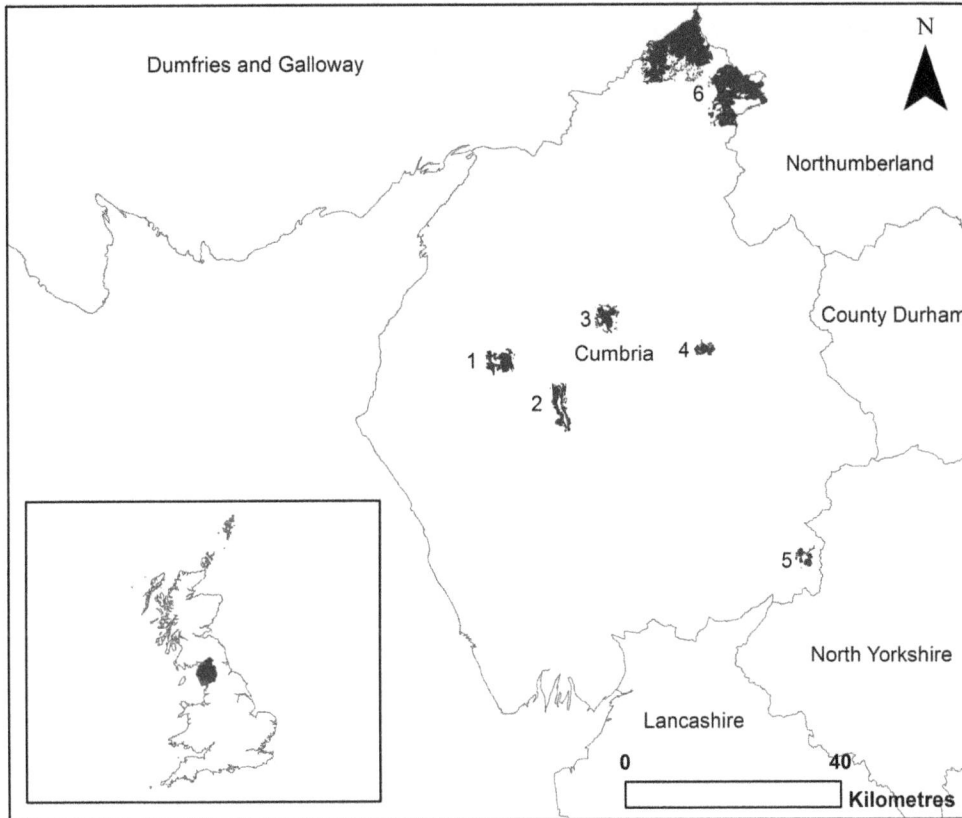

Figure 1. Map of red squirrel reserves in Cumbria and neighbouring counties with reference to its location in the UK. * 1. Whinlatter; 2. Thirlmere; 3. Greystoke; 4. Whinfell; 5. Garsdale/Mallerstang and 6. Kielder (Cumbria proportion of). Boundary lines were obtained through EDINA Digimap Ordnance Survey Service, http://digimap.edina.ac.uk/digimap/home.

grey squirrels for commuting [11,20]. Path, track, road verge, road and railway verge may also be used as commuting corridors, [13], but their use may confer higher mortality risks and therefore they were assigned a relatively high score. Improved/arable/amenity, rough grassland and heath were all attributed higher values still, as squirrel species tend to avoid open habitats [39]. Due to the threat of railways and the difficulty of moving over marsh, water, urban areas, buildings, and rocky areas like cliffs, the high scores assigned in previous studies were maintained.

Least-cost networks were created for each set of resistance values (Table 1) using the least-cost network process outlined in Watts et al. [10]. This network tool analysis utilises ArcView 9.1 and the Spatial Analyst extension (ESRI, Redlands, CA). The first step defined suitable patches of woodland habitat and generated a cost surface raster from the land cover map, by joining the resistance values (Table 1) to the 21 land cover classes. Secondly, the 'cost-distance' function in the Spatial Analyst toolbox was used to create a cost-distance surface between woodland patches. The resulting accumulated cost raster was then reclassified to a standardised maximum dispersal distance of 8 km to ensure comparability between the different resistance sets. The 'region group' function was used to define each discrete network, using an eight-cell rule so that touching cells, either adjacent and diagonally opposite, within the minimum distance of any given patch were considered part of the same network.

Deriving resistance scores from habitat suitability modelling

Records of grey squirrel presence were obtained from Save Our Squirrels (http://www.saveoursquirrels.org.uk/). These consisted of 2,281 verified sightings recorded year-round between 2000 and 2009 given by members of the public from both within woodland habitat (35%) and the wider landscape (65%). The grid references and type of habitat the sightings were recorded and verified by Save Our Squirrels. Sightings that were recorded outside of the grey squirrels known distribution range were also verified by contacting the recorder. The points outside of core woodland habitat are believed to relate to landscape use and movement, rather than indicating suitable foraging, breeding or nesting resources [37,38]. It is these non-woodland records that are used to infer the permeability of the landscape matrix using the habitat suitability modelling software, MaxEnt [40,32]. MaxEnt assigns each raster cell a Habitat Suitability Index (HSI) based on the environmental conditions at locations where a species has been recorded, using the maximum entropy method [41]. There are three output formats given by the MaxEnt programme: raw, cumulative and logistic; the most easily intuitive logistic HSI scores, which indicate the probability of occupancy ranging between 0–1 and assuming that this is 0.5 at an average site [40,32], were used in this study.

Both the species records and environmental data were prepared for modelling with MaxEnt. The squirrel data were filtered to remove locations recorded at a resolution of >100 m. Of the remaining 2,008 points, 842 squirrel presences recorded were

Table 1. Land cover resistance values based on previous least-cost modelling studies and resistance set based on expert opinion.

Land cover classification	Stevenson (2008)	Humphrey et al.(2007)	Williams (2008)	Verbeylen et al (2003) *	Gonzales (2000)	New expert-derived	Land cover ranking
Broadleaf	1	0	0	1	1	1	1
Mixed	1	0	0	1	1	1	2
Coniferous	1	0	0	1	1	1	3
Orchard	30	1	4	300	5	16	4
Scrub	1	2	10	10	5	16	18
Coppice	1	1	4	300	5	16	5
Garden	1	1	20	10	5	11	6
Improved/arable/amenity	30	10	20	800	8	40	20
Rough grassland	30	10	20	800	8	40	15
Heath	22	20	20	300	9	37	19
Path	16	50	20	800	1	27	9
Railway verge	12	10	20	800	4	27	11
Road verge	12	10	20	800	4	27	7
Marsh	76	50	50	800	9	91	14
Water	115	50	100	800	999	130	13
Urban	57	50	50	800	1	72	16
Railway	40	50	50	800	9	55	12
Road	12	50	30	800	9	27	8
Track	12	50	30	800	1	27	10
Building	1000	50	50	1000	999	1000	17
Rock	1000	50	50	1000	999	1000	21

*Note the resistance scores given in Verbeylen et al.'s (2003) study are for red squirrels in an urbanised matrix and are used for comparison.

within the matrix. A categorical land cover raster (gridded data map) was created using the same Ordnance Survey Master Map data and 21 broad land cover categories as previously described (Table 1). However, a coarser resolution of 100 m was used to match the spatial accuracy of the squirrel records. To ensure that linear habitat features were not under-represented, each land cover type was rasterised separately at a 10 m resolution and then aggregated to 100 m using the 'maximum' rule. These rasters were mosaicked using ranks that prioritised the classification of each 100 m square containing more than one land cover type (Table 1). All areas of woodland (core habitat), rocks and buildings (highly impermeable) were removed from the land cover map to prevent their incorporation in the model. In an effort to account for sampling bias towards accessible areas, a well-known and common characteristic of species data collected in an *ad-hoc* or non-systematic way [42], all areas over 500 m from a road, track or path were also removed from the map. This left a total of 665 squirrel records that fell within the remaining areas of the land cover map which were used to train and test the habitat suitability model. Each point (located in the south west of the 100 m grid square) was adjusted by 50 m east and 50 m north to locate each point in the centre of the grid square. This was to ensure that the points matched the 100 m raster landscape i.e. were within one cell, not potentially boarding four.

All models were run in MaxEnt Version 3.3.3k, using primarily default settings (regularisation multiplier = 1; duplicate occurrences removed; maximum number of background points = 10000, as used in Kramer-Schadt et al. [43]). Five-fold cross validation was used to calculate mean Area Under Roc Curve (AUC) and extrinsic omission rates (the average proportion of test points that fall outside the area predicted to be suitable), following use of the occupancy threshold rule that maximises the sum of test sensitivity and specificity (as recommended by Liu et al., [44]). Residual spatial autocorrelation (rSAC) can inflate measures of model performance [45–47] therefore Moran's correlograms were created (1 – predicted HSI for each species record; [48]) using the Spatial Analysis in Macroecology software program (SAM; [49]). Significance of Moran's *I* was calculated using a randomisation test with 9,999 Monte Carlo permutations, correcting for multiple testing.

The response curves, which showed the mean predicted probability of a species' presence (*p*; 0–1 scale) within each land cover type, were used to derive the resistance values for each land cover type. For both the new expert-derived and the HSM-informed values, woodland was given a value of 1, as permeable core habitat, and rock and building given values of 1000, as impermeable land cover types. The remaining land cover type values were inverted and standardised to the same scale as the new expert-derived values, (1–130; using 1-(*p*×130)). These values were then used to identify least-cost networks using the same approach as applied to the new expert-derived resistance scores.

Comparing resistance scores and resulting habitat networks

An area-minimisation methodology was applied to select for the smallest network that captures the majority (≥90%) of the filtered distribution point data (n = 842). This methodology, derived during this study, was based on the principle that when managing invasive species, areas for control must be defined and defensible to provide successful management [50]. As the grey squirrel population continues to expand in the Cumbria study site, it is important that control efforts are targeted to provide effective management. By identifying habitat networks management can be targeted in these specific areas of the landscape. The larger the

habitat networks are the more widespread management would have to be. Therefore, the resistance set which produced networks that include a high proportion of distribution points but a small network area are regarded as the better networks as management can be targeted in these focused areas. In addition a chi square test was used to test whether a significant number of distribution points were within the networks when compared to random points.

The HSM-informed resistance scores and the resulting networks were compared to those created with the new expert-derived set selected by the area-minimisation criteria. A Wilcoxon signed ranks test was used to assess the relative difference between scores. The habitat networks produced were also measured and compared visually and using the distribution points. Distribution points that were within the new expert-derived networks but not within the HSM-informed networks were identified along with the land cover type they were in and vice versa.

Results

Habitat Suitability Model performance

The results from five-fold cross-validation test showed that the models performed well (Training sample size = 532; Test sample size = 133; Training AUC = 0.80±0.001; Test AUC = 0.78±0.04; Test gain = 0.70±0.19; Extrinsic omission rate = 0.23, P<0.001), indicating that land cover type provides useful information on the likelihood of grey squirrel presence. No significant residual spatial autocorrelation was found at any distance lag. Moran's I values were <0.05 and statistically insignificant at each distance lag, indicating that the residuals were not spatially autocorrelated.

Selecting an 'optimal model' from expert-derived resistance sets

There was considerable variation between the previous studies and new expert-derived resistance sets, with network area ranging from 78% to 15% of the total landscape area, containing between 99% and 32% of the squirrel point data (Figure 2). However, when the networks were tested against the occurrence data within the matrix all resulting networks contained significantly more distribution points than expected by chance (n = 842, Stevenson 2008, χ2 = 623, df. = 1, p<0.001; Humphreys et al. 2007, χ2 = 238, df. = 1, p<0.001; Williams 2008, χ2 = 357, df. = 1, p< 0.001; Verbeylen et al. 2003, χ2 = 169, df. = 1, p<0.001; Gonzales 2000, χ2 = 213, df. = 1, p<0.001; new expert-derived, χ2 = 623, df. = 1, p<0.001). Using the area-minimisation methodology, the new expert-derived resistance set was shown to have above 90% of sightings within the networks and the lowest networks area of 49% of the total landscape (Figure 2). This was therefore selected and used for further comparison with the HSM-informed networks.

Comparing expert-derived and HSM-informed networks

The HSM-informed network had significantly more grey squirrel distribution points within it than expected by chance (n = 842, χ2 = 836, df. = 1, p<0.001). However, the new expert-derived network contained significantly more points than the HSM-informed network (n = 842, χ2 = 185, df. = 1, p<0.001). The majority of land cover types were given higher resistance values using the HSM approach compared to those derived from the new expert-derived values, with relative differences ranging from 7–86% (Table 2). These differences were found to be statistically significant (n = 16, Wilcoxon signed ranks test, *p* = 0.002); water and coppice were the only habitats to be assigned an HSM-informed lower resistance value compared to those derived from the new expert-derived resistance set. The

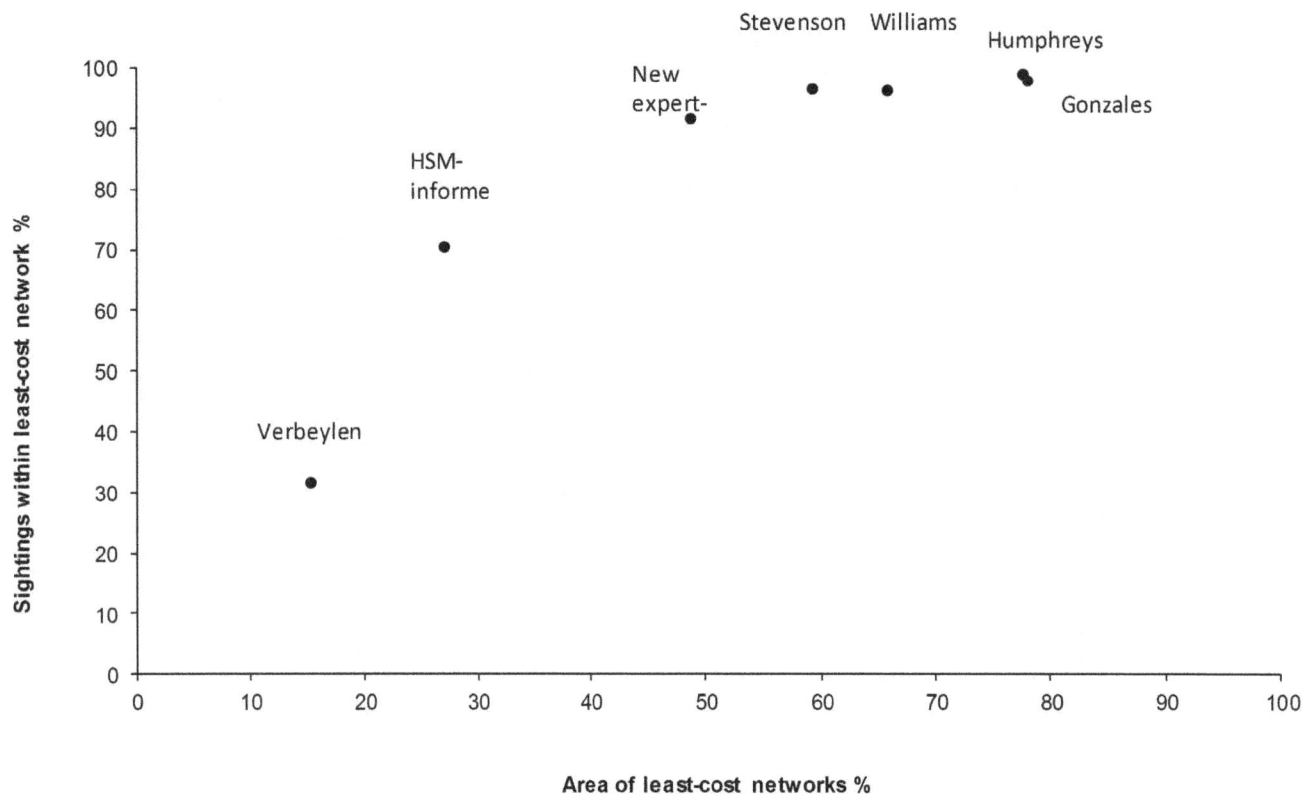

Figure 2. Comparison between expert-derived and Habitat Suitability Model-derived resistance values. Note: values that produce a network with>90% sightings points and the lowest network area is considered the best model for management.

largest differences in the resistance values assigned to habitats by the two approaches were between scrub, tracks, railways and railway verges (Table 2).

The least-cost model parameterised with the new expert-derived resistance values identified 738 discrete networks, although two of these cover substantial areas; habitat network 1 in the north and habitat network 2 in the south (Figure 3). The mean network size was 4.7 km^2 (± 84.4). These networks accounted for 42% of the Cumbrian land cover (3,518 km^2) and appear to be separated by the land cover types within the Cumbrian Mountains. The HSM-informed resistance values generated comparatively smaller and more fragmented networks, owing to the higher resistance scores attributed to most habitat types. This network was 55% the size of the new expert-derived network (1,953 km^2; 34% of land cover) and sat almost entirely inside it, with only 0.2% extending beyond the expert-derived network, over areas of water. The mean network size was 0.3 km^2 (± 5.0) and 5,840 separate networks were identified in Cumbria (Figure 3). Ten of these were relatively large (>20 km^2). The HSM-informed networks also indicated that networks in the north and south of the county were separated by the Cumbrian Mountains range. Both identified Grizdale Forest and surrounding woodlands as a large, well connected grey squirrel habitat network (Figure 3).

The smaller HSM-informed least-cost networks contained 592 (70%) of 842 species records within the habitat network (compared to 772 (92%) using new expert-derived) (Figure 4). As the HSM-informed scores were based upon the actual distribution data it was expected that the resulting networks would include a substantial amount of distribution points. The number of points outside of the HSM-informed least-cost networks was 250; of these

points missed by the HSM-informed network 180 were included within the new expert-derived networks. These 180 points were located in improved/arable/amenity land (77%), gardens (8%), rough grassland (6%), urban (3%), road (2%), road verge (1%), tracks (1%), marshland (1%), scrub (1%) or water (1%). The number of points outside of the new expert-derived networks was 70; of these points none were included within the HSM-informed least-cost networks.

Discussion

When estimating resistance values Beier [19] highlighted three ranked choices. Although using animal movement data, genetic distance or rates of inter-patch movements (option 1) is the preferable option to define resistance values, animal occurrence data (option 2) and/or literature review and expert opinion (option 3) may be the only information available to many researchers and conservationists trying to model functional connectivity in fragmented landscapes. In this study resistance values derived from expert-opinion have been compared to HSM-informed values. Both techniques identified least-cost networks that contained significantly more distribution points than would be expected by chance. However, differences occur between the degree of model assumptions and biases (based on the different types of data), resistance values for certain land cover types and the least-cost networks identified. This has implications for the reliability of using such data in meeting conservation and management objectives.

To derive a set of expert-opinion resistance values it is useful to compare previous resistance values from multiple sources, particularly if the studies have similar species and environmental

Table 2. Average probability of grey squirrel presence according to land cover type.

Habitat type	HSM p score	HSM-resistance score	New expert-derived resistance score	Difference between HSM and Expert-derived resistance scores
Scrub	0.10	117	16	0.86
Track	0.16	109	27	0.75
Railway	0.17	108	27	0.75
Railway verge	0.17	108	27	0.75
Path	0.34	86	27	0.69
Heath	0.17	108	37	0.66
Garden	0.73	32	11	0.66
Road	0.41	77	27	0.65
Improved/arable/ amenity	0.13	113	40	0.65
Rough grassland	0.13	113	40	0.65
Road verge	0.43	74	27	0.64
Orchard	0.77	30	16	0.47
Urban	0.29	92	72	0.22
Marsh	0.17	108	91	0.16
Broadleaf	N/A	1	1	0.00
Coniferous	N/A	1	1	0.00
Mixed	N/A	1	1	0.00
Building	N/A	1000	1000	0.00
Rock	N/A	1000	1000	0.00
Coppice	0.86	15	16	−0.07
Water	0.18	107	130	−0.21

p = mean predicted probability of presence according to habitat type.

conditions. The resistance values given in previous studies were highly variable, resulting in varied least-cost habitat network areas and number of distribution points within networks. Although the land cover resistance values given in these studies were for red or grey squirrels, the studies took place in different countries with different regional environmental conditions and large scale and inevitable differences in landscape composition and structure. This may account for the differences in values given and resulting networks. Verbeylen et al [3] in particular was focused on red squirrels and based in an urban area which is very different to the largely-rural and sparsely populated Cumbria. However by assessing the range of different resistance values given in these studies and additional literature on land cover use, the new expert-derived resistance set was created. The area-minimisation method suggests that these values appear to be the best set for management purposes in this area, capturing a high percentage of distribution points within the smallest network area.

The resistance values for the new expert-derived and HSM-informed least-cost models in this study were significantly different from one another. The HSM-informed model provided higher resistance values for most land cover types. The validity of HSM-informed least-cost models may be limited as the probability of occurrence in a particular land cover type does not always equate to the resistance of that land cover type during species movement [19]. In using distribution/occurrence data, certain land cover types may be undervalued when in reality they are used by the species. Conversely there will be land cover types that are overvalued. A key assumption of presence only modelling is that the data has come from random sampling or is representative of

the whole landscape [51]. It is questionable whether the degree of bias in presence data can be truly known [51]. Squirrels are well known to use scrub habitat and will use this and linear features to aid dispersal [13,52–54], yet scrub and railway verge (a linear feature) were given high HSM-informed resistance values due to a low number of distribution points. Of the distribution points missed by the HSM-informed networks but included within the new expert-derived networks, 77% were within improved/arable/ amenity land cover type. This suggests that the inverted HSM values for this land cover may be too high, and squirrels may be able to cross these hostile areas quickly and undetected. The dispersal distance used for both expert-derived model and the HSM-informed model were set at 8 km. Therefore, it is the higher resistance values given to certain land cover types using the inverted-HSM that led to the identification of smaller and more fragmented networks.

The HSM-informed networks were 45% smaller than the expert-derived networks and were spatially nested inside these networks. The smaller mean size of HSM-informed networks suggests that grey squirrel occurs in a highly fragmented and functionally unconnected landscape. Both models highlight the land cover types of the Cumbrian Mountains as a barrier to movement; the combination of relatively high elevation and intense grazing result in a lack of woodland in the area. Although, some individuals may attempt to cross the barrier, the lack of available habitat will impede dispersal subjecting individuals to high levels of predation and starvation. There are no recorded introductions of the grey squirrel into Cumbria [55,56] and therefore these animals have been able to spread to their present

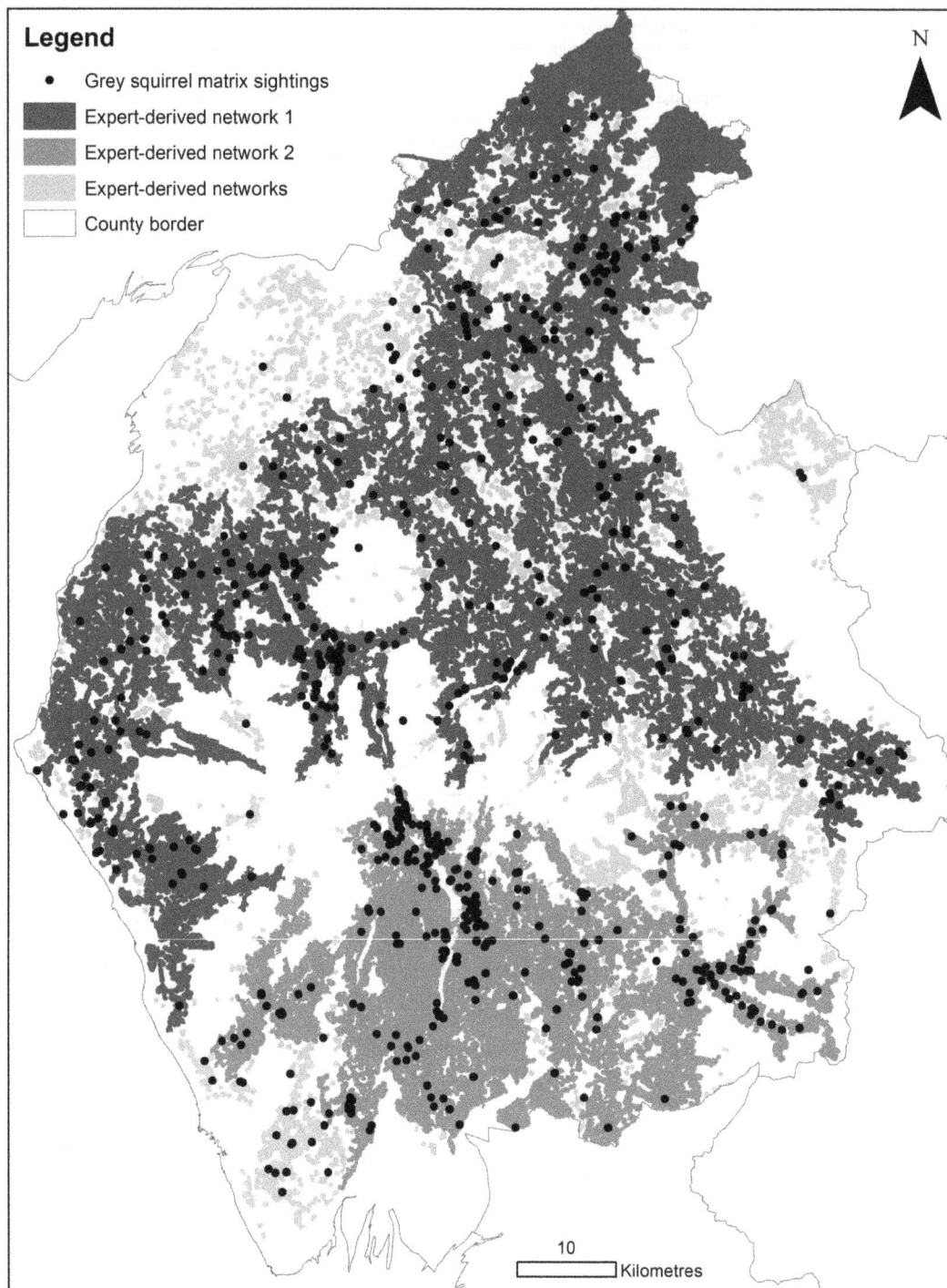

Figure 3. Grey squirrel least-cost habitat networks identified from expert-derived resistance values. Boundary lines were obtained through EDINA Digimap Ordnance Survey Service, http://digimap.edina.ac.uk/digimap/home.

distribution in the north and south of the county by natural means. The expert-derived model identified two large networks, one in the north and one in the south, suggesting a much more connected landscape.

Studies have suggested that expert-opinion based models perform less accurately than models informed by empirical data [24,57,58]. Given that HSM-informed networks are derived from known distribution data, these models could be interpreted as

identifying more precise areas in the landscape that are connected for a species. In comparison, the expert-derived networks include those areas where sighting have not been recorded but are judged by experts as permeable to the species during dispersal. Experts may overestimate the importance of certain land cover types erring on the side of caution and therefore rendering the model less accurate [24]. Where actions might require a more precise approach, such as identifying possible protected areas or sites for

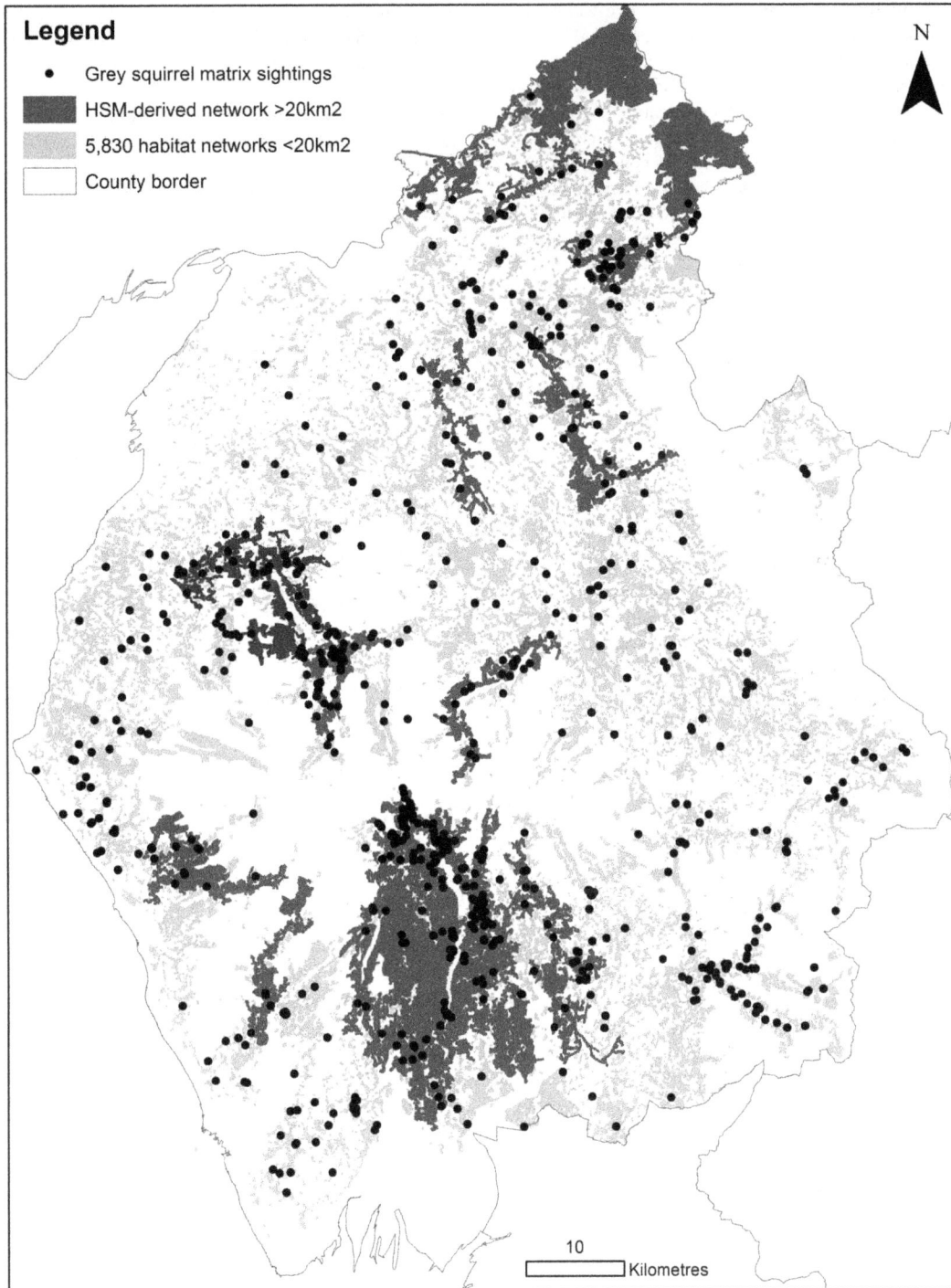

Figure 4. Highly fragmented grey squirrel least-cost habitat networks identified with Habitat Suitability Model-derived resistance values. Boundary lines were obtained through EDINA Digimap Ordnance Survey Service, http://digimap.edina.ac.uk/digimap/home.

an efficient and intensive control program, a HSM modelling approach would be appropriate. However, when assessing invasive species it is not just the most likely areas that a species will disperse to, but the entire possible range that needs identified. In an invasive species context, it may be more appropriate to apply a conservative less precise model, such as the expert-derived model, to enable all possible areas of dispersal to be included within the network.

In the case of invasive species the assessment of potential movement and impact is needed as soon as possible to aid management planning. This method is not dependent upon extensive species distribution data and can therefore be produced relatively quickly. Clevenger et al. [24] found that expert only derived resistance values had a weaker correlation with empirical-derived values than literature-derived values. Systematically collecting expert opinion, as promoted by Eycott et al [59], in combination with published data on land cover usage will enable resistance values to be assigned in the initial stages to give an indication of species movements whilst other empirical data is collected where possible. Adriensen et al [18] suggested that once a 'starter kit' of resistance values has been identified, sensitivity studies can be initiated and multiple alternative resistance sets can be tested [60]. Once species distribution data is collected, HSM-informed least-cost networks can be identified and used to aid the selection of most likely used sites to focus monitoring or eradication programs. It should not be assumed that using distribution data (option 2 in Beier et al. [19]) to identify resistance values is better or worse than using well developed expert-opinion (option 3 in Beier et al. [19]) as the choice of which method to use may depend upon the aims and objectives of the user and the appropriate precision of the approach.

This paper describes the first step towards developing least-cost habitat networks using *ad hoc* species records and a simple, land cover-based habitat suitability model. It is acknowledged, however, that species respond to their surrounding environment over a range of spatial scales and that both local and landscape features will affect both the suitability of the core habitat and the permeability of the surrounding matrix [5,61]. More complex models incorporating multiscale information on the terrain, built environment, and the composition, structure and arrangement of habitat patches are likely to provide more accurate and useful models [45], providing predictions at each location, rather than assuming consistent levels of permeability for a particular land cover type. This spatially explicit technique would enable landscape level decision making, improving our ability to identify important networks of habitat and enabling a targeted and informed approach to both conservation and infrastructural development.

Conclusion

Even though approaches to gather expert opinion are becoming more systematic and robust, it should not be seen as a blanket substitute for empirical data. Empirical data will continue to be important for studies on single species, where there is considerable uncertainty or where there is significant investment in time and money on conservation activities. Conservation planners must be aware of the subjectivity and pitfalls of the different types of data used in least-cost models, without any further validation or sensitivity testing of model values. If expert opinion is the only option available it should be used as a first step by systematically

combining multiple expert opinions and published data, but with the knowledge that further assessment of resistance values through sensitivity analysis and empirical data will be needed. Where distribution data is already available, the type of data collection and the subjective translation issues of over and under valuing land cover types must be assessed with expert knowledge or empirical data and explicitly stated in methodologies [51,62].

This study successfully compared expert-derived and HSM-informed resistance values used in least-cost modelling. Although the results of the models differed, both identified equally useful least-cost networks. For the grey squirrel in Cumbria, both expert-derived and HSM-informed networks have shown that there is a separation between north and south of Cumbria due to the land cover types and lack of habitat of the Cumbrian Mountain range. The expert-derived networks indicate a conservative less precise least-cost network that indicates the potential dispersal range of the grey squirrel and suggests that there may be multiple infiltration routes into the county from the north and south. This conservative expert-derived approach is useful when dealing with invasive or generalist species to identify the potential extend of spread. When assessing endangered or specialist species, or areas that are highly likely to contain target species, the HSM-informed network provides smaller precise networks. These precise networks should be used to inform targeted conservation to increase connectivity for species of conservation concern, or to inform targeted management to prevent the incursion of invasive species. The variable but acceptable precision of both expert-derived and HSM-informed least-cost networks highlights the need to consider data reliability and environmental context when deciding on the most appropriate management of invasive species.

Acknowledgments

The Authors would like to thank Dr Sallie Bailey for advice and comments, Phillip Handley for GIS advice and Simon O'Hare for sightings data. Country and county outlines in Figures 1, 3 and 4 and OSMM data were obtained through EDINA Digimap Ordnance Survey Service, http://digimap.edina.ac.uk/digimap/home.

Author Contributions

Conceived and designed the experiments: CDSH KW OTN ADR. Performed the experiments: CDSH KW CB. Analyzed the data: CDSH KW CB. Contributed reagents/materials/analysis tools: CDSH KW CB OTN ADR. Wrote the paper: CDSH KW CB OTN ADR.

References

1. Worboys G, Francis WL, Lockwood M (2009) Connectivity conservation management: A global guide (with particular reference to mountain connectivity conservation). London: Earthscan/James & James.
2. Taylor PD, Fahrig L, Henein K, Merriam G (1993) Connectivity is a vital element of landscape structure. Oikos 68: 571–573.
3. Verbeylen G, De Bruyn L, Adriaensen F, Matthysen E (2003) Does matrix resistance influence red squirrel (*Sciurus vulgaris* L. 1978) distribution in an urban landscape? Landscape Ecology 18: 791–805.
4. Spear SF, Balkenhol N, Fortin MJ, McRae BH, Scribner K (2010) Use of resistance surfaces for landscape genetic studies: Considerations for parameterization and analysis. Mol Ecol 19: 3576–3591.
5. Ricketts TH (2001) The matrix matters: Effective isolation in fragmented landscapes. The American Naturalist 158: 87–99.
6. Tischendorf L, Fahrig L (2000) On the usage and measurement of landscape connectivity. Oikos 90: 7–19.
7. Crooks KR (2006) Connectivity conservation. Cambridge: Cambridge Univ Pr.
8. Ferreras P (2001) Landscape structure and asymmetrical inter-patch connectivity in a metapopulation of the endangered Iberian lynx. Biol Conserv 100: 125–136.
9. Epps CW, Wehausen JD, Bleich VC, Torres SG, Brashares JS (2007) Optimizing dispersal and corridor models using landscape genetics. J Appl Ecol 44: 714–724.
10. Watts K, Eycott AE, Handley P, Ray D, Humphrey JW, et al. (2010) Targeting and evaluating biodiversity conservation action within fragmented landscapes:

An approach based on generic focal species and least-cost networks. Landscape Ecology 25: 1305–1318.
11. Stevenson CD, Ferryman M, Nevin OT, Ramsey AD, Bailey S, et al. (2013) Using GPS telemetry to validate least-cost modeling of gray squirrel (*Sciurus carolinensis*) movement within a fragmented landscape. Ecology and Evolution 3: 2350–2361.
12. Gonzales EK, Gergel SE (2007) Testing assumptions of cost surface analysis- a tool for invasive species management. Landscape Ecology 22: 1155–1168.
13. Middleton AD (1930) The ecology of the American grey squirrel (*Sciurus carolinensis* gmelin) in the British isles. J Zool, Lond. 100: 809–843.
14. Gurnell J, Wauters LA, Lurz PWW, Tosi G (2004) Alien species and interspecific competition: Effects of introduced eastern grey squirrels on red squirrel population dynamics. J Anim Ecol 73: 26–35.
15. Gurnell J, Mayle B (2003) Ecological impacts of the alien grey squirrel (*Sciurus carolinensis*) in Britain. In Bowen CP, editor. MammAliens – A one day conference on the problems caused by non- native British mammals. London: Peoples Trust for Endangered Species/Mammals Trust UK. pp.40–45.
16. Kenward RE (1983) The causes of damage by red and grey squirrels. Mamm Rev 13: 159–166.
17. Sawyer SC, Epps CW, Brashares JS (2011) Placing linkages among fragmented habitats: Do least-cost models reflect how animals use landscapes? J Appl Ecol 48: 668–678.

18. Adriaensen F, Chardon JP, De Blust G, Swinnen E, Villalba S, et al. (2003) The application of 'least-cost' modelling as a functional landscape model. Landscape and Urban Planning 64: 233–247.

19. Beier P, Majka DR, Spencer WD (2008) Forks in the road: Choices in procedures for designing wildland linkages. Conserv Biol 22: 836–851.

20. Driezen K, Adriaensen F, Rondinini C, Doncaster CP, Matthysen E (2007) Evaluating least-cost model predictions with empirical dispersal data: A case-study using radio tracking data of hedgehogs (*Erinaceus europaeus*). Ecological Modelling 209: 314–322.

21. Stevens VM, Polus E, Wesselingh RA, Schtickzelle N, Baguette M (2005) Quantifying functional connectivity: Experimental evidence for patch-specific resistance in the natterjack toad (*Bufo calamita*). Landscape Ecol 19: 829–842.

22. Eycott AE, Stewart GB, Buyung-Ali LM, Bowler DE, Watts K, et al. (2012) A meta-analysis on the impact of different matrix structures on species movement rates. Landscape Ecol 27: 1263–1278.

23. Zeller KA, McGarigal K, Whiteley AR (2012) Estimating landscape resistance to movement: A review. Landscape Ecol 27: 777–797.

24. Clevenger AP, Wierzchowski J, Chruszcz B, Gunson K (2002) GIS-generated, expert-based models for identifying wildlife habitat linkages and planning mitigation passages. Conserv Biol 16: 503–514.

25. Chardon JP, Adriaensen F, Matthysen E (2003) Incorporating landscape elements into a connectivity measure: A case study for the speckled wood butterfly (*Pararge aegeris* L.). Landscape Ecology 18: 561–573.

26. Wang Y, Yang K, Bridgman CL, Lin L (2008) Habitat suitability modelling to correlate gene flow with landscape connectivity. Landscape Ecol 23: 989–1000.

27. Richards-Zawacki CL (2009) Effects of slope and riparian habitat connectivity on gene flow in an endangered Panamanian frog, *Atelopus varius*. Divers Distrib 15: 796–806.

28. Decout S, Manel S, Miaud C, Luque S (2010) Connectivity loss in human dominated landscape: Operational tools for the identification of suitable habitat patches and corridors on amphibian's population. Landscape International Conference IUFRO, Portugal.

29. Wang IJ, Summers K (2010) Genetic structure is correlated with phenotypic divergence rather than geographic isolation in the highly polymorphic strawberry poison-dart frog. Mol Ecol 19: 447–458.

30. Decout S, Manel S, Miaud C, Luque S (2012) Integrative approach for landscape-based graph connectivity analysis: A case study with the common frog (*Rana temporaria*) in human-dominated landscapes. Landscape Ecol 27: 267–279.

31. Howard A, Bernardes S (2012) A maximum entropy and least cost path model of bearded capuchin monkey movement in Northeastern brazil. Intergraph. Available: https://intergraphgovsolutions.com/assets/white-paper/A_Maximum_Entropy-2_Imagery.sflb.pdf Accessed 2013.

32. Phillips SJ, Anderson RP, Schapire RE (2006) Maximum entropy modeling of species geographic distributions. Ecol Model 190: 231–259.

33. Gonzales EK (2000) Distinguishing between modes of dispersal by introduced eastern grey squirrels (*Sciurus carolinensis*). MSc Thesis, University of Guelph.

34. Humphrey J, Smith M, Shepherd N, Handley P (2007) Developing lowland habitat networks in Scotland: Phase 2. Edinburgh: Forestry Commission.

35. Stevenson CD (2008) Modelling red squirrel population viability under a range of landscape scenarios in fragmented woodland ecosystems on the Solway plain, Cumbria. London: People's Trust for Endangered Species. Available http://insight.cumbria.ac.uk. Accessed 2013.

36. Williams S (2008) Red squirrel strongholds consultation. Edinburgh: Forestry Commission.

37. Lowe VPW (1993) The spread of the grey squirrel (*Sciurus carolinensis*) into Cumbria since 1960 and its present distribution. J Zool, Lond 231: 663–667.

38. Skelcher G (1997) The ecological replacement of red by grey squirrels. In: Gurnell J, Lurz P, editors. The conservation of red squirrels, *Sciurus vulgaris* L. London: People's Trust for Endangered Species. pp.67–78.

39. Nixon CM, McClain MW, Donohoe RW (1980) Effects of clear-cutting on grey squirrels. Journal of Wildlife Management 44: 403–412.

40. Phillips SJ, Dudik M (2008) Modeling of species distributions with maxent: New extensions and a comprehensive evaluation. Ecography 31: 161–175.

41. Phillips SJ, Dudík M, Elith J, Graham CH, Lehmann A, et al. (2009) Sample selection bias and presence-only distribution models: Implications for background and pseudo-absence data. Ecol Appl 19: 181–197.

42. Warton DI, Renner IW, Ramp D (2013) Model-based control of observer bias for the analysis of presence-only data in ecology. Plos One 8: e79168.

43. Schadt S, Knauer F, Kaczensky P, Revilla E, Wiegand T, et al. (2002) Rule-based assessment of suitable habitat and patch connectivity for the Eurasian lynx. Ecol Appl 12: 1469–1483.

44. Liu C, White M, Newell G (2013) Selecting thresholds for the prediction of species occurrence with presence-only data. J Biogeogr 40: 778–789.

45. Bellamy C, Scott C, Altringham J (2013) Multiscale, presence-only habitat suitability models: Fine-resolution maps for eight bat species. J Appl Ecol 50: 892–901.

46. Merckx B, Steyaert M, Vanreusel A, Vincx M, Vanaverbeke J (2011) Null models reveal preferential sampling, spatial autocorrelation and overfitting in habitat suitability modelling. Ecol Model 222: 588–597.

47. Veloz SD (2009) Spatially autocorrelated sampling falsely inflates measures of accuracy for presence-only niche models. J Biogeogr 36: 2290–2299.

48. De Marco P, Diniz-Filho JA, Bini LM (2008) Spatial analysis improves species distribution modelling during range expansion. Biol Lett 4: 577–580.

49. Rangel TF, Diniz-Filho JAF, Bini LM (2010) SAM: A comprehensive application for spatial analysis in macroecology. Ecography 33: 46–50.

50. Zalewski A, Piertney SB, Zalewska H, Lambin X (2009) Landscape barriers reduce gene flow in an invasive carnivore: Geographical and local genetic structure of American mink in Scotland. Mol Ecol 18: 1601–1615.

51. Yackulic CB, Chandler R, Zipkin EF, Royle JA, Nichols JD, et al. (2013) Presence-only modelling using MaxEnt: When can we trust the inferences? Methods in Ecology and Evolution 4: 236–243.

52. Taylor KD, Shorten M, Lloyd HG, Courtier FA (1971) Movements of the grey squirrel as revealed by trapping. J Appl Ecol 8: 123–146.

53. Fitzgibbon CD (1993) The distribution of grey squirrel dreys in farm woodland: The influence of wood area, isolation and management. J Appl Ecol 30: 736–742.

54. Wauters LA, Gurnell J, Currado I, Mazzoglio P (1997) Grey squirrel *Sciurus carolinensis* management in Italy - squirrel distribution in a highly fragmented landscape. Wildlife Biology 3: 117–123.

55. Shorten M (1954) Squirrels. London: Collins.

56. Shorten M (1957) Squirrels in England, Wales and Scotland, 1955. J Animal Ecol 26: 287–294.

57. Pearce J, Cherry K, Whish G (2001) Incorporating expert opinion and fine-scale vegetation mapping into statistical models of faunal distribution. J Appl Ecol 38: 412–424.

58. Seoane J, Bustamante J, Diaz-Delgado R (2005) Effect of expert opinion on the predictive ability of environmental models of bird distribution. Conserv Biol 19: 512–522.

59. Eycott AE, Marzano M, Watts K (2011) Filling evidence gaps with expert opinion: The use of Delphi analysis in least-cost modelling of functional connectivity. Landscape Urban Plann 103: 400–409.

60. Rayfield B, Fortin MJ, Fall A (2010) The sensitivity of least-cost habitat graphs to relative cost surface values. Landscape Ecol 25: 519–532.

61. Wiens JA (1989) Spatial scaling in ecology. Funct Ecol 3: 385–397.

62. Beier P, Majka DR, Newell SL (2009) Uncertainty analysis of least-cost modeling for designing wildlife linkages. Ecological Applications 19: 2067–2077.

11

Thermal Carrying Capacity for a Thermally-Sensitive Species at the Warmest Edge of Its Range

Daniel Ayllón[1], Graciela G. Nicola[2], Benigno Elvira[1], Irene Parra[1], Ana Almodóvar[1]*

1 Department of Zoology and Physical Anthropology, Faculty of Biology, Complutense University of Madrid, Madrid, Spain, 2 Department of Environmental Sciences, University of Castilla-La Mancha, Toledo, Spain

Abstract

Anthropogenic environmental change is causing unprecedented rates of population extirpation and altering the setting of range limits for many species. Significant population declines may occur however before any reduction in range is observed. Determining and modelling the factors driving population size and trends is consequently critical to predict trajectories of change and future extinction risk. We tracked during 12 years 51 populations of a cold-water fish species (brown trout *Salmo trutta*) living along a temperature gradient at the warmest thermal edge of its range. We developed a carrying capacity model in which maximum population size is limited by physical habitat conditions and regulated through territoriality. We first tested whether population numbers were driven by carrying capacity dynamics and then targeted on establishing (1) the temperature thresholds beyond which population numbers switch from being physical habitat- to temperature-limited; and (2) the rate at which carrying capacity declines with temperature within limiting thermal ranges. Carrying capacity along with emergent density-dependent responses explained up to 76% of spatio-temporal density variability of juveniles and adults but only 50% of young-of-the-year's. By contrast, young-of-the-year trout were highly sensitive to thermal conditions, their performance declining with temperature at a higher rate than older life stages, and disruptions being triggered at lower temperature thresholds. Results suggest that limiting temperature effects were progressively stronger with increasing anthropogenic disturbance. There was however a critical threshold, matching the incipient thermal limit for survival, beyond which realized density was always below potential numbers irrespective of disturbance intensity. We additionally found a lower threshold, matching the thermal limit for feeding, beyond which even unaltered populations declined. We predict that most of our study populations may become extinct by 2100, depicting the gloomy fate of thermally-sensitive species occurring at thermal range margins under limited potential for adaptation and dispersal.

Editor: David L. Roberts, University of Kent, United Kingdom

Funding: This study was funded by the Government of Navarra and the Spanish Ministry of Science and Innovation through the research project CGL2008-04257/BOS. I. Parra was funded by a postgraduate contract from the Government of Madrid and the European Social Fund (ESF). Fisheries staff of the Regional Service of Navarra (Wildlife Service) carried out the fish surveys. The funders had not any other role in study design, data collection and analysis, decision to publish, or preparation of the manuscript.

Competing Interests: The authors have declared that no competing interests exist.

* E-mail: aalmodovar@bio.ucm.es

Introduction

Natural and anthropogenic disturbances are impacting global ecological systems and causing elevated rates of population extirpation, so that there is increasing concern that the rate of environmental change may exceed the capacity of populations to persist and maintain their range [1]. A population's extinction risk, persistence time and duration of its final decline to extinction, as well as the probability of evolutionary rescue, strongly depend on initial numbers and population size variability [2–4]. In many systems, imminent extinction will be signalled early by a decreasing rate of recovery from small perturbations. This critical slowing down is typically characterized by an increase in variance or autocorrelation of fluctuations of the system as a tipping point is approached [5–6]. In highly stochastic systems, critical transitions will on the contrary happen far from local tipping points and an increasing variability will reflect the shift to a contrasting regime [7]. Improving wildlife's conservation and management requires therefore a deep comprehension of not only spatial patterns in local species abundance but also the way and rate a population's size changes through time - its trajectory. Since dynamics are driven by the interplay of density-dependent and density-independent aspects of the environment, determining how the strength of density dependence varies with environmental variance remains critical for predicting near-term population trajectories [8–9]; the heart of the matter is then, what limits and regulates the size of natural populations in a fluctuating world?

Theoretically, there is a limit to the maximum number of individuals that can be supported by a system over a period of time for a particular level of resources (i.e., the environment's carrying capacity); and most important, population growth must decrease as the population approaches that limit at a rate dependent on the functional form of density dependence operating on the system [10]. This latter notion has been factually the cornerstone of the management of wildlife populations subject to commercial exploitation (see [11–12]). In conservationist settings, the probability of extinction and the persistence time of a population are a function of the environment's carrying capacity and the amplitude of its fluctuations along time (e.g., [13]). Carrying capacity is however typically set as a static parameter in predictive population dynamics models notwithstanding the fact that levels of resources naturally change through time, and that these changes will be amplified by climate change in many regions. Mechanistic behavioural process-based models provide a useful alternative to simulate carrying capacity dynamics under changing conditions across multiple spatio-temporal scales, contexts and taxa (e.g., [14–16]). Yet most models do not account for social interactions even though the carrying capacity of an environment is greatly determined by how individuals compete over the available resources. This is especially relevant for territorial species because behavioural responses induced by aggressive interactions typically result in reduced exploitation of the limited resource, so that the population stability-enhancing effects of territoriality are paid-off by decreased carrying capacity [17].

In addition to the resources that set the carrying capacity, which are dynamically consumed and may be hence the object of competition, there are scenopoetic variables that are not dynamically affected by the presence of a species but may limit the species' final performance in the other way round. As such, temperature is a primary driver of species' distribution and numbers over the long-term, especially in ectotherm organisms, as their fundamental niche is physiologically bounded by their thermal niche space [18]. Within the temperature range in which survival occurs, there are a series of decreasing ranges for different functions (e.g., feeding, growth, reproduction) so that outside their limits population performance declines (e.g., [19]). Therefore, increasing temperatures may first constrain the carrying capacity of a system for a particular species to a lower thermal capacity and ultimately drive that organism outside its tolerance window.

Alterations in the realized thermal niche resulting from on-going anthropogenic global warming is in fact the underlying cause of the rapid range shifts [20], local and worldwide extinctions [21], and population declines [22] observed in species from a wide variety of taxa. Ominously, range sizes and population numbers of thermally-sensitive species are projected to keep on shrinking along warmer margins (latitudinal or altitudinal), with particularly deleterious impacts on peripheral populations living at the most extreme margins of the species' realized climatic niche (e.g., [23–25]). There is also increasing evidence that the amplitude and probability distribution of environmental variability is changing in response to anthropogenic impacts [9], with the intensification of weather and climate extremes linked to anthropogenic climate change at the far-end of this spectrum (see [26]). This trend can have a substantial influence on population extinction dynamics since increased environmental variability can alter a population's vital rates in several interrelated ways [9]. Understanding then the way and to what extent the internal dynamics of a system responds to temperature fluctuations over time is critical for predicting trajectories of change under future scenarios.

In this article, we address how the carrying capacity and the thermal capacity of the system act and interact to drive spatial patterns and temporal fluctuations in population abundance of thermally-sensitive species. For this purpose, we tracked during 12 years 51 populations of a cold-water fish species, brown trout *Salmo trutta*, living along a temperature gradient at the warmest thermal edge of its range. In this study, we develop a carrying capacity model in which maximum population size is limited by environmentally-driven physical habitat conditions and regulated through habitat selection and territorial behaviour. We test whether the spatial and temporal variations in the numbers of young-of-the-year, juvenile and adult brown trout (1) are

explained by modelled carrying capacities, and (2) are disrupted by thermal conditions. If so, we target on establishing (3) the temperature range within which the thermal capacity of the system is lower than its carrying capacity, and (4) the rate at which carrying capacity declines with temperature within that limiting thermal range.

Materials and Methods

Study populations

The study area was situated in the Iberian peninsula between latitudes 42°29′ and 43°16′N and longitudes 0°43′ and 2°20′W. Brown trout population trajectory was analysed in 37 study sites located in 22 rivers from three major basins (Aragón, Arga and Ega river basins) belonging to the Ebro river basin, a Mediterranean drainage; 14 sites located in 12 Atlantic rivers from the Bay of Biscay drainage were additionally studied. Sampling sites corresponded to first- to fifth-order rivers and were located at an altitude ranging from 40 to 895 m above mean sea level (a.s.l.). Selected sites were chosen to (1) cover the existing variability of environmental (physical habitat, flow, water temperature) conditions, and (2) represent an anthropogenic multiple-stressor gradient within the study area. Location and environmental and physical features of sampling sites can be checked elsewhere (e.g., [27–29]). Brown trout is the dominant fish species throughout the area, and its populations only consist of stream-dwelling individuals.

Brown trout populations were sampled by electrofishing with multiple successive passes every summer from 1993 to 2004. Prior to sampling, each site was blocked upstream and downstream with nets. Trout were placed into holding boxes and were anaesthetised with tricaine methane-sulphonate (MS-222) to both facilitate their manipulation and minimize physiological stress. Individuals were measured (fork length, to the nearest mm) and weighed (to the nearest g), and scales were taken for age determination. Scales were removed from the area between posterior edge of dorsal fin and the lateral line, approximately two scale rows above the lateral line on the left side of the fish. Scales were removed by gently scraping against the grain of the scales with the blade of a clean scalpel or knife. After sampling routines, trout were placed into different holding boxes to recover, being immediately released back into the river after recovery. A grand total of 159,563 individuals were sampled during the study. Population density was used as a measure of population size. Fish density (trout ha^{-1}) with variance was estimated separately for each sampling site by applying the maximum likelihood method [30] and the corresponding solution proposed by Seber [31] for three removals assuming constant-capture effort. Population estimates were carried out separately for each year class.

Ethics statement

The described field study, including electrofishing and all sampling routines, was approved by the Wildlife Regional Service of Navarra (Department of Rural Development and the Environment and Local Administration of the Government of Navarra) accordingly to the current legislation (Ley Foral 2/1993). The study did not require any ethical approval from the corresponding Ethics Committee on Animal Experimentation (Ley Foral 2/1993, article 9; Real Decreto 1201/2005, articles 2–3). Fish surveys were carried out by experienced fisheries staff of the Wildlife Regional Service of Navarra and all sampling procedures complied with the Spanish and European Union legislation on animal welfare. The fish were handled with great care throughout this study to minimize any negative effects. This includes electrofishing and sampling routines such as weighing and length measuring and scale collection. After sampling routines, all fish were returned alive into the river. The study did not involve field sampling during the emergence or spawning critical periods, when trout are more susceptible to undergo potential physiological or behavioural disruptions. The described field studies did not involve endangered or protected species.

Carrying capacity modelling

Carrying capacity dynamics was modelled following the rationale and methodology described in Ayllón et al. [32]. We define carrying capacity as the maximum density of fish a river can naturally support during the period of minimum available physical habitat. In our model, the quantity of suitable physical habitat available for fish of a given age is estimated as a function of stream discharge using physical habitat simulations, and the maximum number of fish that can be sustained is estimated as the area of suitable physical habitat divided by the expected individual territory area for the given aged cohort.

Dynamics of stream physical habitat was modelled by means of the Physical Habitat Simulation system (PHABSIM; [33]). PHABSIM simulations determine the potentially available physical habitat for an aquatic species and its life stages as a function of discharge by coupling a hydraulic model with a Habitat Suitability Model (HSM). The longitudinal distribution of habitat types within the stream is described through transects positioned perpendicular to the channel. Along each transect, measurements of physical habitat variables are made at regular intervals to describe their cross-sectional distributions. As a result, the study site is depicted as a mosaic of cells characterized by their area, structure (substrate and cover) and hydraulics (water depth and velocity), which are a function of discharge [34]. For this work, topographic, hydraulic and channel structure data needed to carry out the physical habitat simulations were collected at each study site during the summer of 2004 following survey methods extensively described for e.g. in Parra et al. [29]. Hydraulic conditions were simulated following procedures set out in Waddle [34]. Finally, the suitability of channel structure and simulated hydraulic conditions for fish is assessed by means of the HSM. In this study, we used the reach-type specific habitat selection curves developed for young-of-the-year (YOY, 0+), juvenile (1+) and adult (>1+) brown trout by Ayllón et al. [27]. Habitat selection represents habitat preference under the prevailing biotic and abiotic conditions in any particular system, so these curves can be seen as operational applications of the realized ecological niche. The curves that relates the Weighted Usable Area (WUA; m^2 WUA ha^{-1}, an index combining quality and quantity of available physical habitat) for each life stage with stream discharge are the final outputs of PHABSIM simulations.

Importantly, we modelled spatial segregation of cohorts to avoid an overestimation of available physical suitable habitat. Since habitat selection patterns overlap among life stages up to a certain degree (see [27]), there is a potential for intercohort competition in some areas of the stream. This results in PHABSIM cells where one life stage has a higher composite suitability index than other life stage, and other cells where the converse holds. We considered that younger life stages would not occupy the shared cells where they are dominated (have less favourable habitat conditions) by older ones, so that this WUA was not added to their total available physical habitat.

Discharge time series for the study period (1993–2004) were obtained at each site from the closest gauging stations. Then, summer (July-September) physical habitat time series for each life stage were calculated by coupling WUA-discharge curves with discharge time series. Mean summer WUA was calculated as the daily average for each life stage and year. Finally, physical habitat time series were transformed to carrying capacity time series by means of an allometric territory size relationship specifically developed for brown trout [35]: $Log_{10} T = (2.64 - 0.96 \cdot age\ category) \cdot Log_{10} L - (2.72 - 0.90 \cdot age\ category)$, where T (m^2 of WUA) is territory size, L (cm) is length and $age\ category$ is 0 for YOY trout or 1 for juvenile and adult trout. Carrying capacity was estimated for every age class (0+, 1+ and >1+), year and site through the following ratio: $K_i = WUA_i / T_i$, where K_i is the carrying capacity of age-class i (trout ha^{-1}), WUA_i is the mean summer WUA of age class i ($m^2 ha^{-1}$) and T_i is the area of the territory used by an individual of average body size of age-class i (m^2 $trout^{-1}$).

Water temperature modelling

We used the maximum mean water temperature during 7 consecutive days from July to September ($T_{max7d-water}$, °C) to study potential limiting effects of physiological stress on brown trout performance. Seven days is the usual standard to estimate thermal tolerance of fish to short-term exposure (e.g., [19]). We developed a regional spatial model to predict extreme water temperatures in the study area during the study period (1993–2004). Since water temperature data were not available, they were estimated from air temperature data. At a first stage, we built a regional air temperature model by regressing annual maximum mean air temperature during seven consecutive days ($T_{max7d-air}$) to latitude and altitude for 48 meteorological stations located at altitudes ranging from 38 to 1344 m a.s.l. Year was included as a random factor to induce an autocorrelation structure among data within the same year and account for yearly differences in the relationship among variables. $T_{max7d-air}$ was significantly related to latitude and altitude following the model: $T_{max7d-air}$ (°C) $= 323.25 - 6.914 \cdot Latitude$ (decimal degree) $- 0.0044 \cdot Altitude$ (m) ($R^2 = 0.85$, $P<0.0001$). At a second stage, we fitted a linear mixed-effects regression model relating $T_{max7d-water}$ to $T_{max7d-air}$ with river basin as a random factor. To do this, water temperature was recorded daily at study sites by means of data-loggers installed from June of 2004 to November of 2005. We employed $T_{max7d-air}$ as the independent variable since weekly averages of stream temperature and air temperatures are typically better correlated with each other than are daily values (e.g., [36]). The resulting model was highly significant ($R^2 = 0.85$, $P<0.0001$) and the within-basin fitted lines were specified by $T_{max7d-water} = 3.372+0.656 \cdot T_{max7d-air}$, $4.688+0.589 \cdot T_{max7d-air}$, $4.171+0.626 \cdot T_{max7d-air}$ for Aragón, Arga-Ega and Bay of Biscay basins, respectively.

Data analyses

Effects of carrying capacity dynamics and competition on population size. We tested whether spatio-temporal variations in the number of individuals of a life stage (YOY, juvenile and adult) were driven by carrying capacity dynamics, levels of crowdedness (i.e., carrying capacity saturation) experienced by these individuals the previous year, and levels of crowdedness experienced by individuals of accompanying life stages. The level of carrying capacity saturation was measured as the relationship between observed density and estimated carrying capacity (D/K ratio) and was used as a proxy for intensity of competition among individuals. We also examined the effects of previous inter-seasonal and inter-annual limiting physical habitat bottlenecks on the performance of a life stage: (1) we used the average discharge (Q_{cm}) and the maximum mean discharge during 7 consecutive days (Q_{max7d}) during emergence time (March-April) as proxies of physical habitat availability during this critical period. Both metrics were standardized by dividing by the historical median daily discharge to make them comparable among rivers significantly differing in discharge magnitude; (2) we used the relative carrying capacity ratio between two consecutive life stages to test whether the relative proportion of habitats available for a cohort along its ontogeny limits its performance. Density of life stage x at year i ($D_{x,i}$), as response variable, was therefore regressed against predictors listed on Table 1. In the case of YOY trout, $D_{0+,i}$ was regressed against the level of carrying capacity saturation experienced by adult trout the previous year and the relative ratio between recruitment and adult stock carrying capacity. We fitted our regression models through the Random Forest algorithm (RF, [37]) implemented in the "randomForest" package [38] within the R environment [39].

RF is a member of Regression Tree Analyses (RTA; [40]). RTA recursively partitions observations of the response variable into successive binary splits, each split being based on the value of a single predictor chosen through an exhaustive search procedure across all available predictors to minimize the unexplained variance of the response while maximizing the differences between the offspring branches. RF models increase prediction accuracy compared to traditional RTA by introducing random variation by growing each tree with a bootstrap sample of the training data and only using a small random sample of the predictors to define the split at each node. In outline, ntree bootstrap samples are randomly drawn with replacement from the training data, each containing 2/3 of the data (in-bag). Then, the RF algorithm searches the best split from a random subset of predictors (mtry variables from the whole set of variables) to construct the decision tree. Independent predictions (i.e., independent of the model-fitting procedure) for each tree are then made for the other 1/3 of the data that were excluded from the bootstrap sample (out-of-bag, or OOB). These predictions are averaged over all trees and the prediction error (OOB error) provides an estimate of the generalization error. Here, we first chose the optimal values of ntree and mtry that minimize the OOB error and then we proceeded to develop the RF model.

We employed RF models because they are free from distributional assumptions and automatically fit non-linear relationships and high-order interactions between predictors. Furthermore, as the number of trees increases, the generalization error always converges, so RF models cannot be over-fitted. Finally, as the OOB error is an unbiased estimate of the generalization error, it is not necessary to test the predictive ability of the model on an external data set [37]. The structure of the RF models can be examined using importance measures and partial dependence plots. Predictor Importance was assessed based on how much worse the OOB predictions can be if the values for that predictor are permuted randomly. The increase in mean of the error of a tree (mean square error, MSE) was used to measure the resulting deterioration of the predictive ability of the model after data permutation. Increase of MSE was computed for each tree and averaged over the forest (ntree trees). In addition, partial plots show the marginal effect of analyzed environmental variables in RF estimates of population size.

Effects of water temperature on population size. We deployed quantile regression (QR, [41]) to describe the limiting effect of water temperature on population size. We used this method because, contrarily to most regression analyses which focus exclusively on changes in the mean response, QR estimates multiple rates of change (slopes) in responses with unequal variation, so that it is especially suited to detect changes in heterogeneous distributions where other influencing factors are unmeasured and unaccounted for [42]. Importantly, QR allows the estimation of the rates of change near the upper and lower edges of responses, the parts of the distribution where limiting effects are typically detected. Therefore, we performed bootstrapped (1000 repetitions) QR estimates of quantiles using the "quantreg" package [43] within the R environment. We used the log-transformation of maximum mean water temperature during 7 consecutive days ($T_{max7d\text{-}water}$) as the independent predictor of the residuals from the previously obtained random forest models [expressed as the log(x+1)-transformation of (observed density-predicted density)/predicted density], the response variable.

We additionally assessed the effects of water temperature on the temporal fluctuations of density within each sampling site. To do this, we fitted linear mixed effects models with the "lme4" package in R [44], using the same predictor and response variable and including site as a random factor (random intercept and slope) to induce a correlation structure between observations within the same site.

Finally, based on the calculated predictive water temperature regional model, we mapped $T_{max7d\text{-}water}$ for the average climate conditions during the study period using ArcGis 9.2 software (ESRI Inc., Redlands, CA, USA). We implemented subsequently the linear mixed effects models previously obtained to map the spatially-explicit distribution of average population thermal carrying capacity across the region during the study period. We eventually projected the amount of thermal suitable habitat ($T_{max7d\text{-}water}$ equal or below 19.4°C; see [24]) and the thermal carrying capacity under warming scenarios based on the air temperature regional projections for the B2 SRES emission scenario presented by Brunet et al. [45].

Table 1. Predictors of YOY, juvenile and adult brown trout density used in Random Forest (RF) regression models.

Generic predictor	Predictors	Description
$K_{x,i}$	K_{yoy}, K_{juv}, K_{adu}	Carrying capacity of life stage x at year i
$D_{x-1,i-1}/K_{x-1,i-1}$	Past D/K_{yoy}, D/K_{juv}, D/K_{adu}	Level of carrying capacity saturation experienced by individuals of age x on year i-1 when they were age x-1
$D_{y,i}/K_{y,i}$	D/K_{yoy}, D/K_{juv}, D/K_{adu}	Level of carrying capacity saturation experienced by accompanying life stage y at year i
$K_{x,i}/K_{x-1,i-1}$	K_{yoy}/K_{adu}, K_{juv}/K_{yoy}, K_{adu}/K_{juv}	Relative carrying capacity ratio experienced by individuals of age x across years i-1 and i
$Q_{em,i}$	Q_{em}	Average discharge during emergence at year i
$Q_{max7d,i}$	Q_{max7d}	Maximum mean discharge during 7 consecutive days during emergence at year i

Results

Effects of carrying capacity dynamics and competition on population size

The out-of-bag estimates of the error rate (OOB error) were used to select the optimum Random Forest parameters ($mtry = 3$, $ntree = 600$ for all models). Compared to older life stages, RF performance was subordinate for YOY trout, in which the model only explained 50% ($P<0.001$) of the observed density variance (Fig. 1). The RF algorithm performed better for juvenile and adult life stages with the models explaining 75% and 76% ($P<0.001$) of total density variance. OOB predictions seemed to be in proper scale (regression slopes ranging from 0.96 to 1.05~1) with slight deviations from observed data (Fig. 1).

Carrying capacity (K) ranked first in importance for all life stages, its contribution to the prediction accuracy of the models being disproportionately higher than the rest of predictors (Fig. 2). Carrying capacity saturation experienced the previous year (past D/K ratio) was an important predictor of density for all life stages. It had considerable importance for juveniles, but appeared less important for YOY trout, whose density variations were strongly driven by carrying capacity (Fig. 2; note also the marked differences in the range of predicted density values across predictors shown in Figure 3). Density of a life stage increased with increasing past D/K ratio up to a threshold where further increases in D/K ratio have either deleterious or no effects on density (Fig. 3). YOY and juvenile trout interacted in an antagonistic manner as density of either life stage decreased with increasing D/K ratio of the other one (Fig. 3). Intercohort interactions did not contribute noticeably to adult trout density though, since neither D/K_{yoy} nor D/K_{juv} ranked among its three most important predictors (Fig. 2). By contrast, the relative ratio between adult and juvenile K was a top-three determinant of adult density (Fig. 2), with maximum performance at a relative ratio close to one and a sharp drop at values greater than two (Fig. 3). Interestingly, Q_{max7d} was only important for YOY trout (Fig. 2), density falling sharply when Q_{max7d} exceeded the historical median daily discharge and the highest negative effects being observed during strong flow events when Q_{max7d} values were over ten times this historical median daily discharge (plot not shown).

Effects of water temperature on population size

Regression quantiles for YOY trout were significant up to the $Q95$, negative deviation from RF model's predicted values increasing with increasing $T_{max7d\text{-}water}$ throughout the whole range of quantiles (Fig. 4A). Slope of the regression quantiles significantly differed across quantiles, lower quantiles having increasingly greater negative slopes (Fig. 4B). By contrast, regression quantiles for juveniles and adults were only significant up to the $Q75$ and regression slopes were not significantly different down to the $Q25$ where negative steepness of slopes significantly increased with lower quantiles (Fig. 4B). Regression slopes for YOY were significantly steeper for any quantile compared to juvenile and adult models (Fig. 4B). All statistical outputs from quantile regressions can be checked in Appendix S1. We observed the existence of a critical temperature threshold (CTT) beyond which no positive residuals existed, and this threshold increased with age, from 1.318 (20.8°C) for YOY to 1.330 (21.4°C) for adult trout (Fig. 4A). Further analyses of the residuals' distribution revealed that most of the data linked to the lowest quantiles ($Q5$–$Q25$) belonged to the sampling sites having also the lowest mean K (lower than the 25th percentile of the K distribution across the whole population of sites) (Fig. 4A). Finally, residuals from the most limiting quantile ($Q5$) were significantly related to anthropogenic disturbance metrics (Appendix S2).

Focusing on temporal trends, we observed that regression lines between temperature and YOY density in sites whose temperature range did not include values beyond the CTT could adopt different patterns, having either positive, negative or no slope (Fig. 5). Nevertheless, there was always a negative relationship between temperature and density in sites where temperature was over the CTT at least at one year during the study period. This pattern was accurately described through a piecewise linear mixed-effects regression model, comprising a non-significant line with a population slope of -1.12 ($N = 509$, $t = -0.68$, $P = 0.50$) and significant random variation across sites ($SD = 3.89$), plus a highly significant negative line with a population slope of -12.80 ($N = 509$, $t = -5.22$, $P<0.001$) allowing for high random variation around it ($SD = 9.21$) (Fig. 5). This pattern was less marked for juveniles and adult trout. The slope of the juvenile's regression model after the breakpoint was not significant (slope $= -1.28$, $SD = 9.17$; $N = 509$, $t = -1.28$, $P = 0.20$), while adult's one was in the boundary of significance (slope $= -2.94$, $SD = 2.39$; $N = 509$, $t = -2.11$, $P = 0.035$) (Fig. 5). Interestingly, the breakpoint was almost the same for the three models (around 19.4°C). When aggregating all life stages together (population model), total population density was not significantly related to temperature (slope $= 1.27$, $SD = 2.87$; $N = 509$, $t = 1.39$, $P = 0.16$) up to 19.42°C when it significantly declined at a rate of -4.11 ($N = 509$, $t = -2.51$, $P = 0.016$) with high random variation in regression slopes across sites ($SD = 6.86$).

Spatial simulations based on the piecewise regression population model and temperatures averaged for the study period showed that the thermal capacity of the environment was permanently lower than its habitat capacity (up to a 36%) in a significant area of the study region (Fig. 6). Projected suitable thermal habitat will decrease down to 7% of total study area by the year 2100 under the B2 SRES emission scenario. By that time and emission scenario, population thermal capacity will be on average 39.9 ± 10.0% (range 1.5–68.1) lower than its habitat capacity, but YOY maximum potential density will be reduced on average a 61.3 ± 19.1% (range 2.4–93.1).

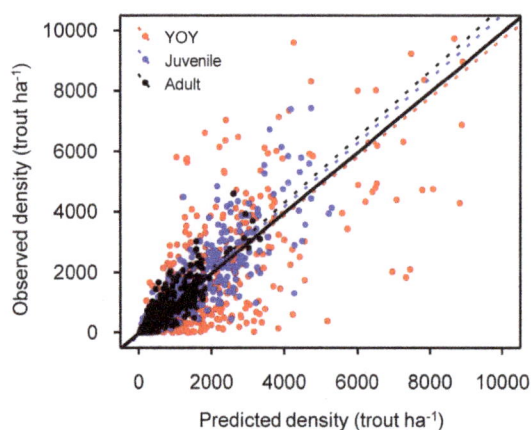

Figure 1. Random Forest performance on density of each life stage. Observed vs. predicted density of YOY, juvenile and adult brown trout. Dotted lines represent fitted linear models (YOY: $y = 13.15 + 0.96 \cdot x$, $R^2 = 0.50$, $P<0.001$; Juvenile: $y = -83.37 + 1.04 \cdot x$, $R^2 = 0.75$, $P<0.001$; Adult: $y = -56.56 + 1.06 \cdot x$, $R^2 = 0.76$, $P<0.001$) while solid line shows perfect match between observed and predicted.

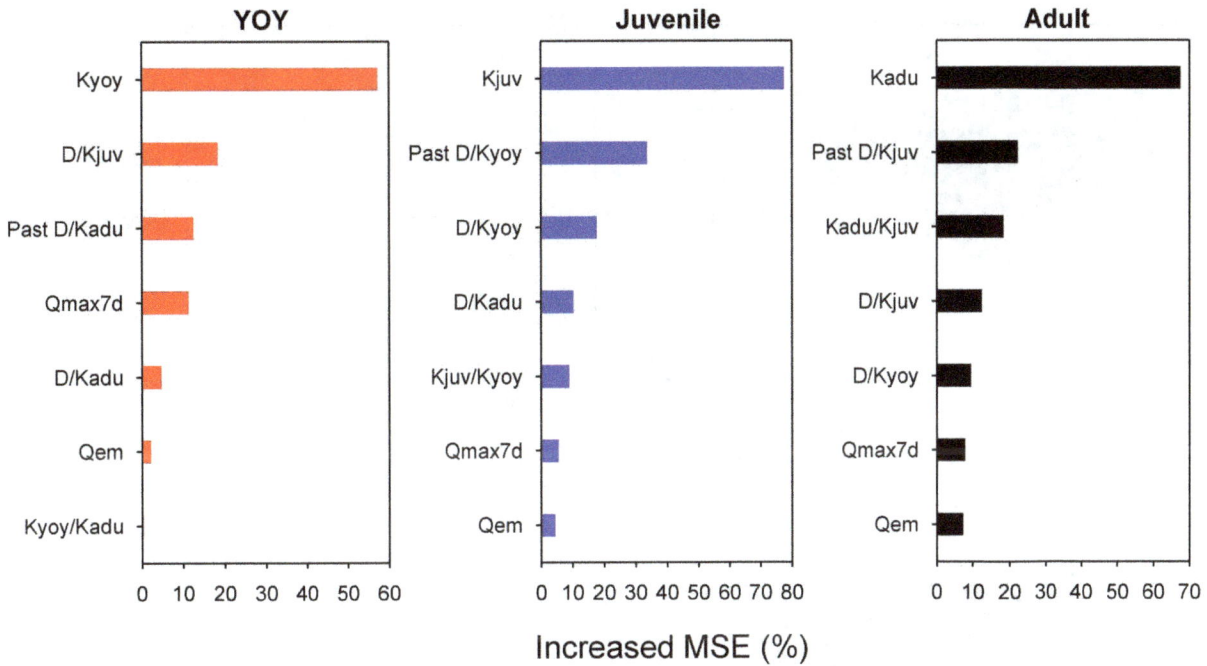

Figure 2. Predictor importance on density of each life stage. The plots show predictor importance measured as the increased mean square error (%), which represents the deterioration of the predictive ability of the model when the data of a predictor are randomly permuted. Higher Increased MSE indicates greater predictor importance. Note that axes are not constant across plots.

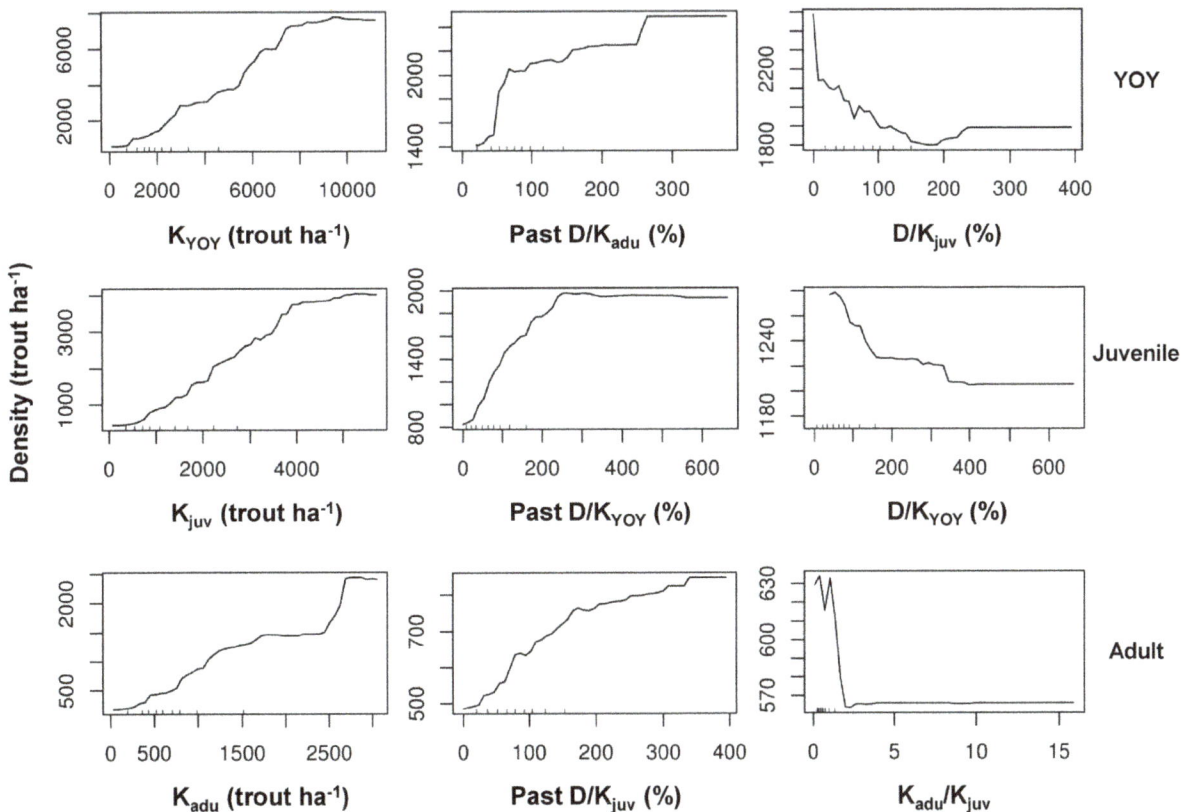

Figure 3. Marginal contribution of most important predictors on density of each life stage. Partial plots representing the marginal contribution of the three most important predictors in the RF models to density of each life stage while averaging out the effect of all the other predictors. Note that in a partial plot of marginal effects, only the range of values (and not the absolute values) can be compared between plots of different predictors.

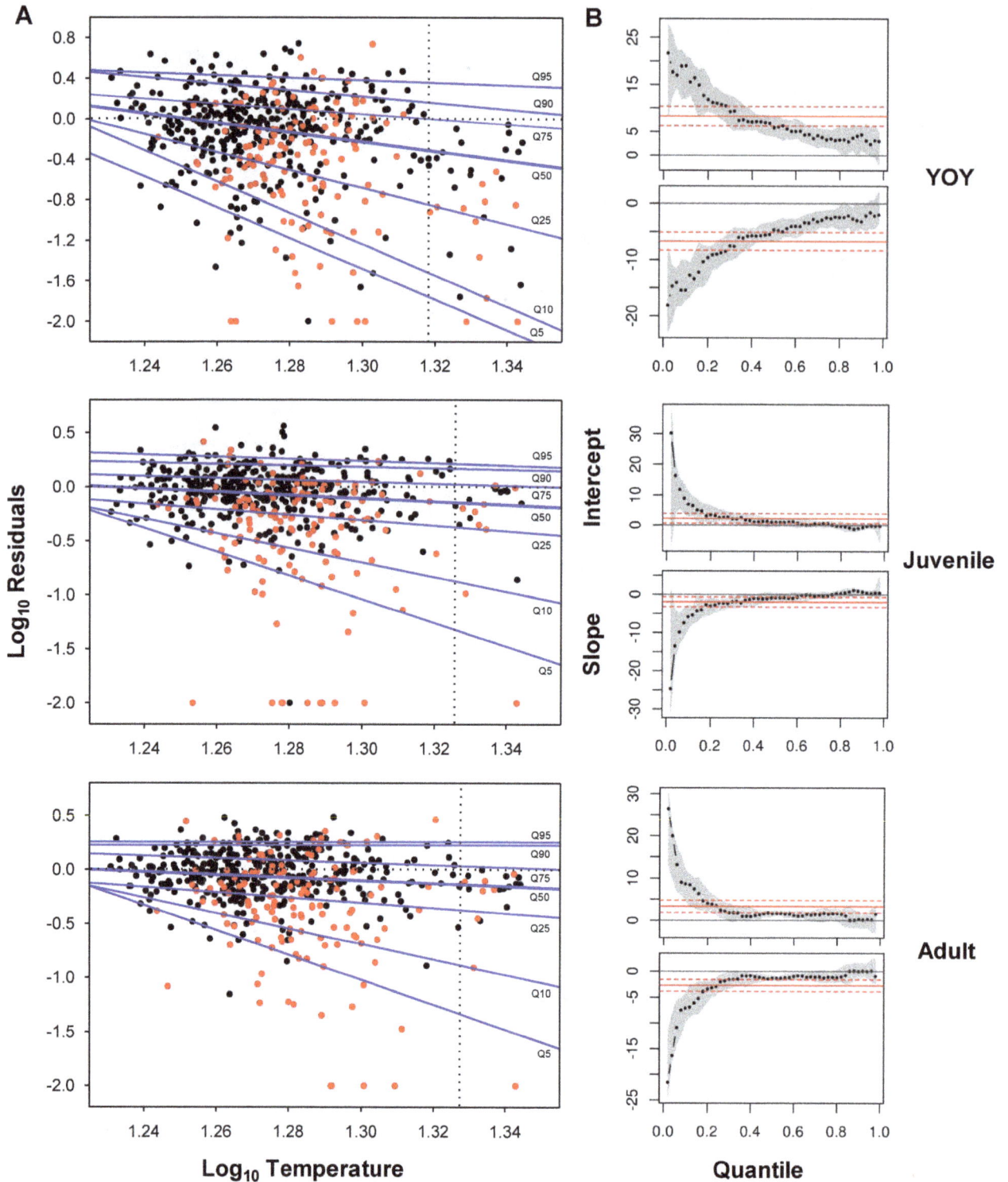

Figure 4. Limiting effects of water temperature on density of each life stage. (A) Quantile Regression (QR) estimates of the 5th, 10th, 25th, 50th, 75th, 90th and 95th quantiles (*Q5, Q10, Q25, Q50, Q75, Q90* and *Q95*) of log(x+1)-transformed residuals from RF models vs. log-transformed maximum mean water temperature during 7 consecutive days. Horizontal dotted lines show perfect match between observed and predicted density from RF models. Vertical dotted lines indicate the critical temperature threshold (CTT) beyond which no positive residuals exist. Red data belong to sampling sites having the lowest mean carrying capacity (below the 25th percentile of the whole distribution); (B) Intercept and slope coefficient estimates with associated confidence intervals for QR across varying quantiles. Mean and confidence interval of the mean are represented in red.

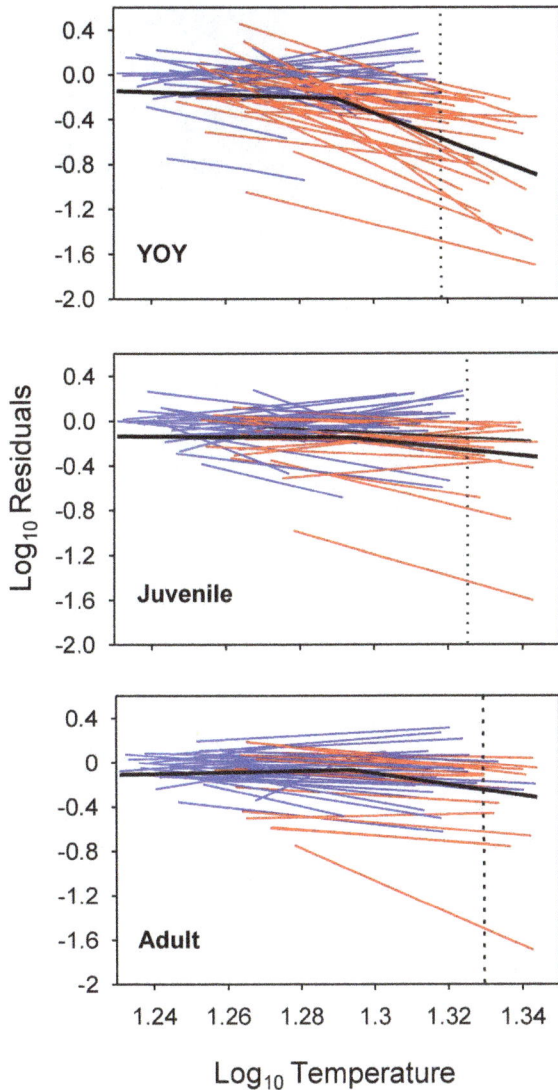

Figure 5. Regression lines for site-specific water temperature vs. density temporal relationships. Red/blue lines are fitted models for sites whose temperature range did include/not include values over the critical temperature threshold (CTT; vertical dotted lines). Piecewise linear mixed-effects regression models for the whole population of sampling sites are shown in black.

Discussion

Recent human-induced species' extinction rates are overwhelmingly greater than at any other time in human history [46] and number of species at the verge of imminent extinction is also increasing at an unparalleled speed [47]; meanwhile, current rates of population extirpation are at least three orders of magnitude higher than species extinction rates [48]. This latter is made evident by the fact that species are shifting their ranges two or three times faster than previously reported [20], especially freshwater fish, which may be responding to global warming at higher rates than terrestrial organism [49]. However, significant population declines of species of high conservation concern may occur before any reduction in range is observed, so that determining and modelling the factors driving population size and trends is crucial to predict their future extinction risk [50]. In

our study, distribution and dynamics of carrying capacity along with emergent density-dependent responses explained up to 76% of spatio-temporal density variability of juvenile and adult brown trout, but only 50% of YOY's. By contrast, YOY trout were highly sensitive to thermal conditions, their performance declining with increasing temperature at a higher rate than older life stages, and disruptions being triggered at lower temperature thresholds.

Carrying capacity (K), primarily based on quantity and quality of available physical habitat, was the strongest and most consistent contributor to density of any life stage; by contrast, most of analyzed habitat and competition predictors just qualified final numbers within narrow ranges around set carrying capacity. This provides empirical support to the theoretical prediction that density-independent factors should predominate over density-dependent ones in setting population numbers when environmental conditions are harsh - such as the ones experienced in distributional margins - [51]. Not only present physical habitat conditions but also previous habitat bottlenecks limited density though. Strong flow events during emergence depressed summer recruitment. Such disturbance events drastically reduce the quantity and quality of suitable physical habitat, which results in high YOY mortality through both direct downstream displacement of subordinate individuals without shelter (e.g., [52]) and delayed carry-over effects on individuals occupying low-quality habitats that affect their performance in the following season (see [53]). We also observed that juvenile physical habitat can limit subsequent adult abundance. Halpern et al. [54] showed that, in stage-structured species, juvenile habitat availability limits adult abundance in a relatively small region of parameter space compared with the regions where recruitment and adult K are limiting. This notion appears to apply for our populations since limitations in adult abundance by juvenile physical habitat seemed to be critical only in populations with low adult K.

Physical habitat quality and quantity is also a resource that, by limiting K, clearly stimulates the operation of density dependence. Intracohort density dependence was the second most important and consistent density predictor, having a large effect on final numbers. The annual realized density of a cohort relative to its K increased with increasing level of K saturation experienced by the cohort the previous year (or by adult stock in the case of YOY). This is in accordance with many model systems which suggest that individuals are strongly affected by both current and past environments, even when the past environments may be in previous generations (reviewed by Benton et al. [8]). This intracohort response is non-linear so that beyond a saturation level further increases in cohort crowdedness have deleterious or no effects on cohort numbers next year. The saturation threshold is well over 100%, indicating that a large proportion of individuals may remain in the stream as non-territorial or floaters. This is consistent with the idea that most animal populations spend more time above than below carrying capacity since population regulation is generally the result of a concave relationship between a population's growth rate and its size [10] (but see [55]). The intracohort density dependence also implies that density disruptions can be transmitted through generations so that constant pressures (either natural or anthropogenic) over time on a population may substantially depress its growth rate, and thus its density at equilibrium, turning the population more prone to become extinct through stochastic events [3]. Furthermore, we found that YOY and juvenile densities were mutually affected by the level of crowdedness experienced by the competing cohort, suggesting a negative density-dependent regulation of each life stage over the other. In stage-structured populations, density-dependent interactions between life stages can affect population

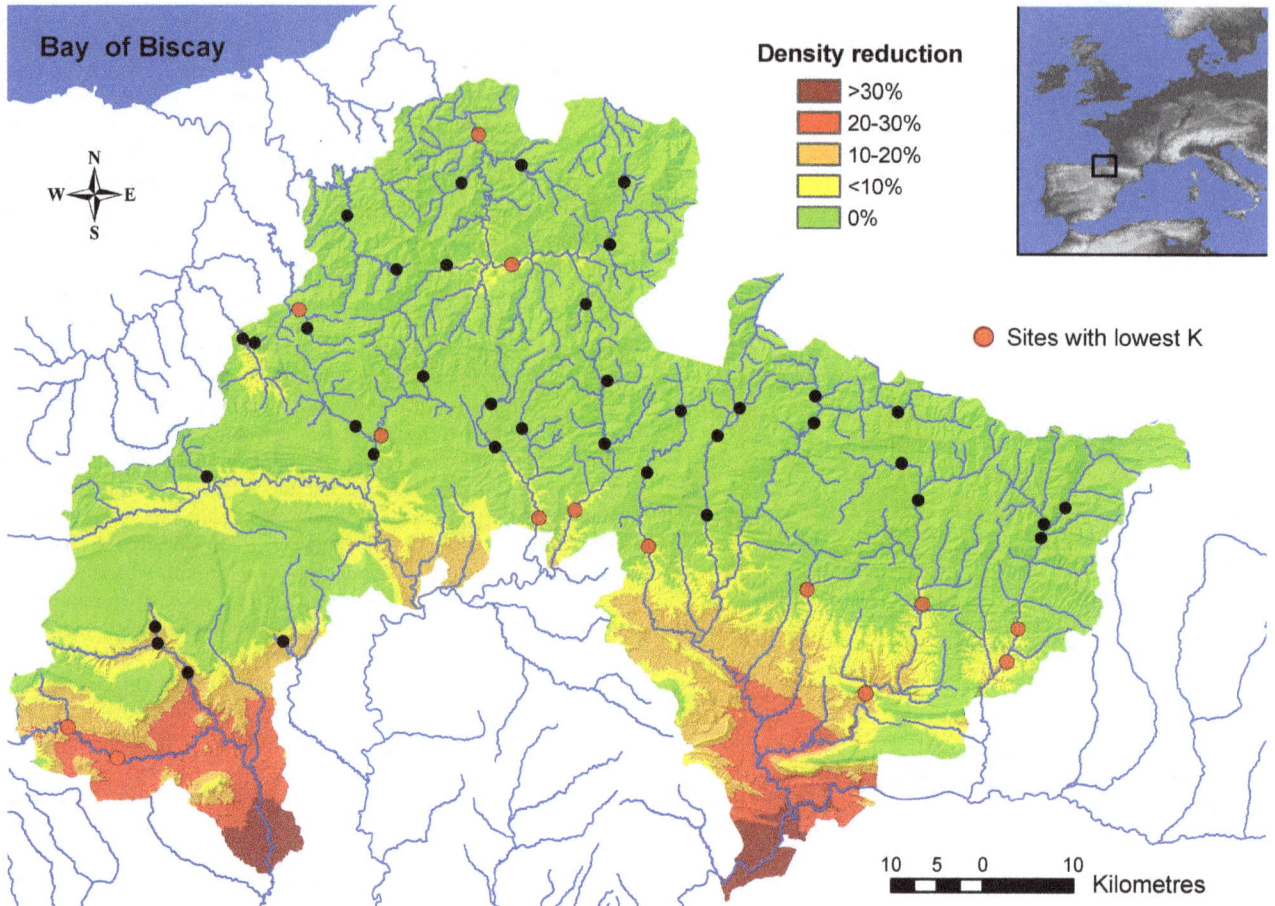

Figure 6. Spatial patterns of density reduction set by the thermal capacity of the environment. Spatially-explicit temperature-driven decrease in potential population numbers predicted by RF model. Effects of temperature on density are estimated from the piecewise linear mixed-effects regression models for the whole population of sampling sites. Dots represent sampling sites, while red dots represent those sites having the lowest mean carrying capacity (below the 25th percentile of the whole distribution: K<2700 trout ha^{-1}).

trajectories and lead to natural selection operating within populations and across life stages (see [56] and references therein). Density-dependent processes may interact in fact with density-independent factors (for e.g., K dynamics) in shaping adaptive landscapes, potentially leading to strong non-additivity in the development of vital rates driving population dynamics [57].

Distribution and abundance of species reflect their specific traits that allow them to pass through multiple environmental filters at hierarchical spatial scales, so that species lacking traits suitable for passing through a large scale filter are limited in abundance at all lower scales [58]. In our study, high summer temperatures restricted or reduced brown trout habitat use from certain areas of study basins where the physical microhabitat was otherwise suitable (see Fig. 6). Our quantile regression models showed that water temperature had a limiting effect on density, this limitation being significantly stronger for YOY trout. This was expected as small fish are more sensitive to temperature fluctuations than larger fish (see [19] for details of underlying mechanisms). Importantly, regression slopes significantly changed across quantiles, the steepest slopes being associated to the lowest quantiles. This means that rising temperatures had an increasingly higher negative effect on density performance as density departs from the maximum potential numbers set by K and density-dependent

dynamics. There is a gradual shift from physical habitat to temperature being the active environmental limiting factor.

The reasons of the shift could be two-fold. First, such changes in regression slopes indicate strong interactions of temperature with unmeasured factors while results reported in Appendix S2 reveal complex synergies among temperature and multiple anthropogenic drivers and stressors. The tight significant relationship between the density-carrying capacity ratio and levels of anthropogenic disturbance previously observed in most of the study populations [28] suggests that the degree of mismatch between densities observed and predicted from random forest models (RF) would be driven by disturbance intensity. In such a case, increasing disturbance would result in physical habitat conditions being no longer an active limiting factor so that density dynamics may get decoupled from K dynamics. On the contrary, the negative effects of increasing temperatures are stronger in populations already disrupted by anthropogenic stressors. Further, temperature impacts on density would be synergistically amplified by disturbance intensity since the predominant anthropogenic drivers in the study area (agricultural land uses and damming) typically imply both a local increase in water temperature and a decrease in energy inputs [59–61], which would affect accordingly fish energy budgets. Second, our data indicate that there is a critical temperature threshold (CTT) beyond which observed density is always lower than predicted by RFs irrespective of disturbance intensity. This CTT, beyond which the thermal capacity of the environment is always lower than its habitat capacity, decreases with age and roughly matches the incipient thermal limit for survival estimated for the different brown trout life stages (see [19]). This CTT is likely to diminish with increasing levels of anthropogenic disturbance, a clear example of how synergies among stressors form self-reinforcing mechanisms that hasten the dynamics of population extinction [2].

The analysis of temporal trends within sampling sites was consistent with such a picture. There is a thermal range within which there are strong spatial variations across populations in the functional relationship between temperature and density fluctuations, the sign of the relationship being dependent on anthropogenic disturbance intensity. However, there is a point beyond which density performance always decline with increasing temperature and at a faster rate than before. This pattern is especially patent in YOY trout, but less marked in older life stages. The breakpoint is however fairly constant across life stages (around 19.4°C), matching the upper thermal limit for feeding, where the starvation zone begins for brown trout (see [19]). This differentiation is important since it entails that population decline may start well below the CTT. In general, individuals without territories may survive either adopting a high-return/high-cost strategy, attempting to maximize energy intake at a cost of increasing interactions with territorial individuals, or a low-return/low-cost strategy, occupying poor feeding positions but minimizing energy costs by avoiding competition [62]. Within the starvation zone, the high-return/high-cost strategy rapidly fails and with increasing temperatures the low-return/low-cost strategy is no longer energetically feasible either. Over the CTT, mortality of individuals holding a territory in high-quality habitat patches could not be buffered by the floater population anymore, so that the population may turn unstable over time.

Two natural compensatory responses are possible against anthropogenic global warming. Given enough time and dispersal capabilities, species may shift to more favourable thermal environments, or they may track climate change through adaptation to avoid demographic collapse and extinction [21]. However, the probability of evolutionary rescue seems to be contingent on low rates of environmental deterioration [63] and there is no empirical evidence of thermal adaptation at the upper temperature limits for either survival or feeding in salmonids [19,64]. We have also provided evidence that anthropogenic disturbances may fasten the rate of population decline under warming, while the vast network of dams in our study basins (see [28]) would additionally prevent upward dispersal to find suitable thermal conditions. Based on our temperature models, we predict that the 93% of our study area would be thermally unsuitable for brown trout and the thermal capacity of the environment for recruitment could be on average 61% lower than its carrying capacity by 2100 under the ecologically friendly B2 SRES emission scenario. In that case, recruitment disruptions would have long-term amplifying downstream effects through density-dependent processes. This is dramatic as populations with the lowest K are located in areas where thermal constraints are predicted to be highest (Fig. 6), so that they are likely to become extinct well before 2100. It is worth noticing that this modelling exercise is somehow burdened by uncertainties inherent to both habitat suitability and climate envelope models (for e.g., see [65–66] for further discussion). Notwithstanding possible uncertainties, for marginal salmonid populations constrained to linear networks, temperature is destiny in a warming world [67]. Drastic reductions of distributional ranges are projected even in core areas [68]. By contrast, Piou and Prévost [69] model predicts that rising river temperatures alone should not lead open anadromous populations to extinction and that such river warming may even bumper the synergistic negative effects of flow regime alteration and ocean conditions deterioration on population persistence.

We acknowledge that this is an oversimplified picture as population trajectories of individual species cannot be scrutinized in isolation considering that climate change can alter multitrophic level interactions so strongly that entire food webs can undergo radical restructuring [70–71]. Warming should lead to a decrease in K and/or a decrease in the mean body mass (to buffer the potential decrease in K) if altered conditions cannot concurrently increase the supply rate of a species' feeding resources [72]. Nevertheless, notwithstanding that our predicted numbers are certainly uncertain, they roughly portray the gloomy fate of thermally-sensitive species occurring at contracting range margins under limited potential for adaptation and/or dispersal.

Acknowledgments

We appreciate the constructive comments provided by E. García-Berthou, M. Vickers and an anonymous reviewer, which considerably improved the quality of the manuscript.

Author Contributions

Conceived and designed the experiments: AA GGN. Performed the experiments: AA GGN BE DA IP. Analyzed the data: DA AA GGN BE. Contributed reagents/materials/analysis tools: AA BE. Wrote the paper: DA. Commented on the manuscript: AA GGN BE IP.

References

1. Bell G, Gonzalez A (2011) Adaptation and Evolutionary Rescue in Metapopulations Experiencing Environmental Deterioration. Science 332: 1327–1330.

2. Brook BW, Sodhi NS, Bradshaw CJA (2008) Synergies among extinction drivers under global change. Trends Ecol Evol 23: 453–460.

3. Griffen BD, Drake JM (2009) Scaling rules for the final decline to extinction. P Roy Soc B 276: 1361–1367.

4. Osmond MM, de Mazancourt C (2012) How competition affects evolutionary rescue. Phil Trans R Soc B 368, 20120085.

5. Drake JM, Griffen BD (2010) Early warning signals of extinction in deteriorating environments. Nature 467: 456–459.

6. Dai L, Vorselen D, Korolev KS, Gore J (2012) Generic indicators for loss of resilience before a tipping point leading to population collapse. Science 336: 1175–1177.

7. Scheffer M, Carpenter SR, Lenton TM, Bascompte J, Brock W, et al. (2012) Anticipating Critical Transitions. Science 338: 344–348.

8. Benton TG, Plaistow SJ, Coulson TN (2006) Complex population dynamics and complex causation: devils, details and demography. P Roy Soc B 273: 1173–1181.

9. Boyce MS, Haridas CV, Lee CT (2006) Demography in an increasingly variable world. Trends Ecol Evol 21: 141–148.

10. Sibly RM, Barker D, Denham MC, Hone J, Pagel M (2005) On the regulation of populations of mammals, birds, fish, and insects. Science 309: 607–610.

11. Sibly RM, Hone J, Clutton-Brock TH (2003) Wildlife Population Growth Rates. Cambridge: Cambridge University Press.

12. del Monte-Luna P, Brook BW, Zetina-Rejon MJ, Cruz-Escalona VH (2004) The carrying capacity of ecosystems. Global Ecol Biogeogr 13: 485–495.

13. Lande R, Engen S, Sæther BE (2003) Stochastic population dynamics in ecology and conservation. New York: Oxford University Press.
14. Goss-Custard JD, Stillman RA, West AD, Caldow RWG, McGrorty S (2002) Carrying capacity in overwintering migratory birds. Biol Conserv 105: 27–41.
15. Hayes JW, Hughes NF, Kelly LH (2007) Process-based modelling of invertebrate drift transport, net energy intake and reach carrying capacity for drift-feeding salmonids. Ecol Model 207: 171–188.
16. Morris DW, Mukherjee S (2007) Can we measure carrying capacity with foraging behavior? Ecology 88: 597–604.
17. López-Sepulcre A, Kokko H (2005) Territorial defense, territory size, and population regulation. Am Nat 166: 317–329.
18. Monahan WB (2009) A Mechanistic Niche Model for Measuring Species' Distributional Responses to Seasonal Temperature Gradients. PLoS ONE 4: e7921.
19. Elliott JM, Elliott JA (2010) Temperature requirements of Atlantic salmon Salmo salar, brown trout Salmo trutta and Arctic charr Salvelinus alpinus: predicting the effects of climate change. J Fish Biol 77: 1793–1817.
20. Chen IC, Hill JK, Ohlemueller R, Roy DB, Thomas CD (2011) Rapid Range Shifts of Species Associated with High Levels of Climate Warming. Science 333: 1024–1026.
21. Sinervo B, Méndez-de-la-Cruz F, Miles DB, Heulin B, Bastiaans E, et al. (2010) Erosion of Lizard Diversity by Climate Change and Altered Thermal Niches. Science 328: 894–899.
22. Parmesan C (2006) Ecological and Evolutionary Responses to Recent Climate Change. Annu Rev Ecol Evol Syst 37: 637–669.
23. Lassalle G, Rochard E (2009) Impact of twenty-first century climate change on diadromous fish spread over Europe, North Africa and the Middle East. Glob Change Biol 15: 1072–1089.
24. Almodóvar A, Nicola GG, Ayllón D, Elvira B (2012) Global warming threatens the persistence of Mediterranean brown trout. Glob Change Biol 18: 1549–1560.
25. Warren R, VanDerWal J, Price J, Welbergen JA, Atkinson I, et al. (2013) Quantifying the benefit of early climate change mitigation in avoiding biodiversity loss. Nature Clim Change 3: 678–682.
26. Jentsch A, Kreyling J, Beierkuhnlein C (2007) A new generation of climate-change experiments: events, not trends. Front Ecol Environ 5: 365–374.
27. Ayllón D, Almodóvar A, Nicola GG, Elvira B (2010) Ontogenetic and spatial variations in brown trout habitat selection. Ecol Freshw Fish 19: 420–432.
28. Ayllón D, Almodóvar A, Nicola GG, Parra I, Elvira B (2012) A new biological indicator to assess the ecological status of Mediterranean trout type streams. Ecol Indic 20: 295–303.
29. Parra I, Almodóvar A, Ayllón D, Nicola GG, Elvira B (2012) Unravelling the effects of water temperature and density dependence on the spatial variation of brown trout (Salmo trutta) body size. Can J Fish Aquat Sci 69: 821–832.
30. Zippin C (1956) An evaluation of the removal method of estimating animal populations. Biometrics 12: 163–189.
31. Seber GAF (1982) The estimation of animal abundance and related parameters. London: Charles Griffin Publications.
32. Ayllón D, Almodóvar A, Nicola GG, Parra I, Elvira B (2012) Modelling carrying capacity dynamics for the conservation and management of territorial salmonids. Fish Res 134–136: 95–103.
33. Milhous RT, Updike MA, Schneider DM (1989) Physical Habitat Simulation System Reference Manual-Version II. Instream Flow Information Paper 26. United States Fish and Wildlife Service, Fort Collins. Available: http://www.fort.usgs.gov/Products/Publications/3912/3912.pdf. Accessed 2013 Oct 23.
34. Waddle TJ (Ed) (2012) PHABSIM for Windows User's manual and exercises. U.S. Geological Survey Open-File Report 2001-340, Fort Collins. Available: http://www.fort.usgs.gov/Products/Publications/pub_abstract.asp?PubId=15000. Accessed 2013 Oct 23.
35. Ayllón D, Almodóvar A, Nicola GG, Elvira B (2010) Modelling brown trout spatial requirements through physical habitat simulations. River Res Appl 26: 1090–1102.
36. Morrill JC, Bales RC, Conklin MH (2005) Estimating stream temperature from air temperature: Implications for future water quality. J Environ Eng 131: 139–146.
37. Breiman L (2001) Random forests. Mach Learn 45: 5–32.
38. Liaw A, Wiener M (2002) Classification and regression by random forest. R News 2: 18–22.
39. R Development Core Team (2012) R: A Language and Environment for Statistical Computing. R Foundation for Statistical Computing: Vienna. Available: http://www.R-project.org/. Accessed 2013 Oct 23.
40. Breiman L, Friedman J, Olshen R, Stone C (1984) Classification and regression trees. Belmont: Wadsworth.
41. Koenker R, Bassett G (1978) Regression quantiles. Econometrica, 46: 33–50.
42. Cade BS, Noon BR (2003) A gentle introduction to quantile regression for ecologists. Front Ecol Environ 1: 412–420.
43. Koenker R (2012) quantreg: Quantile Regression. R package version 4.91. Available: http://cran.r-project.org/web/packages/quantreg/index.html. Accessed 2013 Oct 23.
44. Bates D, Maechler M, Bolker B (2012) lme4: Linear mixed-effects models using S4 classes. R package version 0.999999-0. Available: http://CRAN.R-project.org/package=lme4. Accessed 2013 Oct 23.
45. Brunet M, Casado MJ, de Castro M, et al. (2009) Regional climate change scenarios for Spain (in Spanish). Spanish Meteorological Agency. Available: http://www.aemet.es/es/elclima/cambio_climat/escenarios. Accessed 2013 Oct 23.
46. Pimm SL, Russell GJ. Gittleman JL, Brooks TM (1995) Science 269: 347–350.
47. Ricketts TH, Dinerstein E, Boucher T, Brooks TM, Butchart SHM, et al. (2005) Pinpointing and preventing imminent extinctions. Proc Natl Acad Sci USA 102: 18497–18501.
48. Hughes JB, Daily GC, Ehrlich PR (1997) Population diversity: its extent and extinction. Science 278: 689–692.
49. Comte L, Grenouillet G (2013) Do stream fish track climate change? Assessing distribution shifts in recent decades. Ecography. In press. doi: 10.1111/j.1600-0587.2013.00282.x
50. Renwick AR, Massimino D, Newson SE, Chamberlain DE, Pearce-Higgins JW, et al. (2012) Modelling changes in species' abundance in response to projected climate change. Divers Distrib 18: 121–132.
51. Haldane JBS (1953) Animal populations and their regulation. New Biologist 15: 9–24.
52. Nicola GG, Almodovar A, Elvira B (2009) Influence of hydrologic attributes on brown trout recruitment in low-latitude range margins. Oecologia 160: 515–524.
53. Harrison XA, Blount JD, Inger R, Norris DR, Bearhop S (2011) Carry-over effects as drivers of fitness differences in animals. J Anim Ecol 80: 4–18.
54. Halpern BS, Gaines SD, Warner RR (2005) Habitat size, recruitment, and longevity as factors limiting population size in stage-structured species. Am Nat 165: 82–94.
55. Clark F, Brook BW, Delean S, Reşit Akçakaya H, Bradshaw CJA (2012) The theta-logistic is unreliable for modelling most census data. Methods Ecol Evol 1: 253–262.
56. Samhouri JF, Steele MA, Forrester GE (2009) Inter-cohort competition drives density dependence and selective mortality in a marine fish. Ecology 90: 1009–1020.
57. Einum S, Robertsen G, Fleming IA (2008) Adaptive landscapes and density-dependent selection in declining salmonid populations: going beyond numerical responses to human disturbance. Evol Appl 1: 239–251.
58. Poff NL (1997) Landscape filters and species traits: Towards mechanistic understanding and prediction in stream ecology. J N Am Benthol Soc 16: 391–409.
59. Poff NL, Hart DD (2002) How dams vary and why it matters for the emerging science of dam removal. Bioscience 52: 659–668.
60. Allan JD (2004) Landscapes and riverscapes: The influence of land use on stream ecosystems. Annu Rev Ecol Evol Syst 35: 257–284.
61. Erös T, Gustafsson P, Greenberg LA, Bergman E (2012) Forest-Stream Linkages: Effects of Terrestrial Invertebrate Input and Light on Diet and Growth of Brown Trout (Salmo trutta) in a Boreal Forest Stream. PLoS ONE 7: e36462.
62. Puckett KJ, Dill LM (1985) The energetics of feeding territoriality in juvenile coho salmon (Oncorhyncus kisutch). Behaviour 92: 97–111.
63. Lindsey HA, Gallie J, Taylor S, Kerr B (2013) Evolutionary rescue from extinction is contingent on a lower rate of environmental change. Nature 494: 463–467.
64. Crozier LG, Hendry AP, Lawson PW, Quinn TP, Mantua NJ, et al. (2008) Potential responses to climate change in organisms with complex life histories: evolution and plasticity in Pacific salmon. Evol Appl 1: 252–270.
65. Hirzel AH, Le Lay G (2008) Habitat suitability modelling and niche theory. J Appl Ecol 45: 1372–1381.
66. Araújo MB, Pearson RG, Thuiller W, Erhard M (2005) Validation of species-climate impact models under climate change. Glob Change Biol 11: 1504–1513.
67. Isaak DJ, Rieman BE (2013) Stream isotherm shifts from climate change and implications for distributions of ectothermic organisms. Glob Change Biol 19: 742–751.
68. Filipe AF, Markovic D, Pletterbauer F, Tisseuil C, De Wever A, et al. (2013) Forecasting fish distribution along stream networks: brown trout (Salmo trutta) in Europe. Divers Distrib 19: 1059–1071.
69. Piou C, Prévost E (2013) Contrasting effects of climate change in continental vs. oceanic environments on population persistence and microevolution of Atlantic salmon. Glob Change Biol 19: 711–723.
70. Van der Putten WH, Macel M, Visser ME (2010) Predicting species distribution and abundance responses to climate change: why it is essential to include biotic interactions across trophic levels. Philos Trans R Soc B 365: 2025–2034.
71. Woodward G, Perkins DM, Brown LE (2010) Climate change and freshwater ecosystems: impacts across multiple levels of organization. Philos Trans R Soc B 365: 2093–2106.
72. Daufresne M, Lengfellner K, Sommer U (2009) Global warming benefits the small in aquatic ecosystems. Proc Natl Acad Sci USA 106: 12788–12793.

Similar Processes but Different Environmental Filters for Soil Bacterial and Fungal Community Composition Turnover on a Broad Spatial Scale

Nicolas Chemidlin Prévost-Bouré[1]*, **Samuel Dequiedt**[2], **Jean Thioulouse**[3], **Mélanie Lelièvre**[2],
Nicolas P. A. Saby[4], **Claudy Jolivet**[4], **Dominique Arrouays**[4], **Pierre Plassart**[2], **Philippe Lemanceau**[1],
Lionel Ranjard[1,2]

1 Unité Mixte de Recherche 1347 Agroécologie, Institut National de la Recherche Agronomique-AgroSup Dijon-Université de Bourgogne, Dijon, France, **2** Unité Mixte de Recherche 1347 Agroécologie-Plateforme GenoSol, Institut National de la Recherche Agronomique-AgroSup Dijon-Université de Bourgogne, Dijon, France, **3** Unité Mixte de Recherche 555 Laboratoire de Biométrie et Biologie Evolutive, Université Lyon 1-Centre National de la Recherche Scientifique, Villeurbanne, France, **4** Unité de Services 1106 InfoSol, Institut National de la Recherche Agronomique, Orléans, France

Abstract

Spatial scaling of microorganisms has been demonstrated over the last decade. However, the processes and environmental filters shaping soil microbial community structure on a broad spatial scale still need to be refined and ranked. Here, we compared bacterial and fungal community composition turnovers through a biogeographical approach on the same soil sampling design at a broad spatial scale (area range: 13300 to 31000 km^2): i) to examine their spatial structuring; ii) to investigate the relative importance of environmental selection and spatial autocorrelation in determining their community composition turnover; and iii) to identify and rank the relevant environmental filters and scales involved in their spatial variations. Molecular fingerprinting of soil bacterial and fungal communities was performed on 413 soils from four French regions of contrasting environmental heterogeneity (Landes<Burgundy≤Brittany<<South-East) using the systematic grid of French Soil Quality Monitoring Network to evaluate the communities' composition turnovers. The relative importance of processes and filters was assessed by distance-based redundancy analysis. This study demonstrates significant community composition turnover rates for soil bacteria and fungi, which were dependent on the region. Bacterial and fungal community composition turnovers were mainly driven by environmental selection explaining from 10% to 20% of community composition variations, but spatial variables also explained 3% to 9% of total variance. These variables highlighted significant spatial autocorrelation of both communities unexplained by the environmental variables measured and could partly be explained by dispersal limitations. Although the identified filters and their hierarchy were dependent on the region and organism, selection was systematically based on a common group of environmental variables: pH, trophic resources, texture and land use. Spatial autocorrelation was also important at coarse (80 to 120 km radius) and/or medium (40 to 65 km radius) spatial scales, suggesting dispersal limitations at these scales.

Editor: Maarja Öpik, University of Tartu, Estonia

Funding: This study was granted by ADEME (Energy and Environment Management Agency) and by the French National Research Agency (ANR). RMQS soil sampling and physico-chemical analyses were supported by a French Scientific Group of Interest on soils: the "GIS Sol", involving the French Ministry for Ecology and Sustainable Development (MEDAD), the French Ministry of Agriculture (MAP), the French Institute for Environment (IFEN), the French Agency for Energy and Environment (ADEME), the French Institute for Research and Development (IRD) and the National Institute for Agronomic Research (INRA). This work, through the involvement of technical facilities of the GenoSol platform of the infrastructure ANAEE378 Services, received a grant from the French state via the National Agency for Research under the program "Investments for the Future" (reference ANR-11-INBS-0001), as well as a grant from the Regional Council of Burgundy. The funders had no role in study design, data collection and analysis, decision to publish, or preparation of the manuscript.

* Email: n.chemidlin@agrosupdijon.fr

Introduction

For over two centuries, biogeographical studies have been carried out on macroorganisms and have provided a better understanding of species distribution, extinction and interactions [1–2]. For microorganisms, the first biogeographic postulate was developed by Baas Becking in 1934 [3]: "Everything is everywhere, *but*, the environment selects" suggesting that microbial "species" may be everywhere due to huge dispersal potentials, but that their abundances are constrained by contemporary environmental context, which may be especially true at broad spatial scales (spatial scales larger than 100 km^2 are considered as broad in this study). The number of studies in microbial biogeography has increased exponentially over the past decade thanks to new molecular tools applicable in routine on wide scale sampling networks constituted of several hundreds of samples [4–5]. These studies revealed that soil microorganisms are not strictly cosmopolitan since their distributions are systematically heterogeneous and structured into biogeographical patterns [2], [6–9].

One way to discriminate the spatial processing of microbial diversity is to evaluate either the Taxa-Area Relationship (TAR), *i.e.* the accumulation of new taxa with increasing sampling area, or the Distance-Decay Relationship (DDR), *i.e.* the rate of change in compositional similarity with increasing distance. [10–11]. Although significant TAR and DDR have recently been demonstrated for both soil fungal [12] and bacterial [8], [11], [13–14] communities, the relative importance of the ecological processes shaping these communities is still under debate. Therefore, it needs to be more deeply considered at the community level. According to Vellend [15], four processes are involved in shaping microbial community composition: selection, dispersal, ecological drift and speciation. Speciation is difficult to consider at the community level because the molecular markers used to discriminate microbial taxa mainly target highly conserved regions (*e.g.* ribosomal genes) with low mutation rates. The stochastic demographic processes underlying ecological drift are also difficult to consider since it remains a challenge to fully characterize demographic evolutions within complex microbial communities in environmental samples. Consequently, most biogeographical studies have focused on environmental selection and dispersal limitations, the later leading to a spatial autocorrelation between sites independently of environmental factors. Numerous studies have identified environmental selection as relevant in shaping soil bacterial community composition [8–9], [16–24]. Conversely, dispersal limitation is still under debate regarding the high dispersal potentials of microorganisms and because some environmental variables always remain unmeasured. Nevertheless, recent publications also suggest that bacteria may be dispersal limited [9], [21–23] or that part of soil bacterial communities is endemic [17]. As regards soil fungi, the relevance of environmental selection and dispersal limitations has been demonstrated at the community level and for ectomycorrhizal groups [16], [19], [22], [24–29]. Nevertheless, most of these studies were performed on different sampling designs with different molecular techniques. Only few studies have investigated such processes for both soil fungal and bacterial communities simultaneously to compare their biodiversity turnover. Most of them were performed in particular ecosystems and lead to diverging conclusions: Pasternak et al [18] concluded that bacterial and fungal communities were primarily shaped by environmental selection rather than dispersal limitations at the scale of the Israeli desert. On the contrary, Talbot et al. [27] highlighted a strong endemism for fungi in pine forests and Hovatter [22] suggested that the ecological processes shaping soil bacterial community could differ at a local scale due to the presence/absence of a particular plant. Altogether, this suggests that environmental heterogeneity may determine the relative importance of the ecological processes at work and therefore affect the distance-decay relationship for both soil bacteria and fungi [9–11], also suggested by other macrobial studies [30–31]. The comparison of different microbial communities along different levels of environmental heterogeneity may therefore help to reach a consensus.

Both selection and dispersal are based on the various ecological attributes of soil bacteria and fungi in terms of soil colonization, dispersal forms, trophic requirements, biological interactions and adaptation to environmental conditions, together with stochastic factors. Consequently, studies focusing particularly on soil bacteria or soil fungi have identified numerous environmental filters involved in shaping these particular communities but no consensus could be reached regarding microbial community as a whole on broad spatial scales. The filter most frequently identified for bacteria is soil pH [5], [17], [20], [23–24], [32–34] and it is commonly assumed that this is an important driver for fungal communities [18], [24]. Soil texture and carbon content have also been identified as important filters for bacteria [6], [22–23], [33–

34]. Similarly, the quality of soil organic matter, represented by the C:N ratio, and the amount of N were shown to have a significant effect on the abundance and composition of bacterial and fungal communities [24], [34–35]. These edaphic factors are often considered as the main determinants of bacterial diversity since the importance of climate may vary across biomes at a continental scale [5], [12], [16]. Land-use, agricultural practices, and plant community composition are also important filters for both bacteria and fungi on a wide scale [18], [19], [22–24], [36–38]. Therefore, this suggests that common filters determine the composition of both bacterial and fungal communities, but they still need to be ranked according to their relative importance to reach a consensus. This may be achieved by comparing bacterial and fungal communities over regions contrasted in terms of habitat heterogeneity but also with wide ranges of variations for the identified filters.

The objectives of this study were: i) to examine the spatial structuring of bacterial and fungal communities on a broad spatial scale; ii) to investigate the relative importance of environmental selection and spatial autocorrelation in determining the community composition turnover of these communities; and iii) to identify and rank the relevant environmental filters and scales involved in their spatial variations. To attain these objectives, four regions in the RMQS data set ("Réseau de Mesures de la Qualité des Sols" = French Monitoring Network for Soil Quality, recovering 2,200 soils over the whole of France) were selected along a gradient of environmental heterogeneity, representing a total of 413 soils. This gradient was chosen in order to confront the community composition turnover rates of bacterial and fungal communities to soil habitat heterogeneity [9]. Bacterial and fungal communities were characterized by Automated RISA fingerprinting of soil DNA. Community composition turnover (z) was estimated by means of a similarity DDR using an exponential model as suggested by Harte et al [39–40] for microorganisms. Together with this measure of community composition turnover over broad spatial scales, the initial similarity of communities was evaluated [30]. It represents the variability of community composition at finer spatial scales. High initial similarity corresponds to low local variability. The relative influence of environmental selection and spatial autocorrelation was investigated through a variance partitioning approach involving pedoclimatic characteristics and land-use and spatial variables (geographic coordinates and Principal Coordinates of Neighbour Matrices; PCNM), respectively.

Methods

Soil samples

Soil samples were provided by the Soil Genetic Resource Conservatory (platform GenoSol, http://www.dijon.inra.fr/plateforme_genosol, [41]) and obtained from the soil storage facility of the RMQS ("Réseau de Mesures de la Qualité des Sols" = French Monitoring Network for Soil Quality). The RMQS database consists of observations of soil properties on a 16-km regular grid across the 550000 km^2 French metropolitan territory and was designed to monitor soil properties [42]. The baseline survey consisting of 2,200 sites (each corresponding to a composite soil sample constituted of 25 soil cores) was completed in 2009. The sites were selected at the centre of each 16×16-km cell. In this study, we focused on a subset of 413 sites from the RMQS data set. The samples were organized into four regions: Brittany (131 sites), Burgundy (109 sites), Landes (52 sites) and South-East (121 sites, Fig. 1A) which are contrasted in terms of soil type, land-use (coarse level of the CORINE Land Cover classification; IFEN, http://

www.ifen.fr; 7 classes: forest, crop systems, grasslands, particular natural ecosystems, vineyards/orchards, parkland and wild land), climate and geomorphology (Table S1). Within a region, sites were separated by 16 km at least. For each soil, the pedo-climatic characteristics considered were particle-size distribution, pH in water (pH$_{water}$), organic carbon content (C$_{org}$), N content, C:N ratio, soluble P contents, CaCO$_3$ and exchangeable cations (Ca, K, Mg), sum of annual temperature (°C) and annual rainfall (mm). Physical and chemical analyses were performed by the Soil Analysis Laboratory of INRA (Arras, France) which is accredited for such analyses by the French Ministry of Agriculture.

Bacterial and fungal community fingerprinting

Soil DNA extraction. For each soil sample, the equivalent of 1.5 g of dry soil was used for DNA extraction, following the procedure optimized by platform GenoSol [35]. Briefly, extraction buffer (100 mM Tris pH 8.0, 100 mM EDTA pH 8.0, 100 mM NaCl and 2% (w/v) SDS) was added to the sample in the proportion 3:1 (v/w), with two grams of glass beads (106 µm diameter) and eight glass beads (2 mm diameter) in a bead-beater tube. All beads were acid washed and sterilized. The samples were homogenized for 30 s at 1600 rpm in a mini bead-beater cell disruptor (Mikro-dismembrator, S. B. Braun Biotech International), incubated for 30 min at 70°C in a water bath and centrifuged for 5 min at 7000 g and room temperature. The supernatant was collected, incubated on ice with 1/10 volume of 3 M potassium acetate (pH 5.5) and centrifuged for 5 min at 14000 g. DNA was

precipitated with one volume of ice-cold isopropanol and centrifuged for 30 min at 13000 rpm. The DNA pellet was washed with ice-cold 70% ethanol and dissolved in 100 µl of ultra pure water. For purification, aliquots (100 µL) of crude DNA extracts were loaded onto PVPP (polyvinyl polypyrrolidone) minicolumns (BIORAD, Marne la Coquette, France) and centrifuged for 4 min at 1000 g and 10°C. This step was repeated if the eluate was opaque. The eluate was then collected and purified for residual impurities using the Geneclean Turbo kit as recommended by the manufacturer (Q Biogene, France).

PCR conditions. The bacterial ribosomal IGS was amplified using the PCR protocol described in Ranjard et al [43]. 12.5 ng of DNA was used as the template for PCR volumes of 25 µl. The fungal ribosomal ITS was amplified using the primer set ITS1F/ ITS4-IRD800 (5'- CTTGGTCATTTAGAGGAAGTAA -3'/5'- IRD800-TCCTCCGCTTATTGATATGC -3'). 20 ng of DNA was used as the template for PCR volumes of 25 µl with the following PCR conditions: denaturation at 95°C for 3 min, 35 cycles of 30 s at 95°C, 45 s at 55°C and 1 min at 72°C, and a final elongation of 7 min at 72°C. The primer Tm was the same for bacterial IGS and fungal ITS. Every PCR products were purified using the MinElute Kit (QIAGEN, Courtaboeuf, France) and quantified using a calf thymus DNA standard curve.

ARISA fingerprinting conditions. 2 µL of the PCR product was added to deionized formamide and denatured at 90°C for 2 min. Bacterial and Fungal ARISA fragments were resolved on 3.7% polyacrylamide gels under denaturing conditions as

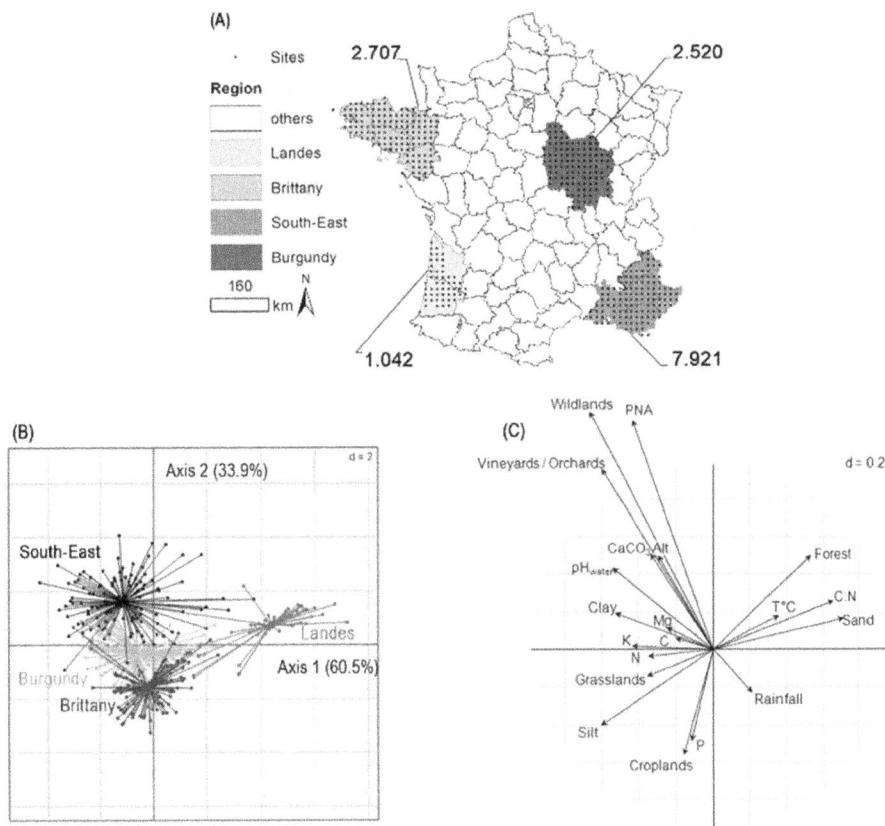

Figure 1. Comparison of the regions considered on the basis of their soil habitats. A. Soil average dissimilarity of soil habitat for the different regions (number linked to the corresponding region) and position of sites; B. Between group analysis of soil habitats according to the region; C. Correlation circle of the variables defining soil habitat in the between group analysis. The length of the arrow corresponds to the Pearson's correlation coefficient for quantitative variables and to the correlation ratio for qualitative variables. Symbols: Alt.: Elevation; T°C: Sum of annual temperatures; P$_{ass}$: Assimilable P; C:N: Carbon to Nitrogen ratio; C$_{org}$: Organic Carbon content.

described in Ranjard et al [9] on a LiCor DNA sequencer (ScienceTec).

Image analysis. The data were analyzed using the 1D-Scan software (ScienceTec), converting fluorescence data into electrophoregrams, where peaks represented PCR fragments (100 to 110 peaks retained per sample, the resolution limit to avoid considering background noise). The height of the peaks was calculated in conjunction with the median filter option and the Gaussian integration in 1D-Scan, and represented the relative proportion of the fragments in the total products. Lengths (in base pairs) were calculated by using a size standard with bands ranging from 200 to 1659 bp. The data were then converted into a contingency table with prepRISA package in R.

Statistical analyses

Characterization of habitat variability and average dissimilarity across regions. Habitats were compared between regions in a Hill & Smith multivariate analysis [44] using the ade4 package in R [45–46]. The analysis was applied to pedoclimatic characteristics, land-use and geomorphology, by centering and scaling the quantitative variables, and converting the qualitative ones into weighted binary variables (weight equal to $1/n$; n is the number of classes for the qualitative variables). Differences between these regions were examined by between group analysis and tested by applying a Monte-Carlo permutation test (1000 permutations). The average dissimilarity between soil habitats was determined by transposing a method based on the dissimilarity matrix for communities [47] to soil habitat. The dissimilarity matrix for soil habitat was derived from the site coordinates in the Hill & Smith analysis, following equation 1 [9]:

$$D_{i,j} = \frac{ED_{i,j}}{ED_{\max}} \qquad (1)$$

Where $D_{i,j}$ and $ED_{i,j}$ are the dissimilarity and the Euclidean distance between sites i and j, respectively. $ED_{\max}$ is the maximum Euclidean distance observed between sites. The average dissimilarity between soil habitats was then calculated as follows [47]:

$$\overline{D}_{habitat} = \frac{1}{n} \sum_{h=1}^{n-1} \sum_{i=h+1}^{n} D_{h,i}^2 \qquad (2)$$

$\overline{D}_{habitat}$ is the average dissimilarity between soil habitats of soil habitat and n the number of sites in the region.

Evaluation of the similarity distance-decay relationship and initial similarity of Bacterial and Fungal community composition. The *similarity distance-decay* relationship was estimated as proposed by Harte et al [39] for organisms with large populations per taxa. From this relationship, the community composition turnover rates (z) for bacterial and fungal communities composition were derived as described in Ranjard *et al.* [9] following the method and the exponential model (equation 3) proposed by Harte et al [39–40] for microorganisms.

$$\log_{10}(\chi_d) = (-2 * z) * \log_{10}(d) + b \qquad (3)$$

Where χ_d is the observed Sørensen's similarity between two soil samples that are d meters apart from each other; b is the intercept of the linear relationship and z the turnover rate of the community composition. The z estimate and its 95% confidence interval were derived from the slope ($-2*z$) of the relationship between similarity and distance by weighted linear regression. The overlap of the 95% confidence intervals was used to test for significant differences

in community composition turnover rates between regions or between bacteria and fungi. The initial similarity of community composition was taken as the average similarity between sites 16 km apart and the 95% confidence intervals of the mean were determined [30].

Variance partitioning of community composition variations according to environmental filters and space. The relative importance of spatial variables, pedoclimatic characteristics and land-use in determining community composition turnover was tested by db-RDA [47–48]. Quantitative data were centered and scaled. Spatial variables were constructed from site coordinates (x, y, elevation) to reveal potential spatial trends at scales larger than the region, and of Principle Coordinates of Neighbour Matrices [49] in each region. The PCNM approach creates independent spatial descriptors that can be introduced in canonical analysis models to consider the spatial autocorrelation between sites in the model [49]. PCNMs with a significant Moran index ($P<0.001$) were selected. Land-use corresponded to the Corinne Land Cover classes recoded into dummy variables. Pedo-climatic characteristics consisted of climate and all the physico-chemical variables except sand. The most parsimonious model was obtained by forward selection from null to full model in two steps: a first step for selecting environmental variables and a second step for selecting the relevant PCNMs. Then, the pure effects of each set of filters or each individual filter were tested with an anova-like permutation test for canonical analyses (anova.cca function in vegan package, [50]). The PCNMs approach does not provide directly the range of the spatial descriptor. Nevertheless, Bellier et al [51] demonstrated that kriging approach could be applied to PCNMs to estimate their spatial range. Consequently, when PCNMs were selected in the most parsimonious model, their ranges were determined by standard kriging techniques (ordinary kriging with a Gaussian model). The hierarchy of these filters must nevertheless be considered with caution due to the small amounts of variance explained by each one. Land-use was not included in the filterranking since it corresponded to a set of categories and a global "land-use" category was already taken into account in the processes section. Maps of soil fungal community structure variations are provided as (Fig. S1, mapping methodology is described in the legend).

Results

Heterogeneity of soil habitat

The four regions were selected for their contrasting environmental heterogeneity as demonstrated by the between group analysis (Fig. 1B) and comparison of the calculated average dissimilarity between soil habitats ($\overline{D}_{habitat}$) (Fig. 1A and C). Multivariate analysis revealed a clear discrimination of the four regions on the first and second axes (Monte-Carlo permutation test, $P<0.001$): Landes was significantly discriminated from Brittany, Burgundy and South-east on the first axis and these three regions were discriminated from each other on the second axis. In addition, the environmental variability strongly differed between the four regions as demonstrated by the dispersal of sites in the factorial map. Sites from Landes were less dispersed on the factorial map than the sites from Brittany or Burgundy, which were less dispersed than sites from South-East. The calculated $\overline{D}_{habitat}$ ranged from 1.042 in the Landes to 7.921 in the South-East, with intermediate values for Burgundy and Brittany (2.520 and 2.707, respectively; Fig. 1A). It provided the same discrimination between regions. Fig. 1C shows that the four regions could mainly be distinguished according to land-use (*e.g.*: 86% of the

Landes sites are forest sites), a restricted set of soil physico-chemical characteristics (sand and silt contents, pH_{water} and $CaCO_3$ content, assimilable P content and organic matter quality as measured by C:N ratio) and by differences in elevation. Climatic conditions did not play a significant role in regional discrimination.

Distance-Decay Relationship for bacterial and fungal soil communities

Bacterial and fungal community similarity decreased with increasing distance in each region (Fig. 2). Community similarity was systematically higher for bacteria than for fungi in all regions at small or large distances. For bacteria, the linear regression model was highly significant in each region ($P<0.001$, Fig. 2) except in Landes where it was just below the significance threshold of 5% ($P<0.02$). For fungi, the linear regression model was highly significant in each region ($P<0.001$) except Landes (Fig. 2).

Community composition turnover rates were derived from the parameters of the linear regression. The community composition turnover rates for the bacterial and fungal communities ranged from 0.006 to 0.013 (Table 1). No significant differences were highlighted between these organisms when the community composition turnover rates were compared within each region.

When the community composition turnover of bacterial or fungal communities was compared between regions, a significant difference was only found between Brittany and the South-East ($P<0.05$) for bacteria.

The initial similarity was always higher for bacterial communities than for fungal communities within each region, ranging respectively from 54.4% to 61.4% and from 39.7% to 43.1%. The initial similarity of the bacterial community in Landes was significantly lower than in the other regions, which did not differ from each other. For fungi, the initial similarities in Landes and Brittany were similar and significantly lower than those in Burgundy and the South-East.

These results were confirmed by a covariance analysis comparing the models between organisms within a region and between regions for a given organism (data not shown).

Variance partitioning of community composition variations

The relative importance of the sets of spatial variables, land-use and pedo-climatic characteristics on variations in bacterial and fungal community composition was tested by db-RDA using the Sørensen index (Fig. 3). The amount of variance in bacterial and fungal community composition explained by the three sets of filters

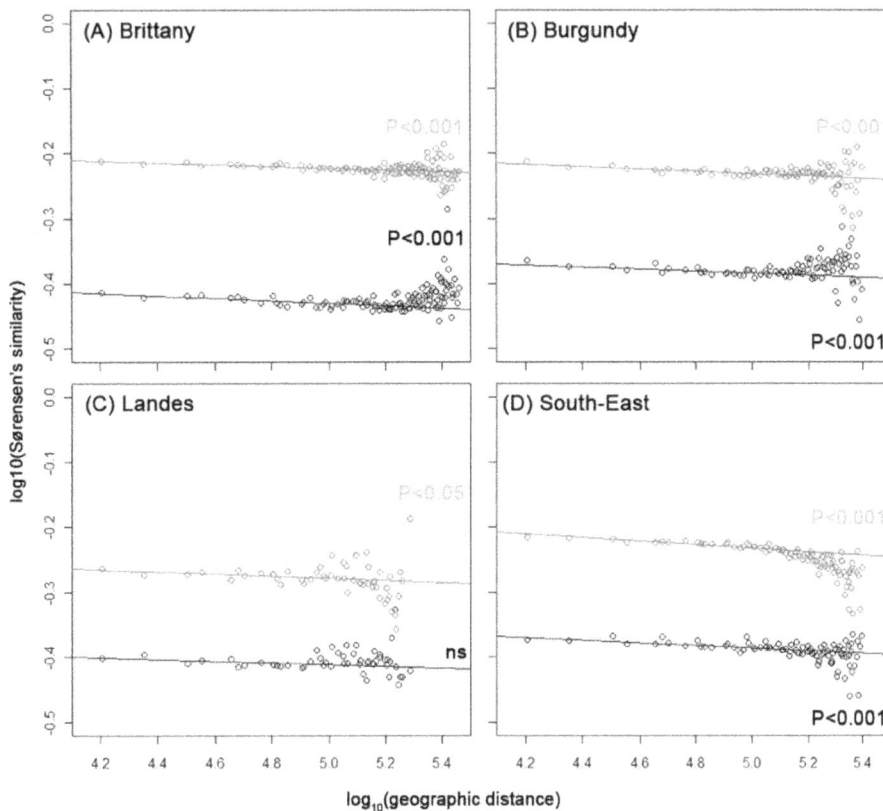

Figure 2. Distance-Decay Relationships for bacteria and fungi. Each panel correspond to a region: Brittany (A), Burgundy (B), Landes (C) and South-East (D) Points the average Sørensen's similarity between sites for each distance class. Lines represent the regression model based on the whole set of paired comparisons; for bacteria (grey) and fungi (black). The equations for the regression models were as follows: (**A**) Brittany: Bacteria***: log10(Sørensen's similarity) = −0.014×log10(geographic distance)−0.156; Fungi***: log10(Sørensen's similarity) = −0.017×log10(geographic distance)−0.350; (**B**) Burgundy: Bacteria*** log10(Sørensen's similarity) = −0.018×log10(geographic distance)−0.144; Fungi***: log10(Sørensen's similarity) = −0.015×log10(geographic distance)−0.316; (**C**) Landes: Bacteria*: log10(Sørensen's similarity) = −0.017×log10(geographic distance)−0.198; Fungi ns: log10(Sørensen's similarity) = −0.012×log10(geographic distance)−0.357; (**D**) South-East: Bacteria***: log10(Sørensen's similarity) = −0.027×log10(geographic distance)−0.101; Fungi***: log10(Sørensen's similarity) = −0.019×log10(geographic distance)−0.298. A graph with points representing all paired-comparisons between sites as points can be found in Figure S2. Significance of the model is indicated as an exponent for each organism: ns: not significant; $P<0.05$: *; $P<0.01$: **, $P<0.001$: ***.

Table 1. Regression parameters of the Distance-Decay Relationships for Bacteria and Fungi.

Region	Parameter	Organism	Estimate	95% Confidence interval
Brittany (131)	Z	Bacteria	0.007	[0.005; 0.009]
		Fungi	0.009	[0.006; 0.011]
	Initial similarity	Bacteria	61.4%	[60.4%; 62.3%]
		Fungi	38.5%	[37.7%; 39.4%]
Burgundy (109)	z	Bacteria	0.009	[0.006; 0.012]
		Fungi	0.008	[0.003; 0.012]
	Initial similarity	Bacteria	61.4%	[60.4%; 62.5%]
		Fungi	43.1%	[42.0%; 44.3%]
Landes (52)	z	Bacteria	0.009	[0.001; 0.016]
		Fungi	0.006	[−0.001; 0.013]
	Initial similarity	Bacteria	54.4%	[52.9%; 56.0%]
		Fungi	39.7%	[38.4%; 40.9%]
South-East (121)	z	Bacteria	0.013	[0.011; 0.016]
		Fungi	0.009	[0.006; 0.013]
	Initial similarity	Bacteria	60.9%	[59.9%; 61.9%]
		Fungi	42.4%	[41.5%; 43.3%]

The number of observations per region is provided in brackets beside the name of the region. The community composition turnover rate (z) and the initial similarity are derived from the slope of the regression ($-2z$) and the mean of similarity at 16 km; respectively. The statistical comparison between region and organism was performed by examining the overlap of the 95% confidence intervals of turnover rates or initial similarities.

ranged from 17% to 32%. The significance of the interactions between the three sets of filters could not be tested but always explained a small amount of the total variance (from 1.2% to 8.4%).

Among the sets of filters, soil pedo-climatic characteristics were the main contributor to variations in bacterial and fungal community composition. Spatial variables systematically explained a lower amount of variance than pedo-climatic characteristics, but a higher amount than land-use for both bacteria and fungi except for bacteria in Landes region where neither spatial variables nor land-use were significant.

The amount of variance explained by pedo-climatic characteristics for bacteria or fungi was always significant, ranging from 5% to 15%, and was similar between regions. Nevertheless, within each region, the amount of variance explained by soil pedo-

climatic characteristics was always higher for bacteria than for fungi (Fig. 3).

Spatial variables significantly explained part of the community composition variations in all regions except for bacteria in Landes region. When significant, spatial variables represented from 3% to 9% of the total variance and were of the same order of magnitude both between bacteria and fungi and between regions (Fig. 3).

Similarly, land-use explained a significant amount of community composition variations in all regions except for bacteria in Landes region. The amount of explained variance ranged from 2.6% to 6.5% of the total variance. Within each region, land-use explained similar amounts of bacterial and fungal community variations when significant. Between regions, amounts of variance explained by land-use were similar for both bacteria and fungi (Fig. 3).

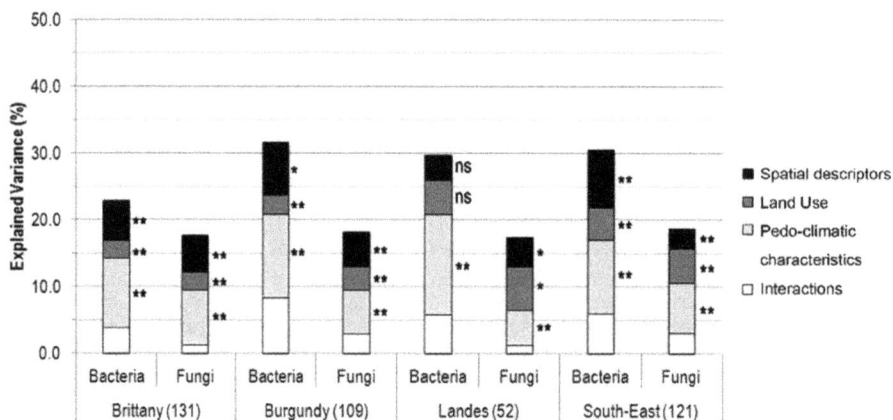

Figure 3. Variance partitioning of bacterial and fungal community composition. The number indicated in brackets corresponds to the number of samples for the region. Significance levels: ns: not significant; P<0.05: *; P<0.01: **, P<0.001: ***.

Hierarchy of environmental filters

Within the sets of soil pedo-climatic characteristics and spatial variables, the pure effects of individual filters on bacterial and fungal community structure are presented in Figure 4. These pure effects account for relatively small proportions of the total variance (from 0.7% to 6.5%) because of the large number of filters explaining the total variance of bacterial or fungal community compositions. Regarding the pedo-climatic characteristics, the significant filters for soil bacterial communities in Brittany were first pH and secondly N content. For fungi, these were the quality of organic matter resource as indicated by the selection of N content, C_{org} content and C:N ratio. The least significant filters corresponded to soil texture and other soil nutrients for bacteria (clay and silt contents, Mg and $CaCO_3$ concentrations) and fungi (clay content, K and P concentrations, Figure 4). In Burgundy, as in Brittany, the soil bacterial and fungal communities were principally affected by pH. Beside pH, bacterial community composition was shaped by the quality of organic matter resource (N content, C_{org} content, C:N ratio) followed by clay content and annual rainfall. The fungal community composition was shaped by C:N ratio only. The only filter that had a significant effect in the Landes region, on both soil bacteria and soil fungi, was the C:N ratio. The pH was the most important filter, followed by clay content and K concentration, for bacteria in the South-East. In this region, N content and K concentration were more important filters for fungi than pH and clay content.

The spatial variables corresponded to the sites coordinates and 16 significant PCNM eigenfunctions, each representing a different spatial scale of analysis (coarse, medium and fine scales, Figure 4). Longitude, latitude or altitude coordinates did not influence community composition except latitude for fungi in Burgundy region (Figure 4). PCNMs representing spatial structures of 80–120 km radius explained significant amounts of variance in the composition of bacterial and fungal communities in Brittany and South-East. This highlighted that these communities were spatially structured at a coarse spatial scale. Similarly, PCNMs representing spatial structures of 40–65 km radius explained significant amounts of variation in the composition of bacterial community

in Brittany and of fungal community in Burgundy, South-East and Landes. This showed the spatial structuration of these communities at medium spatial scales. PCNMs accounting for fine scale variables were neither significant for bacteria nor fungi in any region.

Discussion

The four regions were selected to challenge the hypothesis that different levels of environmental heterogeneity, i.e. different habitat diversity and fragmentation for soil bacteria and fungi, results in different community composition turnovers [9], [14] and to compare their determinism. The multivariate analysis and the calculated $\overline{D}_{habitat}$ both highlighted a significant gradient in environmental heterogeneity following the sequence: Landes< Burgundy≤Brittany<<South-East. The four regions were mainly discriminated by environmental variables already demonstrated to influence soil microbial community abundance and diversity, such as land-use [52] and soil characteristics (texture, pH_{water}, P content and C:N ratio; [6], [14], [33], [35]). Among these four regions, Landes and South-East represent two extremes of environmental heterogeneity. Landes was distinct in having sites with conifer forest on acidic soils and low altitudinal variations. This specificity may turn Landes into an outlier for its environmental variability. On the other hand, South-East is characterized by a strong altitudinal gradient and a mosaic of land-use. In comparison, Burgundy and Brittany were also characterized by a mosaic of land-use but were more marked by croplands than South-East and less by forests. In addition, these two regions presented intermediate soil types regarding Landes and South-East: silty-clay calcic soils and silty acidic soils, respectively; associated to intermediate levels of organic Carbon content. Altogether, these observations highlight that these four regions allow the consideration of a large range of environmental conditions associated to different levels of environmental heterogeneity. Therefore, community similarity turnover rates can be confronted to environmental heterogeneity and their comparison may lead to a consensus regarding the environmental filters shaping soil microbial communities.

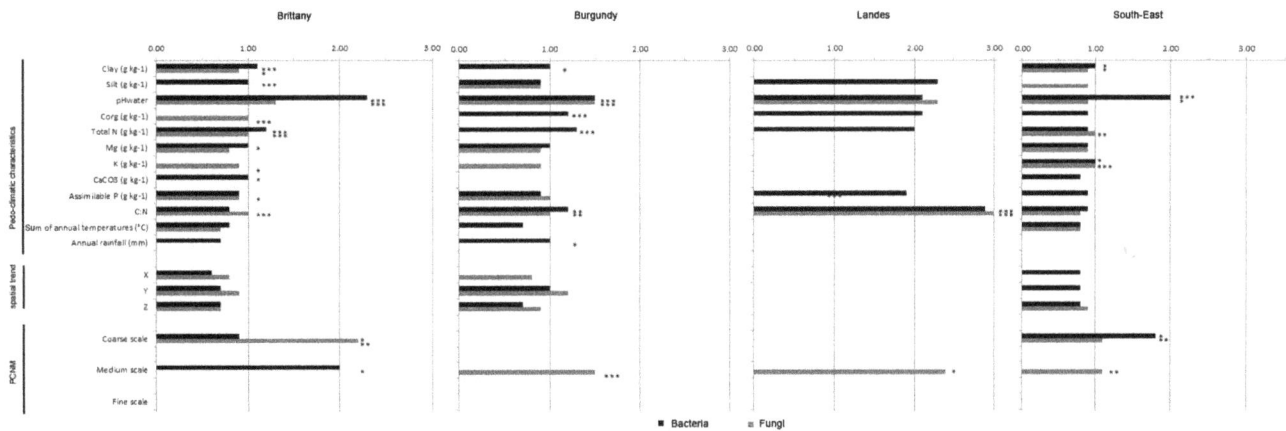

Figure 4. Variations of microbial communities partitionned according to edaphic variables and space. For each organism and region, only variables retained in the most parsimonious model are presented and their pure effect is tested by a permutation test. Significance levels are: P< 0.05: *; P<0.01: **, P<0.001: ***. Missing values or variables indicate that the variable was not retained in the model. Sand was removed prior to model evaluation since it was represented by the opposite of the sum of silt and clay content. Rainfall: Sum of annual rainfall (mm). Temperature: Sum of annual temperature (°C). Spatial components were summarized according to the scale considered: trend (x, y and z coordinates), coarse, medium or fine. The interval in brackets indicates the numbers of PCNMs retained in the model for each scale. The proportion of variance for each scale was determined as the sum of the pure effects of each PCNM when these were significant. Coarse, medium and fine scales correspond to PCNM with a spatial range of 80 to 120 km, 40 to 65 km and less than 40 km; respectively.

In all regions, except Landes, the soil bacterial and fungal communities were spatially structured as indicated by the significant DDR. This supports that the concept of DDR for soil microorganisms may be generalized, as suggested by several other studies [1], [9], [12], [24], [53]. The estimated bacterial and fungal community composition turnover rates ranged from 0.006 to 0.013. This is in agreement with the recent community composition turnover rates observed for soil microorganisms [10], [14], [24]. However, the turnover rates estimates were low. This could be due to technical limitations; particularly the low taxonomic resolution of DNA fingerprinting. Species variations were aggregated by the DNA fingerprints into few dominant bands, which precluded the accumulation of new minor species with increasing distance [54–55]. This would be supported by the higher community composition turnovers observed in Zinger [24] with the high resolution level provided be the pyrosequencing approach. The sampling design could also have led to low estimates of the community composition turnover rate. Indeed, lower composition turnover rates are commonly observed in regions composed of contiguous habitats [13] with gradual variations of habitat characteristics across sites and higher rates of community composition turnover have been observed at finer taxonomic levels and spatial scales [56].

Beside these technical points, community composition turnover rates were very close within each region for bacteria and fungi. This observation was in agreement with maps of bacterial and fungal communities' structures which revealed large patches of ca. 100–140 km radius (see Fig. 2 in [7] for bacteria and Fig. S1 for fungi). Altogether, this indicated a low-level of aggregation of both communities on a broad spatial scale [57]. This observation would suggest that differences in terms of biological and ecological features (habitat characteristics, colonization modes, trophic requirements, biotic interaction, dormancy; [6], [58–59]) would not lead to different rates of community composition turnover on a broad spatial scale. This would be supported by strong biotic interactions between bacterial and fungal communities highlithed by a significant correlation between bacterial and fungal communities' compositions. Nevertheless, this correlation was much lower than that between each community composition and environmental variables (data not shown). This strong dependency to environmental conditions resulted in a trend towards higher turnover rates in regions with higher $\overline{D}_{habitat}$. This would corroborate our first hypothesis that the microbial DDR is positively related to environmental heterogeneity. This observation was more marked and significant for bacterial communities but only a tendency for fungal communities. Although this conclusion is based on analyses of only 4 regions, it is in agreement with Ranjard et al. [9] who demonstrated a correlation between habitat heterogeneity and turnover rate for soil bacterial communities on a broad spatial scale, and with other microbial and macrobial studies [24], [30–31], [60].

The differences between the sensitivities of soil bacteria and fungi to environmental heterogeneity could be due to the ecology of soil fungi [58] at the spatial scale of this study. Nevertheless, according to Peay et al [25–26], ectomycorhizal fungi are spatially structured at very fine spatial scales. This would suggest that these organisms are strongly dependent on environmental conditions but that the grain size of this study (252 km^2) did not allow perceiving this dependence. This hypothesis would be supported by the estimated similarity of the microbial communities at a local spatial scale (initial similarity, [30]). Indeed, the initial similarity was systematically higher for bacterial than for fungal communities. According to Morlon et al [57], this would indicate that fungal communities are more aggregated than bacterial communities, i.e.

more variable than bacterial communities, at a local spatial scale. This would lead to the conclusion that fungi are somehow more dependent on environmental conditions than bacteria at a local scale, in agreement with Peay et al [25–26] and the numerous biological interactions they are involved in.

The observed patterns and significant turnover rates result from different ecological processes [8] which may shape differently bacterial and fungal community composition on a broad spatial scale according to their relative importance. Recent studies have demonstrated that the main processes involved in increasing or decreasing community composition turnover rate are environmental selection and dispersal, respectively [8–9]. Dispersal is commonly supposed, but not empirically demonstrated to be high for microbes [61]. The maintenance of significant turnover rates, as recorded in our study, thus requires environmental selection to be high and dispersal not infinite. This hypothesis was tested by partitioning the β-diversity variations of microbial communities within regions. This partition was made according to filters involved in environmental selection: soil pedo-climatic characteristics and land-use; and to spatial variables characterizing spatial autocorrelation unexplained by the environmental variables. The residual spatial autocorrelation could result from dispersal limitations, but also from unmeasured spatially autocorrelated environmental variables despite the extensive number of environmental variables considered in this study. The amount of explained variance ranged from 23% to 32% for bacteria and from 17.5% to 18.8% for fungi. These values are within the range reported in the literature for the whole communities of bacteria [8], [14] and fungi [25–26], [62].

Soil characteristics accounted for higher amounts of variance, for both bacteria and fungi, than spatial variables. This supports the idea that soil microbial communities are primarily affected by environmental selection and secondly by other processes leading to spatial structuration independently of the environment [2], [7–8], [38]. More precisely, environmental selection was mainly driven by pedo-climatic characteristics, which accounted for larger amounts of variance than land-use. This is in agreement with several studies conducted on a broad spatial scale evidencing the higher dependency of bacterial and fungal communities on soil physico-chemical characteristics [18], [38], [63] than on land-use. Nevertheless, the amount of variance explained by pedo-climatic characteristics was higher for bacteria than for fungi, highlighting the greater effect of soil habitat on shaping the bacterial community. This observation is supported by the higher diversity and reactivity of the soil bacterial community to changes in surrounding conditions which, in turn, leads to a community structure that is better fitted to the habitat characteristics [13], [58]. On the contrary, land-use explained a higher amount of fungal community variance than of bacterial community variance. This latter could be due to plant-soil microbe interactions resulting from the type of vegetative cover [19], [38], [52], [64] as well as from human activities, especially agricultural or industrial practices [35]; [37] potentially affecting fungi more strongly than bacteria. This difference would be in agreement with a highly patchy distribution of soil fungi at a local scale as suggested above by the low level of initial similarity. Altogether, these results support the postulate of Baas-Becking "the environment selects" in summarizing microbial biogeography.

Nevertheless, variance partitioning also highlighted that bacterial and fungal communities are spatially structured independently of environmental characteristics since spatial variables accounted for a significant amount of community variance. A hypothesis in explaining this spatial autocorrelation is that soil bacterial and fungal communities might be dispersal limited, even if this result may also be related to unmeasured, spatially structured, environmental variables. This hypothesis would be supported by the study of Ranjard et al [9] demonstrating significant turnover rates of bacterial diversity in fully homogenous regions and by the consideration of several soil physico-chemical characteristics in the variance partitioning approach of the present study. Nevertheless, this must be confirmed by more in-depth studies since few evidence remains to date on the shaping of bacterial and fungal communities by limited dispersal [8], [17], [26–27], [65]. Comparisons have indicated that fungi tend to be less dependent on spatial variables than bacteria. This result is surprising since fungi are demonstrated to be dispersal limited [27] and bacteria are generally expected to disperse over larger distances, even wider than our regional scale [14], [66]. It was estimated in the literature that 10^{18} viable bacteria were transported annually in the atmosphere between continents [61]. Moreover, particular fungal populations have been demonstrated to disperse over short distances, *e.g.* ectomycorrhiza at the "plant island scale" [25–26], or to be endemic [27]. On the other hand, variance explained by spatial variables was similar in each region for fungi, whereas it followed the environmental heterogeneity gradient: Landes<Brittany≤Burgundy<South-East for bacteria. This observation supports the hypothesis that bacteria are more sensitive to geomorphology and that the presence of natural barriers (*e.g.* mountains), or a higher fragmentation of landscape in the South-East, due to a mosaic of agricultural and natural plots [9] may lead to significant spatial variations.

Our last objective was to identify and rank the environmental filters and the spatial variables involved in environmental selection process and spatial autocorrelation of bacteria and fungi. To do so, the environmental and spatial filters selected in the variance partitioning approach were ranked according to their pure effect on bacterial and fungal community composition turnover. First, environmental selection was mainly based on soil characteristics but not on climatic conditions. This is in accordance with previous studies highlighting the weak influence of climate on soil microbial diversity on a broad spatial scale [6–7], [35], [37], [67], [68]. Nevertheless, recent studies also demonstrated strong differences between bacterial communities across different biomes, especially cold deserts *versus* temperate biomes and hot deserts, suggesting that climate may play a role [32] at very large spatial scales. Overall, the main soil characteristics identified as important filters for soil bacterial and fungal community composition were pH, trophic resources (N, C, K, P contents; C:N ratio) and texture (clay, silt content). These findings are consistent with the literature where i) pH is regularly identified as the main filter for both soil bacteria [6], [16–17] and fungi [25], [62], ii) soil organic C and N contents and the C:N ratio constitute important components of microbial niches [18], [32], [35], [58], and iii) texture determines the size and stability of soil micro-habitats [18], [69], [70]. The hierarchy of these filters depended on the type of organisms and the regions. The above-described hierarchical sequence (pH>trophic resources≥texture) was observed for three of the four regions, but not for Landes. In the Landes region, the filter variability is low and environmental selection is basically driven by organic matter quality (C:N ratio). This is related to the large number of conifer forests sites leading to soil organic matter with a high C:N ratio and strong recalcitrance to microbial decomposition. A strong selection of particular populations with enzymatic ability to transform this organic matter is occurring in such soils [58], which is deeply influencing the corresponding community structure. Comparison of the overall filter hierarchies for bacteria and fungi in a given region did not reveal any discrepancy for the primary drivers. The only differences were observed for secondary filters, including C:N ratio and mineral nutrients (such as K and P) for fungi, and clay and N content for bacteria. This is consistent with the well-known dependency of the soil fungal community on soil P content [34] and soil organic matter quality [58]. As regards bacteria, clay content is positively correlated with the biotic capacity of soil as well as its indigenous bacterial diversity by enhancing the level of protection of the soil habitat and the retention of nutrients [69–70]. Second, regarding spatial autocorrelation, different scales were derived from the hierarchy of spatial variables and their range. The main scale identified was the coarse scale (80 to 120 km radius), for both bacteria and fungi in Brittany and South-east regions. This spatial scale is smaller than the one at which soil habitat changed on the RMQS Network (150 to 470 km, [9]), whereas it is in agreement with the large patches obtained by mapping the bacterial and fungal community structure over these regions [7] and Fig. S1). In addition, finer spatial scales (medium scales; 40 to 65 km radius) were also identified as significant in Burgundy, Landes and South-East for fungi and in Brittany for bacteria. Altogether, this scale dependency would support a hypothesis for the dispersal limitation of bacterial and fungal communities [33]. This scale dependency is in agreement with Martiny et al [14] for bacteria and is supported by the observations of Peay et al [25–26] on soil fungi. As in other studies at continental scales [14], this highlights the importance of considering multiple scales to better understand microbial ecology.

Altogether, our study demonstrated the spatial structuring of soil bacterial and fungal communities on local to coarse scales, which was based on environmental selection and on an unexplained spatial autocorrelation that could be related to limited dispersal. Selection and spatial autocorrelation were shown to have a similar influence on soil bacteria and soil fungi but the filters involved could differ depending on the environmental heterogeneity. Nevertheless, the comparison of bacterial and fungal communities helped to propose a primary consensus regarding the environmental filters shaping soil microbial community composition as a whole: Land-Use and pH are the primary filters, followed by trophic resources quantity (organic carbon content and nitrogen content) and then quality (C:N ratio). The results of this study increase our knowledge on the effects of soil habitat and provide insights in the scale at which dispersal may occur according to the ecological attributes of bacteria and fungi. However further investigations, based on up-scaling approaches, are now required to: i) provide a direct measurement of bacterial cells dispersal which is now crucial to demonstrate limited dispersal of bacteria; ii) identify the filters operating at each spatial scale. Especially, these upscaling approaches could be associated to high-throughput sequencing to achieve a finer resolution on the communities. This would improve our ability to sustainably manage soil biodiversity.

Supporting Information

Figure S1   Maps of interpolated MULTISPATI scores for the first three MULTISPATI axes (columns) and for the four geographical regions (rows). Each map was generated as described in Dequiedt et al (2009) and corresponds to the spatial synthesis of the F-ARISA genetic structure of indigenous fungal communities from the corresponding soils sampled in the four regions of France. Colours on the map are proportional to the score of each soil sample on each MULTISPATI axis following the scale provided at the bottom of the figure. Below each column, the empirical variogram is provided for each MULTISPATI axis.

Figure S2   Distance-Decay Relationship for bacteria and fungi. Each panel correspond to: **(A–D): Bacteria** in Brittany,

Burgundy, Landes and South-East; **(E–H): Fungi** in Brittany, Burgundy, Landes and South-East. Points represent paired-comparisons between sites and line the linear model. The equations for the regression models were as follows: **(A)** log10(Sørensen's similarity) = −0.014×log10(geographic distance)−0.156; **(B)** log10(Sørensen's similarity) = −0.018×log10(geographic distance)−0.144; **(C)** log10(Sørensen's similarity) = −0.017×log10(geographic distance)−0.198; **(D)** (Sørensen's similarity) = −0.027×log10(geographic distance)−0.101; **(E)** log10(Sørensen's similarity) = −0.017×log10(geographic distance)−0.350; **(F)** log10(Sørensen's similarity) = −0.015×log10(geographic distance)−0.316; **(G)** log10(Sørensen's similarity) = −0.012×log10(geographic distance)−0.357; **(H)** log10(Sørensen's similarity) = −0.019×log10(geographic distance)−0.298. Signifi-

cance of the model is indicated as an exponent for each organism: ns: not significant; P<0.05: *; P<0.01: **, P<0.001: ***.

Table S1   Summary statistics of regions characteristics. PNE: Particular Natural Ecosystems, SE: standard error of the mean.

Author Contributions

Conceived and designed the experiments: NCPB SD NPAS CJ DA LR PL. Performed the experiments: NCPB ML. Analyzed the data: NCPB SD JT PP. Contributed reagents/materials/analysis tools: NCPB SD JT ML. Wrote the paper: NCPB LR.

References

1. Horner-Devine MC, Lage M, Hughes JB, Bohannan BJM (2004) A taxa-area relationship for bacteria. Nature 432: 750–753.
2. Martiny JBH, Bohannan BJM, Brown JH, Colwell RK, Fuhrman JA, et al. (2006) Microbial biogeography: putting microorganisms on the map. Nat Rev Microbiol 4: 102–112.
3. Baas Becking LGM (1934) *Geobiologie of inleiding tot de milieukunde*, The Hague, The Netherlands.
4. Maron P-A, Mougel C, Ranjard L (2011) Soil microbial diversity: Methodological strategy, spatial overview and functional interest. C R Biol 334: 403–411.
5. Gilbert J, Meyer F, Jansson J, Gordon J, Pace N, et al. (2010) *The Earth Microbiome Project: Meeting report of the "1st EMP meeting on sample selection and acquisition"* at Argonne National Laboratory October 6th 2010.
6. Fierer N, Jackson RB (2006) The diversity and biogeography of soil bacterial communities. Proc Natl Acad Sci U 103: 626–631.
7. Dequiedt S, Thioulouse J, Jolivet C, Saby NPA, Lelievre M, et al. (2009) Biogeographical patterns of soil bacterial communities. Environ Microbiol Rep 1: 251–255.
8. Hanson CA, Fuhrman JA, Horner-Devine MC, Martiny JBH (2012) Beyond biogeographic patterns: processes shaping the microbial landscape. Nat Rev Microbiol 10: 497–506.
9. Ranjard L, Dequiedt S, Chemidlin Prévost-Bouré N, Thioulouse J, Saby NPA, et al. (2013) Turnover of soil bacterial diversity driven by wide-scale environmental heterogeneity. Nat Commun 4: 1434.
10. Gleason AH (1922) On the relationship between species and area. Ecology (N. Y.), 3: 158–162.
11. Zinger L, Boetius A, Ramette A (2014) Bacterial taxa–area and distance–decay relationships in marine environments. Mol Ecol 23: 954–964.
12. Green JL, Holmes AJ, Westoby M, Oliver I, Briscoe D, et al. (2004) Spatial scaling of microbial eukaryote diversity. Nature 432: 747–750.
13. Prosser JI, Bohannan BJM, Curtis TP, Ellis RJ, Firestone MK, et al. (2007) Essay - The role of ecological theory in microbial ecology. Nat Rev Microbiol 5: 384–392.
14. Martiny JBH, Eisen JA, Penn K, Allison SD, Horner-Devine MC (2011) Drivers of bacterial β-diversity depend on spatial scale. P Natl Acad Sci USA 108: 7850–7854.
15. Vellend M (2010) Conceptual synthesis in community ecology. Q Rev Biol 85: 183–206.
16. Green JL, Bohannan BJM (2006) Spatial scaling of microbial biodiversity. Trends Ecol Evol 21: 501–507.
17. Chu H, Fierer N, Lauber CL, Caporaso JG, Knight R, et al. (2010) Soil bacterial diversity in the Arctic is not fundamentally different from that found in other biomes. Environ Microbiol DOI: 0.1111/j.1462–2920.2010.02277.x.
18. Pasternak Z, Al-Ashhab A, Gatica J, Gafny R, Avraham S, et al. (2013) Spatial and Temporal Biogeography of Soil Microbial Communities in Arid and Semiarid Regions. PLoS ONE 8: e69705. doi:10.1371/journal.pone.0069705.
19. Brodie E, Edwards S, Clipson N (2002) Bacterial Community Dynamics across a Floristic Gradient in a Temperate Upland Grassland Ecosystem. Microb Ecol 44: 260–270.
20. Rousk J, Baath E, Brookes PC, Lauber CL, Lozupone C, et al. (2010) Soil bacterial and fungal communities across a pH gradient in an arable soil. ISME J 4: 1340–1351.
21. Langenheder S, Szekely AJ (2011) Species sorting and neutral processes are both important during the initial assembly of bacterial communities. ISME J 5: 1086–1094.
22. Hovatter SR, Dejelo C, Case AL, Blackwood CB (2010) Metacommunity organization of soil microorganisms depends on habitat defined by presence of Lobelia siphilitica plants. Ecology 92: 57–65.
23. Yergeau E, Bezemer TM, Hedlund K, Mortimer SR, Kowalchuk GA, et al. (2010) Influences of space, soil, nematodes and plants on microbial community composition of chalk grassland soils. Environ Microbiol 12: 2096–2106.
24. Zinger L, Lejon DPH, Baptist F, Bouasria A, Aubert S, et al. (2011) Contrasting Diversity Patterns of Crenarchaeal, Bacterial and Fungal Soil Communities in an Alpine Landscape. PLoS ONE 6: e19950.
25. Peay KG, Bruns TD, Kennedy PG, Bergemann SE, Garbelotto M (2007) A strong species–area relationship for eukaryotic soil microbes: island size matters for ectomycorrhizal fungi. Ecol Lett 10: 470–480.
26. Peay KG, Schubert MG, Nguyen NH, Bruns TD (2012) Measuring ectomycorrhizal fungal dispersal: macroecological patterns driven by microscopic propagules. Mol Ecol 21: 4122–4136.
27. Talbot JM, Bruns TD, Taylor JW, Smith DP, Branco S, et al. (2014) Endemism and functional convergence across the North American soil mycobiome. P Natl Acad Sci USA 111: 6341–6346.
28. Taylor DL, Hollingsworth TN, McFarland JW, Lennon NJ, Nusbaum C, et al. (2013) A first comprehensive census of fungi in soil reveals both hyperdiversity and fine-scale niche partitioning. Ecol Monogr 84: 3–20.
29. Taylor DL, Herriott IC, Stone KE, McFarland JW, Booth MG, et al. (2010) Structure and resilience of fungal communities in Alaskan boreal forest soilsThis article is one of a selection of papers from The Dynamics of Change in Alaska's Boreal Forests: Resilience and Vulnerability in Response to Climate Warming. Can J For Res 40: 1288–1301.
30. Soininen J, McDonald R, Hillebrand H (2007) The distance decay of similarity in ecological communities. Ecography 30: 3–12.
31. Drakare S, Lennon JJ, Hillebrand H (2006) The imprint of the geographical, evolutionary and ecological context on species–area relationships. Ecol Lett 9: 215–227.
32. Fierer N, Leff JW, Adams BJ, Nielsen UN, Bates ST, et al. (2012) Cross-biome metagenomic analyses of soil microbial communities and their functional attributes. Proc Natl Acad Sci USA 109: 21390–21395.
33. Ramette A, Tiedje JM (2007) Multiscale responses of microbial life to spatial distance and environmental heterogeneity in a patchy ecosystem. P Natl Acad Sci USA 104: 2761–2766.
34. Lauber CL, Strickland MS, Bradford MA, Fierer N (2008) The influence of soil properties on the structure of bacterial and fungal communities across land-use types. Soil Biol Biochem 40: 2407–2415.
35. Dequiedt S, Saby NPA, Lelievre M, Jolivet C, Thioulouse J, et al. (2011) Biogeographical patterns of soil molecular microbial biomass as influenced by soil characteristics and management. Global Ecol Biogeog 20: 641–652.
36. Brodie E, Edwards S, Clipson N (2003) Soil fungal community structure in a temperate upland grassland soil. FEMS Microbiol Ecol 45: 105–114.
37. Lienhard P, Tivet F, Chabanne A, Dequiedt S, Lelièvre M, et al. (2013) No-till and cover crops shift soil microbial abundance and diversity in Laos tropical grasslands. Agron Sustain Dev 33: 1–10.
38. Griffiths RI, Thomson BC, James P, Bell T, Bailey M, et al. (2011) The bacterial biogeography of British soils. Environ Microbiol 13: 1642–1654.
39. Harte J, Kinzig A, Green J (1999) Self-Similarity in the Distribution and Abundance of Species. Science 284: 334–336.
40. Harte J, Smith AB, Storch D (2009) Biodiversity scales from plots to biomes with a universal species–area curve. Ecol Lett 12: 789–797.
41. Ranjard L, Dequiedt S, Lelievre M, Maron PA, Mougel C, et al. (2009) Platform GenoSol: a new tool for conserving and exploring soil microbial diversity. Environ Microbiol Rep 1: 97–99.
42. Arrouays D, Jolivet C, Boulonne L, Bodineau G, Saby NPA, et al. (2002) A new projection in France: a multi-institutional soil quality monitoring networkUne initiative nouvelle en France : la mise en place d'un réseau multi-institutionnel de mesure de la qualité des sols (RMQS). Comptes Rendus de l'Académie d'Agriculture de France 88: 93–105.
43. Ranjard L, Poly F, Nazaret S (2000) Monitoring complex bacterial communities using culture-independent molecular techniques: application to soil environment. Res Microbiol 151: 167–177.
44. Hill MO, Smith AEJ (1976) Principal Component Analysis of Taxonomic Data with Multi-State Discrete Characters. Taxon 25: 249–255.

45. Thioulouse J, Dray S (2007) Interactive multivariate data analysisin R with the ade4 and ade4TkGUI packages. J stat softw 22: 1–14.

46. R Development Core Team (2011) *R: A Language and Environment for Statistical Computing*, Vienna, Austria.

47. Legendre P, Borcard D, Peres-Neto PR (2005) Analyzing beta diversity: Partitioning the spatial variation of community composition data. Ecol Monogr 75: 435–450.

48. Legendre P, Fortin MJ (2010) Comparison of the Mantel test and alternative approaches for detecting complex multivariate relationships in the spatial analysis of genetic data. Mol Ecol Resour 10: 831–844.

49. Dray SP, Legendre P, Peres-Neto PR (2006) Spatial modelling: a comprehensive framework for principal coordinate analysis of neighbour matrices (PCNM) Ecol Model 196: 483–493.

50. Oksanen J, Blanchet FG, Kindt R, Legendre P, Minchin PR, et al. (2011) vegan: Community Ecology Package.

51. Bellier E, Monestiez P, Durbec J-P, Candau J-N (2007) Identifying spatial relationships at multiple scales: principal coordinates of neighbour matrices (PCNM) and geostatistical approaches. Ecography 30: 385–399.

52. Drenovsky RE, Steenwerth KL, Jackson LE, Scow KM (2010) Land use and climatic factors structure regional patterns in soil microbial communities. Global Ecol Biogeog 19: 27–39.

53. Bell T, Ager D, Song J-I, Newman JA, Thompson IP, et al. (2005) Larger Islands House More Bacterial Taxa. Science 308: 1884.

54. Woodcock S, Curtis TP, Head IM, Lunn M, Sloan WT (2006) Taxa–area relationships for microbes: the unsampled and the unseen. Ecol Lett 9: 805–812.

55. MacArthur RH, Wilson EO (1967) *The theory of island biogeography/Robert H. MacArthur and Edward O. Wilson*. Princeton University Press, Princeton, N.J.

56. Noguez AM, Arita HT, Escalante AE, Forney LJ, Garcia-Oliva F, et al. (2005) Microbial macroecology: highly structured prokaryotic soil assemblages in a tropical deciduous forest. Global Ecol Biogeog 14: 241–248.

57. Morlon H, Chuyong G, Condit R, Hubbell S, Kenfack D, et al. (2008) A general framework for the distance–decay of similarity in ecological communities. Ecol Lett 11: 904–917.

58. de Boer W, Folman LB, Summerbell RC, Boddy L (2005) Living in a fungal world: impact of fungi on soil bacterial niche development. FEMS Microbiol Rev 29: 795–811.

59. Lennon JT, Jones SE (2011) Microbial seed banks: the ecological and evolutionary implications of dormancy. Nat Rev Micro 9: 119–130.

60. Jobe RT (2008) Estimating ladscape-scale species richness: reconciling frequency- and turnover-based approaches. Ecology 89: 174–182.

61. Fenchel T (2003) Biogeography for Bacteria. Science 301: 925–926.

62. Dumbrell AJ, Nelson M, Helgason T, Dytham C, Fitter AH (2010) Relative roles of niche and neutral processes in structuring a soil microbial community. ISME J 4: 337–345.

63. Hossain Z, Sugiyama S-I (2011) Geographical structure of soil microbial communities in northern Japan: Effects of distance, land use type and soil properties. Eur J Soil Biol 47: 88–94.

64. Bryant JA, Lamanna C, Morlon H, Kerkhoff AJ, Enquist BJ, et al. (2008) Microbes on mountainsides: Contrasting elevational patterns of bacterial and plant diversity. P Natl Acad Sci USA 105: 11505–11511.

65. Monroy F, van der Putten WH, Yergeau E, Mortimer SR, Duyts H, et al. (2012) Community patterns of soil bacteria and nematodes in relation to geographic distance. Soil Biol Biochem 45: 1–7.

66. Cho JC, Tiedje JM (2000) Biogeography and Degree of Endemicity of Fluorescent Pseudomonas Strains in Soil. Appl Environ Microb 66: 5448–5456.

67. Lozupone CA, Knight R (2007) Global patterns in bacterial diversity. Proceedings of the National Academy of Sciences 104: 11436–11440.

68. Auguet J-C, Barberan A, Casamayor EO (2009) Global ecological patterns in uncultured Archaea. ISME J 4: 182–190.

69. Ranjard L, Richaume AS (2001) Quantitative and qualitative microscale distribution of bacteria in soil. Res Microbiol 152: 707–716.

70. Kong AYY, Scow KM, Cordova-Kreylos AL, Holmes WE, Six J (2011) Microbial community composition and carbon cycling within soil microenvironments of conventional, low-input, and organic cropping systems. Soil Biol Biochem 43: 20–30.

Primates Living Outside Protected Habitats Are More Stressed: The Case of Black Howler Monkeys in the Yucatán Peninsula

Ariadna Rangel-Negrín[1], Alejandro Coyohua-Fuentes[1], Roberto Chavira[2], Domingo Canales-Espinosa[1], Pedro Américo D. Dias[1]*

1 Instituto de Neuroetología, Universidad Veracruzana, Xalapa, Veracruz, Mexico, **2** Instituto de Ciencias Médicas y Nutrición Salvador Zubirán, México D.F., Mexico

Abstract

The non-invasive monitoring of glucocorticoid hormones allows for the assessment of the physiological effects of anthropogenic disturbances on wildlife. Variation in glucocorticoid levels of the same species between protected and unprotect areas seldom has been measured, and the available evidence suggests that this relationship may depend on species-specific habitat requirements and biology. In the present study we focused on black howler monkeys (*Alouatta pigra*), a canopy-dwelling primate species, as a case study to evaluate the physiological consequences of living in unprotected areas, and relate them with intragroup competition and competition with extragroup individuals. From February 2006 to September 2007 we collected 371 fecal samples from 21 adults belonging to five groups (two from protected and three from unprotected areas) in Campeche, Mexico. We recorded agonistic interactions within groups and encounters with other groups (1,200 h of behavioral observations), and determined fecal glucocorticoid metabolite (FGM) concentrations with radioimmunoassays. We used linear mixed models and Akaike's information criterion to choose the best model explaining variation in FGM concentrations between protected and unprotected areas calculated from five categorical variables: habitat type (protected vs. unprotected), participation in agonistic interactions, intergroup encounters, sex and female reproductive state, and season. The best model included habitat type, the interaction between habitat type and agonism, and the interaction between habitat type and season. FGM concentrations were higher in unprotected habitats, particularly when individuals were involved in agonistic interactions; seasonal variation in FGM concentrations was only detected in protected habitats. High FGM concentrations in black howler monkeys living in unprotected habitats are associated with increased within-group food competition and probably associated with exposure to anthropogenic stressors and overall food scarcity. Because persistent high GC levels can be detrimental to health and fitness, populations living in disturbed unprotected areas may not be viable in the long-term.

Editor: Antje Engelhardt, German Primate Centre, Germany

Funding: This work was supported by CFE (RGCPTTPUV-001/04), Universidad Veracruzana, CONACyT (Grant Number: 235839; i010/152/2014 & C-133/2014) and Idea Wild. The funders had no role in study design, data collection and analysis, decision to publish, or preparation of the manuscript.

Competing Interests: The authors have declared that no competing interests exist.

* Email: paddias@hotmail.com

Introduction

The effectiveness of protected areas to preserve biodiversity has been questioned [1], despite the prominence of protected areas as a cornerstone of current conservation efforts. Throughout the world, land clearing, logging, fires, hunting and grazing are less pronounced inside protected areas than in their surroundings [2], indicating that protected areas represent refuges for many threatened species and natural ecosystem processes [3]. The factors with the most obvious negative consequences on wildlife persistence in unprotected areas are loss of natural habitat and hunting. However, in these areas wildlife faces numerous deterministic and stochastic threats that affect their biology and behavior, and may lead to population decline and extinction [4]. For instance, habitat disturbance, such as habitat loss, results in shorter breeding seasons in blue tits (*Parus caeruleus*) [5], altered home ranges in greater gliders (*Petauroides volans*) [6], and increased food competition in Tana mangabeys (*Cercocebus galeritus*) [7]. In this context, understanding the physiological mechanisms underlying the responses of wildlife to anthropogenic disturbances may be instrumental to predict its persistence in unprotected areas.

The non-invasive sampling of glucocorticoid (GC) hormones allows researchers to assess the physiological effects of anthropogenic disturbances on wildlife [8]. The modulation of GCs is part of the adaptive physiological stress response, and these hormones are involved in diverse actions [9]. GCs may alter an organism's response to an ongoing stressor and may prepare an organism's response to a subsequent stressor. Independently from the nature of the stressful stimuli, during the stress response, GCs increase circulating glucose through a number of mechanisms, contributing to the depletion of present and, when the action of a stressor is prolonged, future energy stores [9]. When GC levels remain elevated for days or weeks they may become detrimental to health and fitness [10]. Anthropogenic disturbances, such as pollution, exposure to humans, and human-induced habitat transformation, elicit changes in GC levels in a variety of species [11]. For instance, GCs increase in: Magellanic penguins (*Spheniscus magellanicus*) exposed to oil following a petroleum spill [12]; wolves (*Canis lupus*) and elk (*Cervus elaphus*) that live in areas with high vehicle traffic [13]; and in northern spotted owls (*Strix occidentalis caurina*) that live in areas with timber harvesting [14]. However, variation in GC levels of the same species between protected and unprotected areas seldom has been measured, and the available evidence suggests that such variation may depend on species-specific habitat requirements and biology. For instance, whereas in hyenas (*Crocuta crocuta*), lions (*Panthera leo*) and maned wolves (*Chrysocyon brachyurus*) GC levels of individuals increase with increasing anthropogenic pressures in unprotected areas [15–17], there is no statistically significant difference in GC levels between elephants (*Loxodonta africana*) living in protected areas and in community conservation areas where human pastoral settlements and livestock grazing occur [18].

In the present study, we focused on the Yucatan black howler monkey (*Alouatta pigra*; hereafter, black howler monkeys) to evaluate the physiological consequences of living in unprotected areas, and relate them to changes in behavior. Black howler monkeys are tree-dwelling primates with a geographic distribution restricted to the Yucatan Peninsula in Mexico and Belize, and some parts of northern and central Guatemala [19]. In 2003, the International Union for Conservation of Nature conservation status of this species was revised from Insufficiently known to Endangered due to habitat loss and better available information [20]. There is evidence that habitat loss, hunting and continued presence of humans have a negative effect on the biology and behavior of black howler monkeys. In disturbed habitats, black howler monkeys: 1) live at densities up to five times higher than in extensive forests [21]; 2) search for food sources outside their habitats by walking on the ground, where they may be predated by domestic dogs and killed during road crossings [22], [23]; 3) are hunted for food and pet trading [24], [25]; and 4) show increased GC levels [26]. If black howler monkeys living in unprotected habitats show increased GC levels (associated with frequent exposure to anthropogenic stressors) and competition for resources (associated with more individuals living in smaller forests) their long-term presence in these habitats could be compromised due to reduced reproduction, immune function, and survival [27]. Therefore, the assessment of GC levels and competition levels in black howler monkeys may be very informative for the development of conservation policies involving populations of this endangered species.

We compared fecal GC metabolite (FGM) levels between black howler monkeys living in small unprotected forest fragments and black howler monkeys living in nearby extensive protected areas. We hypothesized that individuals living in unprotected forests would have higher FGM levels due to increased physiological stress associated with anthropogenic activities compared to individuals in protected areas. Because in forest fragments population densities are higher and the availability of food resources for howler monkeys is lower [28], we further hypothesized that higher FGM concentrations of black howler monkeys living in unprotected habitats would be associated with increased within- and between-group competition. Besides habitat type, within- and between-group competition, we examined whether the effects of sex, female reproductive state and environmental seasonality on FGM variation (which have been demonstrated to

consistently affect the physiological stress response) [29], [30] varied according to habitat type.

Methods

Our research complied with the Mexican law and was approved by the corresponding authorities (SEMARNAT SGPA/DGVS/ 01273/06 & 04949/07); and complied with the Guidelines for the Treatment of Animals in Behavioral Research and Teaching from the Animal Behavior Society.

Study Area

We focused on populations of black howler monkeys living in the state of Campeche in Mexico. In Campeche the climate is hot and humid, and mean annual rainfall is 1,300 mm, with a drier season from November to May (mean monthly rainfall $\pm$ SD $= 43.7 \pm 25.8$ mm), and a wetter period between June and October (218.9 ± 14.1 mm). Mean annual temperature is 26°C [31].

Campeche has a total area of 57,924 km², from which approximately 40% is protected. There are black howler monkeys in two of the three larger protected areas in Campeche, the Calakmul Biosphere Reserve (18°19'00.28" N, 89°51'28.92" W) and the Laguna de Términos Reserve (18°51'15.38" N, 91°18'41.70" W). Together, these encompass an area of 14,282 km². Although human activities occur in these reserves (e.g., apiculture, fishery), habitat availability is high for black howler monkeys and they face low anthropogenic stressors (e.g., hunting) [32]. We studied one group of black howler monkeys in each of these reserves (hereafter, protected habitat; Table 1). In contrast, the remaining non-urban territory of Campeche consists of highly humanized landscapes, where original habitats have been converted into forest-agricultural mosaics. Black howler monkeys living in these landscapes occupy forest fragments of variable size where they face multiple anthropogenic stressors on a daily basis that are generally absent in protected habitats, such as livestock grazing, predation threat by domestic animals or forest fires associated with slash-and-burn agriculture [23] (Figure 1). We studied three groups of black howler monkeys (Rancho El Álamo: 18°48'45.44" N, 90°58'54.61" W; ejido Chicbul: 18°46'51.66" N, 90°56'13.45" W; ejido General Ignacio Gutiérrez: 18°54'6.58" N, 90°53'37.90" W) living under these circumstances in three different forest fragments with areas <1 km² (hereafter, unprotected habitat; Table 1). Mean ($\pm$SD) group size (7.1 ± 0.5 individuals) and composition (adult males: 1.5 ± 0.5; adult females: 2.3 ± 0.7; immatures: 3.3 ± 1.1) were very similar across groups (Table 1). All groups were habituated to human observers before the beginning of systematic behavioral and fecal sampling. Habituation consisted on the presence of three to four researchers near the group for a total of five days each week for two weeks (*ca.* 30 h). During habituation, researchers performed the same activities that would be performed during the study and observed whether animals reacted to them. We did not observe flights, avoidance, curiosity or displays directed towards researchers during habituation.

Behavioral data collection

To study within-group feeding competition, from February 2006 to September 2007 we recorded all occurrences of agonistic interactions (displacements, threats, chases and fights) in feeding context, defined as any interaction exchanged among adult individuals ($N = 21$; 12 females and nine males) when at least one of the individuals fed in a 10 min period before or after the interaction. When there was a latency of ≤ 5 seconds between two interactions of the same type, only one interaction was recorded.

Figure 1. Black howler monkeys foraging on the ground in a recently burned forest fragment.

The sampling of social interactions was performed by a single observer (A. R.-N.) during complete day follows (i.e., 6:00–7:00 to 17:00–18:00, depending on the time of year). High spatial cohesion of howler monkey groups [33] allows for the observation of all individuals at the same time at any given point in time, and we therefore assume that we were able to sample adequately social interactions that occurred in feeding context. We also recorded encounters (both visual and vocal) with extragroup individuals to study intergroup competition. In each group, behavioral sampling was performed for five days a week (Monday to Friday) during four weeks in each season. A total of 30 sampling hours were collected each week, resulting in 240 sampling hours per group.

Following Dias et al. [34] and Van Belle et al. [35], we classified females as pregnant (defined as the 6 months preceding parturition), lactating (defined by either observations of lactation, or, as starting from the day of parturition until 15 months), or in other reproductive state (neither pregnant nor lactating). We based this classification on observations of births and lactation during the study and during periodical visits to the study groups up to 7 months after the end of the study.

Fecal sample collection and hormone analyses

Fecal samples were collected opportunistically during behavioral samplings whenever they could be matched with individuals. Fresh samples uncontaminated by urine were collected from the forest floor and deposited in polyethylene bags labeled with the identity of each individual. Based on the estimated retention time of the digesta in the gut of *Alouatta pigra* of *ca.* 35 h [36], in order to match the occurrence of social interactions with the FGM levels of *Alouatta pigra* in each sampling week, we collected fecal samples from Tuesdays to Sundays. We analyzed 371 fecal samples (129 from individuals in protected habitats and 242 from individuals in unprotected habitats), with an average (± SD) of 17.7 ± 6.5 samples per individual, 3.2 ± 1.2 samples per week per individual and 9.8 ± 2.9 samples per individual per season. Fecal samples were kept in a cooler with frozen gel packs while in the field and stored at the end of the day in a freezer at $-20°C$ at the field station until extraction was performed following the methods described by Rangel-Negrín *et al.* [37].

FGM assays were conducted at the Instituto de Ciencias Médicas y Nutrición Salvador Zubirán, in Mexico City. We used a radioimmunoassay, a commercial ^{125}I cortisol kit (SIEMENS Coat-a-count Cortisol), and a gamma counter (Cobra 5005, Packard Inc., MI, USA) to measure FGM levels in all samples.

As a biological validation of our assays, we determined the short-term effect of capture (an acute stressor) and anesthesia (ketamine) on the FGM excretion profile of three black howler monkeys (one male and two females) following the samples collection and conservation procedures described above. We collected all fecal samples ($n = 42$; 14 ± 1 SD samples per individual) from 72 h before to 96 h after capture, and compared pre-capture levels with peak concentrations (i.e., the highest post-stressor values that were $\geq 2*SD$ above the mean concentration before capture) with a Wilcoxon signed-rank test. FGM levels peaked at a mean (± SD) of 35.3 ± 9.9 h after capture. Peak FGM levels were significantly higher than pre-capture levels for the three individuals (female 1: $Z_{13} = 2.27$, $p < 0.05$; female 2: $Z_{14} = 3.29$, $p < 0.001$; male: $Z_{15} = 3.41$, $p < 0.001$), indicating that our FGM assays reliably measured adrenal responses of black howler monkeys to stressors. Our capture and handling procedures [38] were approved by Mexican authorities (SEMARNAT, SGPA/ DGVS/01273/06 & 04949/07).

Howler monkeys' pooled fecal extracts, when added to the standard curve points, exhibited an accuracy of $R^2 = 0.99$ ($n = 5$, $p = 0.004$), and serial dilutions of a fecal pool from howler monkeys

Table 1. Characteristics of the groups and habitats of black howler monkeys that were studied in unprotected and protected areas in Campeche, Mexico.

	Unprotected habitat			Protected habitat	
	Álamo	Chicbul	I. Gutiérrez	Calakmul	Términos
Area (km²)	0.96	0.05	0.02	7231.9	7061.5
Distance to nearest village (m)	1,640	771	1,092	27,000	18,100
Population density (ind/km²)	52	120	63	15.2	19.2
Group size *	6.4	7.7	6.7	7.3	7.3
Number of adult males *	1.5	1	2	2	1
Number of adult females *	2	2	3	1.5	3
Sampling months (dry/wet season samplings)	January/August	January/August	March/July	February/August	February/July

*Calculated as mean number of individuals during the two sampling periods (i.e., rainy and dry season).

yielded results that ran parallel to the FGM standards ($R^2 = 0.97$, $n = 5$, $p<0.001$). Samples were run in the order in which they were collected in a total of 58 assays, with a new set of quality controls beginning with assay 27. Cortisol intra-assay variation averaged 6.8% (fecal extract pool, $n = 6$). Inter-assay variation, estimated for the 58 assays from fecal pools with varying levels of cortisol, averaged 19.3% (low), 15.1% (medium), and 7.2% (high). All samples were run in duplicate, and mean FGM values are reported as ng/g (dry feces).

Statistical analyses

We calculated weekly individual participation in agonistic interactions, participation in intergroup encounters (i.e., participated vs. did not participate), and mean FGM levels. Four individuals emigrated from our study groups from the first to the second sampling period (one in El Álamo and three in Calakmul). Therefore, our analyses were performed on 152 individual weeks.

To assess consistency within habitat types in individual variation in FGM levels, we compared the first with the last fecal sample collected for each individual in each season with a nonparametric Wilcoxon signed rank test.

We used linear mixed models (LMM) to assess the effects of five categorical variables on FGM levels: habitat type, participation in agonistic interactions (i.e., within-group competition), encounters with other groups (i.e., between-group competition), sex/female reproductive state and season. Because we were specifically interested in assessing the effects of habitat type on FGM levels, variables were included in the models as their interaction with habitat type (e.g., sex/female reproductive state x habitat type, season x habitat type). As the same individuals were repeatedly sampled, we included individual identity as a random factor in the model with first-order autocorrelation as a covariance structure. We used Akaike's information criterion (AIC) to choose the best model (i.e., lowest AIC: [39]). Post-hoc exploratory analyses were conducted on the effects of the interaction between habitat type and categorical within-group competition, and the interaction between habitat type and seasonality on FGM levels. For these analyses we used LMMs, in which pairwise combinations of habitat type x within-group competition and habitat type x seasonality were included as predictive fixed factors and individual identity was included as a random factor. FGM levels were normalized via logarithmic (ln) transformation. We checked that the assumptions of normally distributed and homogeneous residuals were fulfilled. All LMM analyses were performed in SPSS 22.0 (SPSS, Chicago, Illinois, U.S.A.).

Results

There was no significant difference within seasons between the first and last FGM concentration measured for each individual in protected (rainy, $Z_6 = 0.524$, $p = 0.601$; dry, $Z_9 = 1.01$, $p = 0.314$) and unprotected habitats (rainy, $Z_9 = 1.125$, $p = 0.260$; dry, $Z_9 = 0.415$, $p = 0.678$).

Descriptive data on rates of within-group agonistic interactions and encounters with extragroup individuals are presented in Table 2.

Factors contributing to increased FGM levels

Our final model ($F_{5,143} = 4.683$, $p = 0.001$) included habitat type ($F_{1,18.220} = 7.504$, $p = 0.013$), the interaction between habitat type and agonistic interactions ($F_{2,134.866} = 5.104$, $p = 0.007$) and the interaction between habitat type and season ($F_{2,134.384} = 3.937$, $p = 0.022$; Table 3). The interaction between sex/female repro-

ductive state and habitat type, as well as encounters with extragroup individuals were not selected.

Overall mean FGM levels ($\pm$ SE) of individuals living in unprotected habitats were approximately 20% higher (338.9 ± 19.9 ng/g) than those of individuals living in protected habitats (266.2 ± 20.3 ng/g; Figure 2a).

Agonistic interactions among group members occurred at similar proportions during sampling weeks in both habitat types (protected: 57.1% of individual/weeks; unprotected: 62%). However, when individuals living in unprotected habitats participated in agonistic interactions they had significantly (all post-hoc pairwise LMM $p<0.01$) higher mean ($\pm$ SE) FGM levels (400.7 ± 30.5 ng/g) than when they were not involved in such interactions (292.7 ± 24.6 ng/g), or than individuals living in protected habitats (involved in agonistic behavior: 271.1 ± 30.9 ng/g; non-involved in agonistic behavior: 263.3 ± 26.9 ng/g; Figure 2b).

FGM levels ($\pm$ SE) of individuals living in protected habitats (204.9 ± 18.3 ng/g) were significantly lower (all post-hoc pairwise LMM $p<0.05$) during the rainy season than during the dry season (309.5 ± 30.2 ng/g) and were lower than in both seasons in unprotected habitats (rainy: 315.8 ± 23.9 ng/g; dry: 363.8 ± 32.2 ng/g; Figure 2c).

Discussion

Variation in FGM levels of black howler monkeys was significantly explained by the variables that were examined. In general, variation in FGM concentrations was highly consistent within each habitat type, and as predicted, FGM levels were higher in individuals living in unprotected habitats, particularly when they were involved in agonistic interactions. Additionally, seasonal variation of FGM levels was absent in unprotected habitats (albeit significantly higher overall than in individuals from protected habitats), whereas in protected areas, FGM decreased during the wet season. These results converge with previous evidence documented for this species [26], and suggest that black howler monkeys living in unprotected habitats have higher FGM levels due to increased physiological stress associated with anthropogenic disturbance, within-group competition for resources and possibly food scarcity.

Black howler monkeys living in unprotected habitats in Campeche face numerous anthropogenic stressors. Slash-and-burn agriculture, cattle grazing, hunting, threats and attacks by domestic dogs, and fires were observed during the study in unprotected areas. These stressors have the potential to increase GC levels [11], and none of the events were recorded in protected habitats. In addition to the presence of researchers (usually three people), one of the groups living in protected areas (Calakmul Biosphere Reserve) was visited on several occasions by small groups (three to five people) of tourists, suggesting that either contact with small groups of quiet tourists does not lead to increases in FGM of black howler monkeys living in protected areas, or that we were unable to detect such effect. The fact that a positive relationship between GCs levels and intensity of tourist visitation in protected areas has been found in the majority of studies that have addressed this subject in other species [13], [40], [41] suggests that further research is required to understand the responses of black howler monkeys to the presence of tourists in protected areas [42].

Contrary to our expectation, the patterns of agonism were similar between habitat types, but in unprotected habitats individuals showed stronger FGM output to participation in agonistic interactions than individuals in protected habitats.

Table 2. Mean (± SE) rates of within-group agonistic interactions and encounters with extragroup individuals for groups of black howler monkeys in unprotected and protected areas in Campeche, Mexico.

Habitat type		Weekly rates of agonistic interactions	Weekly rates of encounters with extragroup individuals
Unprotected	Álamo	0.021 (±0.003)	0.033 (±0.039)
	Chicbul	0.045 (±0.009)	0.004 (±0.031)
	I. Gutiérrez	0.052 (±0.013)	0.007 (±0.025)
Protected	Calakmul	0.034 (±0.006)	0 (±0.0)
	Términos	0.057 (±0.009)	0 (±0.0)

Actually, when black howler monkeys living in unprotected habitats were not involved in agonistic interactions, their mean FGM levels, although still higher than in protected habitats, were more similar to FGM levels of individuals from protected habitats than to FGM in weeks with agonism. Therefore, in unprotected habitats black howler monkeys present higher FGM responsiveness to within-group competition. Social interactions have the potential to represent strong stressors, because they may entail both a high degree of unpredictability and an increase in metabolic demands [43], [44]. GCs play a critical role in the stimulation of gluconeogenesis and the mobilization of amino and fatty acids from body stores [45]. Thus, changes in energetic expenditure and energy balance can affect GC production, independently of psychological variables. As a consequence, in unprotected habitats, where in addition to being exposed to anthropogenic stressors, individuals may experience frequent food-deprivation [28], it is possible that acute physical exercise, such as that associated with agonistic interactions (e.g., chases, prolonged threats), increases GC secretion [46], [47]. Still, this result should be interpreted cautiously, because agonistic interactions were overall infrequent (ca. 0.04 interactions/h in both habitat types; see also [26]) limiting the possibility to match temporarily FGM concentrations with behavioral data [48]. Future studies should account for this limitation, and include additional measures of energy expenditure (e.g., C-peptide [49]) to understand the interplay of within-group competition, energy expenditure and FGM secretion.

As observed in many vertebrates [29], [50], the FGM of black howler monkeys varied seasonally, but this variation was only detected in protected habitats, where a decrease in hormone levels occurred in the wet season. Although we did not measure food availability, it has been reported that there is a reduction in the number of fruiting trees (a preferred food resource for this primate species) during the dry season in the forests of the Yucatan Peninsula [51], suggesting that black howler monkeys may face reduced food availability during this period. Also, fruit availability is typically reduced in forest fragments compared to continuous forests [52], [53]. In a closely related howler monkey species (*A. palliata*) [54] as well as in other primate and non-primate species (e.g., elephants: [46]; Sykes' monkeys, *Cercopithecus mitis albogularis*: [55]), when food availability decreases individuals increase energy expenditure, leading to higher FGM levels. A similar effect could explain the results found in protected habitats. FGM concentrations remained high throughout the year for black howler monkeys living in unprotected habitats. Therefore, whereas in extensive protected habitats individuals may experience seasonal increases in metabolic stress, black howler monkeys living in unprotected habitats may suffer from long-term stress due to comparatively constant food scarcity and exposure to anthropo-

genic stressors. Because sex was not a significant predictor of variation in GC concentrations between protected and unprotected areas and a similar number of females were lactating and pregnant in both habitats, it is unlikely that these results were affected by seasonal variation in female reproductive state [54], [56].

In Campeche, all protected areas are large (mean = 3,255.4 km^2), whereas few large forest remnants exist elsewhere [57]. As a consequence, in this study all protected habitats corresponded to extensive forests and all unprotected habitats were small forest fragments; and in our results the effects of habitat type on behavior and FGM cannot be separated from those of habitat size. In red howler monkeys (*Alouatta seniculus*) FGM levels vary significantly between fragments, but neither size nor human impact (logging, hunting) predict such variation, suggesting that other factors besides habitat size and human activities may be more important to understand the physiological stress responses of howler monkeys living in disturbed habitats [58]. For black howler monkeys, it remains for future research to determine if, independently from habitat size, habitat protection, accompanied by a decrease in anthropogenic pressures, is sufficient to prevent increased FGM levels. A recent study conducted in the Lacandona rainforest, however, suggests otherwise. Population composition and structure of black howler monkey groups are more strongly affected by local-scale habitat metrics, such as habitat size, than by landscape-scale variables [59]. Furthermore, forest fragment size has been proposed to be the main factor constraining populations of howler monkeys living in fragmented habitats, probably because fragment size is positively related to food availability, and negatively related to anthropogenic pressures, physiological stress and parasite loads [28]. Therefore, as for other large canopy-dwelling Neotropical mammals, including primates [60], [61], the conservation of black howler monkey populations will probably depend on the protection and maintenance of large forest tracts.

In conclusion, black howler monkeys living in unprotected areas have high FGM compared to individuals living in protected areas. As persistent high GC levels may be detrimental to health and fitness [9], black howler monkey populations living in disturbed unprotected habitats may not be viable in the long-term because there is an extinction debt to be paid in these habitats [62], [63]. To test this prediction, future research should concentrate on quantifying the effects of anthropogenic disturbances on population structure and dynamics to determine population viability. Because such data will be difficult to get, as it requires long-term studies, it may be too late for the conservation of this species by the time the data would be available. Therefore, the present study supports the idea that conservation measures may need to be taken earlier.

Table 3. Parameter estimates for the best model explaining variation in FGM concentrations between groups of black howler monkeys living in protected and unprotected habitats in Campeche, Mexico.

Response variable	Estimate	SE	d.f.	t ratio	P	Confidence intervals	
						Upper limit	Lower limit
Intercept	2.42	0.05	56.95	51.177	<0.001	2.32	2.51
Habitat type	0.00	0.06	64.81	0.006	0.995	−0.13	0.13
Habitat type [non-protected]*agonism	0.15	0.05	140.36	3.112	0.002	0.06	0.25
Habitat type [protected]*agonism	0.05	0.07	129.79	0.723	0.471	−0.08	0.18
Habitat type [non-protected]*season	−0.04	0.05	127.39	−0.718	0.474	−0.13	0.06
Habitat type [protected]*season	−0.17	0.06	142.19	−2.713	0.007	−0.29	−0.05

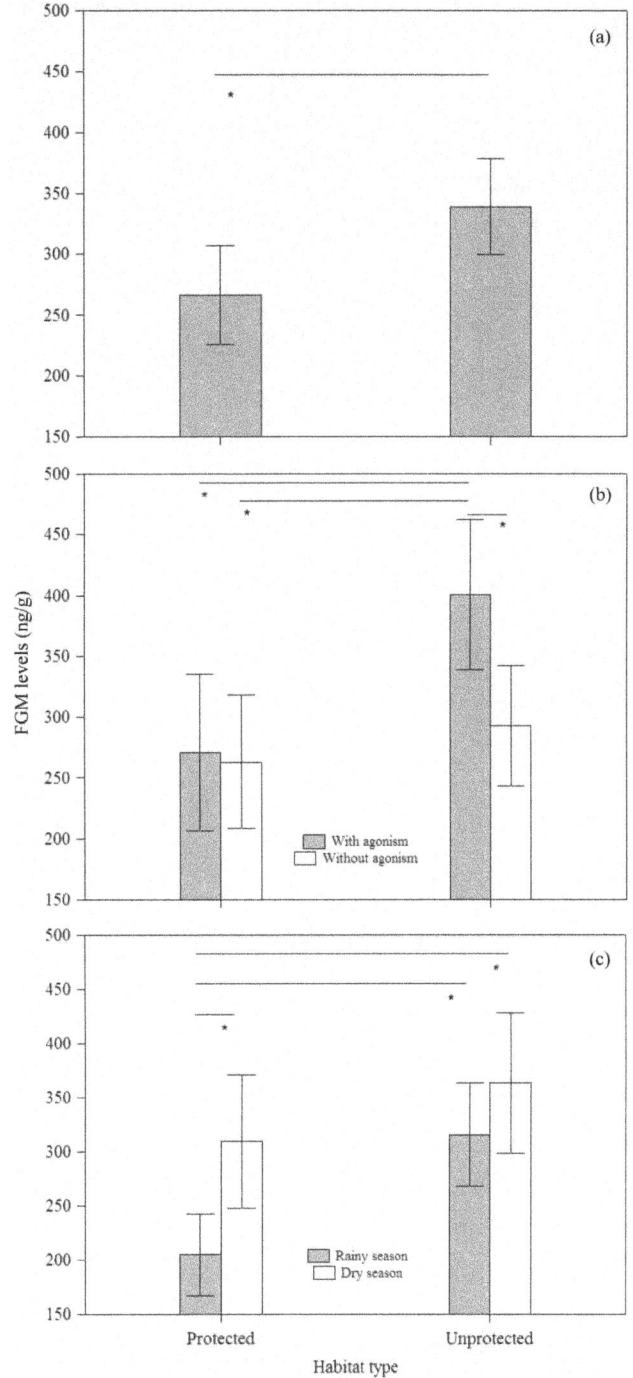

Figure 2. Variation in mean (±SE) FGM (fecal glucocorticoid metabolite) levels as a function of: a) habitat type; b) participation in agonistic interactions per habitat type; c) season per habitat type. Significant differences are denoted by an asterisk.

Acknowledgments

We thank all the students and volunteers that helped during fieldwork. The following people and institutions granted permission to work in their properties and facilitated our fieldwork: Comisarios Ejidales de Abelardo Domínguez, Calax, Conhuas, Plan de Ayala, and Carmén Gómez; Ayuntamiento de Calakmul; Lic. C. Vidal and Lic. L. Álvarez, INAH Campeche; Biól. F. Durand Siller, Reserva de la Biósfera Calakmul,

CONANP; Ing. V. Olvera, El Álamo. We thank C.M. Schaffner and J. Dunn for very useful comments on pre-submission versions of the manuscript; and A. Engelhardt, J. Lynch Alfaro and two anonymous referees for encouraging and positive comments, which helped greatly improve the manuscript. A.R.N. and P.A.D.D. thank Mariana for her continued inspiration to study primate behavior.

Author Contributions

Conceived and designed the experiments: ARN PADD. Performed the experiments: ARN ACF RC DCE PADD. Analyzed the data: ARN RC PADD. Contributed reagents/materials/analysis tools: ARN RC DCE PADD. Wrote the paper: ARN PADD.

References

1. Ghimire KB, Pimbert MP (1997) Social change and conservation: environmental politics and impacts of national parks and protected areas. London: Earthscan. 352 p.

2. Bruner AG, Gullison R, Rice R, da Fonseca G (2011) Effectiveness of parks in protecting tropical biodiversity. Science 291: 125–128.

3. Laurance WF, Carolina Useche C, Rendeiro J, Kalka M, Bradshaw CJA, et al. (2012) Adverting biodiversity collapse in tropical forest protected areas. Nature 489: 290–294.

4. Fischer J, Lindenmayer DB (2007) Landscape modification and habitat fragmentation: a synthesis. Global Ecol Biogeogr 16: 265–280.

5. Hinsley SA, Rothery P, Bellamy PE (1999) Influence of woodland area on breeding success in Great Tits Parus major and Blue Tits Parus caeruleus. J Avian Biol 30: 271–281.

6. Pope ML, Lindenmayer DB, Cunningham RB (2004) Patch use by the Greater Glider in a fragmented forest ecosystem. I. Home range size and movements. Wildl Res 31: 559–568.

7. Wieczkowski J (2005) Examination of increased annual range of a Tana mangabey (Cercocebus galeritus) group. Am J Phys Anthropol 128: 381–388.

8. Palme R, Rettenbacher S, Touma C, El-Bahr SM, Möstl E (2005) Stress hormones in mammals and birds: comparative aspects regarding metabolism, excretion, and noninvasive measurement in fecal samples. Ann N Y Acad Sci 1040: 162–171.

9. Sapolsky RM, Romero LM, Munck AU (2000) How do glucocorticoids influence stress responses? Endocrinol Rev 21: 55–89.

10. Breuner CW, Patterson SH, Hahn TP (2008) In search of relationships between the acute adrenocortical response and fitness. Gen Comp Endocrinol 157: 288–295.

11. Busch DS, Hayward LS (2009) Stress in a conservation context: a discussion of glucocorticoid actions and how levels change with conservation-relevant variables. Biol Conserv 142: 2844–2853.

12. Fowler GS, Wingfield JC, Boersma PD (1995) Hormonal and reproductive effects of low level of petroleum fouling in Magellanic penguins (Spheniscus magellanicus). The Auk 112: 382–389.

13. Creel S, Fox JE, Hardy A, Sands J, Garrott B, et al. (2002) Snowmobile activity and glucocorticoid stress responses in wolves and elk. Conserv Biol 16: 809–814.

14. Wasser SK, Bevis K, King G, Hanson E (1997) Noninvasive physiological measures of disturbance in the northern spotted owl. Conserv Biol 11: 1019–1022.

15. Van Meter PE, French JA, Dloniak SM, Watts HE, Kolowski JM, et al. (2009) Fecal glucocorticoids reflect socio-ecological and anthropogenic stressors in the lives of wild spotted hyenas. Horm Behav 55: 329–337.

16. Spercoski KM, Morais RN, Morato RG, de Paula RC, Azevedo FC, et al. (2012) Adrenal activity in maned wolves is higher on farmlands and park boundaries than within protected areas. Gen Comp Endocrinol 179: 232–240.

17. Creel S, Christianson D, Schuette P (2013) Glucocorticoid stress responses of lions in relationship to group composition, human land use, and proximity to people. Conserv Physiol 1: 10.1093/conphys/cot021.

18. Ahlering MA, Maldonado JE, Eggert LS, Fleischer RC, Western D, et al. (2013) Conservation outside protected areas and the effect of human-dominated landscapes on stress hormones in savannah elephants. Conserv Biol 27: 569–575.

19. Horwich RH, Johnson ED (1986) Geographical distribution of the black howler (Alouatta pigra) in Central America. Primates 27: 53–62.

20. Marsh LK, Cuarón AD, Cortés-Ortiz L, Shedden A, Rodríguez-Luna E, et al. (2008) Alouatta pigra. IUCN 2014 Red List of Threatened Species. Version 2013.1. Available: http:www.iucnredlist.org. Accessed 11 June 2014.

21. Van Belle S, Estrada A (2006) Demographic features of Alouatta pigra populations in extensive and fragmented forest. In: Estrada A, Garber PA, Pavelka MSM, Luecke L, editors. New perspectives in the study of Mesoamerican primates: distribution, ecology, behavior and conservation. New York: Kluwer. pp. 121–142.

22. Candelero-Rueda R, Pozo-Montuy G (2011) Mortalidad de monos aulladores negros (Alouatta pigra) en paisajes altamente fragmentados de Balancán, Tabasco. In: Gama-Campillo L, Pozo-Montuy G, Contreras-Sánchez WM, Arriaga-Weiss SL, editors. Perspectivas en Primatología Mexicana. Villahermosa: UJAT. pp. 289–317.

23. Rangel-Negrín A, Dias PAD, Canales-Espinosa D (2011a) Impact of translocation on the behavior and health of black howlers. In: Gama-Campillo L, Pozo-Montuy G, Contreras-Sánchez WM, Arriaga-Weiss SL, editors. Perspectivas en Primatología Mexicana. Villahermosa: UJAT. pp. 271–288.

24. Watts ES, Rico-Gray V, Chan C (1986) Monkeys in the Yucatan peninsula, Mexico: preliminary survey of their distribution and status. Primate Conserv 7: 17–22.

25. Jones CB, Young J (2004) Hunting restraint by Creoles at the Community Baboon Sanctuary, Belize: a preliminary survey. J Appl Anim Welfare Sci 7: 127–141.

26. Martínez-Mota R, Valdespino C, Sánchez-Ramos MA, Serio-Silva JC (2007) Effects of forest fragmentation on the physiological stress response of black howler monkeys. Anim Conserv 10: 374–379.

27. Wingfield JC, Romero LM (2001) Adrenocortical responses to stress and their modulation in free-living vertebrates. In: McEwen BS, Goodman HM, editors. Handbook of Physiology; Section 7: The Endocrine System. Coping With the Environment: Neural and Endocrine Mechanisms. New York: Oxford University Press. pp.211–234.

28. Arroyo-Rodríguez V, Dias PAD (2010) Effects of habitat fragmentation and disturbance on howler monkeys: a review. Am J Primatol 72: 1–16.

29. Romero ML (2002) Seasonal changes in plasma glucocorticoid concentrations in free-living vertebrates. Gen Comp Endocrinol 128: 1–24.

30. Goymann W (2012) On the use of non-invasive hormone research in uncontrolled, natural environments: the problem with sex, diet, metabolic rate and the individual. Methods Ecol Evol 3: 757–765.

31. Dias PAD, Rangel-Negrín A, Coyohua-Fuentes A, Canales-Espinosa D (2014) Factors affecting the drinking behavior of black howler monkeys (Alouatta pigra). Primates 55: 1–5.

32. Escamilla A, Sanvicente M, Sosa M, Galindo-Leal C (2000) Habitat mosaic, wildlife availability, and hunting in the tropical forest of Calakmul, Mexico. Conserv Biol 14: 1592–1601.

33. Di Fiore A, Link A, Campbell C (2011) The Atelines. In: Campbell C, Fuentes A, MacKinnon K, Bearder S, Stumpf R, editors. Primates in Perspective, 2nd edn. New York: Oxford University Press. pp.155–188.

34. Dias PAD, Rangel-Negrín A, Canales-Espinosa D (2011) Effects of lactation on the time-budgets and foraging patterns of female black howlers (Alouatta pigra). Am J Phys Anthropol 145: 137–146.

35. Van Belle S, Estrada A, Ziegler TE, Strier KB (2009) Sexual behavior across ovarian cycles in wild black howler monkeys (Alouatta pigra): male mate guarding and female mate choice. Am J Primatol 71: 153–164.

36. Edwards MS, Ullrey DE (1999) Effect of dietary fiber concentration on apparent digestibility and digesta passage in non-human primates. II. Hindgut- and foregut-fermenting folivores. Zoo Biol 18: 537–549.

37. Rangel-Negrín A, Dias PAD, Chavira R, Canales-Espinosa D (2011b) Social modulation of testosterone levels in male black howlers (Alouatta pigra). Horm Behav 59: 159–166.

38. Canales-Espinosa D, Dias PAD, Rangel-Negrín A, Aguilar-Cucurachi S, García-Orduña F, et al. (2011) Translocación de primates mexicanos. In: Dias PAD, Rangel-Negrín A, Canales-Espinosa D, editors. La conservación de los primates en México. Xalapa: Covecyt. pp.77–122.

39. Burnham KP, Anderson DR (2010) Model selection and multi-model inference: a practical information-theoretic approach. New York: Springer. 488 p.

40. Muehlenbein MP, Ancrenaz M, Sakong R, Ambu L, Prall S, et al. (2012) Ape conservation physiology: fecal glucocorticoid responses in wild Pongo pigmaeus morio following human visitation. PLoS ONE 7: e33357.

41. Zwijacz-Kozica T, Selva N, Barja I, Silván G, Martínez-Fernández L, et al. (2013) Concentration of fecal cortisol metabolites in chamois in relation to tourist pressure in Tatra National Park (South Poland). Acta Theriol 58: 215–222.

42. Behie AM, Pavelka MSM, Chapman CA (2010) Sources of variation in fecal cortisol levels in howler monkeys in Belize. Am J Primatol 71: 1–7.

43. Creel S (2001) Social dominance and stress hormones. Trends Ecol Evol 16: 491–497.

44. Muller MN, Wrangham RW (2004) Dominance, cortisol and stress in wild chimpanzees (Pan troglodytes schweinfurthii). Behav Ecol Sociobiol 55: 332–340.

45. Miller WL, Chrousos GP (2001) The adrenal cortex. In: Felig P, Frohman LA, editors. Endocrinology and metabolism. New York: McGraw-Hill. pp. 387–524.

46. Foley CAH, Papageorge S, Wasser SK (2001) Noninvasive stress and reproductive measures of social and ecological pressures in free-ranging African elephants. Conserv Biol 15: 1134–1142.

47. Girard I, Garland T (2002) Plasma corticosterone response to acute and chronic voluntary exercise in female house mice. J Appl Physiol 92: 1553–1561.

48. Edwards KL, Walker SL, Bodenham RF, Ritchie H, Shultz S (2013) Associations between social behaviour and adrenal activity in female Barbary macaques: consequences of study design. Gen Comp Endocrinol 186: 72–79.

49. Emery-Thompson M, Muller MN, Wrangham RW (2012) The energetics of lactation and the return to fecundity in wild chimpanzees. Behav Ecol 23: 1234–1241.

50. Rangel-Negrín A, Alfaro JL, Valdéz RA, Romano MC, Serio-Silva JC (2009) Stress in Yucatan spider monkeys: effects of environmental conditions on fecal cortisol levels in wild and captive populations. Anim Conserv 12: 496–502.

51. Schaffner CM, Rebecchini L, Aureli F, Ramos-Fernandez G, Vick LG (2012) Spider monkeys cope with the negative consequences of hurricanes through changes in diet, activity budget and fission-fusion dynamics. Int J Primatol 33: 922–936.

52. Putz FE, Leigh EGJ, Wright SJ (1990) Solitary confinement in Panama. Garden 2: 18–23.

53. Arroyo-Rodríguez V, Mandujano S (2006) Forest fragmentation modifies habitat quality for *Alouatta palliata*. Int J Primatol 27: 1079–1096.

54. Dunn JC, Cristóbal-Azkarate J, Schulte-Herbrüggen B, Chavira R, Veà JJ (2013) Travel time predicts fecal glucocorticoid levels in free-ranging howlers (*Alouatta palliata*). Int J Primatol 34: 246–259.

55. Foerster S, Monfort SL (2010) Fecal glucocorticoids as indicators of metabolic stress in female Sykes' monkeys (*Cercopithecus mitis albogularis*). Horm Behav 58: 685–697.

56. Gómez-Espinosa E, Rangel-Negrín A, Chavira R, Canales-Espinosa D, Dias PAD (2013) Glucocorticoid response in mantled howlers (*Alouatta palliata*) in relation to energetic and psychosocial stressors. Am J Primatol 76: 362–373.

57. Gobierno del Estado de Campeche (2012) Sistema Estatal de Áreas Naturales Protegidas del Estado de Campeche. Campeche: Gobierno del Estado de Campeche & Pronatura Península de Yucatán A.C. Available: www.smaas.campeche.gob.mx/anp/. Accessed 11 June 2014

58. Rimbach R, Link A, Heistermann M, Gómez-Posadas C, Galvis N, et al. (2013) Effects of logging, hunting, and forest fragment size on physiological stress levels of two sympatric ateline primates in Colombia. Conserv Physiol 1: 10.1093/conphys/cot031.

59. Arroyo-Rodríguez V, González-Pérez IM, Garmendia A, Solà M, Estrada A (2013) The relative impact of forest patch and landscape attributes on black howler monkey populations in the fragmented Lacandona rainforest, Mexico. Landscape Ecol 28: 1717–1727.

60. Fahrig L (2003) Effects of habitat fragmentation on biodiversity. Annu Rev Ecol Evol Syst 34: 487–515.

61. Harcourt AH, Doherty DA (2005) Species-area relationships of primates in tropical forest fragments: a global analysis. J Appl Ecol 42: 630–637.

62. Cowlishaw G (1999) Predicting the pattern of decline of African primate diversity: an extinction debt from historical deforestation. Conserv Biol 13: 1183–1193.

63. Metzger JP, Martensen AC, Dixo M, Bernacci LC, Ribeiro MC, et al. (2009) Time-lag in biological responses to landscape changes in a highly dynamic Atlantic forest region. Biol Conserv 142: 1166–1177.

Predation Limits Spread of *Didemnum vexillum* into Natural Habitats from Refuges on Anthropogenic Structures

Barrie M. Forrest[1]*, **Lauren M. Fletcher**[1], **Javier Atalah**[1], **Richard F. Piola**[1,2], **Grant A. Hopkins**[1]

1 Coastal and Freshwater Group, Cawthron Institute, Nelson, New Zealand, **2** Maritime Division, Defence Science and Technology Organisation, Melbourne, Victoria, Australia

Abstract

Non-indigenous species can dominate fouling assemblages on artificial structures in marine environments; however, the extent to which infected structures act as reservoirs for subsequent spread to natural habitats is poorly understood. *Didemnum vexillum* is one of few colonial ascidian species that is widely reported to be highly invasive in natural ecosystems, but which in New Zealand proliferates only on suspended structures. Experimental work revealed that *D. vexillum* established equally well on suspended artificial and natural substrata, and was able to overgrow suspended settlement plates that were completely covered in other cosmopolitan fouling species. Fragmentation led to a level of *D. vexillum* cover that was significantly greater than was achieved as a result of ambient larval recruitment. The species failed to establish following fragment transplants onto seabed cobbles and into beds of macroalgae. The establishment success of *D. vexillum* was greatest in summer compared with autumn, and on the underside of experimental settlement plates that were suspended off the seabed to avoid benthic predators. Where benthic predation pressure was reduced by caging, *D. vexillum* establishment success was broadly comparable to suspended treatments; by contrast, the species did not establish on the face-up aspect of uncaged plates. This study provides compelling evidence that benthic predation was a key mechanism that prevented *D. vexillum*'s establishment in the cobble habitats of the study region. The widespread occurrence of *D. vexillum* on suspended anthropogenic structures is consistent with evidence for other sessile invertebrates that such habitats provide a refuge from benthic predation. For invasive species generally, anthropogenic structures are likely to be most important as propagule reservoirs for spread to natural habitats in situations where predation and other mechanisms do not limit their subsequent proliferation.

Editor: Philippe Archambault, Université du Québec à Rimouski, Canada

Funding: This research was funded by NIWA under Coasts and Oceans Research Programme 4– Marine Biosecurity (2012/13 SCI) and the New Zealand Ministry of Business Innovation and Employment (Contract C01X0502, Effective Management of Marine Biodiversity and Biosecurity). The funders had no role in study design, data collection and analysis, decision to publish, or preparation of the manuscript.

Competing Interests: The authors have declared that no competing interests exist.

* E-mail: barrie.forrest@cawthron.org.nz

Introduction

Human activities in the marine environment have led to the creation of extensive areas of artificial habitat (e.g. floating pontoons, wharf piles, aquaculture structures) along coastal margins [1,2]. As non-indigenous species (NIS) can be dominant fouling organisms in such habitats, the role of anthropogenic structures in facilitating NIS establishment and spread is increasingly being recognized [3–6]. An important question that remains poorly understood is the extent to which populations of NIS on artificial structures act as reservoirs for invasion into natural systems [2,7]. Although the community composition and dominant species inhabiting artificial structures often differs greatly to that on the adjacent natural seabed [7–11], there are a number of non-indigenous fouling organisms that are also highly invasive in natural systems [3,12–16]. Among these, the ascidians are a group that often dominate the fouling biomass on suspended artificial structures, but whose invasiveness into natural habitats is highly variable among species, and within species among different geographic regions [7,17,18].

The colonial ascidian *Didemnum vexillum* Kott [19] has been widely described as forming highly invasive populations in natural habitats. The most dramatic reported example of invasiveness in this species (hereafter referred to as *Didemnum*) is an extensive population covering 230 km^2 of pebble and gravel habitat (up to 65 m water depth) on Georges Bank off the northeastern coast of the United States [16,20]. In the last 10–15 years, non-indigenous populations of *Didemnum* have also been reported from many other regions, including: the west coast of the United States (including southern Alaska), British Columbia (Canada), the United Kingdom, Ireland, northern France, the Netherlands, northern Italy and New Zealand [19,21–26]. Although *Didemnum* is most prevalent on artificial structures in these locations, it has also been described from natural rocky substrata, macroalgal beds, seagrass habitats, tide pools, estuaries, lagoons, and open coastal areas [27–29].

In contrast to these examples, New Zealand is one of few regions where *Didemnum* is prolific on artificial structures (including vessel hulls), but appears to have only a limited ability to establish in natural habitats [30]. Where invasion of such habitats occurs in

New Zealand, it is evident only in situations where *Didemnum* colonies are not in direct contact with natural seabed substrata. In the Marlborough Sounds region where *Didemnum* is most widespread, the species occurs in natural habitats only where there is biogenic structure such as that provided by horse mussels and macroalgal canopies, or where debris such as submerged logs are present (Table 1). However, such populations are not common, and have only been recorded in localities that have infestations on adjacent marine farms and other structures. In contrast, regional surveys conducted during a *Didemnum* management program ending in 2008, detected populations of *Didemnum* on approximately 123 artificial structures, but did not record the species from rocky (primarily cobble) subtidal habitats that were adjacent.

Previous studies with ascidians (both invasive and indigenous species) have highlighted competition [31–33], predation [34–38] and the diversity of benthic assemblages [33,39], as factors that can reduce the establishment and persistence of populations; limited evidence suggests these factors may also be important in the case of *Didemnum* [40,41]. Predation in particular has the potential to not only limit, but also prevent, some non-indigenous species from establishing in natural habitats. This is in contrast to suspended structures, which may provide a refuge from predation by benthic invertebrates [42]. For example, a study from Chile showed that the non-indigenous solitary ascidian *Ciona intestinalis* was able to establish abundant populations on suspended structures, but could not establish in adjacent natural habitats unless predators were excluded using cages [17].

The present paper describes a range of experiments designed to examine processes that affect the establishment of *Didemnum* on suspended and rocky subtidal habitats in New Zealand's Marlborough Sounds (41°13′S 174°07′E; Fig. 1). Based on experimental evidence, as well as field observations, we consider the potential for *Didemnum* to establish in natural seabed habitats, and the role of reservoir populations on artificial structures in the spread and establishment of the species.

Materials and Methods

Ethics Statement and Data Availability

No permits were required for the described study, which complied with all relevant regulations. No locations were privately-owned or protected in any way and the studies did not involve endangered or protected species. All data are available for inspection at the Cawthron Institute.

Establishment in Relation to Resident Fouling and Substratum Type

Didemnum is a colonial species that has the capacity to establish from planktonic larval recruitment and via the reattachment of fragmented colony tissue [43–45]. Two initial experiments evaluated the relative importance of recruitment and fragment inoculation, and tested the hypothesis that *Didemnum* could establish equally well on different substratum types (Table 2). The first of these experiments investigated the influence of resident fouling communities. Roughened black Perspex settlement plates (20×20 cm, n = 3 per treatment as described below) were deployed in a horizontal orientation approximately 1 m beneath the sea surface for three months at a location free of *Didemnum*, until their underside had developed a complete two-dimensional fouling layer. These pre-fouled plates, consisting almost exclusively of colonial ascidians (*Botrylloides leachi*, *Botryllus schlosseri*, *Diplosoma listerianum*) and the encrusting bryozoan *Watersipora subtorquata*, were subsequently suspended among an extensive *Didemnum* population on a floating salmon aquaculture cage at Ruakaka Bay (Fig. 1), along with bare plates. Triplicate pre-fouled and bare plates had a small (~20×40 mm) fragment of *Didemnum* colony attached to them (secured by rubber band), and were randomized among triplicate pre-fouled and bare plates that had no *Didemnum* attached. Hence, whereas all plates were exposed to ambient larval recruitment, half of them were additionally subjected to a single fragment inoculation event. Plates were suspended horizontally facing down (0.5–1.5 m depth) for three months (17 January to 21 April 2008) during the peak summer period of *Didemnum* larval recruitment in the study region [46]. At monthly intervals during the three month period, the plates were photographed and colony percentage cover subsequently estimated using ImageJ software [47]. Any *Didemnum* colonies present after three months were considered to be successfully established, on the assumption that new colonies could have reproduced in that time. Reproductive maturity in *Didemnum* can be reached two weeks after inoculation with fragments [48], and occurs within two months after larval settlement in the case of other colonial ascidians [49,50].

In the second experiment (see Table 2), the cover of *Didemnum* developing from ambient larval inoculation and fragments was compared between replicate (n = 4) settlement plates and natural greywacke cobbles of comparable surface area, which were collected from the study site. The selected cobbles were devoid of visible macrofouling, and were centre-drilled so that they could be hung in a similar manner to the plates. Plates and cobbles were suspended horizontally facing down (0.5–1.5 m depth) for three

Table 1. Habitats in Marlborough Sounds study region on which *Didemnum* has been recorded or has established.

Habitat	Description
Artificial structures floating or elevated off seabed	
Floating structures	Widespread and abundant on mussel farms, salmon farms, floating pontoons and mooring lines
Fixed elevated structures	Widespread and abundant on wharf piles
Biogenic features and seabed structures or debris	
Erect sessile epifauna	Present on horse mussels, finger sponges and hydroid trees adjacent to infected artificial structures
Erect macroalgae	Present on canopy-forming macroalgae (*Carpophyllum flexuosum* & *Macrocystis pyrifera*), adjacent to infected artificial structures
Mobile epibiota	Decorator crabs adjacent to infected artificial structures
Organic/inorganic debris	Common on submerged logs, cables and other debris adjacent to infected artificial structures
Artificial rock walls	Abundant in one location on rip-rap beneath heavily infected wharf piles

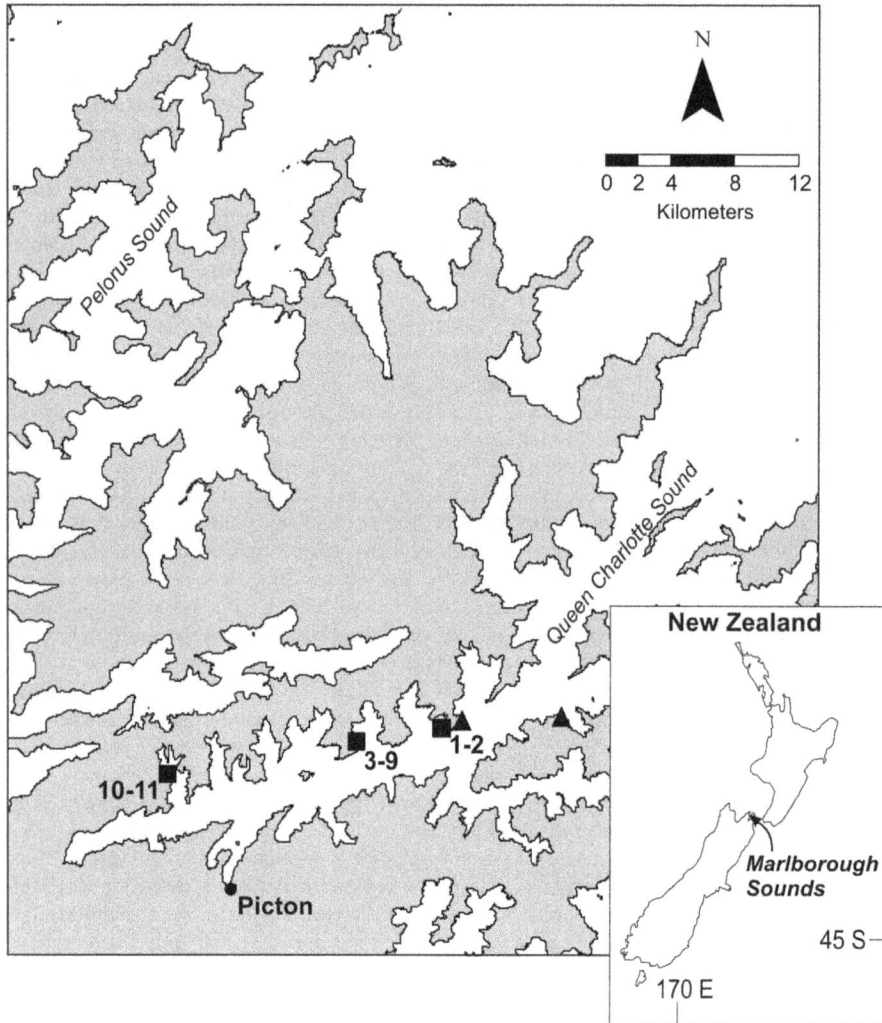

Figure 1. Sites used for *Didemnum vexillum* experiments. Transplants of *Didemnum* fragments were conducted along 11 transects (1–11) across three main locations (filled squares) also used for small scale experiments: Ruakaka Bay (1–2), Blackwood Bay (3–9) and Onahau Bay (10–11). Filled triangles indicate locations of fragment transplants into macroalgal (*Carpophyllum flexuosum*) beds.

Table 2. Description of experiments conducted with *Didemnum*.

Experimental component	Substratum and position	*Didemnum* inoculation
A) Effect of resident fouling and substratum		
Resident fouling	Suspended bare vs pre-fouled settlement plates	Fragments vs ambient larvae
Substratum	Suspended settlement plates vs cobbles	Fragments vs ambient larvae
B) *Didemnum* transplant to seabed habitats		
Regional fragment transplant to natural cobbles	Seabed cobbles	Fragments
Fragment transplant to algal beds (*Carpophyllum flexuosum*)	Mid-canopy & canopy base, seabed within canopy & canopy edge	Fragments
C) Seabed invasion resistance/caging experiments		
Blackwood Bay summer 2008	Suspended, seabed caged, seabed uncaged	Fragments vs ambient larvae
Blackwood Bay autumn 2008	Suspended, seabed caged, seabed uncaged	Fragments vs pre-established *Didemnum*
Onahau Bay summer 2009	Suspended, seabed uncaged	Fragments vs ambient larvae

All treatments in components A and C were exposed to ambient larval inoculation from adjacent reproductive *Didemnum* populations. Invasion resistance experiments in component C also included a comparison of face-up and face-down orientations. See Fig. 1 for experimental locations.

months at Ruakaka Bay (Fig. 1). The cover of *Didemnum* was determined at the end of the three month deployment, and was expressed as surface area (cm^2) to account for the differences in cobble sizes.

Transplant of *Didemnum* to Seabed Habitats

Two transplant experiments were conducted in January 2008 (austral summer) to examine the regional-scale establishment potential of *Didemnum* in the Marlborough Sounds (Table 2). Firstly, large colony fragments (~100×200 mm) were attached to cobbles and transplanted to subtidal cobble habitats along each of 11 transects (1–16 m depth, 10 m long, n = 102 in total) in three main locations: Ruakaka Bay, Blackwood Bay and Onahau Bay (Fig. 1). The transects were revisited over the following 24 hours, and after 1 month, and the presence/absence of *Didemnum* colonies was determined.

Secondly, fragments were transplanted into canopies of a fucoid macroalgal species (*Carpophyllum flexuosum*) that *Didemnum* had previously been observed to establish on (see Table 1). It was hypothesized that the erect biogenic structure provided by the algal canopy would elevate *Didemnum* from the seabed and provide a refuge from predation. Two plots (a treatment and reference plot) of *C. flexuosum* were identified in each of two locations (Fig. 1), each covering an area of at least 10×10 m at approximately 2 m depth. Ten *Didemnum* fragments (~30×100 mm) were transplanted into the treatment plots in each of four positions: (i) within canopy approximately 1 m from the seabed, (ii) within canopy at the base of the stipe of each *C. flexuosum*, (iii) within canopy on the seabed, and (iv) at the edge of each canopy on the seabed (i.e. at the distinct boundary between the canopy and barren rock habitat). For each of the ten canopy transplants, fragments were attached to ten separate tagged *C. flexuosum* plants. The plots were revisited over the following 24 hours, and after 1 month, and the presence/absence of *Didemnum* colonies was determined.

Experimental Evaluation of Seabed Invasion Resistance

A series of field caging experiments was undertaken to test the hypothesis that *Didemnum* could establish on (or survive transplant to) the seabed, if benthic predation was reduced (Table 2). The overall experimental design compared establishment success among settlement plates (20×20 cm) that were deployed in two horizontal orientations (face-up and face-down), and which were either: (i) uncaged on the seabed, (ii) fully caged on the seabed using a 10 mm stainless mesh screen (to exclude larger predators), or (iii) suspended at comparable depths from adjacent suspended structures that were infected with *Didemnum*. Seabed cages and plates were attached to gridded (150×150 mm) steel reinforcing mesh. Previous studies in the region have shown that the cage design and method does not introduce experimental artefacts [44].

Caging experiments were first conducted for three months (17 January –11 April 2008) during the recruitment peak at the study site in Blackwood Bay (Fig. 1). Seabed caged and uncaged plates were deployed at approximately 10 m depth, and arranged 25 m distance from a floating pontoon (dimensions ~2×4 m) that had been pre-seeded with approximately 350 kg of reproductive *Didemnum* [51]. Uncaged plates were also suspended beneath the pontoon at a similar depth. A fragment of *Didemnum* was attached to one set of plates (n = 6) and colony percentage cover measured after three months. The ambient larval supply from the pontoon was simultaneously measured on an additional set of plates (n = 3 suspended, n = 6 seabed caged and uncaged) that were deployed for the same period, and involved counting the density of recruits in the laboratory using a binocular microscope. Daily mean water temperature during the experiment were determined from hourly

measurements with a TidbiT v2 Temperature Logger, and ranged from 17.0 to 19.6°C.

The fragment inoculation component of the caging experiment was repeated during April to June 2008, covering a two month autumn to mid-winter period in New Zealand. Daily mean water temperatures of 13.1 to 17.1°C during the second experimental period were within the species optimal range [25,29], and *Didemnum* was still actively recruiting throughout this period. Simultaneous with the fragment inoculation experiment, the persistence of pre-established *Didemnum* colonies was assessed using the same experimental design (n = 6 per treatment). For this component, the colonies were pre-established on settlement plates using fragments. The resultant colony cover on the plates at the beginning of the experiment ranged from 50 to 92% (mean 72.1% ±1.7SE) and plates were randomly allocated across treatments. Results were expressed as colony survival, reflected as the change in percentage cover at the completion of the experiment.

The fragment inoculation and recruitment experiments conducted in summer 2008 were repeated at a different location (Onahau Bay, see Fig. 1) in summer 2009, with the exception that the cage treatment was not included (the suspended versus seabed contrast was of primary interest) (see Table 2). The infected structure at the second location was a suspended pontoon of similar size to that in Blackwood Bay; however, site constraints meant that seabed plates were deployed on shallow cobbles (2–3 m depth) immediately adjacent to the pontoon. *Didemnum* percentage cover (from fragments) and recruit density on suspended and seabed plates (n = 4) was assessed after three months, as described above.

Statistical Analyses

A two-way repeated-measures analysis of variance (RM-ANOVA) was used to test for differences through time in the percentage cover of *Didemnum* on bare and pre-fouled plates at the Ruakaka site. The experimental design included 'Inoculation' (two levels: Fragments and Recruits) and 'Fouling' (two levels: Bare and Pre-fouled) as fixed orthogonal factors. Data from the substratum type experiment were analyzed using a two-way ANOVA with 'Inoculation' (two levels: Fragments and Recruits) and 'Substratum' (two levels: 'Plate' and 'Cobble') as fixed orthogonal factors.

The Blackwood Bay caging experiments involved two-way ANOVA analyses with 'Treatment' (three levels: Cage, No Cage, Suspended) and 'Orientation' (two levels: face-up, face-down) as fixed orthogonal factors. Data from the summer and autumn experiments resulted in a total of three response variables (recruit density, colony cover and colony survival), which were analyzed separately in the same manner. For the summer 2008 recruit density data, one of the 'Suspended-Down' and 'Suspended-Up' replicates was lost during the experimental phase, so the analysis for this experiment was consequently unbalanced. Recruit density and colony cover data from the Onahau Bay experiments were analyzed separately using the same design, with the exception that there were only two treatment levels: 'Suspended' and 'No Cage'.

Due to data over-dispersion and a high proportion of zero values in some treatments, all ANOVA analyses were performed using a permutational approach [52], with the PERMANOVA add-in of PRIMER 6 software [53]. Analyses used resemblance matrices based on Euclidean distance, and the partial sum of squares (Type III) with 9999 permutations of residuals under a reduced model. Data were log (x+1) transformed to achieve approximate unimodal symmetry and to avoid right skewness. Significant terms were then investigated using *a posteriori* pair-wise comparisons with the PERMANOVA *t*-statistic and 999 permutations.

Results

Establishment in Relation to Resident Fouling and Substratum Type

Didemnum established equally well on settlement plates, irrespective of whether they were bare or pre- fouled; however, inoculation with fragments led to a colony cover that was considerably greater than that arising from ambient recruitment alone (Fig. 2A). On pre-fouled plates, microscopical observation revealed new recruitment and colony growth on top of the existing fouling. After three months, plates inoculated with fragments had a mean colony cover of 59.2 to 78.3% (±23.7 and 7.3%, respectively) compared with <3% cover on plates exposed to larval recruitment alone. Hence, there was a statistically significant effect of inoculation type, which was consistent over time (Table 3A). In contrast, there was no overall effect of substratum, but the substratum effect was variable over time (Table 3A, Substratum×Month, p<0.05). Even after one month, the *Didemnum* cover resulting from fragment inoculation (mean 36.3±6.1 and 18.3±1.6% for bare and pre-fouled plates, respectively) was already considerably greater than achieved from ambient recruitment. The comparison of plates and cobbles yielded a similar pattern, with significantly greater colony cover arising from fragment inoculation compared to larval recruitment, but no significant effect of substratum type (Fig. 2B, Table 3B).

Transplant of *Didemnum* to Seabed Habitats

The regional-scale transplant of large *Didemnum* fragments along 11 transects in cobble habitats failed to lead to any colonies establishing. In fact, within hours of transplant, fragments were usually covered by benthic predators, which subsequently consumed intact colony tissue (i.e. not just cut edges). Most prevalent among the predators observed to actively consume the fragments were small (up to 60 mm diameter) cushion stars, *Patiriella regularis*, and sea urchins, *Evechinus chloroticus* (Fig. 3).

Table 3. Effects of inoculation and substratum on *Didemnum* cover.

Source	df	MS	F	Sig
A) Bare and pre-fouled plates				
Inoculation	1	92.60	397.9	**
Substratum	1	0.00	0.0	
Inoculation×Substratum	1	0.39	1.7	
Month	2	1.95	28.3	***
Inoculation×Month	2	0.24	3.4	
Substratum×Month	2	0.35	5.1	*
Inoculation×Substratum×Month	2	0.17	2.5	
Residual	16	0.07		
B) Plates and cobbles				
Inoculation	1	7.83	13.28	**
Substratum	1	1.33	2.25	
Inoculation×Substratum	1	0.05	0.09	
Residual	12	0.59		

A) RM-ANOVA comparing percentage cover from fragment and larval inoculation on bare and pre-fouled plates; B) Permutational ANOVA comparing percentage cover from fragment and larval inoculation on plates and cobbles. Statistical significance denoted as follows: * = p<0.05, ** = p<0.01, *** = p<0.001.

However, large eleven-arm sea stars (*Coscinasterias muricata*), top shells (*Turbo smaragdus*), chitons (*Cryptoconchus porosus*) and hermit crabs (Paguridae) were also observed on the fragments and were presumed to be eating them.

Didemnum also failed to establish in the *Carpophyllum flexuosum* transplant experiment. In the case of the seabed transplants

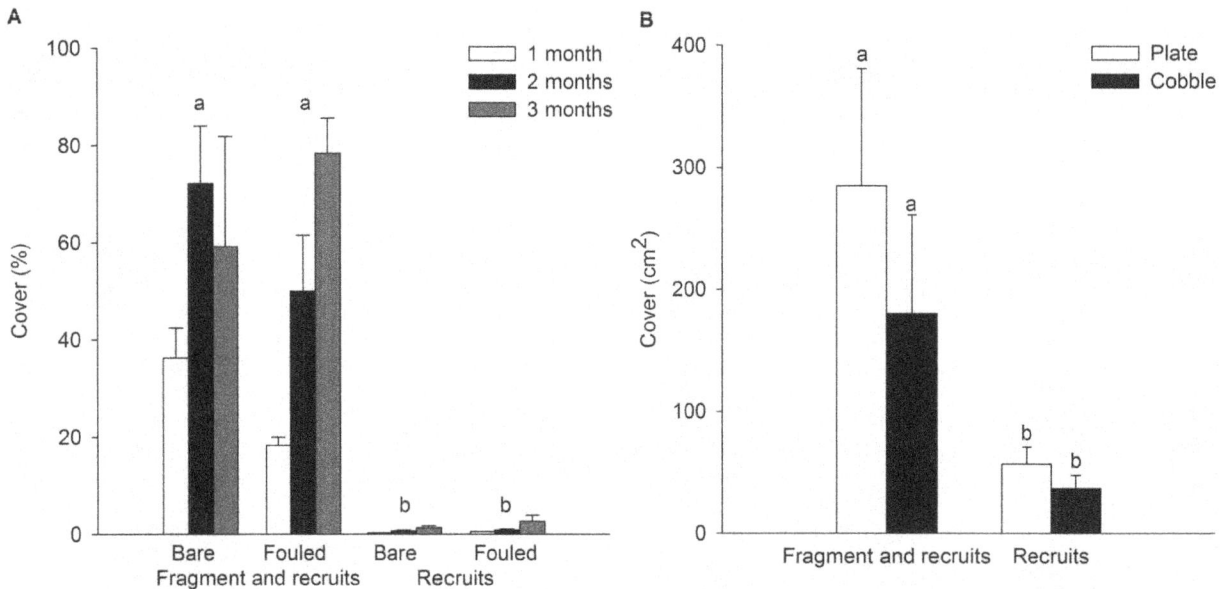

Figure 2. Establishment of *Didemnum* on different substrata. In two separate experiments of three months duration, different substrata were exposed to ambient larval recruitment, with half of them also inoculated with a *Didemnum* fragment. A) Mean *Didemnum* cover (% cover ±1SE) on bare settlement plates compared with plates that were pre-fouled with other sessile species. B). Mean *Didemnum* cover (cm²±1SE) on bare settlement plates compared with cobbles that were collected from the adjacent intertidal zone. In both experiments, substrata were suspended between 0.5 and 1.5 m deep. Different letters (a,b) indicate significant differences as indicated by the PERMANOVA post-hoc *t*-test.

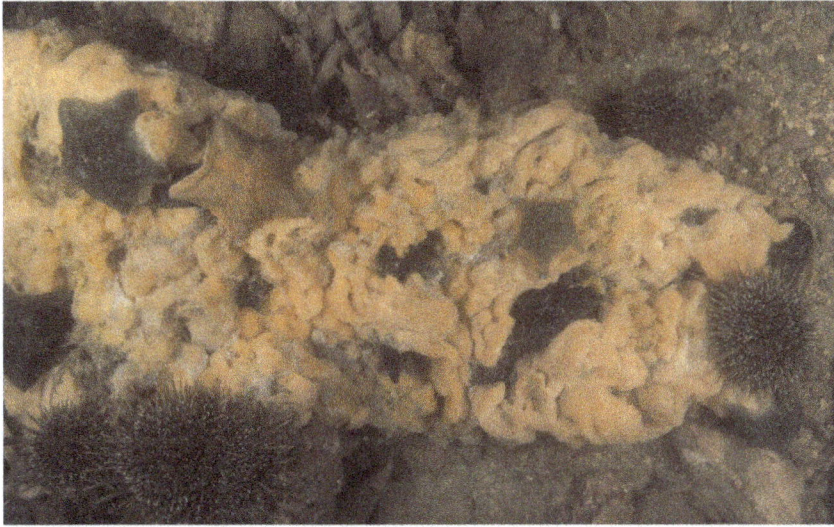

Figure 3. Predation on *Didemnum* following transplant to the seabed. Image shows predation by cushion stars, *Patiriella regularis*, and sea urchins, *Evechinus chloroticus*, on a large (~20×35 cm) fragment of *Didemnum*. Predation by these and/or other species was observed following fragment inoculation into beds of the macroalga *Carpophyllum flexuosum*, and when settlement plates with fragments or pre-established *Didemnum* colonies were placed on the seabed.

(adjacent to and at the edge of the *C. flexuosum* canopy), this result was consistent with the regional-scale transplant experiment, with predation on the fragments observed within the first day of the experiment. However, *Didemnum*'s failure to establish following transplant to the basal stipe of *C. flexuosum*, and especially the canopy, was contrary to our expectations. The seastar *C. muricata* was observed to consume one of the basal transplants, and one of the canopy transplants had *T. smaragdus* on it. One month later, neither the transplanted fragments nor their remains were visible.

Seabed Invasion Resistance

During the summer of 2008 in Blackwood Bay, *Didemnum* established a greater colony cover and recruited at greater densities on face-down plates compared to face-up ones (Orientation, p<0.01, Table 4A, Figure 4A). Additionally, there was a significant Treatment effect (Table 4A, Figure 4A), with Suspended plates having greater cover than the seabed No Cage treatment (PERMANOVA, pair-wise comparison p<0.05), but not the seabed Cage treatment. On face-down plates, fragment inoculation led to 34.2 (±12.8) and 39.8% (±17.9) mean cover in Suspended and Cage treatments, respectively, compared with 8% cover (±3.4) on No Cage plates. On face-up plates, *Didemnum* failed to establish without caging, and cover was relatively low but comparable on suspended and caged plates. Patterns in recruitment density largely mimicked the fragment inoculation patterns, with a significant effect of Orientation and Treatment (Table 4B, Figure 4B). No recruits were recorded on the face-up plates unless they were caged or suspended (Fig. 4B).

In autumn 2008 at Blackwood Bay, *Didemnum* establishment by fragments was poor compared with the previous summer (Fig. 5A), but was consistent with the fact that the persistence of pre-established colonies was simultaneously low (Fig. 5B). In the fragment inoculation experiment, a significant Treatment×Orientation interaction (Table 5A) reflected the significantly greater *Didemnum* cover on face-down suspended plates (6% cover) than in all other treatments in which cover was <1% (Fig. 5A, PERMANOVA, pair-wise comparison p<0.05). On No Cage plates, no *Didemnum* was recorded in the face-up orientation

(Fig. 5A), nor did any pre-established *Didemnum* persist in that orientation (Fig. 5B). By contrast, survival was significantly greater in seabed Cage and face-down Suspended treatments (Fig. 5B, PERMANOVA, pair-wise comparison p<0.05). Whereas mean survival was comparable for the two orientations in the Cage treatments (19.5±5.8 and 31.2±21.1% for face-down and face-up plates, respectively), the face-up Suspended plates became covered in a layer of filamentous algae and *Didemnum* survival was poor (Fig. 5B).

The experiment conducted in summer 2009 at Onahau Bay generally confirmed the importance of orientation, and the greater establishment (Fig. 6A) and recruitment (Fig. 6B) of *Didemnum* in suspended compared with seabed habitats. In the fragment inoculation experiment, a significant Treatment×Orientation interaction (Table 6A) indicated an orientation effect for Suspended, but not Seabed plates (see Fig. 6A, PERMANOVA, pair-wise comparison p<0.05). A similar general pattern was evident for larval recruitment (Fig. 6B), although significant effects of both Treatment and Orientation were detected (Table 6B).

Discussion

Invasiveness of *Didemnum*

Substratum type and the presence of an established fouling assemblage were unimportant as explanatory variables for the invasion success of *Didemnum*. The species established equally well on settlement plates and cobbles, and was able to recruit to, and subsequently overgrow, cosmopolitan fouling taxa such as colonial ascidians and encrusting bryozoans. The latter finding is consistent with field observations and manipulative experiments from a number of localities world-wide, which indicate the ability of *Didemnum* and other ascidians to overgrow other sessile organisms [16,18,54,55]. A single exception for *Didemnum* is evident from a recent experimental study (~10 weeks duration) in Long Island Sound, which suggested that invasion success was linked to the availability of unoccupied space [41].

Didemnum revealed a greater ability to establish as a result of fragment inoculation than ambient larval recruitment. This was

Table 4. Effects on colony percentage cover and recruit density at Blackwood Bay in summer 2008.

Source	df	MS	F	Sig
A) Cover				
Treatment	2	7.5	6.1	*
Orientation	1	27.7	22.3	**
Treatment×Orientation	2	0.2	0.2	
Residual	28	1.2		
B) Recruit density				
Treatment	2	7.0	4.5	*
Orientation	1	52.2	33.5	**
Treatment×Orientation	2	0.2	0.2	
Residual	24	1.6		

Results show effects of treatment (Suspended, Cage, No Cage) and orientation, from permutational ANOVA. Statistical significance denoted as follows: * = p<0.05, ** = p<0.01, *** = p<0.001.

evident in experiments that compared different substrata (see Fig. 2), and was also observed in the field during the Blackwood and Onahau Bay experiments. In fact, at Blackwood Bay, the cover of *Didemnum* resulting from ambient recruitment was not measurable by image analysis. The important role of fragments in the establishment of *Didemnum* [43] and other colonial ascidians [56] has previously been recognized. Natural fragmentation in the species arises from the tendril-like lobes that colonies often exhibit [57], and fragments can remain viable for as long as four weeks when artificially kept in suspension [45]. Although *Didemnum* can reattach from fragments as small as 5×3 mm, survivorship

increases with fragment size under field conditions [44]; thus, relatively large fragments were used in the present study to maximize the likelihood that they would be viable and able to reattach.

The non-measurable cover resulting from recruitment at Blackwood Bay is perhaps explained by the fact that the experimental plates were up to 25 m from the *Didemnum* larval source, whereas plates were only 1–3 m from the larval source in the other experiments, and less subjected to dilution processes. This hypothesis is consistent with parallel research in the study region that described an exponential decline in *Didemnum* recruitment within tens of meters from reproductive populations [51], and also consistent with the notion that invasion success can depend on 'propagule pressure' [58,59]. However, a propagule pressure hypothesis does not explain why successful recruits (or divided zooids) did not continue to grow at Blackwood Bay, as the three month experimental duration was more than sufficient for this to occur [60,61]. Competition with other recruiting fouling species seems an unlikely explanation, as the diversity and prevalence of other sessile species at Blackwood Bay was quite low (e.g. consisted of occasional small barnacles, spirorbid and serpulid polychaetes, hydroids, and encrusting bryozoans), and comparable to the other locations. Further investigation is needed to ascertain the extent to which site-specific factors (e.g. environmental conditions, post-settlement processes) may have contributed to differences in recruitment and subsequent colony establishment success among locations.

While the results suggests a more important role for fragmentation in the initial stages of establishment, localized ambient larval recruitment may lead to a comparable level of establishment to fragments in the longer-term (>12 months) [54]. Thus, the apparent advantage in establishment by fragmentation may diminish over time. The findings highlight that despite the

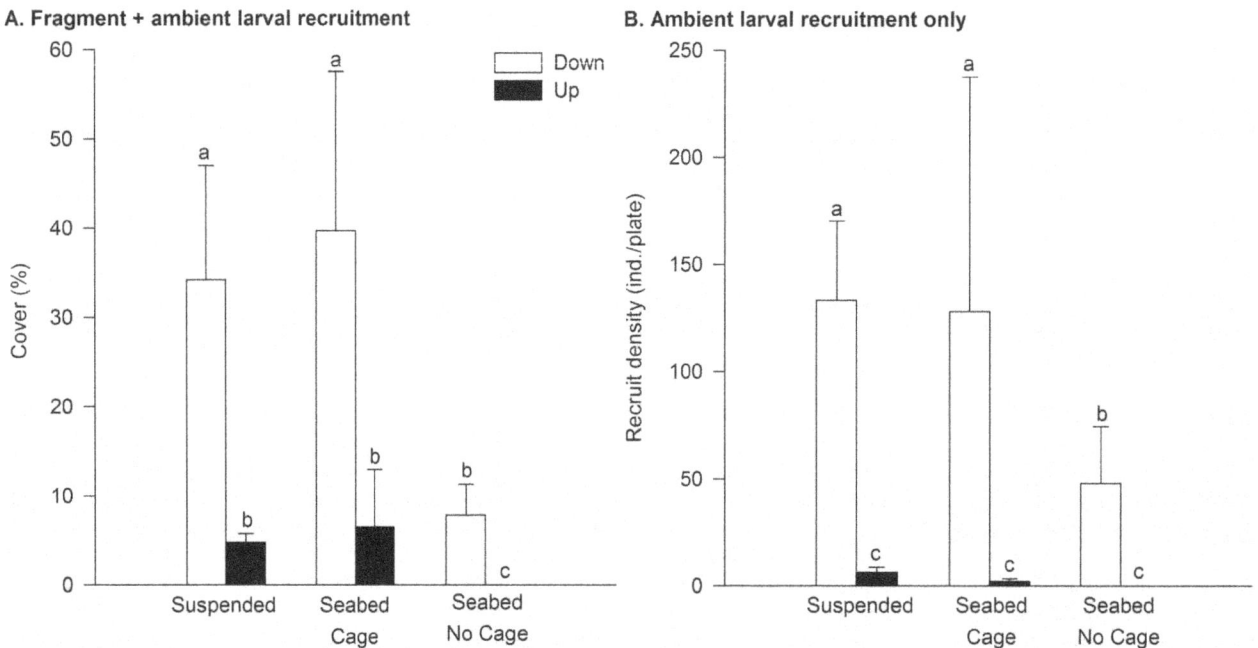

Figure 4. Establishment of *Didemnum* in suspended and seabed treatments in summer 2008 at Blackwood Bay. Perspex plates were positioned in face-up and face-down orientations, with the seabed treatment including a Cage and No Cage comparison. All plates were exposed to larval recruitment from a floating pontoon approximately 25 m away, for a three month experimental duration. A). Mean *Didemnum* cover (% cover ±1SE) on plates inoculated with *Didemnum* fragments. B). Mean density of *Didemnum* recruits (±1SE) on plates subjected only to ambient larval supply. Different letters (a,b,c) indicate significant differences as indicated by the PERMANOVA post-hoc *t*-test.

A. Fragment + ambient larval recruitment

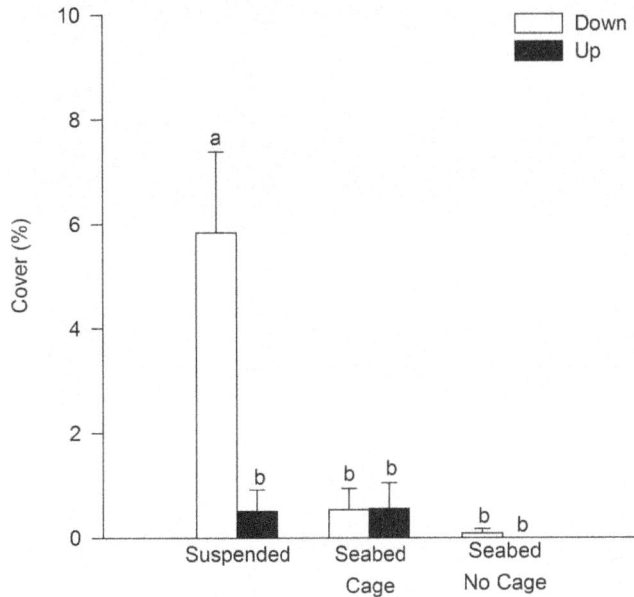

B. Pre-established colony + ambient larval recruitment

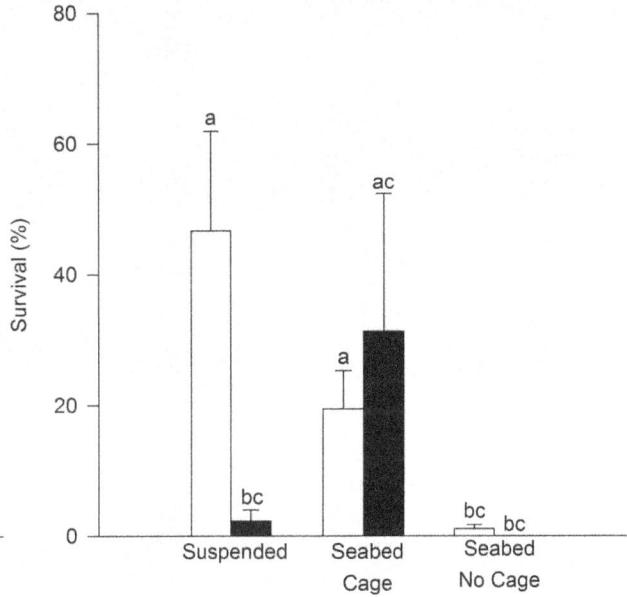

Figure 5. Establishment of *Didemnum* in suspended and seabed treatments in autumn 2008 at Blackwood Bay. Settlement plates were positioned in face-up and face-down orientations, with the seabed treatment including a Cage and No Cage comparison. All plates were exposed to an ambient larval supply from a suspended pontoon approximately 25 m away, for a two month experimental duration. A). Mean *Didemnum* cover (% cover ±1SE) on plates inoculated with *Didemnum* fragments. B). Mean survival of *Didemnum* (% survival ±1SE) colonies on plates for which a colony cover (mean ~73%) was pre-established at the start of the experiment. Different letters (a,b,c) indicate significant differences as indicated by the PERMANOVA post-hoc *t*-test.

potential for dispersal in this species over scales of hundreds of meters and perhaps a few kilometers [51], low level larval recruitment at the outer limits of the dispersal range may not immediately lead to the development of extensive new populations of *Didemnum*, even on suspended artificial structures which generally appear optimal for establishment.

The decline in establishment success of fragments in the autumn experiment at Blackwood Bay (compared with summer) was accompanied by poor survival of pre-established colonies at the time of the mid-winter cessation of the experiment. Such

Table 5. Effects on colony percentage cover and survival at Blackwood Bay in autumn 2008.

Source	df	MS	F	Sig
A) Cover				
Treatment	2	3.1	14.6	
Orientation	1	2.5	11.5	
Treatment×Orientation	2	2.1	9.7	***
Residual	30	0.2		
B) Survival				
Treatment	2	14.8	10.6	
Orientation	1	13.4	9.6	
Treatment×Orientation	2	5.3	3.8	*
Residual	30	1.4		

Results show effects of treatment (Suspended, Cage, No Cage) and orientation, from permutational ANOVA. Statistical significance denoted as follows: * = p<0.05, ** = p<0.01, *** = p<0.001.

findings are consistent with a seasonal decline in success due to decreasing water temperatures [61], as opposed to a failure of the inoculation method. Thus, even though *Didemnum* can persist and still recruit until mid-winter in the study region [46], the period of greatest risk for establishment and proliferation is clearly during warmer summer months. Despite an apparent seasonal decline in invasion success in cooler months, the fact that *Didemnum* still persists means that established populations will provide a reservoir from which extensive colonies may develop the following summer.

Invasion Resistance in Natural Habitats

The combination of regional scale transplants, smaller spatial scale experiments, and associated field observations, provided compelling evidence that benthic predation is a key mechanism that limits *Didemnum*'s establishment in the natural cobble habitats of the study region. Simultaneously, the widespread occurrence of the species in habitats that are elevated off the seabed (see Table 1), is consistent with evidence for other sessile marine invertebrates that both anthropogenic and biogenic structures can provide a refuge from benthic predation [17,37,62,63]. However, most studies of *Didemnum* and other colonial ascidians have shown that predation acts to reduce but not prevent establishment in natural habitats [36–38,40]. By contrast, the present study indicates that benthic predation may completely exclude *Didemnum* from certain habitat types. This result is consistent with recent analysis showing that, unlike other didemnid ascidians that contain compounds that are thought to reduce predation [64], *Didemnum* does not contain potent secondary metabolites [26].

In the caging experiments, the establishment success of *Didemnum* was greatest on suspended plates, from which benthic predators would have been completely excluded. Where predation

A. Fragment + ambient larval recruitment

B. Ambient larval recruitment only

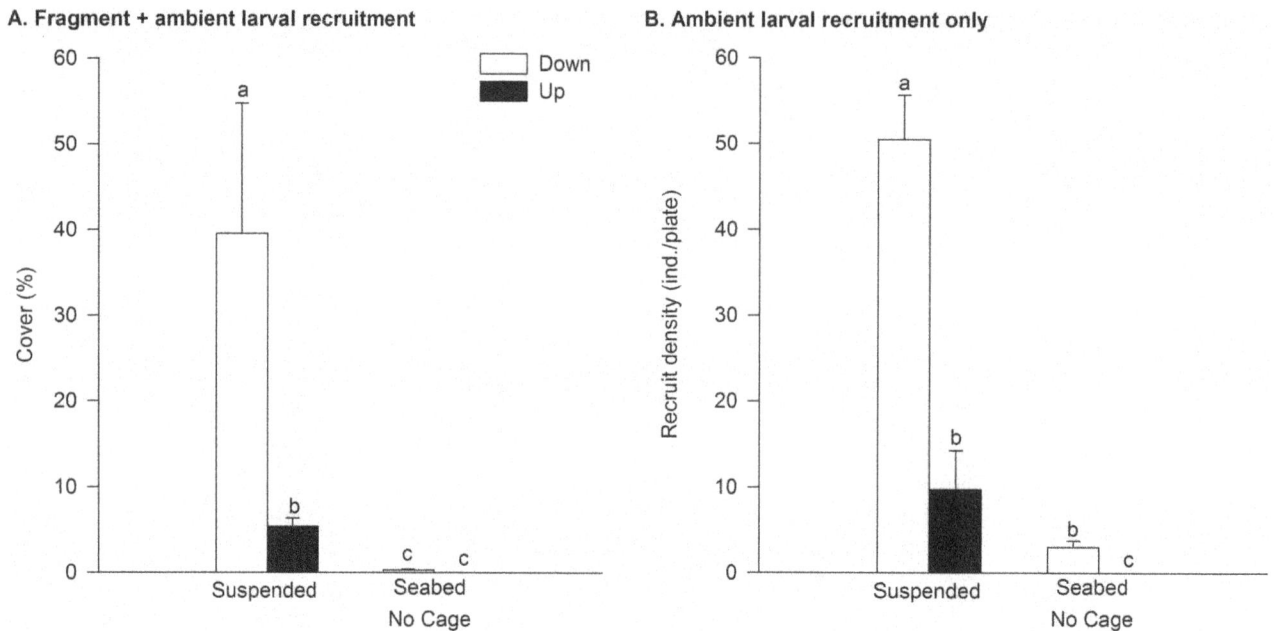

Figure 6. Establishment of *Didemnum* in suspended and seabed treatments in summer 2009 at Onahau Bay. Settlement plates were positioned in face-up and face-down orientations, with suspended plates compared to uncaged seabed plates. All plates were exposed to an ambient larval supply from a suspended pontoon 2–3 m away, for a three month experimental duration. A). Mean *Didemnum* cover (% cover ±1SE) on plates inoculated with *Didemnum* fragments. B). Mean density of *Didemnum* recruits (±1SE) on plates subjected only to ambient larval supply. Different letters (a,b,c) indicate significant differences as indicated by the PERMANOVA post-hoc *t*-test.

pressure was reduced by caging, *Didemnum* fragment establishment and recruitment yielded broadly similar patterns to suspended treatments. Greater recruitment to the face-down aspect of plates was consistent with the negative phototaxis exhibited by ascidian larvae immediately prior to settlement [61,65]. Marked variability in establishment success in some caged treatments was evident (see Fig. 4), which possibly reflected incursion into some cages of small predators (in particular juvenile cushion stars, *Patiriella regularis*) that were not excluded by the 10 mm mesh. The low level recruitment and colony cover on uncaged face-down seabed plates was not anticipated, but likely reflects reduced predation pressure. The face-down orientation would have restricted predator access

into the spaces among the cobbles, especially in the case of larger-bodied urchins and cushion stars.

The consistent failure of *Didemnum* to establish in any of the seabed transplants, or on the face-up orientation of the No Cage treatments, suggests that predators had relatively unlimited access. In contrast, colonies on face-up plates within cages persisted to a similar extent to that observed on suspended treatments, thus illustrating the invasion potential of *Didemnum* in the absence of benthic predation pressure. Such results are of interest, as a face-up orientation approximates a common scenario in which *Didemnum* forms high-biomass colonies on suspended structures, which can slough off and provide a considerable supply of

Table 6. Effects on colony percentage cover and recruit density at Onahau Bay in summer 2009.

Source	df	MS	F	Sig
A) Cover				
Treatment	1	25.7	146.0	
Orientation	1	3.7	20.8	
Treatment×Orientation	1	2.1	11.9	**
Residual	12	0.2		
B) Recruit density				
Treatment	1	21.7	89.3	***
Orientation	1	10.2	41.7	***
Treatment×Orientation	1	0.2	1.0	
Residual	12	0.2		

Results show effects of treatment (Suspended, No Cage) and orientation, from permutational ANOVA. Statistical significance denoted as follows: * = p<0.05, ** = p<0.01, *** = p<0.001.

fragments to the adjacent seabed [30]. However, irrespective of whether *Didemnum* propagules colonize suspended structures or the seabed, subsequent establishment and invasiveness on face-up surfaces may be less than face-down, and subject to a greater suite of limiting factors than predation alone, such as over-settlement by seasonal algae (this study) and smothering by sediment [44].

In addition to these limiting factors, considerable small-scale variability was often evident within the same treatment (e.g. among plates in the same treatment separated by only tens of centimeters), which was not always explainable by extrinsic factors such as differential predation pressure. This variability mirrors that previously observed across small spatial scales in the recruitment and establishment of *Didemnum* and other ascidians [18,66]. Such findings suggest a stochastic element to successful establishment that may in part be driven by inherent variability in *Didemnum* (e.g. differential viability of fragments from the same or different colonies).

Propagule Pressure and Invasion Potential of *Didemnum*

Given the apparent stochastic nature of establishment, and the design of our experiments (e.g. short-term with one-off inoculation by fragments) a pertinent question raised by this study is whether increased propagule pressure, such as that provided by a sustained larval and fragment release from an infected anthropogenic structure, could overcome invasion resistance and enhance *Didemnum*'s invasiveness in adjacent natural habitats? This phenomenon has been described theoretically [67,68], and in empirical studies from a range of aquatic and terrestrial systems that highlight the important interaction between propagule supply and factors that alter mechanisms of invasion resistance [17,69–72].

In the study area, it is apparent that propagule supply alone may not overwhelm invasion resistance, as rocky habitats devoid of *Didemnum* lie adjacent to numerous heavily infected structures throughout the region. However, there is a single location where *Didemnum* has extensively colonized a large ($\sim$10,000 m^2) area of artificial rip-rap wall (comprising quarry rock) beneath heavily infected piles at an international shipping wharf (see Table 1), yet the invasion has not spread to natural rocky habitats that are contiguous with the rip-rap. A feature of the rip-rap compared with adjacent natural areas is its construction from relatively large angular rocks with many spaces among them, which appeared impoverished in terms of predators and other mobile benthic species (BF, pers. obs.). The reasons are unknown, but the topography of the rip-rap (especially the size of the rocks) may not provide a favorable habitat for mobile species to access or move across. Hence, in this instance, it is conceivably reduced invasion resistance, rather than propagule pressure, that has enabled *Didemnum* to flourish. This hypothesis is consistent with the 'Persistent Pressure Scenario' conceptualized in a recent study of the solitary ascidian *Ciona intestinalis* [17]. That scenario proposed that despite a continuous propagule supply from a fouled suspended structure, the establishment of certain marine benthic species may be prevented by predation, unless predation pressure was reduced by processes such as disturbance.

In the present study, predation pressure on *Didemnum* was attributable to a range of species. As these are common species around the New Zealand coastline [73], it may be the case that predation explains the apparent absence of *Didemnum* from rocky seabed habitats elsewhere in the country, despite its increasing prevalence on artificial structures. In general, while anthropogenic structures may function as significant propagule reservoirs that facilitate the spread of invasive species into adjacent natural habitats, proliferation in natural systems may only occur in situations where predation and other mechanisms of invasion resistance are not limiting. That being so, the risk to natural habitats may be exacerbated in situations where a high connectivity of structures (e.g. by natural dispersal) enables them to function as 'stepping-stones' for NIS spread [1,5,74]. Such spread will increase the number of propagule reservoirs, possibly leading to increased likelihood that invadable natural habitats will eventually be encountered. In order to better understand and effectively manage such risks, there is a need for a greater understanding of the interactions between propagule supply and invasion resistance that preclude or enhance the establishment of invasive species.

Acknowledgments

We are grateful to the Aaron Pannell, Terry Schwass, Mark Gillard and James Brodie for assistance with field logistics, and to anonymous referees whose helpful comments improved this manuscript.

Author Contributions

Conceived and designed the experiments: BMF GAH RFP. Performed the experiments: BMF GAH RFP LF. Analyzed the data: JA BMF. Wrote the paper: BMF.

References

1. Bulleri F, Chapman MG (2010) The introduction of coastal infrastructure as a driver of change in marine environments. Journal of Applied Ecology 47: 26–35.

2. Ruiz GM, Freestone AL, Fofonoff PW, Simkanin C (2009) Habitat distribution and heterogeneity in marine invasion dynamics: the importance of hard substrate and artificial structure. In: Rilov G, Crooks JA, editors. Biological Invasions in Marine Ecosystems: Ecological, Management, and Geographic Perspectives. Berlin Heidelberg: Springer-Verlag. pp. 321–332.

3. Bulleri F, Airoldi L (2005) Artificial marine structures facilitate the spread of a non-indigenous green alga, *Codium fragile* ssp. *tomentosoides*, in the north Adriatic Sea. Journal of Applied Ecology 42: 1063–1072.

4. Dafforn KA, Johnston EL, Glasby TM (2009) Shallow moving structures promote marine invader dominance. Biofouling 25: 277–287.

5. Glasby TM, Connell SD, Holloway MG, Hewitt CL (2007) Nonindigenous biota on artificial structures: could habitat creation facilitate biological invasions? Marine Biology 151: 887–895.

6. Tyrrell MC, Byers JE (2007) Do artificial substrates favor nonindigenous fouling species over native species? Journal of Experimental Marine Biology and Ecology 342: 54–60.

7. Simkanin C, Davidson IC, Dower JF, Jamieson G, Therriault TW (2012) Anthropogenic structures and the infiltration of natural benthos by invasive ascidians. Marine Ecology 33: 499–512.

8. Bulleri F (2005) Role of recruitment in causing differences between intertidal assemblages on seawalls and rocky shores. Marine Ecology Progress Series 287: 53–65.

9. Connell S (2000) Floating pontoons create novel habitats for subtidal epibiota. Journal of Experimental Marine Biology and Ecology 247: 183–194.

10. Connell SD, Glasby TM (1999) Do urban structures influence local abundance and diversity of subtidal epibiota? A case study from Sydney Harbour, Australia. Marine Environmental Research 47: 373–387.

11. Glasby TM (1999) Differences between subtidal epibiota on pier pilings and rocky reefs at marinas in Sydney, Australia. Estuarine, Coastal and Shelf Science 48: 281–290.

12. Britton-Simmons KH (2004) Direct and indirect effects of the introduced alga *Sargassum muticum* on benthic, subtidal communities of Washington State, USA. Marine Ecology Progress Series 277: 61–78.

13. Currie DR, McArthur MA, Cohen BF (2000) Reproduction and distribution of the invasive European fanworm *Sabella spallanzanii* (Polychaeta: Sabellidae) in Port Phillip Bay, Victoria, Australia. Marine Biology 136: 645–656.

14. Forrest BM, Taylor MD (2002) Assessing invasion impact: survey design considerations and implications for management of an invasive marine plant. Biological Invasions 4: 375–386.

15. Goulletquer P, Bachelet G, Sauriau G, Noel P (2002) Open Atlantic coasts of Europe - a century of introduced species into French waters. In: Leppäkoski E, Gollasch S, Olenin S, editors. Invasive Aquatic Species of Europe: Distribution,

Impacts and Management: Kluwer Academic Publishers, The Netherlands. pp. 276–290.

16. Valentine PC, Collie JS, Reid RN, Asch RG, Guida VG, et al. (2007) The occurrence of the colonial ascidian *Didemnum* sp. on Georges Bank gravel habitat: ecological observations and potential effects on groundfish and scallop fisheries. Journal of Experimental Marine Biology and Ecology 342: 179–181.

17. Dumont CP, Gaymer CF, Thiel M (2011) Predation contributes to invasion resistance of benthic communities against the non-indigenous tunicate *Ciona intestinalis*. Biological Invasions 13: 2023–2034.

18. Osman RW, Whitlatch RB (2007) Variation in the ability of *Didemnum* sp. to invade established communities. Journal of Experimental Marine Biology and Ecology 342: 40–53.

19. Kott P (2002) A complex didemnid ascidian from Whangamata, New Zealand. Journal of Marine Biology Association United Kingdom 82: 625–628.

20. Lengyel NL, Collie JS, Valentine PC (2009) The invasive colonial ascidian *Didemnum vexillum* on Georges Bank: ecological effects and genetic identification. Aquatic Invasions 4: 143–152.

21. Beveridge C, Cook EJ, Brunner L, MacLeod A, Black K, et al. (2011) Initial response to the invasive carpet sea squirt, *Didemnum vexillum*, in Scotland. Scottish Natural Heritage Commissioned Report No. 413. 32 p.

22. Bishop JDD, Wood CA, Yunnie ALE (2012) Surveys for the ascidian *Didemnum vexillum* in the Dart and Kingsbridge-Salcombe estuaries, Devon, in October 2009. Internal Report for Department of the Environment, Food and Rural Affairs (DEFRA). 15 p.

23. Cohen CS, McCann L, Davis T, Shaw L, Ruiz G (2011) Discovery and significance of the colonial tunicate *Didemnum vexillum* in Alaska. Aquatic Invasions 6: 263–271.

24. Tagliapietra D, Keppel E, Sigovini M, Lambert G (2012) First record of the colonial ascidian *Didemnum vexillum* Kott, 2002 in the Mediterranean: Lagoon of Venice (Italy). BioInvasions Records 1: 247–254.

25. Gittenberger A (2007) Recent population expansions of non-native ascidians in The Netherlands. Journal of Experimental Marine Biology and Ecology 342: 122–126.

26. Lambert G (2009) Adventures of a sea squirt sleuth: unraveling the identity of *Didemnum vexillum*, a global ascidian invader. Aquatic Invasions 4: 5–28.

27. Carman MR, Grunden DW (2010) First occurrence of the invasive tunicate *Didemnum vexillum* in eelgrass habitat. Aquatic Invasions 5: 23–29.

28. Mercer JM, Whitlatch RB, Osman RW (2009) Potential effects of the invasive colonial ascidian (*Didemnum vexillum*) on pebble-cobble bottom habitats in Long Island Sound, USA. Aquatic Invasions 4: 133–142.

29. Valentine PC, Carman MR, Blackwood DS, Heffron EJ (2007) Ecological observations on the colonial ascidian *Didemnum* sp. in a New England tide pool habitat. Journal of Experimental Marine Biology and Ecology 342: 109–121.

30. Coutts ADM, Forrest BM (2007) Development and application of tools for incursion response: lessons learned from the management of the fouling pest *Didemnum vexillum*. Journal of Experimental Marine Biology and Ecology 342: 154–162.

31. Keough MJ, Butler AJ (1979) The role of asteroid predators in the organization of a sessile community on pier pilings. Marine Biology 51: 167–177.

32. Osman RW, Whitlatch RB (1995) The influence of resident adults on larval settlement: experiments with four species of ascidians. Journal of Experimental Marine Biology and Ecology 190: 199–220.

33. Stachowicz JJ, Byrnes JE (2006) Species diversity, invasion success, and ecosystem functioning: disentangling the influence of resource competition, facilitation, and extrinsic factors. Marine Ecology Progress Series 311: 251–262.

34. Castilla JC, Guiñez R, Caro AU, Ortiz V (2004) Invasion of a rocky intertidal shore by the tunicate *Pyura praeputialis* in the Bay of Antofagasta, Chile. Proceedings of the National Academy of Sciences of the United States of America 101: 8517–8524.

35. Keough MJ, Downes BJ (1986) Effects of settlement and post-settlement mortality on the distribution of the ascidian *Trididemnum opacum*. Marine Ecology Progress Series 33: 279–285.

36. Osman RW, Whitlatch RB (2004) The control of the development of a marine benthic community by predation on recruits. Journal of Experimental Marine Biology and Ecology 311: 117–145.

37. Simkanin C, Dower JF, Filip N, Jamieson G, Therriault TW (2013) Biotic resistance to the infiltration of natural benthic habitats: examining the role of predation in the distribution of the invasive ascidian *Botrylloides violaceus*. Journal of Experimental Marine Biology and Ecology 439: 76–83.

38. Whitlatch RB, Osman RW (2009) Post-settlement predation on ascidian recruits: predator responses to changing prey density. Aquatic Invasions 4: 121–131.

39. Stachowicz JJ, Whitlatch RB, Osman RW (1999) Species diversity and invasion resistance in a marine ecosystem. Science 286: 1577–1579.

40. Epelbaum A, Pearce C, Barker D, Paulson A, Therriault T (2009) Susceptibility of non-indigenous ascidian species in British Columbia (Canada) to invertebrate predation. Marine Biology 156: 1311–1320.

41. Janiak DS, Osman RW, Whitlatch RB (2013) The role of species richness and spatial resources in the invasion success of the colonial ascidian *Didemnum vexillum* Kott, 2002 in eastern Long Island Sound. Journal of Experimental Marine Biology and Ecology 443: 12–20.

42. Rocha RM, Kremer LP, Baptista MS, Metri R (2009) Bivalve cultures provide habitat for exotic tunicates in southern Brazil. Aquatic Invasions 4: 195–205.

43. Bullard SG, Sedlack B, Reinhardt JF, Litty C, Gareau K, et al. (2007) Fragmentation of colonial ascidians: differences in reattachment capability among species. Journal of Experimental Marine Biology and Ecology 342: 166–168.

44. Hopkins GA, Forrest BM, Piola RF, Gardner JPA (2011) Factors affecting survivorship of defouled communities and the effect of fragmentation on establishment success. Journal of Experimental Marine Biology and Ecology 396: 233–243.

45. Morris J Jr, Carman M (2012) Fragment reattachment, reproductive status, and health indicators of the invasive colonial tunicate *Didemnum vexillum* with implications for dispersal. Biological Invasions 14: 2133–2140.

46. Fletcher LM, Forrest BM, Atalah J, Bell JJ (2013) Reproductive seasonality of the invasive ascidian *Didemnum vexillum* in New Zealand and implications for shellfish aquaculture. Aquaculture Environment Interactions 3: 197–211.

47. Abramoff MD, Magalhaes PJ, Ram SJ (2004) Image processing with ImageJ. Biophotonics International 11: 36–42.

48. Fletcher LM (2013) Ecology of biofouling and impacts on mussel aquaculture: a case study with *Didemnum vexillum*. PhD thesis, Victoria University of Wellington, New Zealand. 234 p.

49. Boyd HC, Brown SK, Harp JA, Weissman IL (1986) Growth and sexual maturation of laboratory-cultured Monterey *Botryllus schlosseri*. Biological Bulletin 170: 91–109.

50. Cloney RA (1987) Phylum Urochordata, Class Ascidiacea. In: Strathmann MF, editor. Reproduction and development of marine invertebrates of the Northern Pacific coast. University of Washington Press, Seattle. pp. 607–646.

51. Fletcher LM, Forrest BM, Bell JJ (2013) Natural dispersal mechanisms and dispersal potential of the invasive ascidian *Didemnum vexillum*. Biological Invasions 15: 627–643.

52. Anderson MJ (2001) A new method for non-parametric multivariate analysis of variance. Austral Ecology 26: 32–46.

53. Anderson MJ, Gorley RN (2007) PERMANOVA+ for PRIMER: Guide to Software and Statistical Methods. Plymouth, UK: PRIMER-E.

54. Fletcher LM, Forrest BM, Bell JJ (2013) Impacts of the invasive ascidian *Didemnum vexillum* on aquaculture of the New Zealand green-lipped mussel, *Perna canaliculus*. Aquaculture Environment Interactions 4: 17–30.

55. Kay AM, Keough MJ (1981) Occupation of patches in the epifaunal communities on pier pilings and the bivalve *Pinna bicolor* at Edithburgh, South Australia. Oecologia 48: 123–130.

56. Agius BP (2007) Spatial and temporal effects of pre-seeding plates with invasive ascidians: Growth, recruitment and community composition. Journal of Experimental Marine Biology and Ecology 342: 30–39.

57. Reinhardt J, Gallagher K, Stefaniak L, Nolan R, Shaw M, et al. (2012) Material properties of *Didemnum vexillum* and prediction of tendril fragmentation. Marine Biology 159: 2875–2884.

58. Campbell ML (2009) An overview of risk assessment in a marine biosecurity context. In: Rilov G, Crooks JA, editors. Biological Invasions in Marine Ecosystems: Ecological, Management, and Geographic Perspectives. Berlin Heidelberg: Springer-Verlag pp. 353–374.

59. Lockwood JL, Cassey P, Blackburn TM (2009) The more you introduce the more you get: the role of colonization pressure and propagule pressure in invasion ecology. Diversity & Distributions 15: 904–910.

60. Fletcher LM, Forrest BM (2011) Induced spawning and culture techniques for the invasive ascidian *Didemnum vexillum* (Kott, 2002). Aquatic Invasions 6: 457–464.

61. Valentine PC, Carman MR, Dijkstra J, Blackwood DS (2009) Larval recruitment of the invasive colonial ascidian *Didemnum vexillum*, seasonal water temperatures in New England coastal and offshore waters, and implications for spread of the species. Aquatic Invasions 4: 153–168.

62. Chapman MG (2003) Paucity of mobile species on constructed seawalls: effects of urbanization on biodiversity. Marine Ecology Progress Series 264: 21–29.

63. Hunt HL, Scheibling RE (1997) Role of early post-settlement mortality in recruitment of benthic marine invertebrates. Marine Ecology Progress Series 155: 269–301.

64. Vervoort HC, Pawlik JR, Fenical W (1998) Chemical defense of the Caribbean ascidian *Didemnum conchyliatum*. Marine Ecology Progress Series 164: 221–228.

65. Svane I, Young CM (1989) The ecology and behaviour of ascidian larvae. Oceanography and Marine Biology: An Annual Review 27: 45–90.

66. Bullard SG, Lambert G, Carman MR, Byrnes J, Whitlatch RB, et al. (2007) The colonial ascidian *Didemnum* sp. A: current distribution, basic biology and potential threat to marine communities of the northeast and west coasts of North America. Journal of Experimental Marine Biology and Ecology 342: 99–108.

67. Lockwood JL, Cassey P, Blackburn T (2005) The role of propagule pressure in explaining species invasions. Trends in Ecology and Evolution 20: 223–228.

68. Ruiz GM, Fofonoff PW, Carlton JT, Wonhom MJ, Hines AH (2000) Invasion of coastal marine communities in North America: apparent patterns, processes and biases. Annual Review of Ecology and Systematics 31: 481–531.

69. Chadwell TB, Engelhardt KAM (2008) Effects of pre-existing submersed vegetation and propagule pressure on the invasion success of *Hydrilla verticillata*. Journal of Applied Ecology 45: 515–523.

70. Clark G, Johnston EL (2009) Propagule pressure and disturbance interact to overcome biotic resistance of marine invertebrate communities. Oikos 118: 1679–1686.

71. Hollebone AL, Hay ME (2007) Propagule pressure of an invasive crab overwhelms native biotic resistance. Marine Ecology Progress Series 342: 191–196.

72. Von Holle B, Simberloff D (2005) Ecological resistance to biological invasion overwhelmed by propagule pressure. Ecology 86: 3212–3218.

73. Morton J, Miller M (1973) The New Zealand Sea Shore. Collins, London. 653 p.

74. Airoldi L, Abbiati M, Beck MW, Hawkins SJ, Jonsson PR, et al. (2005) An ecological perspective on the development and design of low-crested and other hard coastal defence structures. Coastal Engineering 52: 1073–1087.

Great Apes and Biodiversity Offset Projects in Africa: The Case for National Offset Strategies

Rebecca Kormos[1]*, Cyril F. Kormos[2], Tatyana Humle[3], Annette Lanjouw[4], Helga Rainer[5], Ray Victurine[6], Russell A. Mittermeier[7], Mamadou S. Diallo[8], Anthony B. Rylands[9], Elizabeth A. Williamson[10]

1 Department of Integrative Biology, University of California, Berkeley, California, United States of America, 2 The WILD Foundation, Berkeley, California, United States of America, 3 Durrell Institute of Conservation and Ecology, School of Anthropology and Conservation, University of Kent, Canterbury, United Kingdom, 4 Strategic Initiatives and Great Apes Program, The Arcus Foundation, Cambridge, United Kingdom, 5 Conservation Program, The Arcus Foundation, Cambridge, United Kingdom, 6 Business Conservation Initiative and Conservation Finance, Wildlife Conservation Society, Bronx, New York, United States of America, 7 Conservation International, Arlington, Virginia, United States of America, 8 Guinée Écologie, Conakry, Republic of Guinea, 9 Conservation International, Arlington, Virginia, United States of America, 10 Scottish Primate Research Group, School of Natural Sciences, University of Stirling, Scotland, United Kingdom

Abstract

The development and private sectors are increasingly considering "biodiversity offsets" as a strategy to compensate for their negative impacts on biodiversity, including impacts on great apes and their habitats in Africa. In the absence of national offset policies in sub-Saharan Africa, offset design and implementation are guided by company internal standards, lending bank standards or international best practice principles. We examine four projects in Africa that are seeking to compensate for their negative impacts on great ape populations. Our assessment of these projects reveals that not all apply or implement best practices, and that there is little standardization in the methods used to measure losses and gains in species numbers. Even if they were to follow currently accepted best-practice principles, we find that these actions may still fail to contribute to conservation objectives over the long term. We advocate for an alternative approach in which biodiversity offset and compensation projects are designed and implemented as part of a National Offset Strategy that (1) takes into account the cumulative impacts of development in individual countries, (2) identifies priority offset sites, (3) promotes aggregated offsets, and (4) integrates biodiversity offset and compensation projects with national biodiversity conservation objectives. We also propose supplementary principles necessary for biodiversity offsets to contribute to great ape conservation in Africa. Caution should still be exercised, however, with regard to offsets until further field-based evidence of their effectiveness is available.

Editor: Clinton N. Jenkins, Instituto de Pesquisas Ecológicas, Brazil

Funding: Funding for part of this study was provided by the Arcus Foundation grant no. 1104-38 (http://www.arcusfoundation.org/). The funders had no role in study design, data collection and analysis, decision to publish. They did contribute to discussions of these ideas and preparation of the manuscript.

Competing Interests: The authors have declared that no competing interests exist.

* Email: rebeccakormos@yahoo.com

Introduction

Great apes–gorillas, chimpanzees, and bonobos–are distributed across 21 countries on the African continent [1]. Their conservation is important in several respects. Their geographic ranges are strongly associated with the tropical forests that harbor some of the richest biodiversity in the world and overlap extensively with those of many endemic species [2]. Great apes have large home ranges [3], and thus protection of their habitat will also bring many other species under protection. Great apes are keystone species, playing important roles in maintaining the health and diversity of their ecosystems through their seed dispersal [4–6]. In addition, they act as physical ecosystem engineers [7,8] shaping the forest structure by trampling, bending and breaking vegetation as they travel, forage and build nests [9,10]. Apes and ape habitat are important for people; protecting ape habitats protects important water catchment areas. In Rwanda, for example, the Volcanoes National Park provides much of that country's water [11], and the Fouta Djallon in Guinea is the source of a number of

West Africa's major rivers, including the Niger. Tourism with great apes also provides significant income to local communities through revenue sharing and local businesses, thus providing livelihood opportunities for local people [12].

All great ape taxa are listed as either Endangered (EN) or Critically Endangered (CR) by the International Union for Conservation of Nature (IUCN) [13]. Threats to great apes include habitat loss, hunting and disease [13]. These threats are exacerbated by large-scale development activities such as hydro-electric projects, roads, and extractive industries [14], all of which result in the destruction of large areas of ape habitat and provide access to remote areas, facilitating bushmeat hunting [15]. Industrial development projects result in large influxes of people, exposing great apes to human diseases that can be fatal to them [16]. Human presence and activity can cause apes to leave their habitual ranges, which can result in competition, conflict and stress, with long-term consequences for the health and reproduction of the population [17,18]. Mortalities are likely to occur if chimpanzees are forced into an area that is already occupied by conspecifics because chimpanzees are highly territorial and often attack intruders [19]. The effects of development projects are intensified for great apes because of their reproductive biology and slow maturation [14]. Even low levels of disturbance to ape populations can result in declines that require decades for their subsequent recovery [20].

Industrial development is proliferating throughout Africa [21,22]. The interface between development projects and great ape conservation will, therefore, intensify in coming decades. Most countries in Africa where great apes occur rank high on the United Nations poverty index and are undergoing intensive infrastructure development [23]. A mining boom is occurring [24,25], which will result in the expansion of transportation infrastructure [25]. More than 50% of the range of chimpanzees and gorillas in Western Equatorial Africa has been allocated to logging concessions [26]. The extensive overlap between the distribution of commodities, biodiverse areas and great ape ranges means that companies will increasingly need to mitigate the negative impacts of their projects on great ape populations [14,27].

Options for mitigating impacts on great apes are limited. Relocation is risky [28] and can lead to mortalities [29,30], and restoring habitat is not feasible on a time-scale meaningful to great apes [31]. Unless great ape habitat is avoided entirely, in most cases mitigation is unlikely to prevent great ape losses and most projects will result in some population decline.

"Biodiversity offsets" are increasingly used worldwide to compensate for the negative impacts of development and private sector projects on biodiversity [31–40]. The Business and Biodiversity Offset Program (BBOP), a broad consortium including civil society and private sector organizations, financial institutions, governments, and intergovernmental organizations, defines biodiversity offsets as "measurable conservation outcomes resulting from actions designed to compensate for significant residual adverse biodiversity impacts arising from project development after appropriate prevention and mitigation measures have been taken" [36]. According to BBOP, the goal of biodiversity offsets is "to achieve no net loss and preferably a net gain of biodiversity on the ground with respect to species composition, habitat structure, ecosystem function and people's use and cultural values associated with biodiversity" [36]. Offsets are distinct from the broader category of biodiversity compensation projects, which mitigate impacts but do not follow a mitigation hierarchy or comply with other offset requirements [36].

Seventeen countries worldwide have national policies requiring biodiversity offsets, and more than 29 countries have national policies that suggest or enable the use of offsets [41]. No countries in the range of great apes in West and Central Africa, however, currently have policies guiding or requiring offsets [41]. Biodiversity offsets are therefore guided by private sector internal standards or those of lenders, rather than by government policy [42–45]. Several international organizations such as BBOP, the International Council on Mining and Metals (ICMM) and IUCN have proposed best practices for biodiversity offsets [36,46]. BBOP best practices include a list of 10 guiding "principles" for the design and implementation of biodiversity offsets. These principles are reflected in many national offset policies around the world [34], in the scientific literature [31–35,37–39,47], in private sector internal guidelines [42], and in other international best practices [46], and are reflected in the lending standards and principles of many of the largest international banks [45,46].

Despite this new emphasis on biodiversity offsets, there is little empirical evidence from the field to demonstrate that they are achieving conservation objectives over the long term. This is a result of the lack of standardized evaluation criteria, limited monitoring of projects, under-reporting of projects that are not working [31], and lack of access to information on projects due to confidentiality of reports. In addition, many offsets projects are still in the design or early implementation phase and do not yet have results to report. Biodiversity offset policies are still in their infancy [38], and more reviews of field projects are urgently needed before encouraging their wider use as a conservation tool [37,38].

To assess how well field projects adhere to international best practice principles, we examine four projects in Africa where private sector or development projects are impacting great apes and their habitat and are seeking to use offsets and compensation projects to counterbalance these impacts. We measure them against the six BBOP principles related to biological criteria: (1) limits to what can be offset, (2) adherence to the mitigation hierarchy, (3) additional conservation outcomes, (4) landscape context, (5) no net loss, and (6) long-term outcomes. We also assess whether full compliance to these principles would be sufficient to generate conservation benefits for great apes in light of their EN and CR status, their shrinking habitat, and their vulnerability to disturbance. Although this paper focuses on great apes, we believe the results of this study will also apply to many other EN and CR taxa.

Case Studies

We examine four projects in Africa that are investigating either biodiversity offsets or compensation for residual impacts to great apes and their habitat, or have attempted such projects (Fig. 1). They are: (1) the Simandou Project in the Republic of Guinea, (2) the Global Alumina Project (GAP) in the Republic of Guinea, (3) the Bumbuna Hydroelectric Project (BHP) in Sierra Leone, and (4) the Lom Pangar Dam in Cameroon. For each project, we researched all publicly available documents that included information on measuring and mitigating impacts on great apes and their habitat. Table 1 provides a list of those documents, available as of May 2014. We are also aware of other projects that have considered, or are considering, the use of offsets to compensate for residual damage to ape habitat in Africa. We focus on these four case studies because they are those of which the authors have the most direct experience. It will be important to eventually expand this type of analysis to include a larger set of projects. The current analysis, however, is an important first step. Table 2 provides a summary of predicted impacts on great apes and proposed mitigation measures for each project.

The Simandou Project, Republic of Guinea

Simfer is a Guinean-registered company and holder of an iron-ore mining concession called the "Simandou Project" in the Simandou mountains of Southeast Guinea. The "Simandou Project" partners include the Republic of Guinea, Rio Tinto, Aluminium Corporation of China ("Chinalco"), and the International Finance Corporation ("IFC"). The proposed Simandou Project includes: (i) an open-pit iron-ore mine in the Simandou mountain range; (ii) approximately 670 km of railway across Guinea to transport ore to the coast; (iii) a new port facility; and (iv) associated infrastructure, such as housing, roads, quarries, and power generation and distribution.

The Simandou Project has been collecting data on chimpanzees in the Pic de Fon Classified Forest since 2007 to guide the development and implementation of a mitigation plan–the Pic de Fon Management Plan and the Simandou Project Biodiversity Offsets Strategy. The study has identified an estimated 36–46 western chimpanzees (*Pan troglodytes verus*) living in the forest in

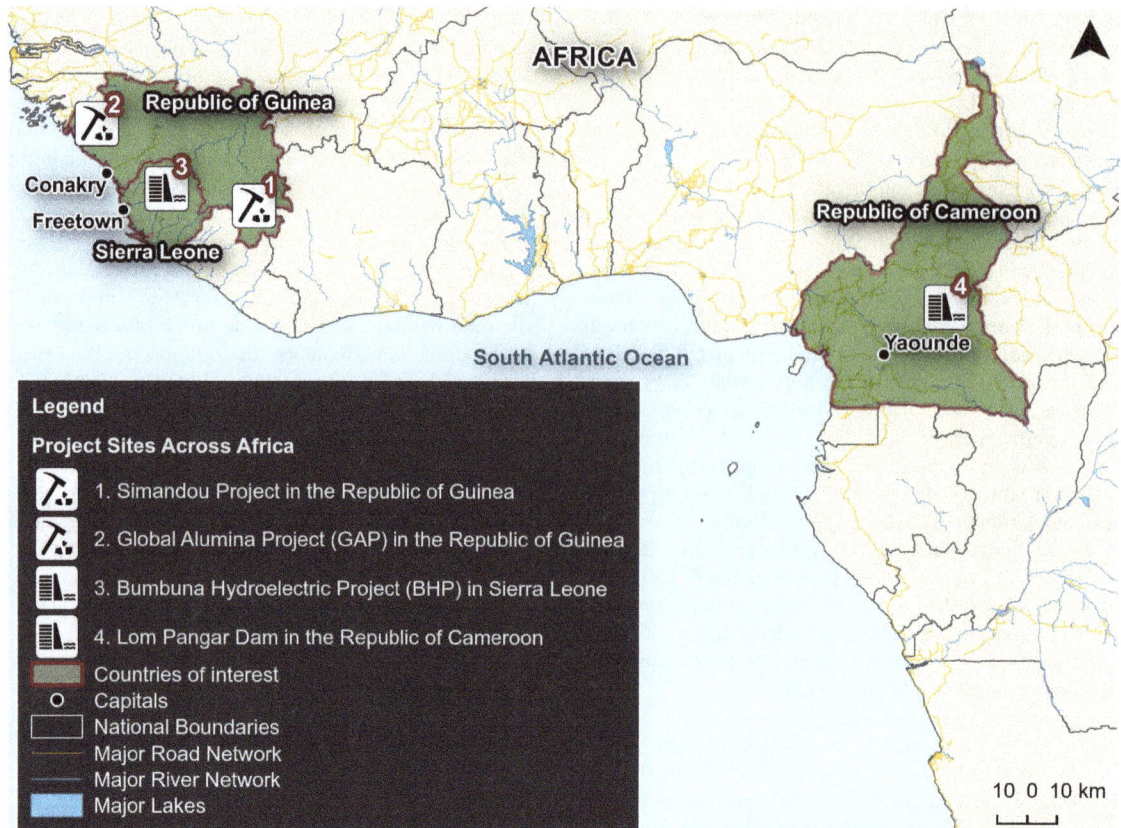

Figure 1. Sites in Africa where private sector or development projects are seeking to use offsets and compensation projects to counterbalance residual negative impacts to great apes and their habitat. Sites include (1) the Simandou Project in the Republic of Guinea, (2) the Global Alumina Project (GAP) in the Republic of Guinea, (3) the Bumbuna Hydroelectric Project (BHP) in Sierra Leone, and (4) the Lom Pangar Dam in the Republic of Cameroon.

the Simandou range dispersed across one or two communities. The number of chimpanzees along the railway is unknown, but 2,750 chimpanzee nests were recorded along the rail study area. The Social and Environmental Impact Assessment (SEIA) outlines the range of impacts on chimpanzees caused by mining activities and forecasts "best-case" and "worst-case" scenarios, dependent on the number of communities of chimpanzees and the relationships between them. Assuming that there are two separate communities, the worst-case scenario predicts a high degree of chimpanzee mortality when the communities are forced together as they lose habitat and move away from mining activities. An estimated 25% of the core of the chimpanzees' range would be permanently and irrecoverably lost to mining.

Mitigation proposed for the chimpanzees in the Simandou mountains includes controlling hunting, protecting habitat currently within the chimpanzee's range that will not be lost to mining, and creating additional habitat for chimpanzees both prior to and during mining activities. The SEIA predicts that, despite mitigation efforts, the sub-montane forest habitat where chimpanzees are living will be impacted, and the project is therefore investigating an offset site to compensate for residual damage to this unique habitat and other species living there. Simfer has formed a technical group called the Simandou Offsets Working Group with representatives from Simfer, the Environment Ministry, and the NGO Guinée Ecologie.

On 26th May 2014, the Government of Guinea, Rio Tinto, Chinalco and the IFC, signed an Investment Framework for

Blocks 3 and 4 of Simandou, which now makes it the largest combined iron-ore and infrastructure project ever developed in Africa.

Global Alumina Project (GAP), Republic of Guinea

The "Guinea Alumina Project" (GAP) is a 690-km² bauxite-mining concession in northwest Guinea that was developed in 2008 by the Global Alumina Corporation–a joint venture of BHP Billiton, Global Alumina, Dubai Aluminum Company Ltd., and the Mubadala Development Company. The initial exploitation zone was approximately 100 km², to be operational for 16 years. GAP includes a mine, an alumina refinery, a steam and power plant, a port facility, additional infrastructure on the concession, and a port facility 82 km from the refinery. GAP expected to employ about 12,000 workers during the 4-year construction period and more than 2,100 employees thereafter.

GAP originally hired the company Bechtel to conduct a chimpanzee Critical Habitat study, but the IFC requested that the study be redone. A second Critical Habitat study was conducted, although only two weeks were allowed for the fieldwork. The second study estimated that a minimum of 50 chimpanzees was living in the area surveyed [48]. GAP then hired the Wild Chimpanzee Foundation (WCF) to conduct a longer-term survey of chimpanzees covering the entire concession, to assist with the design and implementation of a conservation management plan, and to look for conservation sites to offset the expected decrease in the chimpanzee population caused by the

Table 1. Publicly available documentation for each site.

Project	Country	Document type	Source
Simandou Project	Guinea	Social and Environmental Baseline Study 2010	http://ifcext.ifc.org/ifcext/spiwebsite1.nsf/0/A87B7EA570082C41852578E700569CED/$File/Vol.%20D_Biodiversity%20baseline_FINAL.pdf
		Social and Environmental Impact Assessment Chapter 12. Biodiversity	http://www.riotintosimandou.com/documents/Mine/M_Ch12_TerrBiodiv_EN.pdf
		Social and Environmental Impact Assessment Annex12.E. West African Chimpanzee - Supplementary Baseline and Impact Assessment Information	http://www.riotintosimandou.com/documents/Mine/M_An12E_Chimp_EN.pdf
		Environmental and Social Action Plan July 2013	http://ifcext.ifc.org/ifcext/spiwebsite1.nsf/0/A87B7EA570082C41852578E700569CED/$File/Simandou%20Project%20ESAP%20July%202013%20FINAL.pdf
Global Alumina Project	Guinea	Social and Environmental Assessment	http://ifcext.ifc.org/IFCExt/spiwebsite1.nsf/78e3b305216fcdba85257a8b0075079d/8a0ee1048673cb16852576ba000e2cac?opendocument
		Critical Habitat Assessment 2008	http://ifcext.ifc.org/ifcext/spiwebsite1.nsf/0/8A0EE1048673CB16852576BA000E2CAC/$File/Guinea%20Critical%20Habitat%20Assessment%20Report.pdf
		Action Plan 2008	http://ifcext.ifc.org/ifcext/spiwebsite1.nsf/0/8A0EE1048673CB16852576BA000E2CAC/$File/Post%20Comm%20Action%20Plan%20FINAL%20FINAL290808.pdf
Bumbuna	Sierra Leone	Environmental Impact Assessment 2005	http://www-wds.worldbank.org/external/default/WDSContentServer/WDSP/IB/2005/03/10/000012009_20050310135611/Rendered/PDF/E10930V.02.pdf
		2005	http://bumbuna.sl/admin/images/news/ESAP%20M4%20Draft%20Final%20Report%20for%20transmission%20JJ%2005.03.09.pdf
		Environmental and Social Advisory Panel Report 2010	http://www.bumbuna.sl/admin/images/news/ESAP%20M1%20-%20Final%20Report%2010.11.04.pdf
		ESAP mission report 2010	http://www.bumbuna.sl/admin/images/news/ESAP%20M5%20Draft%20Final%20Report%20ver%202%20JJ%2003%2011%2010.pdf
Lom Pangar	Cameroon	Social and Environmental Impact Assessment 2010	http://www.edc-cameroon.org/IMG/pdf/sde/ANNEXE%204%20PNDD%20projet%20110111.pdf
		Environmental and Social Assessments 2011	http://www-wds.worldbank.org/external/default/WDSContentServer/WDSP/IB/2012/03/07/000350881_20120307112131/Rendered/PDF/673550BR0P11400ffiicial0Use0Only090.pdf

project. Information on the predicted habitat and population decline as a result of project activities and on suggested mitigation for mining activities on chimpanzees is not publicly available.

The Global Alumina Corporation (GAC) is now a branch of the newly formed Emirates Global Aluminum (EGA) founded by Mubadala and DUBAL, and predicted to become the fifth-largest aluminum company in the world by production in 2014 [49].

The Bumbuna Hydroelectric Project (BHP), Sierra Leone

The Bumbuna Hydroelectric Project (BHP) Phase I is a 50-MW, water regulation and hydropower facility located on the Seli River near Bumbuna, Sierra Leone. The project consists of an 88-m high asphalt concrete-faced rock-fill dam, a water intake structure, two spillways with associated tunnels, an above-ground powerhouse with two 25-MW turbo-generator units and a 30-km wide, Y-shaped reservoir. The USD 91.8 million project was funded by the African Development Bank, the Government of Italy, Organization of the Petroleum Exporting Countries

(OPEC), the Netherlands Clean Development Facility, the World Bank, UK DFID, and the Government of Sierra Leone.

An Environmental Impact Assessment (EIA) was completed in 2005 [50]. An Environmental Management and Mitigation Plan (EMP) was designed to include the construction phase, and the mitigation and monitoring activities associated with the first component, and includes the preparation and initiation of the Bumbuna Watershed Management Plan. In 2004, the World Bank assembled an Environmental and Social Advisory Panel (ESAP) to review these studies. In 2006, a full biodiversity assessment was conducted as a follow-up to the BHP EIA. As part of this, the size, distribution and socio-ecology of the chimpanzee population in the immediate catchment area was studied to determine the impact on chimpanzees of the filling of the reservoir and the associated loss of riparian forest [51]. The 2006 study estimated that four communities totaling 33–58 chimpanzees used the area to be flooded and that the main impacts would be (1) loss of the chimpanzees' natural resources, (2) an increase in human-wildlife

Table 2. Summary of predicted project impacts on great apes and proposed mitigation measures.

Project	Country	Type of project	Main Project components	Estimated # of apes possibly affected by project activities	Main predicted impacts on apes	Main proposed mitigation measures for ape conservation	Proposed offset site
Simandou	Guinea	Iron-ore mining	(i) An open-pit iron-ore mine in the Simandou mountain range; (ii) approximately 670 km of railway across Guinea to transport ore to the coast; (iii) a new port facility; (iv) associated infrastructure, incl. housing, roads, quarries, power generation and distribution	<50 at the mine site and unknown number along the railway	Open-pit mining will remove an area approximately 6–8 km long, 1–1.5 km wide and 300 m deep. One quarter of the core of the chimpanzees' range would be permanently and irrecoverably lost to mining	(i) Improving control of hunting; (ii) protecting habitat within the chimpanzee's current range; iii) creating additional habitat for chimpanzees both prior to and during mining activities	Undecided
Global Alumina Corporation	Guinea	Bauxite mining	(i) Mine sites in the 690-km^2 concession in northwest Guinea; (ii) an alumina refinery; (iii) a steam and power plant; (iv) a port facility 82 km from the refinery; (v) additional infrastructure on the concession	50–>100 in 247.6 km^2 of 2008 Critical Habitat survey area	Information not available	Information not available	Undecided
Bumbuna	Sierra Leone	Hydroelectric dam	(i) An 88-m high asphalt concrete-faced rock-fill dam; (ii) a water intake structure; (iii) two spillways with associated tunnels; (iv) an above-ground powerhouse with two 25-MW turbo-generator units; (v) a 30-km wide, Y-shaped reservoir	33–58	(i) loss of natural resources for the estimated four communities of about 33–58 chimpanzees using the area to be flooded; (ii) an increase in human-wildlife conflict as farmers and wildlife are forced closer together by reduction in available land; (iii) the prevention of movement of chimpanzees across the Seli River and between chimpanzee communities on each side of the river resulting in a decrease in their genetic viability	(i) Initiation of monitoring and awareness programs throughout the catchment area; (ii) incorporation of specific conservation activities into a Watershed Management Plan, including hunting controls, environmental awareness, fire control and zonation to give more protection to the most important remaining forest patches; (iii) creation of a Wildlife Conservation Area within the catchment	Loma Mountains National Park

Table 2. Cont.

Project	Country	Type of project	Main Project components	Estimated # of apes possibly affected by project activities	Main predicted impacts on apes	Main proposed mitigation measures for ape conservation	Proposed offset site
Lom Pangar	Cameroon	Hydroelectric dam	(i) A 50-m high dam located on the River Lom in Cameroon's East Region; (ii) a 610-km² reservoir area; (iii) a hydroelectric power plant; (iv) a transmission line; (v) a rural electrification scheme along this transmission line.	About 990 gorillas are located in the greater Deng Deng area, with over 50% of this population resident in Deng Deng NP itself; the others are located in the Forest Management Unit UFA 10-65 and in the Belabo Forest.	The key impacts identified by the project include: (i) loss of natural habitat due to flooding and infrastructure footprint; (ii) risk of reducing the viability of a distinct population of gorillas and other Red-Listed species; (iii) the risk that construction activities will induce significant loss of natural habitat; and (iv) the risk of future habitat loss due to increased human pressures in the region.	Minimization: an adjustment to the pipeline route to avoid central Deng Deng and other dense forest areas and an offset site. Compensation: i) Extension of Deng Deng National Park by 9,000 ha, and creation of protection corridors to allow movement of gorilla populations in and out of Deng Deng National Park; ii) Conservation management programs in two Forest Management Units; iii) Creation of community forests (e.g. Belabo Forest) to support sustainable livelihoods and gorilla protection; iv) Development of mechanisms to control access to the region, especially during construction phase to prevent illegal logging; v) development of a financing mechanism to support conservation management and maintenance of ecoguards.	Forest areas around Deng Deng National Park perimeter

conflict as farmers and wildlife are forced closer together by reduction in available land, and (3) the prevention of movement of chimpanzees across the Seli River and between chimpanzee communities on each side of the river. The study predicted that these effects could result in reduced viability of the chimpanzee population in the BHP catchment over the long term due to genetic isolation [51]. Recommendations for the mitigation and offset of these impacts included initiating a monitoring and awareness program in the catchment area, and incorporating conservation activities such as hunting controls, environmental awareness, fire control and zonation to give more protection to the most important remaining forest patches in a Watershed Management Plan. The 2006 study also suggested the establishment and management of a Wildlife Conservation Area in the catchment area, to be called the Bumbuna Conservation Area (BCA), to preserve and protect biodiversity in the Bumbuna watershed and to serve as a biodiversity offset. The ESAP's view was that this BCA would not be very effective for chimpanzees, given the small size of the area and the many people living and farming within it. It therefore recommended creating an offset conservation area outside the catchment area. It also recommended that the Loma Mountain Non-Hunting Forest Reserve, an even more diverse 396-km^2 forest, 50 km from the dam with a population of approximately 1,065 chimpanzees [52], be upgraded to national park status as additional compensation.

The dam construction was completed and the reservoir area flooded in 2009. In 2012, Loma Mountains National Park (LMNP) was proclaimed, although parliamentary approval is still pending. Measures for the park have included: (a) posting and training staff; (b) providing equipment and materials; (c) conducting consultations on the park boundary, (d) resurveying and physically demarcating the boundary; (e) completing Reserve Settlement Courts Sittings; (f) completing a Process Framework, including a socio-economic baseline study; (g) developing a Management Plan; and (h) developing a provisional 5-year budget. These activities were funded by the original loan guarantee turned into a grant to support environmental and social mitigation at Bumbuna, as well as a GEF project that has been building government capacity to conserve national biodiversity. Sustainable financing for the LMNP has not yet been secured [53]. Since no long-term monitoring of chimpanzees was undertaken, the responses of the chimpanzees to the flooding remain unknown. A new Bumbuna Phase II project has now been launched that will expand the Bumbuna hydroelectric station and flood an even greater area.

Lom Pangar Hydropower Project (LPHP), Cameroon

The Lom Pangar Hydropower Project (LPHP) consists of a regulating, 50-m high dam located on the River Lom in Cameroon's East Region, a 610-km^2 reservoir area, a hydroelectric power plant, a transmission line, and a rural electrification scheme along this transmission line. Estimated costs for the LPHP are USD 393 million from the African Development Bank (AfDB), the Central African States Development Bank (BDEAC), the European Investment Bank (EIB), the French Agency for Development (AFD), the Government of Cameroon, and the World Bank.

The Environmental and Social Assessment noted that the main impact would be the flooding of natural forest, that none of the flooded habitat was critical, but that the dam site was located next to portions of the Deng Deng forest that included critical habitats and populations of the Critically Endangered (CR) Western lowland gorilla (*Gorilla gorilla gorilla*) and the Endangered (EN) Central chimpanzee (*Pan troglodytes troglodytes*). About 990 gorillas are located in the greater Deng Deng area, with over

50% of this population resident in Deng Deng National Park itself; the others are located in the Forest Management Unit UFA 10–65 and in the Belabo Forest [54]. Suggested mitigation included an adjustment to the pipeline route to avoid the central Deng Deng and other dense forest areas and control of access to the area to prevent illegal logging during the construction phase. Nonetheless, the project was predicted to have significant impacts on natural habitats that should be compensated. Studies funded by the AFD and carried out by the Wildlife Conservation Society (WCS), estimated a population size of several hundred gorillas in the greater Deng Deng area, which included a communal forest and a logging concession. As a result, a decision was made that the company would provide a high-level compensation package designed to strengthen protection of the newly designated 580-km^2 Deng Deng National Park, extend its boundary, and establish a corridor to other forest areas [55]. A third-party analysis indicated that an annual payment of USD 700,000 per year would be required to provide sufficient financing to meet the conservation management needs in the region. The activities in Lom Pangar are considered compensation rather than an offset since no efforts were made to quantify losses or gains to biodiversity and offset them to achieve no net loss.

Analysis

In the following, we analyze to what extent these field projects follow international best-practice principles for biodiversity offsets and whether these principles adequately generate conservation benefits for great apes in light of their EN and CR status, shrinking habitats, and vulnerability to disturbance. Table 3 provides a summary of our findings.

Limits to what can be offset

It is generally accepted that there are limits to what can be offset [36,38,39]. Some residual impacts on biodiversity cannot be fully compensated for by a biodiversity offset given the irreplaceability or vulnerability of the biodiversity affected [36]. Ecological criteria used to make decisions on where biodiversity offset limits should be drawn include: "levels of conservation concern", "magnitude of the estimated residual impact", and "opportunity and feasibility of offsets on the ground" [38,39].

All great apes are listed as EN or CR [13]. They are highly vulnerable to habitat disturbances due to their life history, behavior and susceptibility to human diseases [9,56]. They are important seed dispersers and play a role in shaping forest structure [57,58]. Due to the high degree of their "impactability" and low degree of their "offsetability", (i.e. their vulnerability and irreplaceability), these ecological considerations advocate in favor of an extremely high threshold for offsetting apes.

In addition to these ecological factors, offsetting great apes raises serious ethical questions. Great apes are our closest living relatives [59,60], exhibit many of the same emotions as humans [61,62], practice tool-use [63], hunt cooperatively [64,65], and show evidence of culture and traditions [62,66] as well as a capacity for language [68]. Some human communities in the region have religious, cultural and traditional taboos against hunting and eating great apes because of their close resemblance to humans [69–72]. We contend that from an ethical standpoint also, offsets for apes should require an extremely high threshold.

The IFC [44] recognizes both ecological and ethical values, and theoretically sets high standard thresholds for offsetting apes and ape habitat. IFC Guidance Note 6 divides Critical Habitat into two tiers with the likelihood of project investment in a Tier 1 habitat substantially lower than in a Tier 2 habitat. A footnote in

Table 3. Summary of project's implementation of international best practice principles for biodiversity offsets with respect to great ape conservation.

Project	Limits to what can be offset	Adherence to the mitigation hierarchy	Additional conservation outcomes	Landscape context	No net loss	Long-term outcomes
Simandou	Chimpanzees are not considered beyond the limit to what can be offset.	All possible mitigation on site and changes to the mining plan will theoretically be followed before offsets are considered for residual damage to chimpanzees and their habitat.	Unknown.	The potential offset sites being considered are priority areas for chimpanzees.	No net loss may occur in medium-term but there will probably be a short-term and potentially long-term loss of chimpanzees.	Dependent on financing mechanisms, coordination with other offset projects and whether commitment to offset site is in perpetuity. Long-term outcomes therefore unknown at this point.
Global Alumina Corporation	Chimpanzees are not considered beyond the limit to what can be offset.	Some infrastructure and original mining plan already in place before 2008 Critical Habitat study was conducted.	Unknown	Unknown	Unknown because offset site is still not certain.	Dependent on financing mechanisms, site selection, coordination with other projects. Long-term outcomes are therefore unknown at this point.
Bumbuna	Chimpanzees are not considered beyond the limit to what can be offset.	Site selection and infrastructure were already in place before the ESAP was engaged.	Yes, if assumption of baseline decline of chimpanzees in Loma Mountains is correct. No, if assumption is incorrect. Yes, if Loma Mountains protected in perpetuity, but no if not.	The Loma Mountains National Park is a priority area for chimpanzees.	Unknown. Specific calculations on losses and gains of individual chimpanzees were not made.	No
Lom Pangar	Chimpanzees and gorillas are not considered beyond the limit to what can be offset.	No. Mitigation hierarchy was not specifically applied.	Yes, if assumption of baseline decline of chimpanzees in Deng Deng is correct. No, if assumption is incorrect. Yes, if Deng Deng National Park protected in perpetuity, and the adjacent areas are effectively managed, but no if not.	Deng Deng is a priority area for gorillas and chimpanzees and studies indicated that an area larger than Deng Deng forest needed protection and management and the compensation project was designed to address that.	Unknown. Specific calculations on losses and gains of individual chimpanzees and gorillas were not made. Key concern was to maintain gorilla population in the region.	Dependent on both company compliance with financing commitments (30 year annual payments) and securing longer-term financing, and whether Deng Deng will be protected in perpetuity.

Guidance Note 6 states that "special consideration should be given to great apes given their anthropological and evolutionary significance in addition to ethical considerations. Where populations of CR and EN great apes exist, a Tier 1 habitat designation is probable" [44].

In practice, however, the presence of apes in a project area does not seem to have deterred companies, governments or funders from investing in activities that will be detrimental to great ape habitat and likely to result in their decline. Both the GAP and the Simandou project are located in areas of Critical Habitat for chimpanzees, and both have received funding from the IFC. The BHP and the Lom Pangar Dam projects were financed by the World Bank, and both are considering or have implemented offsets or compensation projects for negative impacts to great ape

habitat. In summary it seems that none of these projects have considered great apes to be beyond the limits of what can be offset.

Adherence to the mitigation hierarchy

BBOP Principles [36] emphasize that biodiversity offsets are only appropriate after compliance with the mitigation hierarchy; that is, after avoidance, minimization and on-site rehabilitation measures have been exhausted [36]. Biodiversity offsets are therefore a mechanism of last resort [36,43,46]. In two of the projects profiled above, mitigation measures were not designed until they were already under way, resulting in the need for more off-site compensation than if mitigation measures had been included from the onset. For example, much of the BHP infrastructure was already completed when the ESAP was engaged, and the ESAP concluded that mitigation options for chimpanzees impacted by the dam were limited given that the dam site had already been selected [51]. Similarly in the case of the GAP, a Critical Habitat study for chimpanzees was not requested until the IFC was approached for a loan in 2008 [48], when much of the infrastructure already existed. The concept of "avoidance" was better integrated into the Simandou and Lom Pangar projects. For Simandou, the mining infrastructure development and plans for the sequence of exploitation activities were adapted to reduce impacts on chimpanzees. For Lom Pangar, costly mitigation measures were undertaken, including rerouting the pipeline to avoid the center of Deng Deng.

Additional conservation outcomes

BBOP states that a biodiversity offset should achieve additional conservation outcomes beyond results that would have occurred had the biodiversity offset not taken place [36]. The most common form of "additionality" in countries with offset policies is habitat restoration [31]. Biodiversity offset best practices also state that biodiversity offsets can achieve "additionality" by protecting areas where there is imminent or projected loss of biodiversity [36,73]. Two of the projects we examined based additionality on "averted loss" by updating the protected status of an area; the Loma Mountains National Park in Sierra Leone for BHP, and the Deng Deng National Park in Cameroon for the Lom Pangar Dam, and in the case of Lom Pangar extending areas for conservation around the park to protect great ape habitat. Both GAP and the Simandou project have provided a short list of potential offsets sites, indicating that their offsets will also consider "averted loss" as the counterbalance to actual loss on site. In these cases "no net loss" is working from an assumption of a pre-existing baseline rate of loss, assuming that habitat in the offset site is under threat or will be in the future [74]. This may be true given how many forests are threatened throughout the range of great apes. The result remains nonetheless a net loss of habitat against the extent and condition of that habitat at the time the project is implemented [73].

Maron *et al.* [31] emphasize that calculating the expected benefit of a conservation action–such as the purchase of a new reserve–requires "explicit estimation of the change in conservation value (e.g., population size of a threatened species) both with and without the action taking place, and calculation of the difference between these two scenarios. It is difficult to accurately (1) ascertain a baseline number of apes, (2) estimate the magnitude of change in ape numbers that would have occurred without project activities, (3) predict the magnitude of a population decline resulting from project activities; and (4) determine how much compensation is appropriate based on (1), (2) and (3). It is easier to estimate numbers of apes than of some other species given that their conspicuous nests can be used as indices of abundance. However, it is still extremely difficult to estimate numbers accurately [75]. Few studies have assessed the long-term impacts of extractive industries and other forms of habitat disturbance on apes [14]. As a result, it is very difficult for projects to measure losses, gains and additionality. Nevertheless, such estimations are necessary [36], and there should be consistency as to how they are generated and at what scale they are being assessed.

The projects we examined approached the challenge of measuring "additionality" in different ways, without common standards for measuring losses or gains. For the GAP, researchers from WCF proposed a mathematical formula to predict losses of chimpanzees, and gave a dollar value that companies should pay to compensate for this loss [76–78]. For BHP, it was predicted that all 33–58 chimpanzees would be impacted, but it was not specified

whether "impact" would result in their death. The Loma Mountains compensation was assumed to be far greater than the loss [51]. In the case of the Lom Pangar project, the project appraisal indicated that the project would have significant and irreversible environmental impacts, including the loss of natural habitat and the risk of reducing the viability of a distinct population of gorillas and other Red-Listed species. Bolstering the protection of the national park and designating new areas for conservation management were considered additional.

Another challenge in estimating losses, gains and additionality under the current framework is that methodologies do not take into account the cumulative impacts of multiple projects, which can be far greater than the sum of the impact of individual projects. This lack of information on cumulative impacts is in part a result of offsets being funded, designed and implemented on a project-by-project basis.

Since great apes are distributed across equatorial Africa, they are likely present in or around many mining concessions. When concessions are adjacent to each other, there will be few available locations for apes to escape the mining activities. For example, the GAP concession is adjoined by a concession held by Compagnie des Bauxites de Guinée (CBG) to the east and a concession held by Russian Aluminium (RUSAL) to the north. Chimpanzees fleeing noise and other human disturbance in the GAP concession may not have accessible undisturbed habitat to move into. These projects state that they have conducted cumulative impact assessments, but these generally refer only to the direct effects on the environment from their own activities and not their impacts in combination with the activities of other companies. Tools exist to aid such an analysis. They include the Cumulative Impact Assessment (CIA), the Regional Cumulative Impact Assessment (RCIA), and the Strategic Environmental Assessment (SEA). The IFC recognizes that the "CIA should be an integral component of a good environmental and social impact assessment (ESIA) or a separate stand-alone process". They also recognize, however, that the "CIA is evolving and there is no single accepted state of global practice". In addition, the IFC Performance Standard 1 "does not expressly require, or put the sole onus on, private sector clients to undertake a CIA" [79]. Without better coordination and accounting for cumulative impacts, the risk is that offsets and compensation projects will be insufficient to offset the total cumulative loss of EN and CR species nationally or regionally over time, leading to overall species loss.

Landscape context

Several authors have encouraged biodiversity offsets to be designed in a landscape context [38,39]; however, there is little guidance regarding how this may be accomplished. The Simandou Project SEIA explicitly stated that the project would ensure that offsets were aligned with national biodiversity priorities. Both the Simandou Project and the GAP are considering sites identified in an IUCN action plan for West African chimpanzees [80]. Studies conducted in the Loma Mountains and Deng Deng forests determined both these areas to be important for apes and biodiversity in general. Thus these biodiversity offset and compensation sites appear to be located in national priority sites for these species and seem to be complying with this best practice principle. However, when offset projects are designed on a project-by-project basis without coordination or integration with other offset or compensation projects or other conservation initiatives, opportunities for aggregating sites are missed. Aggregating protection of larger areas of habitat, or connected forest patches, would have a better chance of maintaining viable populations of apes over the long term.

Operating on a project-by-project basis does not rule out placing the offset location into a larger landscape context but could still result in the protection of multiple small, isolated and vulnerable sites, impacting the 'additionality' potential of the offset project.

No net loss

One of the reasons that offset design and implementation continue to be *ad hoc* is that there are differing interpretations even as to the meaning of "no net loss" [37,38]. The IFC defines no net loss as: "the point at which project-related impacts on biodiversity are balanced by measures taken to avoid and minimize the project's impacts, to undertake on-site restoration and finally to offset significant residual impacts, if any, on an appropriate geographic scale (e.g., local, landscape-level, national, regional)". As discussed above, the challenges of accurately estimating the losses and gains of individual apes are enormous. Even if this were possible, and even if a project could result in an increase in ape numbers at a particular site, a small local increase would not necessarily contribute to the viability of the population as a whole. The worst-case scenario would be that the criterion of no net loss would merely result in many isolated offset projects protecting isolated individuals, groups or communities. While that may counterbalance losses of individual apes from the activities of individual projects, it does not necessarily contribute to protecting viable populations that would survive in the long term. This is again due to offsets being designed and implemented on a project-by-project basis.

By not coordinating with other projects, the projects we examined may have missed opportunities to aggregate offsets and create larger, more robust offset areas. Species viability in forest patches depends on many factors, including the size and shape of habitat patches and connectivity between patches. Not only does fragmentation disrupt the distribution of the species, it also affects the ecological processes that are part of the ecosystem [81]. Designing biodiversity offsets on a project-by-project basis could indeed temporarily result in a "no net loss" or "net gain", but in the long term, species viability could be eroded if offset sites exist in isolation from each other.

Long-term outcomes

BBOP addresses the need for long-term protection by emphasizing in their best practice principles that the outcomes of a biodiversity offset should last at least as long as the project's impacts, and preferably in perpetuity [36]. The ability of a biodiversity-offset project to deliver long-term outcomes depends on both biological and financial factors. Above, we have already discussed the biological factors affecting long-term outcomes. When we examined the financial sustainability of these projects, we found no consistency in the way offsets and compensation projects for apes are being funded in Sub-Saharan Africa.

In the BHP and Lom Pangar projects, the intended source of conservation funding is revenue from electricity production, which would be disbursed to specific conservation projects on an annual basis. However, disruption of either dam's operations would threaten this funding. In the case of BHP, problems with tariffs and the distribution system have curtailed profits and caused operational disruptions. In the case of Lom Pangar, the annual conservation payments were designed to last 30 years, but there are no specific payment guarantees to ensure the consistency of payments or any identified system for recourse if the project does not comply. This lack of financial security places compensation projects at risk, as well as the great ape populations whose security is dependent on effective conservation management. No informa-

tion is currently available on how the GAP or Simandou offsets will be funded.

Discussion

Adherence to Biodiversity Offset Principles

Our review of these four projects that are investigating or already employing great ape offset or compensation projects demonstrates that the degree to which international best practices are applied is mixed. Despite the vulnerability of apes to disturbances, the long time it takes ape populations to recover from disturbances, their EN and CR status, and the ethical questions surrounding offsetting apes, all four projects assumed that great apes and their habitats could be at least compensated if not totally offset to achieve a no net loss. Following the mitigation hierarchy in most circumstances appears to have been challenging because project sites had already been selected using non-biodiversity criteria, and in some cases infrastructure development had already taken place, decreasing the options for mitigation and thus increasing residual impacts. Predicting losses, gains and "additionality" for apes is challenging, and most projects avoided making this calculation altogether. In all cases additionality was based on "averted losses". The projects might achieve "no net loss" or even support small local increases in numbers ("net gain") in isolated locations, but long-term outcomes of the two projects that have already been implemented are questionable given the uncertainty of the financing mechanisms and the fact that all offset or compensation projects were being designed on a case-by-case basis. Offset and compensation sites might be placed in priority locations, but without an overall offset strategy, opportunities are being missed for aggregating offsets in time and location and integrating them with species conservation objectives and other national biodiversity priorities. In summary, even though the projects we examined may result in temporary no net loss or even a net gain in species numbers, the current trajectory for great ape biodiversity offset projects is unlikely to result in no net loss over the long term. This indicates that, even if these projects adhere closely to international biodiversity offset principles, this will not ultimately generate a meaningful conservation outcome, and will not be sufficient to protect great ape populations over the long term, or contribute to species recovery.

Our assessment revealed three challenges that limit the effectiveness of efforts to compensate for impacts on great apes even if adherence to biodiversity offsets principles is improved. The first is that current great ape offset and compensation projects fail to account for cumulative development impacts at larger scales. The second is that great ape offset projects are less likely to make a meaningful contribution to great ape conservation if conducted on a project-by-project basis that results in multiple, isolated projects. This could occur even if offsets are placed in a larger, landscape context, since each location within the larger area could be disconnected from the other. The third is that biodiversity offset principles do not require offset projects to be fully operational and delivering the required biodiversity compensation before impacts from the development project occur. We suggest that a fundamental shift in the way offsets are designed and implemented is needed.

National Offset Strategies as an alternative trajectory

The most effective means of ensuring that biodiversity offset projects adhere to existing international best practice principles and contribute more effectively to great ape conservation is to develop National Offset Strategies, supported by conservation trust funds with non-wasting endowments. This new trajectory would have the following advantages.

National Offset Strategies would provide a framework for managing biodiversity offsets in a coordinated and transparent manner, consistent with national biodiversity strategies, including national protected area system plans and species recovery plans. Coordinating offsets and compensation projects through a National Offset Strategy would help ensure that sites are selected strategically, providing synergies with other conservation areas and between offset sites, and ensuring contribution to a landscape-level approach to great ape conservation. It would also help to ensure compliance with the mitigation hierarchy at the outset and that offset sites are prioritized in terms of timing of investment and location.

National Offset Strategies would maximize conservation benefits, for example, by establishing connectivity, buffering existing conservation areas, creating larger areas by aggregating offsets, or by selecting sites in different parts of the country, especially where several distinct and unconnected areas may be needed to buffer against the spread of disease such as Ebola–a major threat to humans and great apes in Africa [82].

In addition to increasing the conservation benefits, National Offset Strategies would also benefit project developers in a number of ways. They could establish a common set of rules, leveling the playing field, helping protect companies from reputational risk, and raising standards of industrial development projects. They would not limit offsets to those companies applying for funding from financial institutions with offset standards or companies with internal offset standards. National Offset Strategies would also allow companies to entrust offset management and implementation to a permanent entity, such as a conservation trust fund, with the responsibility of funding and implementing the strategy, rather than assuming the burden and liability of managing offsets in perpetuity (conservation banks in the U.S. provide a similar function). Investing in aggregated offset sites identified by National Offset Strategies could also decrease transaction costs by pooling resources for formulating offset methodologies, biological surveys, priority setting, or conservation trust fund development, which would otherwise be incurred by developers. With coordinated planning, companies could also share infrastructure or take advantage of efficiencies that lead to greater avoidance and minimization of impacts before offsets are even considered.

National Offset strategies should be developed as a result of a science-based, multi-stakeholder process and should include the following components:

A species recovery goal for EN and CR species rather than no net loss. Given the worsening global species extinction crisis [83–85], the international goal of biodiversity offsets should not just be "no net loss" but rather to make a measurable contribution towards *recovery* of EN- or CR-listed species. The idea that biodiversity offsets should contribute to species recovery has precedent in the U.S. where biodiversity offsets originated. A key aspect of offset policy in the U.S. is that the Endangered Species Act (ESA) seeks to ensure that a species listed under the ESA "recovers", that its conservation status improves to the point where it is no longer endangered (i.e. "in danger of extinction throughout all or a significant portion of its range") or "threatened" (i.e. "likely to become an endangered species within the foreseeable future throughout all or a significant portion of its range"). Thus, the United States Fish and Wildlife Service (USFWS) develops "recovery plans" for endangered species listed under the ESA to guide conservation decisions for that species. A federal permit allowing an impact on endangered species is only

granted if the development activity will not "appreciably reduce" prospects for both the survival and recovery of the species [86].

Articulation of and adherence to no-go zones. Some biodiversity values are not offsetable, either because of their vulnerability and irreplaceability or because of their location. Industrial activities, for example, should not be permitted in World Heritage Sites. National Offset Strategies could play a key role in helping to determine which locations and which species should be off limits because their biodiversity value cannot be offset.

Cumulative impact assessments Sectoral EIAs. The need for projects to take into account the cumulative impacts of neighboring projects impacting EN and CR species when designing offsets is a growing concern [25,38,87,88]. We suggest that a nationwide assessment of the collective and cumulative impacts of planned and ongoing projects impacting EN or CR species should be part of the national offset design process, We also believe that assessment of cumulative impacts *before* offsets are designed should be an additional international principle as this could affect the magnitude of the offset required.

An opportunity for aggregating offsets. The challenge is to develop mechanisms to help ensure that biodiversity offset and compensation projects are implemented in the context of larger frameworks for endangered species conservation rather than on a project-by-project basis. "Conservation banking" was pioneered in the U.S. in part to address the problem of isolated biodiversity offset projects [89]. Conservation banks are lands that are conserved and permanently managed for threatened species. In exchange for permanently protecting and managing the land for the species of concern, the U.S. Fish and Wildlife Service (USFWS) agrees on a set number of habitat or species credits that bank owners may sell for that area. Developers whose projects have unavoidable adverse impacts on that species may purchase the credits from conservation bank owners to offset these impacts [90]. Conservation banks are expected to support a viable population of a species or contribute to the maintenance of a population by expanding an area managed for the species [89]. They provide an aggregated approach to offsets rather than a series of smaller, less viable projects. Conservation banks are permitted by USFWS, the same agency that also develops species recovery plans. Similar opportunities for aggregating offsets should be made available in National Offset Strategies.

Offset implementation to be complete before development occurs. The timing of biodiversity offsets can be an important factor in determining their success [38,73,91]. If development takes place before the offset is implemented, biodiversity may be lost [38], including loss of resources that are key to an EN or CR species' survival (e.g., a tree that must reach maturity to bear fruit necessary for ape survival). Allowing development to proceed before the offset is complete also creates a significant risk to biodiversity if the offset project fails. To increase the likelihood of project success, it is therefore important that offsets are in place before development occurs. Bekessy *et al.* [73] argue for a biodiversity "savings bank" approach rather than a "lending bank" in which the public "owns all the risk of failure", suggesting that "the biodiversity value of offsets should be realized before assets are liquidated", and that assets can only be traded once it has been demonstrated that they have matured (reached ecological equivalence with whatever losses they are being traded against) ([73], p.153).

There is precedence for this approach in the U.S., where conservation banks must demonstrate measurable conservation benefits before issuing credits. The conservation benefit is then far less speculative than if offset activities are concurrent with development, an approach that experts in the U.S. have found usually results in biodiversity losses [88,89,92].

Because restoration of tropical rainforests is a very lengthy process [31], and because great ape population increases would not be apparent for many years given their slow reproductive rates, some may argue that demonstrating an increase in great ape numbers before development can occur is not realistic. At a minimum, however, we suggest that areas proposed for higher protected area status, should at least have been created and already have appropriate levels of trained staff, necessary equipment and secure long-term funding before any development is allowed to occur.

Long-term funding. Industrial projects impacting EN and CR species should also ensure guaranteed, permanent funding for offset projects specifically targeting EN and CR species. We propose that Conservation Trust Funds (CTFs) with non-wasting, or permanent, endowments (which already exist in many countries and could accommodate multiple offset projects) could be the most effective mechanism. CTFs provide independent and permanent entities that can assume responsibility and liability for financing, managing and evaluating offsets and compensation projects in perpetuity. This is an important function, as conservation project management is not part of the core business of most companies. CTFs for countries where trust creation is not possible legally, or where governance is weak, can be located offshore, while maintaining in-country management and operations to ensure multi-stakeholder representation and promote civil society.

Offset insurance sites. Due to the high risk involved in biodiversity offsets, particularly in countries with weak conservation capacity and governance concerns, but also as a result of natural disasters or other unforeseen causes for project failure [93], we propose that National Offset Strategies also identify "insurance" sites. Developers would then be required to invest in both offset and "insurance" sites in case an offset fails.

Conclusion

Developing and enforcing a National Offset Strategy would help promote adherence to current best practice principles for biodiversity offsets and would be more likely than the current trajectory to contribute to species recovery objectives. Because offsets for many large-scale development projects need to be permanent, and because mechanisms for ensuring permanent protection on private lands may be lacking, the default is likely to be the creation of new protected areas, or the expansion or improvement of the management of existing protected areas. National Offset Strategies would, therefore, be likely to contribute to the establishment of well-managed national protected area systems containing a representative cross section of a country's biodiversity.

Globally, the current rate of species extinctions is 1,000 times the predicted background rate of extinction [83]. Habitat degradation and conversion remain a leading cause of biodiversity decline [21]. A new approach to their design and implementation is needed if offsets are to be a useful tool for great ape recovery.

Here we have presented a new framework for designing and implementing biodiversity offsets that we believe has a greater likelihood of protecting EN and CR species in the long term. We stress, however, that caution should be exercised in the use of offsets as a tool for ape conservation until further field-based evidence of their effectiveness is available. Biodiversity offsets remain an unproven mechanism, and the risk of failure is magnified in countries with poor governance and recent histories of civil conflict. More time is needed to gauge whether offsets are truly contributing to no-net-loss objectives and to progress towards more "evidence based" approaches to offset design [94], even in developing countries where biodiversity offsets have been in use for some time [37,72,88,91,95,96]. The international conservation community, development organizations, the private sector, and international lending banks should ensure not only a precautionary approach to the use of conservation offsets, but also that offsets are designed and implemented in the context of National Offset Strategies.

Acknowledgments

We are most grateful to Deborah Mead, John F. Oates, Chris Ransom, and Roger Fotso for reviewing the manuscript, and to anonymous reviewers for their comments and suggestions.

Author Contributions

Conceived and designed the experiments: RK CFK. Performed the experiments: RK CFK TH MSD. Analyzed the data: RK CFK TH MSD. Contributed to the writing of the manuscript: RK CFK TH AL HR RV RAM MSD AR EAW.

References

1. Mittermeier RA, Rylands AB, Wilson DE, editors (2013) Handbook of the Mammals of the World. Volume 3: Primates. Barcelona, Spain: Lynx Edicions. 951 p.
2. Dinerstein E, Varma K, Wikramanayake E, Lumpkin S (2010) Wildlife Premium Market+REDD: Creating a financial incentive for conservation and recovery of endangered species and habitats. Available: http://www.hcvnetwork.org/resources/folder.2006-09-29.6584228415/Wildlife_Premium-REDD%20Oct%2013%202010%20-2-%20-2.pdf. Accessed 2014 Jun 4.
3. Williamson EA, Maisels F, Groves CP (2013) Hominidae. In: Mittermeier RA, Rylands AB, Wilson DE, editors. Handbook of the Mammals of the World. Volume 3: Primates. Barcelona, Spain: Lynx Edicions. 792–843.
4. Tutin CEG, Williamson EA, Rogers ME, Fernandez M (1991) Gorilla dispersal of *Cola lizae* in the Lopé Reserve, Gabon. J Trop Ecol 7: 181–199.
5. Voysey BC, McDonald KE, Rogers ME, Tutin CEG, Parnell RJ (1999) Gorillas and seed dispersal in the Lopé Reserve, Gabon. I: Gorilla acquisition by trees. J Trop Ecol 15: 39–60.
6. Ancrenaz M, Lackman-Ancrenaz I, Elahan H (2006) Seed spitting and seed swallowing by wild orang-utans (*Pongo pygmaeus morio*) in Sabah, Malaysia. J Trop Biol Conserv 2: 65–70.
7. Jones CG, Lawton JH, Shachak M (1997) Positive and negative effects of organisms as physical ecosystem engineers. Ecology 78: 1946–1957.
8. Boogert NJ, Paterson DM, Laland KN (2006) The implications of niche construction and ecosystem engineering for conservation biology. BioScience 56: 570–578.
9. Plumptre AJ (1995) The effects of trampling damage by herbivores on the vegetation of the Parc National des Volcans, Rwanda. Afr J Ecol 32: 115–129.
10. Rogers ME, Voysey BC, McDonald KE, Parnell RJ, Tutin CEG (1998) Lowland gorillas and seed dispersal: the importance of nest sites. Am J Primatol 45: 45–68.
11. Weber AW (1987) Ruhengeri and Its Resources: an Environmental Profile of the Ruhengeri Prefecture. Kigali, Rwanda: ETMA/USAID. 171 p.
12. Macfie EJ, Williamson EA (2010) Best Practice Guidelines for Great Ape Tourism. Gland, Switzerland: IUCN/SSC Primate Specialist Group. 78 p.
13. IUCN (2014) IUCN Red List of Threatened Species. Version 2014.2. Available: http://www.iucnredlist.org. Accessed 2014 Sep 20.
14. Arcus Foundation (2014) State of the Apes: Extractive Industries and Ape Conservation. Cambridge, UK: Cambridge University Press. 377 p.
15. Poulsen JR, Clark CJ, Mavah G, Elkan PW (2009) Bushmeat supply and consumption in a tropical logging concession in northern Congo. Conservation Biology 23: 1597–1608.
16. Köndgen S, Kühl H, N'Goran PK, Walsh PD, Schenk S, et al. (2008) Pandemic human viruses cause decline of endangered chimpanzees. Curr Biol 18: 260–264.
17. Emery Thompson M, Kahlenberg SM, Gilby IC, Wrangham RW (2007) Core area quality is associated with variance in reproductive success among female chimpanzees at Kibale National Park. Anim Behav 73: 501–512.
18. Kahlenberg SM, Thompson ME, Muller MN, Wrangham RW (2008) Immigration costs for female chimpanzees and male protection as an immigrant counterstrategy to intrasexual aggression. Anim Behav 76: 1497–1509.
19. Mitani JC, Watts DP, Amsler SJ (2010) Lethal intergroup aggression leads to territorial expansion in wild chimpanzees. Curr Biol 20: R507–R508.

20. Ryan SJ, Walsh PD (2011) Consequences of non-intervention for infectious disease in African great apes. PLoS One 6: e29030.
21. Butler RA, Laurance WF (2008) New strategies for conserving tropical forests. Trends Ecol Evol 23: 469–472.
22. African Development Bank (2009) Oil and Gas in Africa. New York: Oxford University Press. 233 p.
23. UNDP (2013) The 2013 Human Development Report – The Rise of the South: Human Progress in a Diverse World. United Nations Development Programme.
24. Weng L, Boedhihartono A, Dirks PHGM, Dixon J, Lubis MI, et al. (2013) Mineral industries, growth corridors and agricultural development in Africa. Global Food Security 2: 195–202.
25. Edwards DP, Sloan S, Weng L, Dirks P, Sayer J, et al. (2014) Mining and the African environment. Conserv Lett 7: 302–311.
26. Morgan D, Sanz C (2007) Best Practice Guidelines for Reducing the Impact of Commercial Logging on Great Apes in Western Equatorial Africa. Gland, Switzerland: IUCN/SSC Primate Specialist Group. 48 p. Available: http://www.primate-sg.org/best_practice_logging/Accessed 2014 May 29.
27. Wich SA, Garcia-Ulloa J, Kuehl HS, Humle T, Lee JSH, et al. (2014) Will oil palm's homecoming spell doom for Africa's great apes? Curr Biol 24: 1659–1663.
28. Hockings K, Humle T (2009) Best Practice Guidelines for the Prevention and Mitigation of Conflict between Humans and Chimpanzees. Gland, Switzerland: IUCN/SSC Primate Specialist Group. 50 p.
29. Goossens B, Setchell JM, Tchidongo E, Dilambaka E, Vidal C, et al. (2005) Survival, interactions with conspecifics and reproduction in 37 chimpanzees released into the wild. Biol Conserv 123: 461–475.
30. Humle T, Colin C, Laurans M, Raballand E (2011) Group release of sanctuary chimpanzees (Pan troglodytes) in the Haut Niger National Park, Guinea, West Africa: ranging patterns and lessons so far. Int J Primatol 32: 456–473.
31. Maron M, Hobbs RJ, Moilanen A, Matthews JW, Christie K, et al. (2012) Faustian bargains? Restoration realities in the context of biodiversity offset policies. Biol Conserv 155: 141–148.
32. Kiesecker JM, Copeland H, Pocewicz A, Nibbelink N, McKenney B, et al. (2009) A framework for implementing biodiversity offsets: selecting sites and determining scale. BioScience 59: 77–84.
33. Madsen B, Carroll N, Moore Brands K (2010) State of biodiversity markets: offset and compensation programs worldwide. Washington, DC: Forest Trends Association. Available: http://www.forest-trends.org/publication_details.php?publicationID=2388. Accessed 2014 May 16.
34. McKenney B, Kiesecker JM (2010) Policy development for biodiversity offsets: a review of offset frameworks. Environ Manag 45: 165–176.
35. Suding KN (2011) Toward an era of restoration in ecology: successes, failures and opportunities ahead. Ann Rev Ecol Evol Syst 42: 465–487.
36. BBOP (2013) To No Net Loss and Beyond: An Overview of the Business and Biodiversity Offsets Programme (BBOP). 20 p. Available: http://www.forest-trends.org/documents/files/doc_3319.pdf. Accessed 2014 May 24.
37. Bull JW, Suttle KB, Gordon A, Singh NJ, Milner-Gulland EJ (2013) Biodiversity offsets in theory and practice. Oryx 47: 369–380.
38. Gardner TA, von Hase A, Brownlie S, Ekstrom JM, Pilgrim JD, et al. (2013) Biodiversity offsets and the challenge of achieving no net loss. Conserv Biol 27: 1254–1264.
39. Pilgrim JD, Brownlie S, Ekstrom JMM, Gardner TA, von Hase A, et al. (2013) A process for assessing the offsetability of biodiversity impacts. Conserv Lett 6: 376–384.
40. Quétier F, Regnery B, Levrel H (2014) No net loss of biodiversity or paper offsets? A critical review of the French no net loss policy. Environ Sci Pol 38: 120–131.
41. The Biodiversity Consultancy (2013) Government Policies on biodiversity offsets. http://www.thebiodiversityconsultancy.com/wp-content/uploads/2013/07/Government-policies-on-biodiversity-offsets3.pdf. Accessed 2014 May 16.
42. Rio Tinto (2008) Rio Tinto and Biodiversity: Biodiversity Offset Design. Rio Tinto plc and Rio Tinto Ltd.
43. IFC (2012a) Performance Standard 6: Biodiversity Conservation and Sustainable Management of Living Natural Resources. Washington, DC: International Finance Corporation.
44. IFC (2012b) Guidance Note 6: Biodiversity Conservation and Sustainable Management of Living Natural Resources. Washington, DC: International Finance Corporation.
45. Equator Principles (2013) Available: http://equator-Principles.com/resources/equator_Principles_III.pdf. Accessed 2014 May 16.
46. ICMM IUCN (2012) Independent report on biodiversity offsets. Prepared by The Biodiversity Consultancy. Available: http://www.icmm.com/biodiversity-offsets. Accessed 2014 May 20.
47. Kiesecker JM, Copeland H, Pocewicz A, McKenney B (2010) Development by design: blending landscape level planning with the mitigation hierarchy. Front Ecol Environ 8: 261–266.
48. Ecology, Environment Inc., Kormos R (2008) Critical Habitat Assessment Report Guinea Alumina Corporation Report. Guinea, West Africa. 182 p. Available: http://ifcext.ifc.org/ifcext/spiwebsite1.nsf/0/8A0EE1048673CB16852576BA000E2CAC/$File/Guinea%20Critical%20Habitat%20Assessment%20Report.pdf Accessed 2014 May 25.
49. Reuters (2014) CORRECTED-Emirates Global aluminium project clears Guinea hurdle. Available: http://www.reuters.com/article/2014/06/25/guinea-aluminium-emirates-idUSL6N0P62AE20140625. Accessed 2014 Sep 17.

50. World Bank (2005) Sierra Leone – Completion of the Bumbuna Hydroelectric Project – Environmental Impact Assessment and Resettlement Action Plans. Washington, DC: World Bank. Available: http://documents.worldbank.org/curated/en/2005/01/5653720/sierra-leone-completion-bumbuna-hydroelectric-project-environmental-impact-assessment-resettlement-action-plans. Accessed 2014 May 16.
51. ZSL (2006) Baseline Primate Survey and Monitoring Programme. Final Report. Annexe 8 to: Baseline Biodiversity Surveys, Completion of the Bumbuna Hydroelectric Project. Berkshire, UK: Nippon Koei UK Co. Ltd.
52. Brncic TM, Amarasekaran B, McKenna A (2010) Sierra Leone National Chimpanzee Census. Tacugama Chimpanzee Sanctuary, Freetown, Sierra Leone. 115 p. Available: http://tacugama.com/downloads/SLNCCP_Final_Report_TCS_Sep2010.pdf. Accessed 2014 May 24.
53. Haider SW (2013) Sierra Leone – SL Bumbuna Hydroelectric Environmental and Social Management Project: P086801 – Implementation Status Results Report: Sequence 19. Washington, DC: World Bank. Available: http://documents.worldbank.org/curated/en/2013/06/17822689/sierra-leone-sl-bumbuna-hydroelectric-environmental-social-management-project-p086801-implementation-status-results-report-sequence-19. Accessed 2014 May 16.
54. Maisels F, Strindberg S, Ambahe R, Ambassa E, Yara CN, et al. (2013) Deng Deng National Park and UFA 10-065, Republic of Cameroon. Wildlife and Human Impact Survey 2012. Wildlife Conservation Society. 37 p.
55. Maisels F, Ambahe R, Ambassa R, Fosso B, Puomegne J-B, et al. (2011) Gorilla population in Deng Deng National Park and a logging concession. Gorilla J 4: 18–19.
56. White A, Fa JE (2014) The bigger picture: indirect impacts of extractive industries on apes and ape habitat. In: Arcus Foundation, editor. State of the Apes 2013: Extractive Industries and Ape Conservation. Cambridge: Cambridge University Press. 197–225.
57. Plumptre AJ (1995) The effects of trampling damage by herbivores on the vegetation of the Parc National des Volcans, Rwanda. Afr J Ecol 32: 115–129.
58. Rogers ME, Voysey BC, McDonald KE, Parnell RJ, Tutin CEG (1998) Lowland gorillas and seed dispersal: the importance of nest sites. Am J Primatol 45: 45–68.
59. Chen FC, Li WH (2001) Genomic divergences between humans and other hominoids and the effective population size of the common ancestor of humans and chimpanzees. Am J Hum Genet 68: 444–456.
60. Scally A, Dutheil JY, Hillier LW, Gregory E, Jordan GE, et al. (2012) Insights into hominid evolution from the gorilla genome sequence. Nature 483: 169–175.
61. Warren Y, Williamson EA (2004) Transport of dead infant mountain gorillas by mothers and unrelated females. Zoo Biol 23: 375–378.
62. Anderson JR (2011) A primatological perspective on death. Am J Primatol 73: 410–414.
63. McGrew WC (1992) Tool-use by free-ranging chimpanzees: the extent of diversity. J Zool Lond 228: 689–694.
64. Boesch C (1994) Cooperative hunting in wild chimpanzees. Anim Behav 48: 653–667.
65. Boesch C (2002) Cooperative hunting roles among Taï chimpanzees. Hum Nature 13: 27–46.
66. Whiten A, Goodall J, McGrew WC, Nishida T, Reynolds V, et al. (1999) Cultures in chimpanzees. Nature 399: 682–685.
67. van Schaik CP, Ancrenaz M, Borgen G, Galdikas B, Knott C, et al. (2003) Orangutan cultures and the evolution of material culture. Science 299: 102–105.
68. Gardner A, Gardner BT (1980) Comparative psychology and language acquisition. In: TA Sebok, J Umiker-Sebok, editors. Speaking of Apes: A Critical Anthology of Two-way Communication with Man. New York: Plenum Press. 287–329.
69. Thompson-Handler N, Malenky RK, Reinartz GE, editors (1995) Action Plan for Pan paniscus: Report of Free-ranging Populations and Proposals for Their Preservation. Milwaukee (WI): Zoological Society Milwaukee County. 105 p.
70. Ham R (1998) Nationwide Chimpanzee Census and Large Mammal Survey: Republic of Guinea. Republic of Guinea: European Union. 115 p.
71. Rijksen HD, Meijaard E (1999) Our Vanishing Relative: The Status of Wild Orangutans at the Close of the Twentieth Century. Wageningen: Tropenbos Publications. 480 p.
72. Ancrenaz M, Dabek L, O'Neil S (2007) The cost of exclusion: recognizing a role for local communities in biodiversity conservation. PLOS Biol 5: 2443–2448.
73. Gibbons P, Lindenmayer DB (2007) Offsets for land clearing: no net loss or the tail wagging the dog? Ecol Manag Restor 8: 26–31.
74. Bekessy S, Wintle B, Lindenmayer DB, McCarthy M, Colyvan M, et al. (2010) The biodiversity bank cannot be a lending bank. Conserv Lett 3: 151–158.
75. Kühl H, Maisels F, Ancrenaz M, Williamson EA (2008) Best Practice Guidelines for Surveys and Monitoring of Great Ape Populations. Gland, Switzerland: IUCN/SSC Primate Specialist Group. 36 p.
76. WCF (2011) Biodiversity Management Plan for the Guinea Alumina Concession in Sangaredi, Guinea, Report 2008–2010. Germany: Wild Chimpanzee Foundation. 63 p.
77. WCF (2012a) Etat de la faune et des menaces dans les aires protégées terrestres et principals zones de forte biodiversité de Republique de Guinée. Germany: Wild Chimpanzee Foundation. 83 p.
78. WCF (2012b) Offset Strategy for the Guinea Alumina Project. Germany: Wild Chimpanzee Foundation.

79. IFC (2013) Good Practice Handbook. Cumulative Impact Assessment and Management: Guidance for the Private Sector in Emerging Markets'' Washington, DC: International Finance Corporation. 82 p.

80. Kormos R, Boesch C, Bakarr MI, Butynski, TM editors (2003). West African Chimpanzees: Status Survey and Conservation Action Plan. Gland, Switzerland and Cambridge, UK: IUCN. IUCN/SSC Primate Specialist Group. 219 p.

81. Leader-Williams N, Dublin H (2000) Charismatic megafauna as 'flagship species'. In: Entwistle A, Dunstone N, editors. Priorities for the Conservation of Mammalian Diversity: Has the Panda had its Day? Cambridge, UK: Cambridge University Press. 53–81.

82. Walsh PD, Abernethy KA, Bermejo M, Beyersk R, De Wachter P, et al. (2003) Catastrophic ape decline in western equatorial Africa. Nature 422: 611–614.

83. Pimm SL, Ayres M, Balmford A, Branch G, Brandon K, et al. (2001) Can we defy nature's end? Science 293: 2207–2208.

84. Pimm SL, Jenkins CN, Abell R, Brooks TM, Gittleman JL, et al. (2014) The biodiversity of species and their rates of extinction, distribution, and protection. Science 344: 1246752 [DOI:10.1126/science.1246752].

85. Butchart SHM, Walpole M, Collen B, van Strien A, Scharlemann JPW, et al. (2010) Global biodiversity: indicators of recent declines. Science 328: 1164–1168.

86. Endangered Species Act (ESA) of 1973 As Amended through the 108th Congress Department of the Interior U.S. Washington, DC: Fish and Wildlife Service. 44 p. Available: http://www.nmfs.noaa.gov/pr/pdfs/laws/esa.pdf. Accessed 2014 May 24.

87. Brownlie S, Botha M (2009) Biodiversity offsets: Adding to the conservation estate, or "no net loss"? Impact Assessment and Project Appraisal 27: 227–231.

88. Kormos R, Kormos C (2011) Towards a strategic national plan for biodiversity offsets for mining in the Republic of Guinea, West Africa with a focus on chimpanzees. Report to the Arcus Foundation. Available: http://www.bicusa.org/wp-content/uploads/2013/07/GuineaBiodiversityOffsetKormosKormos2011.pdf. Accessed 2014 May 24. 92 p.

89. Carroll N, Fox J, Bayon R, editors (2008) Conservation and Biodiversity Banking: A Guide to Setting Up and Running Biodiversity Credit Trading Systems. London, UK and Sterling, VA: Earthscan.

90. USFWS (2003) Guidance for the Establishment, Use, and Operation of Conservation Banks. Memorandum to Regional Directors, Regions 1–7 and Manager, California Nevada operations. May 2. Washington, DC: US Fish and Wildlife Service.

91. Bendor T (2009) A dynamic analysis of the wetland mitigation process and its effects on no net loss policy. Landscape Urban Plan 89: 17–27.

92. Bean M, Kihslinger R, Wilkinson J (2008) Design of U.S. Habitat Banking Systems to Support the Conservation of Wildlife Habitat and At-Risk Species. Washington, DC: Environmental Law Institute. 978-1-58576-136-4, ELI Project No.0629-01. Available: http://www.eli.org/sites/default/files/eli-pubs/d18_02.pdf. Accessed 2014 May 24.

93. Stratus Consulting Inc. (2003) A Nationwide Survey of Conservation Banks. Prepared for the Northwest Fisheries Science Center. Seattle, WA: NOAA Fisheries.

94. Sutherland WJ, Pullin AS, Dolman PM, Knight TM (2004) The need for evidence-based conservation, Trends Ecol Evol 19: 305–308.

95. Kate K ten, Bishop J, Bayon R (2004) Biodiversity Offsets: Views, Experience, and the Business Case. Gland, Switzerland, Cambridge, UK and London, UK: IUCN and Insight Investment.

96. Fox J, Nino-Murcia A (2005) Status of species conservation banking in the United States. Conserv Biol 19: 996–1007.

16

Patch Size and Isolation Predict Plant Species Density in a Naturally Fragmented Forest

Miguel A. Munguía-Rosas*, Salvador Montiel

Departamento de Ecología Humana, Centro de Investigación y de Estudios Avanzados del Instituto Politécnico Nacional (CINVESTAV), Mérida, Yucatán, México

Abstract

Studies of the effects of patch size and isolation on plant species density have yielded contrasting results. However, much of the available evidence comes from relatively recent anthropogenic forest fragments which have not reached equilibrium between extinction and immigration. This is a critical issue because the theory clearly states that only when equilibrium has been reached can the number of species be accurately predicted by habitat size and isolation. Therefore, species density could be better predicted by patch size and isolation in an ecosystem that has been fragmented for a very long time. We tested whether patch area, isolation and other spatial variables explain variation among forest patches in plant species density in an ecosystem where the forest has been naturally fragmented for long periods of time on a geological scale. Our main predictions were that plant species density will be positively correlated with patch size, and negatively correlated with isolation (distance to the nearest patch, connectivity, and distance to the continuous forest). We surveyed the vascular flora (except lianas and epiphytes) of 19 forest patches using five belt transects (50×4 m each) per patch (area sampled per patch = 0.1 ha). As predicted, plant species density was positively associated (logarithmically) with patch size and negatively associated (linearly) with patch isolation (distance to the nearest patch). Other spatial variables such as patch elevation and perimeter, did not explain among-patch variability in plant species density. The power of patch area and isolation as predictors of plant species density was moderate (together they explain 43% of the variation), however, a larger sample size may improve the explanatory power of these variables. Patch size and isolation may be suitable predictors of long-term plant species density in terrestrial ecosystems that are naturally and anthropogenically fragmented.

Editor: Eric Gordon Lamb, University of Saskatchewan, Canada

Funding: This study was funded by Consejo NacionaL de Ciencia y Tecnología (project CB-2012-177680), funds to MAM-R. The funder had no role in study design, data collection and analysis, decision to publish, or preparation of the manuscript

Competing Interests: The authors have declared that no competing interests exist.

* Email: munguiarma@mda.cinvestav.mx

Introduction

MacArthur and Wilson's Equilibrium Theory of Island Biogeography (ETIB) postulates that the number of species in oceanic islands can be predicted by island size and isolation. This is firstly because of sample area effects (in any region, larger sample areas will contain more species than smaller samples) and secondly, but more importantly, because of the interplay between immigration and extinction that leads, over time, to a size and isolation-dependent equilibrium number of species (species relaxation) [1,2]. Therefore, the main testable predictions of this theory are that the larger and less isolated the island, the higher the species number at which it should reach equilibrium. The ETIB reached paradigmatic status in biogeography and ecology, and also had an enormous impact on conservation biology [3]. The ETIB has also provided a theoretical framework for understanding habitat fragmentation and making predictions about its effect on biodiversity [3,4,5,6,7].

Since the publication of original monograph of the ETIB in 1967 [2], the authors themselves have suggested that the principles and predictions of the ETIB may apply not only to oceanic islands but also to terrestrial ecosystems that are naturally and anthropogenically fragmented ([2], pp 3–4). However, the current literature is divided about the effects of patch size and isolation on species richness and other measures of biodiversity, such as species density [8,9,10], in terrestrial ecosystems [7,11,12]. In fact, quantitative reviews have shown that patch area and isolation alone are poor predictors of species occupancy in terrestrial forest fragments [13,14]. While the low predictive power of patch area and isolation for biodiversity could be due to several confounding factors associated with anthropogenic disturbance (i.e., patches differ in habitat matrix, elevation, human activity, etc. [15,16,17]), experimental studies in which these factors are controlled have also shown a remarkable lack of consistency in their results, especially with regard to number of species relative to patch size [18,19,20]. A basic condition of the ETIB is that it is only when colonization and emigration rates have reached equilibrium that species number can be accurately predicted by habitat size and isolation [1,2]. Owing to the long life cycle of some plants [21], it may take hundreds or thousands of years after fragmentation to reach a new equilibrium in forest patches [22,23]. Studies addressing the effect of patch and landscape traits on plant biodiversity in anthropogenic forest fragments usually work with forests that have been fragmented for decades or just a few hundred years [e.g., 9, 24]. It is therefore possible that most of the studies of anthropogenic forest fragments have failed to record an effect of patch size and isolation on biodiversity because most of these studies have been conducted in recently formed forest patches, where a new equilibrium has not yet been reached. Therefore, the effects of patch size and isolation on the biodiversity of terrestrial fragmented ecosystems can be more effectively assessed in forests that have been fragmented for a long time.

Naturally fragmented forests offer an excellent opportunity for testing the effects of patch size, isolation and other spatial variables on biodiversity because they have been fragmented for thousands or millions of years [25,26], and consequently can be assumed to have reached equilibrium. The main goal of this study was to assess whether plant species density is explained by patch size and isolation in a naturally fragmented forest on the Yucatan Peninsula. We used species density (number of species in equal-sized samples) as a response variable, as it is a measure of biodiversity that is less influenced by sample area effects than species richness [27,28]. This forest is characterized by the presence of several forest patches (locally known as Petenes, singular Petén) that are roughly circular, variable in size [25,29,30,31] and grow on Quaternary geological formations ca. 1.7 My old [26,32]. Scattered near the coast in a wetland matrix [29,30,31], the Petenes are landscape units that aptly reflect the habitat patch concept as their spatial boundaries naturally contain or delimit populations and communities of plants [12,25]. Therefore, the study area allowed us to test the effect of patch area and isolation while controlling for important confounding factors such as patch shape and matrix. In addition to the effects of patch size and isolation we tested the effect of elevation and patch perimeter. Patch elevation is negatively related to the level of salt water during the rainy season and this affects plant distribution [31,33], while patch perimeter is positively correlated with edge effects [34]. We predicted that plant species density would be positively correlated with patch size and negatively correlated with patch isolation. Patch perimeter would be negatively correlated with plant species density and elevation, positively correlated.

Materials and Methods

Study area

The study area is the Petenes-Celestún-El Palmar biological corridor (19° 53'–21° 11′ N, 90°28'–90°17'W) located along the northwest coast of the Yucatan Peninsula, which has an area of about 240,000 ha [35] (Fig. 1). The weather is tropical subhumid with summer rains, precipitation is 1000–1200 mm y^{-1} and mean temperature 26.1–27.8°C [36]. Our study focused on naturally formed forest patches of semi-evergreen tropical forest, sometimes mixed with tall mangrove species [30,31]. These patches are more abundant in a narrow belt (ca.10 km wide) beside the sea and their abundance decreases toward the mainland where the forest becomes continuous (Fig. 1). Most of these forest patches are roughly circular [30], and those that are amorphous are far less frequent [36]. The forest patches grow on Quaternary geological formations that are approximately 1.7 My old [26,32], and are characterized by taller, more diverse vegetation relative to that of the matrix [30]. It is believed that the presence of these patches is explained by their higher elevation relative to the matrix and the permanent supply of fresh water from one or more sink holes [30]. Tree species such as *Manilkara zapota*, *Metopium brownei*, *Bursera simaruba*, *Laguncularia recemosa* and *Avicennia germinans* dominate the canopy; *Bravaisia tubiflora* and *Sabal yapa* dominate the understorey [26,31,37]. The matrix surrounding the forest patches is dominated by short mangrove species (*Rizophora mangle*, *Conocarpus erectus*), sedges (*Eleocharis cellulose*, *Cladium jamaicense*) and cattails (*Typha dominguensis*) [31,37].

Sampling

From January 2013 to May 2014 we recorded vascular plants in 19 forest patches using belt transects (50×4 m). It was not possible to select these forest patches randomly owing to insurmountable access difficulties. Epiphytes and lianas were not recorded during the vegetation survey owing to the difficulties associated with assessing their presence and abundance (forest canopy height: 26 m [36]). We recorded woody plants with a girth greater than 5 cm (dbh>1.6 cm) and non woody plants taller than 20 cm in five transects per patch. We did not record small seedlings (dbh≤ 1.6 cm) because while propagules do arrive in the forest patches, some species cannot establish because of the water level and increased salinity during the rainy season [33]; recording seedlings might have led to an overestimation of real species density. Following the advice of previous studies [9,27,28,38,39], the area sampled was kept constant in all 19 patches (total sampled area per patch 0.1 ha) to reduce sample area effects. In each patch, the first transect was placed using a random point and the remaining transects were placed systematically: 20 m apart and in a previously defined direction. For all plants, dbh and life form were recorded *in situ*. Plants were identified with the help of field guides [40,41] and expert advice. Unidentified species were morphotyped (N = 2).

The leaves, flower or fruit of plants of uncertain identity were taken to the laboratory for later identification. Biological material was collected under a permit issued by the Mexican Ministry of the Environment (reg. SEMARNAT 31D8B-00780/1303) and field work was conducted with the permission of the authorities of the protected areas (reg. F00.9.DRBRC.002/13 for Ria Celestún Nature Reserve and reg. F00.9.DRBLP.04/13 for Los Petenes Nature Reserve).

Spatial configuration

For each patch sampled we measured its area (ha), perimeter (km) and the elevation at its centre (m a.s.l.). As measures of isolation we used distance to the nearest patch (edge to edge in km), distance to the continuous forest (edge to edge in km) and a connectivity index. The latter was an area-based index weighted by distance, calculated as follows:

$$\text{Connectivity} = \sum_{j}^{N} \frac{A_j}{Dist_j},$$

where A is the area of the focal patch (ha) and Dist is the edge to edge distance (km) between the focal patch and another patch in the surroundings. To calculate this index we used the three patches closest to each sampled patch (Fig. 1). For this index, patch connectivity increases as patch isolation decreases (all values >0). All spatial variables were obtained from digital cartography available in Google Earth Pro 7.2. We chose the most recent images available (2009–2011) and verified that the images represented the current landscape in the field. Spatial variables for all 19 forest patches are available in Appendix S1.

Data analysis

To assess inventory completeness in the patches sampled, we compared the number of species observed in samples to the number of species predicted by three nonparametric estimators: Chao1, ACE and Bootstrap [42]. To minimize potential over- or underestimation of species number by any estimator we used the mean of species predicted by the three estimators as a reference [26]. As previous studies suggest [43,44], we considered an inventory completeness ≥80% as representative; therefore, only

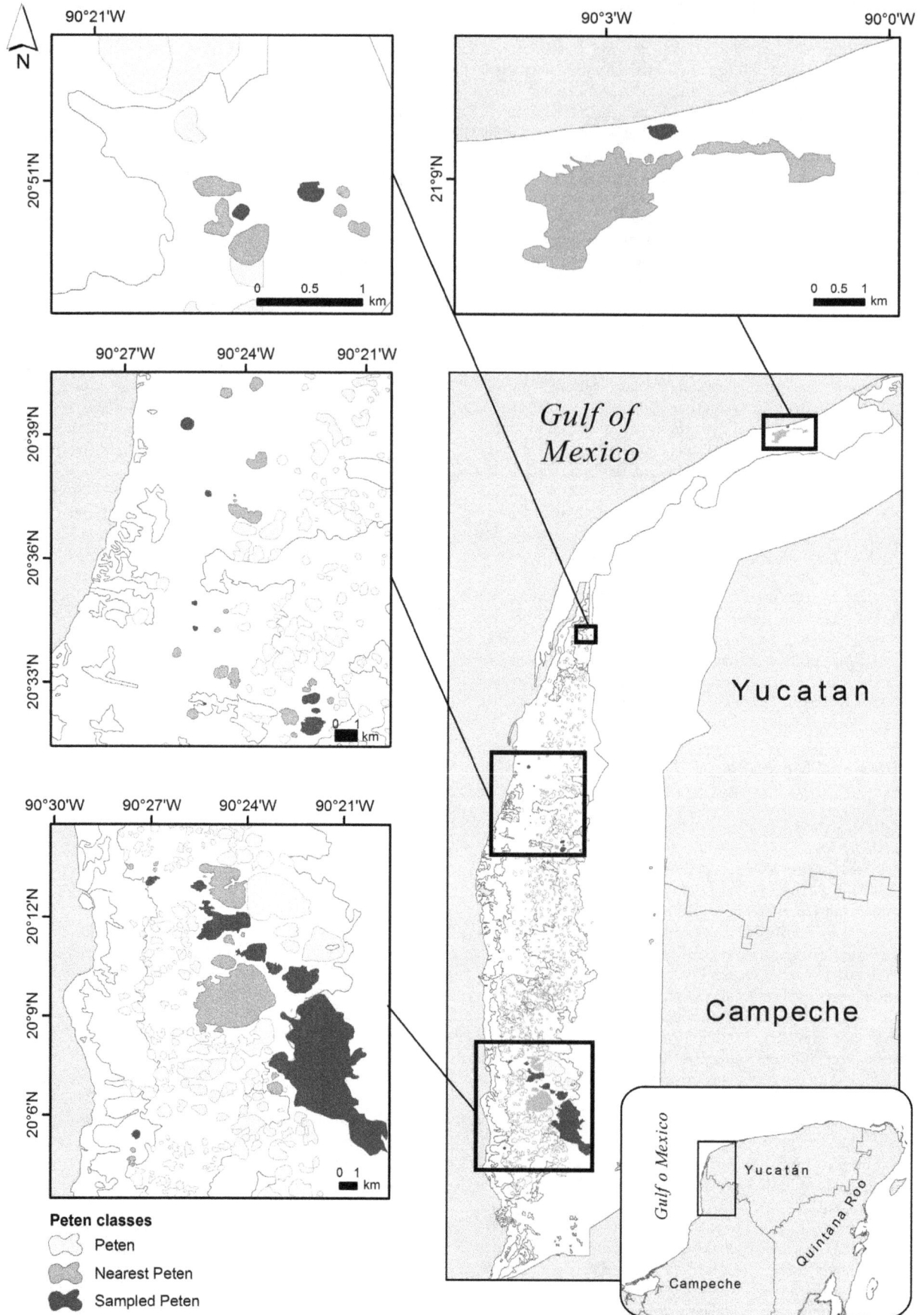

Peten classes

- Peten
- Nearest Peten
- Sampled Peten

Figure 1. Study area map. Forest patches sampled are in black, and gray patches were used to calculate a connectivity index or the distance to the nearest patch. The white area is the terrestrial portion of the Petenes-Celestún-El Palmar biological corridor. The small rectangle in the inset at the bottom right indicates the position of the study area on the Yucatan Peninsula. All bars represent 1 km.

forest patches that met this criterion were included in the analyses (Table 1).

We used plant species density (number of species in 0.1 ha) as the response variable in a model where the predictors were: patch area (ha) on a logarithmic scale, distance to the nearest patch (km), edge to edge distance from the focal patch to the continuous forest (km), patch connectivity index (see "Spatial configuration" above), patch central elevation (m a.s.l.) and patch perimeter (km). This model was fitted to a generalized linear model with a Poisson error distribution and log link function [45]. To determine the minimal adequate model, the complete model was simplified using the AIC criterion and intermediate models were compared using ANOVAs [45] (Table 2). Examination of the residuals of the minimal adequate model indicated a good fit and no evidence of overdispersion (dispersion parameter = 1).

Results

We recorded 55 different plant species in the 19 forest patches sampled (Appendix S2). Species density per forest patch was 6 to 20 species per 0.1 ha (mean = 14 ± 1.44 species; hereafter mean ±1 standard error) (Table 1). The most abundant species in all of the forest patches we sampled was the shrub *Bravaisia tubiflora* (Acanthaceae). The representativeness of our inventory relative to the mean number of species predicted by the three estimators was, in all cases, equal to or greater than 80% (range = 80%–100%; Table 1).

Forest patch size averages 180.15 ha (range: 1.65–2472.82 ha) and average patch perimeter is 4.14 km (range: 0.51–35.21 km), patch altitude is 8.58 m a.s.l. (range: 3–18 m a. s. l.), focal patch to the continuous forest distance is 14.89 km (range: 3.82–89.98 km), focal patch to the nearest patch distance is 3.8 km (range: 0.03–1.62 km) and the connectivity index is 1794.35 (range: 10.85–12030.91) (Appendix S1).

For the model proposed to explain variation in plant species density, of the six variables initially in the complete model, only two (patch size and distance to the nearest patch) were retained in the minimal adequate model (Table 2). The minimal adequate model had the lowest AIC and did not differ statistically from the complete model ($\chi_4 = 3.11$, $P = 0.54$). The minimal adequate model explained 43% of the total deviance (Table 2). Both of the variables in the minimal adequate model –patch size ($\chi_1 = 19.68$, $P = 0.02$) and distance to the nearest patch ($\chi_1 = 14.78$, $P = 0.03$) – significantly affected plant species density. The relationship between plant species density and patch size is positive and logarithmic ($y = 1.125 \log x +10.501$; Fig. 2A), while the relationship between plant species density and the distance to the nearest patch is negative and linear ($y = -5.7668 x +16.123$; Fig. 2B). Patch size and distance to the nearest patch explained 23% and 20% of the deviance of plant species density, respectively.

Table 1. Number of species observed in 0.1 ha samples (S) and predicted number of plant species using three nonparametric estimators (Chao1, ACE and Bootstrap) for 19 forest patches on the Yucatan Peninsula.

Patch	S	Chao1	ACE	Bootstrap	Average	Completeness (%)
1	6	6±1.3	7±0.11	7±0.1	6.7±0.3	90
2	9	10±1.4	10±1.7	10±0.1	10±0	90
3	16	16±0.5	16±21	17±0.1	16.3±0.3	98
4	17	20±2.3	19±1.9	19±1.5	19.3±0.3	88
5	19	19±1.1	20±2.2	22±2.1	20.3±0.9	93
6	16	16±1.9	17±1.2	18±1.9	17±0.6	94
7	19	26±2.4	22±2.1	22±1.7	23.3±1.3	81
8	13	13±1.3	14±1.8	14±0.9	13.7±0.3	95
9	16	16±1.8	18±21	18±1.5	17.3±0.7	92
10	18	20±5.3	23±2.4	20±2.1	21±1	86
11	15	15±3.7	16±1.9	17±1.7	16±0.6	94
12	10	13±2	14±1.8	11±0.9	12.6±0.8	80
13	20	22±5.3	23±2.4	23±2.1	22.7±0.3	88
14	11	12±1.4	13±1.4	12±1.2	12.3±0.3	89
15	11	11±3.7	12±1.6	12±1.3	11.7±0.3	94
16	8	8±0.1	8±1.41	9±0.7	8.3±0.3	96
17	10	10±1.4	10±1.4	10±0.5	10±0	100
18	20	26±2.3	24±2.4	23±3.6	24±1.1	83
19	13	13±1.4	13±1.6	13±0.7	13±0	100

Average is the mean number of species predicted by the three estimators. The percent completeness of our inventory relative to the average predicted number of species is shown (Completeness). In all cases, errors represent one standard error of the mean.

Table 2. Log linear models proposed to explain the variation in plant species density (S) in 19 forest patches on the Yucatan Peninsula.

Model	Model description	D²	AIC
1	log S = log Size+D. nearest patch+D. continuous forest+Connectivity+Elevation+Perimeter	0.53	110
2	log S = log Size+D. nearest patch+D. continuous forest+Connectivity+Elevation	0.53	108
3	log S = log Size+D. nearest patch+D. continuous forest+Connectivity	0.50	107
4	log S = log Size+D. nearest patch+D. continuous forest	0.47	106
5	log S = log Size+D. nearest patch	0.43	105

Model 1 represents the complete model with six explanatory variables: patch size (Size, on a logarithmic scale), distance to the nearest patch (D. nearest patch), distance to continuous forest (D. continuous forest), patch connectivity index (Connectivity), patch elevation (Elevation) and patch perimeter (Perimeter). Model 5 represents the minimal adequate model and models 2–4 are intermediate steps during model simplification. The Akaike information criterion (AIC) and the proportion of explained deviance for each model (D²) are also shown.

Discussion

Although previous studies have obtained contrasting results for the effects of patch size and isolation on species density [e.g. 8, 9, 10], our results clearly show that patch isolation and patch size predict plant species density in our study area on the Yucatán Peninsula. Unlike those of previous studies our study system has been fragmented for a very long time and important confounding factors such as habitat matrix and shape were kept relatively constant in this study. Therefore, we suggest that patch size and isolation are important determinants of species density in fragmented forests, although in some previous studies confounding effects and, more importantly, little elapsed time since fragmentation (i.e. no equilibrium has been reached) have obscured their relevance.

In addition to patch size and isolation, we included other variables in our analysis that were relevant to our study system: patch elevation (negatively correlated with the level of saltwater, which most of the plants in forest patches cannot tolerate; [33]) and patch perimeter (positively correlated with the length of edge in contact with the harsh habitat matrix). However, only patch size and isolation (measured as distance to the nearest patch) significantly explained the variability in species density among forest patches. We believe that patch elevation was not significant because it varies little among patches (3–18 m a.s.l.). Additionally, the internal fresh water supply from the sinkholes that are typically found in petenes [32] may counteract the influence of saltwater from the matrix. Perimeter was probably not a good predictor of species density because most of the patches are roughly circular in shape and therefore variation in area is a better predictor of species density.

In continental islands isolation is clearly the distance to the mainland [1,2], but it seems that this does not apply to terrestrial ecosystems. In our study system we took distance to the continuous forest to be analogous to the distance to the mainland in oceanic islands. However, this variable did not explain the variation in plant density among patches. We suggest that this is because, unlike oceanic islands where the mainland is the primary species pool for colonization, in most forest fragments immigration occurs predominantly from habitats in the vicinity of the patch, rather than from a common mainland [12]. This notion is reinforced by the fact that the distance to the nearest patch significantly explained variation in species density among patches in our study and in other previous studies [46]. Another metric of patch isolation which was not significant in our analysis was a measure of habitat amount weighted by distance to the focal patch (connectivity index). Therefore, our results do not support the suggestion made in a recent study [12] that habitat amount is a better predictor of the number of species in a focal patch than the distance to the nearest patch is.

The most remarkable result of our study was that patch area and isolation significantly predicted species density. However, we must recognize that the predictive value of these variables is moderate (43%; patch area = 23%, patch isolation = 20%). In a previous study [13], the authors evaluated the predictive power of species-area equations and found that these equations usually explain less than a half (49%) of the variation in species number.

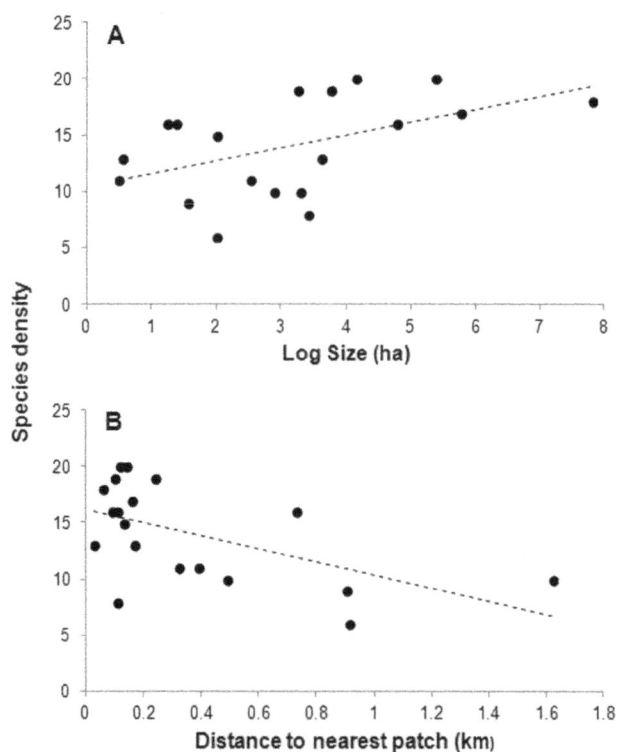

Figure 2. Relationship between plant species density (number of species in 0.1 ha) and patch size (Size) (A) and between plant species density and the distance to the nearest patch (B) for a group of 19 forest patches on the Yucatan Peninsula. Regression lines for A and B are also shown. The scale on the "x" axis for A is logarithmic.

Based on this, the authors suggested that predictions by these models are unreliable. We find their position very conservative. Møller and Jennions [47] addressed the question of how much variance can be explained by ecologists. They examined several quantitative reviews of empirical studies and determined that, on average, ecologists can only account for a small portion of variance in their studies (2.5–5.4%) owing to the randomness and noise typical of ecological systems. Therefore, for the results obtained in our study, we think that values of 20 and 23% explained variance are actually quite good. Previous research has suggested that several confounding factors may reduce or nullify the effects of patch size and isolation on species number. For instance, Cook [48] showed that diversity patterns conform better to the prediction of ETIB when matrix species are removed from the patch samples, suggesting that the habitat matrix is creating a confounding effect. However, the results of experimental studies where several confounding effects were controlled also produced inconsistent results with respect to the effects of patch size and isolation on species number [19], suggesting that the time a forest has been fragmented could be even more important than environmental factors in determining the predictive capacity of these variables. Our study system has been fragmented for a period on a scale of geological rather than ecological time and other potential confounding factors were either considered in the model (patch elevation and perimeter) or kept constant (habitat matrix, shape, human disturbance). Therefore, we think that sample size and not being able to choose the forest patches randomly limited the explanatory power of the model.

Another explanation of the contrasting results of previous studies that evaluated the effect of patch size and isolation on biodiversity may lie in how biodiversity was measured. For example, in her review of the effect of habitat fragmentation on biodiversity, Fahrig [7] examined studies where biodiversity was measured as species abundance, species density, species richness, species incidence, genetic diversity, species interactions, species extinction and species turnover. With this kind of variation in the way biodiversity has been measured it is not surprising to find contrasting results and a vague general pattern. In our study we selected the number of species in equal-sized samples (species density) as the response variable. Some authors have preferred species density (obtained from equal-sized samples) over species richness to evaluate species-area relationships [9,27,28,38,39] because this way patch area is not confounded with the sampled area, as occurs when greater sampling effort is invested in larger patches relative to smaller patches.

The most frequently cited mechanisms to explain the association between species richness and habitat size are sample area effects [49], habitat heterogeneity [50], and immigration-extinction balance [1,2]. However, in our study some of these mechanisms may not apply because we used species density as a response variable instead species richness. Sample area effects cannot account for the species density-area relationship found in this study because the area sampled was kept constant. As habitat heterogeneity is usually correlated with area [1,2], habitat heterogeneity is not likely to be the underlying mechanism of this species density-area relationship either. Previous research has suggested that, as occurs with species richness [1,2], variation in species density can be partially explained by demographic processes such as colonization and extinction [51]. Therefore, we suggest that the most likely mechanism underlying the association between species density and patch area or isolation is a size/isolation dependant equilibrium between colonization and extinction; however, we also recognize that further work is needed to unequivocally identify the mechanism.

In conclusion, species density in fragmented terrestrial ecosystems can be predicted by patch area and isolation. The fact that the study area has been fragmented for a very long period of time may have contributed to this result as this ecosystem has very probably reached equilibrium. We encourage other researchers to test for the effect of patch size and isolation on species number in other naturally fragmented forests to assess whether it has greater predictive power in terrestrial ecosystems that have been fragmented for geological periods of time and are likely to be in equilibrium.

Acknowledgments

A Rojas, SG Jurado-Dzib, CR Mezeta-Cob and A Rivas-Arellano provided invaluable assistance in the field. B Magaña elaborated the map in Fig. 1. L Salinas-Peba, L Arias and the staff from the "U Najil Tikin Xiw" herbarium (CICY) helped with plant identification. B Delfosse revised the English.

Author Contributions

Conceived and designed the experiments: MAMR SM. Performed the experiments: MAMR SM. Analyzed the data: MAMR. Contributed reagents/materials/analysis tools: MAMR SM. Contributed to the writing of the manuscript: MAMR SM.

References

1. MacArthur RH, Wilson EO (1963) An equilibrium theory of insular zoogeography. Evolution 17: 373–387.
2. MacArthur RH, Wilson EO (1967) The Theory of Island Biogeography. New Jersey: Princeton University Press. 203p.
3. Haila Y (2002) A conceptual genealogy of fragmentation research: from island biogeography to landscape ecology. Ecol Appl 12: 321–334.
4. Janzen DH (1968) Host plants as islands in evolutionary and contemporary time. Am Nat 102: 592–595.
5. Henney LR (2000) Dynamic equilibrium: a long term, large scale perspective on the equilibrium model of island biogeography. Global Ecol Biogeogr 9: 59–74.
6. Laurance WF (2008) Theory meets reality: how habitat fragmentation research has transcended island biogeographic theory. Biol Conserv 141: 1731–1744.
7. Fahrig L (2003) Effects of habitat fragmentation on biodiversity. Annu Rev Ecol Evol Syst 34: 487–515.
8. Wettstein W, Schmid B (1999) Conservation of arthropod diversity in montane wetlands: Effect of altitude, habitat quality and habitat fragmentation on butterflies and grasshoppers. J Appl Ecol 36: 363–373.
9. Arroyo-Rodríguez V, Pineda E, Escobar F, Benítez-Malvido J (2008) Values of small patches in the conservation of plant-species diversity in highly fragmented rain forest. Conserv Biol 23: 729–739.
10. Farmilo BJ, Melbourne BA, Camac JS, Morgan JW (2014) Changes in plant species density in an experimentally fragmented forest landscape: Are the effects scale-dependent? Austral Ecol 39: 416–423.
11. Gilbert FS (1980) The equilibrium theory of island biogeography: fact or fiction? J Biogeogr 7: 209–235.
12. Fahrig L (2013) Rethinking patch size and isolation effects: the habitat amount hypothesis. J Biogeogr 40: 1649–1663.
13. Boecklen WJ, Gotelli NJ (1984) Island biogeographic theory and conservation practice: species-area or specious-area relationships? Biol Conserv 29: 63–80.

14. Prugh LR, Hodges KE, Sinclair ARE, Brashares JS (2008) Effect of habitat area and isolation on fragmented animal populations. Proc Natl Acad Sci USA 105: 20770–20775.

15. Walter HS (2004) The mismeasure of islands: implication for biogeographical theory and the conservation of nature. J Biogeogr 31: 177–197.

16. Ewers RM, Bidham RK (2006) Confounding factors in the detection of species responses to habitat fragmentation. Biol Rev Camb Philos Soc 81: 117–142.

17. Hu G, Gu J, Feeley KJ, Xu G, Yu M (2013) The effects of landscape variables on the species-area relationship during late-stage habitat fragmentation. Plos One 7: e43894.

18. Robinson GR, Holt RD, Gaines MS, Hamburg SP, Johnson ML, et al. (1992). Diverse and contrasting effects of habitat fragmentation. Science 257: 524–526.

19. Debinski DM, Holt RD (2000) A survey and overview of habitat fragmentation experiments. Conserv Biol 14: 342–355.

20. Collins CD, Holt RD, Foster BL (2009) Patch size effects on plant species decline in an experimentally fragmented landscape. Ecology 90: 2577–2588.

21. Martínez-Ramos M, Alvarez-Buylla ER (1998) How old are tropical rain forest trees? Trends Plant Sci 3: 400–405.

22. Helm A, Hanski I, Pärtel M (2006) Slow response of plant species richness to habitat loss and forest fragmentation. Ecol Lett 9: 72–77.

23. Johansson V, Snäll T, Ranius T (2013) Estimates of connectivity reveal non-equilibrium epiphyte occurrence patterns almost 180 years after habitat decline. Oecologia 172: 607–615.

24. Santos B, Arroyo-Rodríguez V, Moreno CE, Tabarelli M (2010) Edge-related loss of tree phylogenetic diversity in the severely fragmented Brazilian Atlantic Forest. Plos One 5: e12625.

25. Mas JF, Correa J (2000) Analysis of landscape fragmentation in the "Los Petenes" protected area, Campeche, Mexico. Investigaciones Geográficas 43: 42–59.

26. Munguía-Rosas MA, Jurado-Dzib SG, Mezeta-Cob C, Montiel S, Rojas A, et al. (2014) Continuous forest has greater taxonomic, functional and phylogenetic plant diversity than an adjacent naturally fragmented forest. J Trop Ecol 30: 323–333.

27. Yamahura Y, Kawahara T, Iida S, Ozaki K (2008) Relative importance of the area and shape to the diversity of multiple taxa. Conserv Biol 22: 1513–1522.

28. Rosati L, Fipaldini M, Marignani M, Blasi C (2010) Effect of forest fragmentation on vascular plant diversity in a Mediterranean forest archipelago. Plant Biosyst 144: 38–46.

29. Montiel S, Estrada A, León P (2006) Bat assemblages in a naturally fragmented ecosystem in the Yucatan Peninsula, Mexico: species richness, diversity and spatio-temporal dynamics. J Trop Ecol 22: 267–276.

30. Barrera A (1982) Los petenes del noroeste de Yucatán: su exploración ecológica en perspectiva. Biótica 2: 163–169.

31. Rico-Gray V (1982) Estudio de la vegetación de la zona costera inundable del noreste del estado de Campeche, México: Los Petenes. Biótica 7: 171–188.

32. CONANP-SEMARNAT (2006) Programa de conservación y manejo de la Reserva de la Biosfera Los Petenes. Mexico City: CONANP. 206 p.

33. Rico-Gray V, Palacios-Rios M (1996) Salinidad y nivel del agua como factores en la distribución de la vegetación en la ciénega del NW de Campeche, México. Acta Bot Mex 34: 53–61.

34. Murcia C (1995) Edge effects in fragmented forest: implications for conservation. Trends Ecol Evol 10: 58–62.

35. Costa-Lugo E, Parra DA, Andrade-Hernández M, Castillo-Tzab D, Chablé-Santos J, Durán R, Espadas C, et al. (2010). Plan de conservación de la ecoregión Petenes-Celestún-Palmar. Mérida, Yucatán: Pronatura. 177 p.

36. Durán R (1987) Descripción y análisis de la estructura y composición de la vegetación de los petenes del noroeste de Campeche, México. Biótica 12: 191–198.

37. Durán R (1987). Lista florística de la región de los petenes Campeche, México. Biótica 12: 199–208.

38. Arroyo-Rodríguez V, Cavender-Bares J, Escobar F, Melo F, Tabarelli M, et al. (2012) Maintenance of tree phylogenetic diversity in a highly fragmented rainforest. J Ecol 100: 702–711.

39. Arroyo-Rodríguez V, Rös M, Escobar F, Melo F, Santos B, Tabarelli M, et al. (2013) Plant β diversity in fragmented rain forest: Testing floristic homogenization and differentiation hypotheses. J Ecol 101: 1499–1458.

40. Pennington TD, Sarukhán J (2005) Arboles tropicales de México: manual para la identificación de las principales especies. Mexico City: UNAM-Fondo de Cultura Económica. 523 pp.

41. Brokaw N, Bonilla N, Knapp S, MacVean A, Ortíz JJ, et al. (2011). Arboles del mundo maya. Mérida, Yucatán: Natural History Museum. 263 pp.

42. Gotelli NJ, Colwell RK (2011) Estimating species richness. In Magurran AE, McGill BJ, editors. Biological diversity: frontiers in measurement and assessment. New York: Oxford University Press. 39–54.

43. Castillo-Campo G, Halffter G, Moreno C (2008) Primary and secondary vegetation patches as contributors to floristic diversity in a tropical deciduous forest landscape. Biodivers Conserv 17: 1701–1714.

44. Barragán F, Moreno C, Escobar F, Halffter G, Navarrete D (2011) Negative impacts of human land use on dung beetle functional diversity. PLoS ONE 6: e17976.

45. Crawley MJ (2013) The R book. 2nd ed. Chichester: John Wiley and Sons Ltd. 1051 p.

46. Prugh LR (2009) An evaluation of patch connectivity measures. Ecol Appl 19: 1300–1310.

47. Møller AP, Jennions MD (2002) How much variance can be explained by ecologists and evolutionary biologists? Oecologia 132: 492–500.

48. Cook WM, Kurt TL, Foster BL, Holt RD (2002) Island theory, matrix effects and species richness patterns in habitat fragments. Ecol Lett 6: 619–623.

49. Connor EF, McCoy ED (1979) The statistics and biology of species-area relationship. Am Nat 113: 791–833.

50. Tews J, Brose U, Grimm V, Tielbörger K, Wichmann MC, et al. (2004) Animal species diversity driven by habitat heterogeneity/diversity: the importance of keystone structures. J Biogeogr 31: 79–92.

51. Grace JB (1999) The factor controlling species density in herbaceous plant communities: an assessment. Perspect Plant Ecol Evol Syst 2: 1–28.

17

Contrasting Regeneration Strategies in Climax and Long-Lived Pioneer Tree Species in a Subtropical Forest

Haiyang Wang[1], Hui Feng[2], Yanru Zhang[2], Hong Chen[1,2]*

1 Institute of Landscape Ecology of Montane Horticulture, Southwest University, Chongqing, 400716, China, **2** Department of Botany, College of Horticulture and Landscape Architecture, Southwest University, Chongqing, 400716, China

Abstract

1: This study investigated 15 coexisting dominant species in a humid subtropical evergreen broad-leaved forest in southwest China, consisting of long-lived pioneers and climax species occurring in natural and disturbed regimes. The authors hypothesized that there would be non-tradeoff scaling relationships between sprouting and seed size among species, with the aim of uncovering the ecological relationship between plant sprouting and seed characteristics in the two functional groups.

2: The sprouting variations of the species were initially examined using pairwise comparisons between natural and disturbed habitats within and across species and were noted to show a continuum in persistence niches across the forest dominants, which may underlie the maintenance of plant diversity. Second, a significantly positive, rather than tradeoff, relationship between sprout number and seed size across species within each of the two functional groups was observed, and an obvious elevational shift with a common slope among the two groups in their natural habitat was examined. The results indicate the following: 1) the relationship of seed size vs. sprouts in the natural habitat is more likely to be bet-hedging among species within a guild in a forest; 2) climax species tend to choose seeding rather than sprouting regeneration, and vice versa for the long-lived pioneers; and 3) the negative correlation between sprouting and seed dispersal under disturbed conditions may imply a tradeoff between dispersal and persistence *in situ* during the process of plant regeneration.

3: These findings may be of potential significance for urban greening using native species.

Editor: Norman W. H. Mason, Landcare Research, New Zealand

Funding: This research was funded by Special Research Program for Public-Welfare Forestry (201004064) and Southwest University (SWU110032). The funders had no role in study design, data collection and analysis, decision to publish, or preparation of the manuscript.

Competing Interests: The authors have declared that no competing interests exist.

* Email: chenh.swu@gmail.com

Introduction

Plant regeneration is an important ecological process for a forest ecosystem in which the sprouting and seeding of woody species is involved [1]. As a mechanism for shaping community dynamics [2], plant regeneration is important for the succession of forest communities and for the stability and restoration of vegetation in a forest ecosystem following various disturbances. As new architectural units, species sprouts [3] have been widely perceived as key functional traits [4–6] linked to plant life-history strategies [7,8], functional types [9,10] or population persistence [11,12]. Using sprouts, some damaged trees can occupy their original space niche rapidly and eventually reach the forest canopy again [13].

Furthermore, sprouts, as one of the two plant regenerative traits [4], have been reported to correlate in part to seed size. For example, sprouters would most likely be expressed in seed fecundity [14]. Moreover, it has been argued that there is a tradeoff between sprouts and seed size [15,16] due to competition for resources between vegetative and reproductive growth [14] relative to starch-tissue content [17]. Plants that sprout vigorously tend to be poorer seed recruiters than non-sprouters [8], and non-sprouters produce more seeds than congeneric sprouters growing at the same sites [8,17]. Nevertheless, others have found no correlation between the two variables (e.g., [18]). In brief, few significant associations have been reported thus far.

In contrast, because both seeding and resprouting are all ascribed to plant regeneration [4], we may infer that seed size correlates with sprouting [19,20]. In addition, in fluctuating environments, seeding shows bet-hedging [21,22], which remains correlated with seed size among community species [23–25] during the process of seed germination as an evolutionarily stable strategy to control the plant population (or ESS, e.g., [26]). With respect to resprouting, Nzunda *et al.* [16] recently suggested a bet-hedging model between occasional events and fixed interference in the storage reserve in good resprouting species. Based on these reports, we may hypothesize that plant sprouting may show a pattern of bet hedging in response alternating seeding, which eventually results in a positive scaling relationship, instead of a tradeoff, between plant sprout and seed size.

Moreover, interspecific and intraspecific differences in sprouting ability have been observed [2,7] that were due to various ecological factors [27–30] or the biological properties of the plants themselves [31–33]. For example, as an example of the latter, light-demanding species generally show enhanced resprout-ing abilities compared with shade-tolerant species [18], and species tend to be more widely distributed in less-productive sites [17]. It is reasonable to infer that the connections between sprouting and seed traits may be intrinsically variable in different functional plants (i.e., guilds). An examination of these properties may be vital to understand the regenerative ecology of particular species.

However, direct evidence on the relationship between sprouts and seed size in woody species remains insufficient. Furthermore, much of the existing research on regeneration following distur-bances has focused on the role of pioneer species (e.g., [2,34–36]). There have been few comparative studies among long-lived pioneers (LP) and climax species (CS; however, see [37]), which may be of practical value for forestry (see the summary in [38]).

The current study employed a subtropical evergreen broad-leaved forest in Mt. Jinyun Nature Reserve in China. We monitored the sprouting performance of 15 dominant species with varying seed size living in both natural and disturbed areas, and we analyzed intraspecific variations in sprouting responses to disturbances for the 15 species by examining the relationship between species individual sprout number and seed size among the two groups. Finally, we examined the associations among life-history attributes of the species, such as plant height, starch amount in current-year shoots, individual seed size, and append-ages per seed carried (a parameter characterizing seed-dispersal ability that may be correlated in part with sprouting, see [39]). The primary objectives of this study were the following: 1) compare the sprouting performance within species across disturbance gradients; 2) test the expected ecological relationships between species' sprouting behaviors and seed traits among CS and LP.

Materials and Methods

Study site and vegetation

The study area is located in the Jinyunshan National Nature Reserve (29°50′N, 106°24′E) in southwestern China, covering about 14 km² in total. The elevation within the reserve ranges from 180 to 951.5 m. The climate is of a typical subtropical monsoon type, characterized by mid-subtropical monsoons with a rainy, hot summer and a dry, warm winter. The mean annual rainfall is 1143.1 mm, which primarily falls in summer; thus leading to a relatively high mean annual relative humidity of c. 85%. The annual mean temperature is 17°C, with annual effective accumulated temperature (>10°C) of c. 6000°C, a maximum annual frost-free period of 334 d, and a mean annual sunshine time of only 1160 h. The most common soil type is acid yellow (mean pH value is 4.36) developed from Triassic quartz sandstone, carbonaceous shale and argillaceous sandstone, which contains a total content of 0.0985% nitrogen, 0.040% phosphorus, 1.413% kalium and 2.099% organic matter [40,41]. In addition, Mt. Jinyun is both a nature reserve and a famous scenic area where there are many signs of human habitation, such as roads, farmhouses, and hotels. The plants near these sites are greatly influenced by human disturbance.

With respect to zonal vegetation, there is a humid subtropical evergreen broad-leaved forest with secondary succession of vegetation in varying stages that primarily result from a forest fire (~300 years ago), mining (~50 years ago), and an abandoned tea garden (~20 years ago). The forest shows an obvious vertical structure, primarily comprising Fagaceae, Lauraceae, Theaceae, Symplocaceae, and Elaeocarpaceae [42]. The first tree layer is generally 15–20 m high with a few trees taller than 25 m. This tree layer is chiefly dominated by Fagaceae and Theaceae species that average 15 m in height and range from 25–35 cm in diameter

at breast height, interspersed with large Elaeocarpaceae, Laur-aceae and Symplocaceae trees that also influence the forest community. The second shrub layer is typically dominated by large shrubs or saplings that are often less than 3 m in height [40] and include species that are common in the sunny bare soil or shaded understory.

Materials and study design

Fifteen dominant or co-dominant species in the tree layer and shrub layer were selected as target species that represented, to some extent, the characteristics of the evergreen broad-leaved forest. The 15 species included three Symplocaceae species, four Theaceae species, three Elaeocarpaceae species, two Lauraceae species, two Fagaceae species, and one Rubiaceae species. They are well-layered in the forest community, and *Castanopsis fargesii, Castanopsis carlesii, Elaeocarpus duclouxii, Machilus nanmu, Symplocos laurina* and *Gordonia acuminata* belong to the first tree layer, *Symplocos setchuanensis, Adinandra bockiana, Neolitsea aurata, Adinandra bockiana* and *Sloanea leptocarpa* belong to the second, and *Eurya loguiana, Camellia tsofuii, Aidia cochinchi-nensis* and *Symplocos lancifolia* belong to the third [41,43]. We designated these selected species as either climax species or long-lived pioneers based on their role in the forest community; the dividing approach for climax species and long-lived pioneers followed criteria used in the literature [43,44] and personal observations.

The field investigations were performed between April 2011 and May 2011 using stratified random sampling to arrange five sample plots (covering about 15 ha in total that were separated from each other by 400–600 m) in a sampling area of about 100 ha. The habitats of individual plants were divided into two types: disturbed habitat (DH; defined as a forest edge within 10 m of buildings, squares, or main roads), which was more strongly affected by human activities, and natural habitat (NH; greater than 50 m from the sources of interference), which was little influenced by human. The plots in the two types of habitats were kept as uniform as possible with respect to altitude, slope position and aspect based on target plants and plant habitat category.

This paper used sprout number to measure sprouting capability, which is the most commonly adopted method (c.f. [45,46]). Twenty to forty trees per plot were selected randomly to record sprout number per individual in the field. For the purposes of the census, we recorded "distinguished sprout" by specifically referring to the living shoots originating from the trunk [46], and the main stems branching from the trunk were not included.

Ethics statement

The field studies for each site were permitted by the staff of the Jinyunshan National Nature Reserve, and this study did not involve endangered or protected species.

Data analyses

In view of the variation in sprout data of interspecifics and intraspecifics under varying habitats, in prior to analyses of the measured traits we conducted nested ANOVAs to test variability. The results showed that both species and habitat influences sprout number markedly (F = 45.630, p<0.001 and F = 6.078, p = 0.014, respectively). Moreover, as a variable, species contributed the largest variance (87.7%) of sprouting data, following by habitat (11.5%) and the individual plant of the same species under the same habitat contributed the least to variation (0.8%). These results showed that sprout data may be compared among species, regardless of their intraspecific variations in different habitats. Data were meaned across individuals of the same species living in

Table 1. The species properties and the mean sprout number per individual (mean±SE) in natural habitat and in disturbed habitat for the 15 species studied.

Species	Family	Height	Group	Abbr.	NH	DH	DH/NH
Castanopsis carlesii	Fagaceae	Mt	CS	Cc	3.37±0.39 ***	7.05±2.06	2.09
Eurya loquiana	Theaceae	Ls	LP	El	3.61±0.75 ***	6.11±1.89	1.70
Neolitsea aurata	Lauraceae	St	CS	Na	3.24±0.41 ***	5.44±1.08	1.68
Elaeocarpus japonicus	Elaeocarpaceae	Mt	CS	Ej	3.00±0.46 ***	5.00±0.92	1.66
Aidia cochinchinensis	Rubiaceae	Ls	CS	Ac	1.67±0.29 *	2.56±0.67	1.52
Sloanea leptocarpa	Elaeocarpaceae	Mt	CS	Slc	2.80±0.75 ***	4.02±3.00	1.43
Castanopsis fargesii	Fagaceae	Lt	CS	Cf	2.80±0.42 ***	3.82±0.71	1.36
Adinandra bockiana	Theaceae	St	LP	Ab	6.35±1.54 ***	8.57±2.72	1.35
Symplocos lancifolia	Symplocaceae	St	LP	Slf	6.56±2.12 **	8.64±3.33	1.32
Symplocos setchuanensis	Symplocaceae	Mt	CS	Ss	3.23±1.51 ns	4.07±1.61	1.26
Elaeocarpus duclouxii	Elaeocarpaceae	Lt	CS	Ed	5.48±1.14 ns	5.29±2.33	0.96
Camellia tsofuii	Theaceae	Ls	CS	Ct	3.46±0.93 ns	3.24±0.36	0.94
Gordonia acuminata	Theaceae	Mt	LP	Ga	10.50±2.18 ***	5.83±1.21	0.55
Machilus nanmu	Lauraceae	Lt	LP	Mn	9.70±3.77 ***	5.17±1.89	0.53
Symplocos laurina	Symplocaceae	Mt	LP	Sla	9.44±2.78 ***	2.16±0.31	0.23

* denotes the p-level of student's t-tests for the species in the two habitats. * = $p < 0.05$, ** = $p < 0.01$, *** = $p < 0.001$, ns = non-significant. Ss = small shrub (≤ 0.5 m), Ms = middle shrub (0.5–2 m), Ls = large shrub (2–5 m), St = small tree (5–8 m), Mt = middle tree (8–25 m), Lt = large tree (≥ 25 m); NH & DH = sprout number in natural & in disturbed habitat, respectively; Abbr. = the abbreviation of the species studied. CS denotes climax species and LP does long-lived pioneer. The species are sorted according to the column of DH/NH.

the same habitat and $\log_{10}$-transformed to improve normality prior to analyzing for the scaling relationships.

The scaling relationships between sprout number and seed size (i.e., seed mass, referring to individual seed dry mass, as in [47]) were analyzed using a Model Type II regression method, with allometric slopes of particular interest calculated as Standardized Major Axes (SMA; [48]). The heterogeneity of regression slopes and the common slopes were tested following the methods of [49], and the shifts between lines fitted to groups sharing a common slope (y-intercept) were examined using ANOVAs. The above allometric parameters were conducted using (S)MATR [48,50].

Although sprouting capability showed little phylogenetic conservatism [5], we nevertheless performed phylogenetic independent contrasts (PICs) analyses for the species used in the current study to determine whether a correlation between seed size and sprouts was biased by phylogenetic signal. The phylogenetic trees were constructed following Phylomatic (version 3; [51]), and PICs were conducted using the phylogenetic comparative methods of COMPARE (4.6b; [52]) for the LP and CS species.

In addition, to assess differences within species in the two habitat conditions, t-tests were conducted using Statistica 6.0 (Statsoft I 2001: Tulsa, Oklahoma). To detect broad correlations in variation among sprouting traits and plant seed traits, we performed a principal component analysis (PCA). Six statistics were employed to perform the PCA for 15 species. With the exception of two sprouting variables, the remaining four parameters were seed size (a life-history trait related to successional trade-offs), plant height (a parameter that has been shown to be strongly associated to the coexistence of pioneer and shade-tolerant tree species by [37]; sprouters are typically short plants, such as shrubs or bushes, whereas non-sprouters are commonly tall trees [11,53]), seed appendage per seed carried (taken from [39]), and starch content in shoot pith (estimated roughly using a scale of 1–5, denoting the least to the most amount based on anatomical sections of current-year shoots observed under a microscope). DCA (detrended correspondence analysis) was performed prior to PCA to determine the ordination model (i.e., unimodal or linear) to be adopted following the requirement of [54], using Canoco for Windows 4.5 software. Because all lengths of the gradient were less than 3 (the largest being 0.863), a linear multivariate PCA approach was chosen to examine the associations among the multiple variables including sprout number in NH and DH, seed size, seed appendage per seed carried and starch content in shoot pith and plant height. Of these variables, the first two were assigned as environmental variables and the remaining as species variables for the PCA. For the PCA, the species data were log-transformed, and the focus of scaling was on inter-species correlations. The species scores were divided by the standard deviations, and sample data were centered and standardized, whereas species data were centered by species. The PCA ordination biplot was created using CanoDraw 4.0.

Results

Sprouting characteristics under disturbed and natural habitats

The mean numbers of sprouts per trunk for the 15 species in DH and in NH are summarized in Table 1. Five of the species (*Gordonia acuminate, Machilus nanmu, Symplocos laurina, Camellia tsofuii, Elaeocarpus duclouxii*) showed more sprouts in NH than in DH; in particular, the former three were significantly different (t-test, p<0.001) and showed sprouting numbers that were 1.8-, 1.9-, and 5.9-fold greater than in the DH. The remaining 10 species showed fewer sprouts in NH than in DH.

The sprout numbers for different tree species within the same habitat were also different. The biggest sprout number in the DH was that of *Symplocos lancifolia* (8.64), followed by *Adinandra bockiana* (8.57); the lowest two were *Symplocos laurina* and *Aidia cochinchinensis*, with only 1.6 and 2.56, respectively.

With the exception of 3 species (t-test, p>0.05; *Elaeocarpus duclouxii, Camellia tsofuii* and *Symplocos setchuanensis*), there were notable differences in paired sprouts for the 12 species between NH and DH (t-test, p<0.05; Table 1). Therefore, human activities affect sprouting capacity of trees in general; however, the effect of this interference differs with species.

Scaling relationship between sprouting capability and species seed size

Species sprout number in NH did not generally correlate with seed size in pooled data across all 15 species (p>0.05). However, when dividing the 15 species into two groups (CS group and LP group; Table 1), the sprout number closely scaled with seed mass across congeners within a group ($r^2 = 0.714$, p = 0.034, slope = 5.884 [2.959 11.702] for the LP group; $r^2 = 0.907$, p< 0.001, slope = 6.609 [5.049 8.653] for the CS group) with a significant elevation increase (y-intercept$_{LP} = -4.157$; y-intercept$_{CS} = -1.479$) fitting a common slope (= 6.476; Figure 1). Second, the sprout number in DH did not correlate with seed mass across species (p>0.05) even when the 15 species were divided into CS and LP (p>0.05). In addition, the above tendencies were maintained in PICs, with the exception of a weak correction in pooled data across the 15 species in DH (Table 2).

These results show that the relationship between sprouting capability and species seed size varies considerably between the two species groups in natural habitats and remained when phylogeny signals were removed. Furthermore, the apparent difference in y-intercept (increase) between the CS and LP groups indicates that the species in the CS group produced larger seeds than those of the LP group for given sprouts.

The broad trends among sprouting capability, stem and seed traits, and habitats

The PCA graph summarizes the general correlations (Figure 2). The PCA revealed that the first and second PCA axes explained 58.4% and 33.2% of the variability in species data, respectively; 91.6% in total. The variability of NH sprouts paralleled the direction of seed size (Figure 2) but opposed the direction of height and starch content; however, the DH variable was orthogonal to seed size but conversely parallel to the direction of seed appendage. Thus, the NH variable was more closely and positively associated with seed size and correlated negatively with height and starch content. The DH variable correlated strongly with seed appendage per seed carried but was independent of seed size.

Discussion

We performed intraspecific comparisons of sprouting performance for the dominant species in a subtropical community and tested our predictions on the interspecific relationships between sprouting and seed traits for LP and CS. We observed consistently significant relationships between species sprouting and seed traits; however, the details of these relationships varied based on species and habitat type.

1) Contrasting sprouting patterns within and across species between natural and disturbed habitats

The species sprouting patterns differed in conspecifics in different habitat types (consistently showed using t-tests) and also

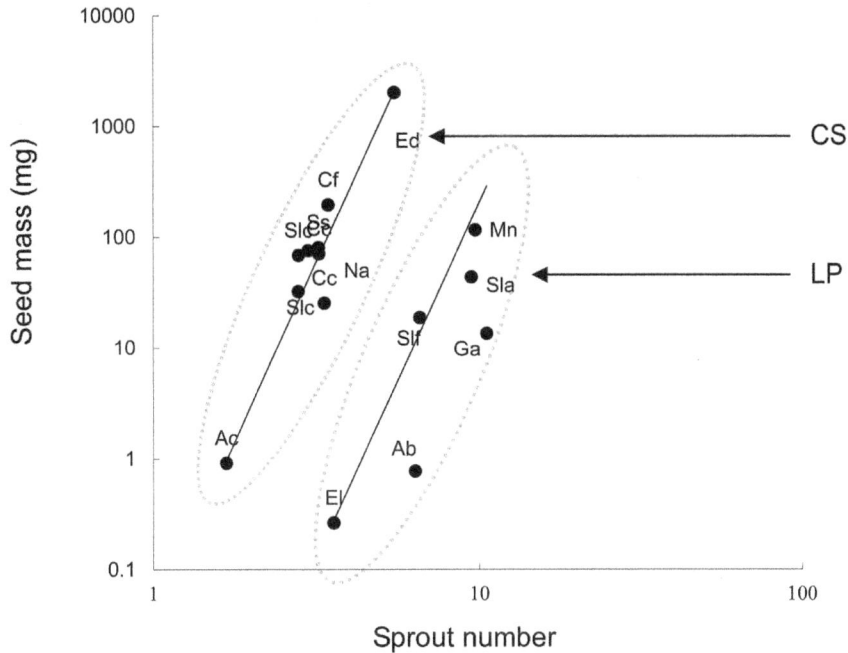

Figure 1. The relationship of seed size vs. sprouts. The bivariate relationship between seed size and trunk sprouting number among climax species (CS) and long-lived pioneers (LP) in natural habitats in a subtropical broad-leaved forest in southwestern China. The two lines in the graph denote the fitted lines to the two functional groups (slope = common slope shared by the two functional groups) using the standardised major axis (SMA); broken line, CS species group; solid line, LP group.

in interspecifics. Some species sprouted vigorously in natural habitats; however, others showed enhanced sprouting in disturbed habitats. The intraspecific sprouting pattern shows that, among the species measured, the majority of species were sensitive to disturbance, indicating that disturbance affects sprouting in plants.

Moreover, the differentiation among dominant species in a forest may be of particular significance due to their leader functions in a forest when considering spatial coexistence among species [55] and the dynamics of the community structure [56]. Some species sprouted more in DH whereas others performed better in NH, as mentioned above, we believe this diverse sprouting property in dominant species forms a vegetative regenerative spectrum, which lays the foundation for them to cope with various natural- or human-derived disturbances, and presumably, it also reflects a bet-hedging sprouting strategy at the species level within a community (c.f. [57]). Thus, the marked differences in sprouting performance may be critical for the dominant species to colonize gaps [58] in a complementary way and may enable maintenance of tree diversity following disturbances, based on the "alternate trait axis" argument [2].

Nevertheless, it is difficult to explain the origin of the difference in sprouting behaviors of some species that sprouted less in disturbed habitats, as observed here in this study. Further research such as investigation for underground clone structure is suggested.

2) Relationship between seed size and sprout number

Larger-sized species tend to produce more sprouts among species within a functional group in an undisturbed habitat; however, this relationship did not reach significance when all species were pooled, indicating that natural guild species may evolve the potential to reproduce vegetatively (bet-hedge rather than reproduce sexually) in response to environmental cues [57]. We believe that this result is reasonable based on two lines of reasoning. First, the observed sprouting continuum across dominant species along the disturbed gradient provided a necessity for species hedging their evolutionary bets in varying environments [59]. Second, woody plants are perennial and not annual. The sprouts on trunks were the sprouts saved for years that eventually become the biological "bet" and enable a positive relationship between seed size and sprouts; hence, there is no need

Table 2. Summary of phylogenetically independent comparative analysis (PICs) for the relationships between sprout number and seed size of the climax species (CS) and long-lived pioneers (LP) studied, and for the pooled data (Pooled) of the two groups.

Habitat	Group	r^2	p	a	b
NH	CS	0.807	0.002	6.395	0.057
NH	LP	0.722	0.002	3.326	−0.284
NH	Pooled	0.702	<0.001	4.508	−0.125
DH	Pooled	0.365	0.029	−2.187	−0.670

Letters a and b are the regressive coefficient and the intercept of the linear regressive equation $Y = ax+b$, where Y and x are seed mass and sprout number, respectively.

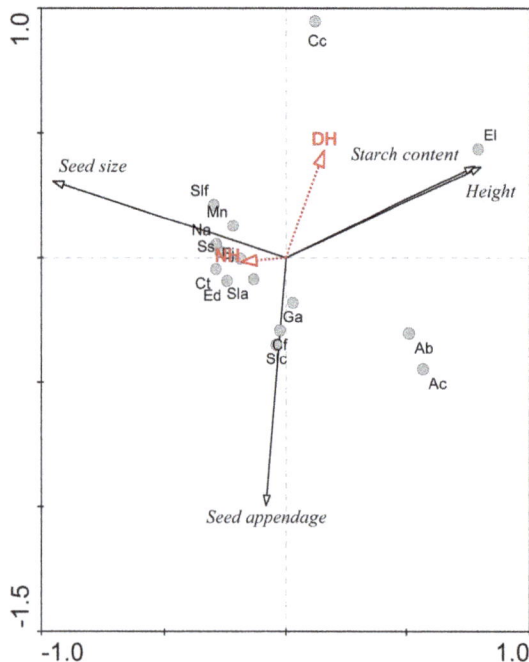

Figure 2. The PCA biplot. Principal component analysis (PCA) based on trunk sprouts in natural sites (NH) and disturbed sites (DH), seed size, seed appendage, starch content in shoot and plant height variables. Circles show sampled species, with abbreviations being as for Table 1.

3) Associations among NH & DH sprouts, seed dispersal, carbon storage and plant height

PCA showed a clear positive correlation of sprouts in NH with seed size, further verifying the relationship between these two variables (Figure 2). NH sprouting was also negatively correlated with starch content and plant height, showing that good sprouters in natural sites may store starch less and include relatively more short species. The DH sprouts showed contrasting results, supporting some previous reports, such as [60], demonstrating that sprouters show higher starch concentrations in roots.

The DH sprouting variable was inversely correlated with seed appendage (Figure 2), suggesting that good sprouters in DH may be weak seed dispersers and vice versa. This relationship may imply a tradeoff between the space occupied by the plant between far (i.e., seed dispersal) and near (i.e., sprouting). The ecological relationship between seed dispersal and plant sprouting may be an interesting topic that deserves further discussion.

4) Ecological and practical implications of the sprouting strategy

The sprouts of dominant species in the natural ecosystem may be very useful when managing forests for products including timber, firewood, edible fruits, and landscape plants, which are some of the different uses of the species observed. Thus, our results are likely to be helpful for silvicultural treatments. For example, the difference in seed size between CS and LP implies different regeneration niches in a subtropical forest and lays a good foundation for a more resilient forest ecosystem apart from human disturbances using complementary mechanisms of responding to a more or less disturbed habitat. Still, long-lived pioneers show some excellent features, such as their fast growth and relative shade tolerance, compared with pioneers, and they have light but strong timber (characteristics proposed as part of the long-lived pioneer syndrome by [38] (pp246), which is superior not to the slow-growing climax species but to the strongly r-selected pioneer. Therefore, LP is an issue of particular concern to foresters when applying "Miyawaki's Method" for reestablishing an urban forest [61].

Conclusions

We tested and confirmed our hypothesis on the ecological correlation of sprout vs. seed size. Dominant forest species differ in their sprouting capability, build divergent sprouting patterns along disturbed gradients and display contrasting relationships of sprouts vs. seed size among guilds, manifesting in diverse sprouting strategies across woody plants and suggesting a strong selection for species life-histories under varying conditions. These results provide evidence for the evolutionary mechanism and consequences of variations in plant sprouting behavior, which may contribute to the ecology of regeneration in a forest community.

Acknowledgments

We thank Yanfang Liu for analysis of the anatomy of plant shoots in the laboratory and Danrong Wang for preparing the images used in this paper. We also appreciate the two anonymous reviewers for their constructive comments on early versions of this manuscript.

Author Contributions

Conceived and designed the experiments: HW. Performed the experiments: HF HW. Analyzed the data: HC. Contributed reagents/materials/analysis tools: HC. Wrote the paper: HC YZ.

for to trade-off their limited resources accumulated within one year. Thus, this strategy may be viewed as the integration of bet-hedging between sprouting and seeding during long-term regenerative evolution.

Furthermore, based on the equal sprouting number on trunks, climax species tended to show larger individual seeds than long-lived pioneers, revealing a difference between the two species groups with respect of life history strategy under natural conditions. This tendency held true in PIC analyses, revealing its ecological significance. The CS guild may win out in a forest by evolving larger seeds, enabling seedlings to settle and establish in a darker habitat. In contrast to CS, the LP species tended to adopt a vegetative means to enable plant regeneration in brighter sites. They produce more sprouts and smaller seeds under natural conditions; thus, enhanced sprouting ability but weaker seeds.

On one hand, this result is inconsistent with authors who argue that vegetative sprouting should require a tradeoff against sexual reproduction in general and consequently, that poor resprouters should show higher seed mass compared with good resprouters (e.g., [15,60]). This inconsistency may result from the differences in the type of environment studied (e.g., natural vs. disturbed habitat), the species type (the species focused upon were woody CS & LP) or the variation in seed size due to seed size/number tradeoff (e.g., [39]). On the other hand, the greater mean number of sprouts in LP compared with CS under natural conditions, as observed in our study, indicates a pattern of smaller-seeded species with more sprouts [the LP group in NH showed markedly more sprouts (= 7.695) than the CS group (= 3.232; t-test, t = −4.661, p<0.001)], which may also represent a tradeoff to some extent between seed size and sprouts. In addition, this relationship did not hold true for the species in disturbed habitats, as shown both in the PCA graph (Figure 2) and in regression analysis (p>0.05), which is consistent with a previous report [18].

References

1. Pratt RB, Jacobsen AL, Hernandez J, Ewers FW, North GB, et al. (2012) Allocation tradeoffs among chaparral shrub seedlings with different life history types (Rhamnaceae). American Journal of Botany 99: 1464–1476.
2. Dietze MC, Clark JS (2008) Changing the gap dynamics paradigm: vegetative regeneration control on forest response to disturbance. Ecological Monographs 78: 331–347.
3. Turnbull CG (2005) Plant architecture and its manipulation: Blackwell.
4. Cornelissen J, Lavorel S, Garnier E, Diaz S, Buchmann N, et al. (2003) A handbook of protocols for standardised and easy measurement of plant functional traits worldwide. Australian Journal of Botany 51: 335–380.
5. Vesk PA, Westoby M (2004) Sprouting ability across diverse disturbances and vegetation types worldwide. Journal of Ecology 92: 310–320.
6. Clarke PJ, Lawes MJ, Midgley JJ, Lamont BB, Ojeda F, et al. (2013) Resprouting as a key functional trait: how buds, protection and resources drive persistence after fire. New Phytologist 197: 19–35.
7. Bellingham PJ, Sparrow AD (2000) Resprouting as a life history strategy in woody plant communities. Oikos 89: 409–416.
8. Bond WJ, Midgley JJ (2003) The evolutionary ecology of sprouting in woody plants. International Journal of Plant Sciences 164: S103–S114.
9. Poorter H, Niklas KJ, Reich PB, Oleksyn J, Poot P, et al. (2012) Biomass allocation to leaves, stems and roots: meta-analyses of interspecific variation and environmental control. New Phytologist 193: 30–50.
10. Weiher E, Werf A, Thompson K, Roderick M, Garnier E, et al. (1999) Challenging Theophrastus: a common core list of plant traits for functional ecology. Journal of Vegetation Science 10: 609–620.
11. Hodgkinson KC (1998) Sprouting success of shrubs after fire: height dependent relationships for different strategies. Oecologia 115: 64–72.
12. Houssard C, Escarré J (1995) Variation and covariation among life-history traits in Rumex acetosella from a successional old-field gradient. Oecologia 102: 70–80.
13. Bellingham PJ, Kohyama T, Aiba S (1996) The effects of a typhoon on Japanese warm temperate rainforests. Ecological Research 11: 229–247.
14. Lamont BB, Wiens D (2003) Are seed set and speciation rates always low among species that resprout after fire, and why? Evolutionary Ecology 17: 277–292.
15. Nzunda E, Lawes M (2011) Costs of resprouting are traded off against reproduction in subtropical coastal dune forest trees. Plant Ecology 212: 1991–2001.
16. Nzunda EF, Griffiths ME, Lawes MJ (2014) Resource allocation and storage relative to resprouting ability in wind disturbed coastal forest trees. Evolutionary Ecology 28: 735–749.
17. Bond WJ, Midgley JJ (2001) Ecology of sprouting in woody plants: the persistence niche. Trends in Ecology & Evolution 16: 45–51.
18. Shibata R, Shibata M, Tanaka H, Iida S, Masaki T, et al. (2014) Interspecific variation in the size-dependent resprouting ability of temperate woody species and its adaptive significance. Journal of Ecology 102: 209–220.
19. West GB, Brown JH, Enquist BJ (1999) The fourth dimension of life: fractal geometry and allometric scaling of organisms. Science 284: 1677–1679.
20. Lohier T, Jabot F, Meziane D, Shipley B, Reich PB, et al. (2014) Explaining ontogenetic shifts in root-shoot scaling with transient dynamics. Annals of Botany 114: 513–524.
21. Cohen D (1966) Optimizing reproduction in a randomly varying environment. Journal of theoretical biology 12: 119–129.
22. Philippi T (1993) Bet-hedging germination of desert annuals: variation among populations and maternal effects in Lepidium lasiocarpum. American Naturalist: 488–507.
23. Pake CE, Venable DL (1996) Seed banks in desert annuals: implications for persistence and coexistence in variable environments. Ecology 77: 1427–1435.
24. Fenner M (2000) Seeds: the ecology of regeneration in plant communities: CABI Publishing.
25. Fenner M, Thompson K (2005) The ecology of seeds: Cambridge University Press.
26. Clauss M, Venable D (2000) Seed germination in desert annuals: an empirical test of adaptive bet hedging. American Naturalist 155: 168–186.
27. Salk CF, McMahon SM (2011) Ecological and environmental factors constrain sprouting ability in tropical trees. Oecologia 166: 485–492.
28. Van Bloem SJ, Murphy PG, Lugo AE (2007) A link between hurricane-induced tree sprouting, high stem density and short canopy in tropical dry forest. Tree Physiology 27: 475–480.
29. Zhu W, Xiang J, Wang S, Li M (2012) Resprouting ability and mobile carbohydrate reserves in an oak shrubland decline with increasing elevation on the eastern edge of the Qinghai Tibet Plateau. Forest Ecology and Management 278: 118–126.
30. Moreira B, Tormo J, Pausas JG (2012) To resprout or not to resprout: factors driving intraspecific variability in resprouting. Oikos 121: 1577–1584.
31. Burrows GE (2008) Syncarpia and Tristaniopsis (Myrtaceae) possess specialised fire-resistant epicormic structures. Australian Journal of Botany 56: 254–264.
32. Meier AR, Saunders MR, Michler CH (2012) Epicormic buds in trees: a review of bud establishment, development and dormancy release. Tree Physiology 32: 565–584.
33. Wu L, Shinzato T, Chen C, Aramoto M (2008) Sprouting characteristics of a subtropical evergreen broad-leaved forest following clear-cutting in Okinawa, Japan. New Forests 36: 239–246.
34. Lusk CH (1999) Long-lived light-demanding emergents in southern temperate forests: the case of Weinmannia trichosperma (Cunoniaceae) in Chile. Plant Ecology 140: 111–115.
35. Lasso E, Engelbrecht BMJ, Dalling JW (2009) When sex is not enough: ecological correlates of resprouting capacity in congeneric tropical forest shrubs. Oecologia 161: 43–56.
36. Miura M, Yamamoto S (2003) Structure and dynamics of a Castanopsis cuspidata var. sieboldii population in an old-growth, evergreen, broad-leaved forest: the importance of sprout regeneration. Ecological Research 18: 115–129.
37. Gutiérrez AG, Aravena JC, Carrasco-Farías NV, Christie DA, Fuentes M, et al. (2008) Gap-phase dynamics and coexistence of a long-lived pioneer and shade-tolerant tree species in the canopy of an old-growth coastal temperate rain forest of Chiloé Island, Chile. Journal of Biogeography 35: 1674–1687.
38. Turner IM (2001) The ecology of trees in the tropical rain forest: Cambridge University Press.
39. Chen H, Felker S, Sun S (2010) Allometry of within-fruit reproductive allocation in subtropical dicot woody species. American Journal of Botany 97: 611–619.
40. Zhong Z (1988) Ecological study on evergreen broadleaved forest: Chongqing Southwest China Normal University Press (in Chinese with English abstracts). Pp. 696.
41. Xiong J, Guo S, Pan T (2005) The plant index of Mt. Jinyun Chongqing: Southwest China Normal University Press (in Chinese).
42. Zeng J, Liu Y (1995) An floristic analysis for the xylophyta in Sichuan. Journal of Southwest China Normal University (Natural Science Edition; in Chinese with English abstracts) 20: 686–692.
43. Liu Y (1988) Vegetation summary of Jinyun Mountain Natural Reserve. In: Zhong Z, editor. Ecology study of evergreen broad-leaved forest. Chongqing: Southwest China Normal University Press (in Chinese with English abstracts).
44. Da L, Yang Y, Song Y (2003) Population structure and regeneration types of dominant species in an evergreen broadleaved forest in Tiantong National Forest Park, Zhejiang Province, Eastern China. Acta Phytoecological Sinica 28: 376–384 (in Chinese with English abstract).
45. Sakai A, Ohsawa T, Ohsawa M (1995) Adaptive significance of sprouting of Euptelea polyandra, a deciduous tree growing on steep slopes with shallow soil. Journal of Plant Research 108: 377–386.
46. Del Tredici P (2001) Sprouting in temperate trees: a morphological and ecological review. Botanical Review 67: 121–140.
47. Chen H, Niklas KJ, Yang D, Sun S (2009) The effect of twig architecture and seed number on seed size variation in subtropical woody species. New Phytologist 183: 1212–1221.
48. Warton DI, Wright IJ, Falster DS, Westoby M (2006) Bivariate line-fitting methods for allometry. Biological Reviews 81: 259–291.
49. Warton DI, Weber NC (2002) Common slope tests for bivariate errors-in-variables models. Biometrical Journal 44: 161.
50. Falster DS, Warton DI, Wright IJ (2006) SMATR: standardised major axis tests and routines ver 2. 0 ed. Available: http://www.bio.mq.edu.au/ecology/SMATR/Accessed 2014 Aug 1.
51. Webb CO, Donoghue MJ (2005) Phylomatic: tree assembly for applied phylogenetics. Molecular Ecology Notes 5: 181–183; Available: http://phylodiversity.net/phylomatic/version 3 Accessed 2014 Aug 1.
52. Martins EP (2004) COMPARE: Computer programs for the statistical analysis of comparative data, 4.6b. Computer program and documentation distributed by the author; Available: http://compare.bio.indiana.edu Accessed 2014 Aug 1
53. Kruger L, Midgley J, Cowling R (1997) Resprouters vs reseeders in South African forest trees; a model based on forest canopy height. Functional Ecology 11: 101–105.
54. Lepš J, Šmilauer P (2003) Multivariate analysis of ecological data using CANOCO: Cambridge university press.
55. Amarasekare P (2003) Competitive coexistence in spatially structured environments: a synthesis. Ecology Letters 6: 1109–1122.
56. Wang X, Kent M, Fang X (2007) Evergreen broad-leaved forest in Eastern China: its ecology and conservation and the importance of resprouting in forest restoration. Forest Ecology and Management 245: 76–87.
57. Venable DL (2007) Bet hedging in a guild of desert annuals. Ecology 88: 1086–1090.
58. Bullock JM (2000) Gaps and seedling colonization. In: Fenner M, editor. Seeds: the ecology of regeneration in plant communities: CABI Publishing. pp. 375–395.
59. Starrfelt J, Kokko H (2012) Bet-hedging – a triple trade-off between means, variances and correlations. Biological Reviews 87: 742–755.
60. Knox KJE, Clarke PJ (2005) Nutrient availability induces contrasting allocation and starch formation in resprouting and obligate seeding shrubs. Functional Ecology 19: 690–698.
61. Miyawaki A (1999) Creative ecology: restoration of native forests by native trees. Plant Biotechnology 16: 15–25.

18

Habitat Loss, Not Fragmentation, Drives Occurrence Patterns of Canada Lynx at the Southern Range Periphery

Megan L. Hornseth[1*¤], Aaron A. Walpole[2], Lyle R. Walton[3], Jeff Bowman[2], Justina C. Ray[4], Marie-Josée Fortin[5], Dennis L. Murray[6]

1 Environmental and Life Sciences, Trent University, Peterborough, Canada, 2 Wildlife Research & Monitoring Section, Ontario Ministry of Natural Resources and Forestry, Peterborough, Canada, 3 Regional Operations Division, Ontario Ministry of Natural Resources, South Porcupine, Canada, 4 Wildlife Conservation Society Canada, Toronto, Canada, 5 Department of Ecology and Evolutionary Biology, University of Toronto, Toronto, Canada, 6 Department of Biology, Trent University, Peterborough, Canada

Abstract

Peripheral populations often experience more extreme environmental conditions than those in the centre of a species' range. Such extreme conditions include habitat loss, defined as a reduction in the amount of suitable habitat, as well as habitat fragmentation, which involves the breaking apart of habitat independent of habitat loss. The 'threshold hypothesis' predicts that organisms will be more affected by habitat fragmentation when the amount of habitat on the landscape is scarce (i.e., less than 30%) than when habitat is abundant, implying that habitat fragmentation may compound habitat loss through changes in patch size and configuration. Alternatively, the 'flexibility hypothesis' predicts that individuals may respond to increased habitat disturbance by altering their selection patterns and thereby reducing sensitivity to habitat loss and fragmentation. While the range of Canada lynx (Lynx canadensis) has contracted during recent decades, the relative importance of habitat loss and habitat fragmentation on this phenomenon is poorly understood. We used a habitat suitability model for lynx to identify suitable land cover in Ontario, and contrasted occupancy patterns across landscapes differing in cover, to test the 'threshold hypothesis' and 'flexibility hypothesis'. When suitable land cover was widely available, lynx avoided areas with less than 30% habitat and were unaffected by habitat fragmentation. However, on landscapes with minimal suitable land cover, lynx occurrence was not related to either habitat loss or habitat fragmentation, indicating support for the 'flexibility hypothesis'. We conclude that lynx are broadly affected by habitat loss, and not specifically by habitat fragmentation, although occurrence patterns are flexible and dependent on landscape condition. We suggest that lynx may alter their habitat selection patterns depending on local conditions, thereby reducing their sensitivity to anthropogenically-driven habitat alteration.

Editor: Brock Fenton, University of Western Ontario, Canada

Funding: This study was funded by the Wildlife Conservation Society Kaplan Awards Program (now the Panthera Foundation, given to MLH), the Ontario Ministry of Natural Resources North East Science and Information Section, a Strategic Natural Sciences and Engineering Research grant to DLM, and a Natural Sciences and Engineering Research Industrial Postgraduate Scholarship with the Wildlife Conservation Society Canada to Megan Hornseth. The funders had no role in study design, data collection and analysis, decision to publish, or preparation of the manuscript.

Competing Interests: The authors have declared that no competing interests exist.

* Email: mhornseth@gmail.com

¤ Current address: Centre for Northern Forest Ecology Research, Ontario of Ministry of Natural Resources and Forestry, Lakehead University, Thunder Bay, Canada

Introduction

Populations occurring at the periphery of a species' geographic range often occupy habitats that are of lower overall quality, leading to reduced survival, reproduction and population density, compared to populations in the core of the range [1]. In addition, peripheral populations tend to be more sensitive to environmental variability than those in the core, which can promote increased demographic stochasticity and lower resilience [2–4]. As a result, individuals in the range periphery may be more sensitive to the processes of habitat loss and fragmentation. Alternatively, animals may respond with more flexible habitat selection patterns, enabling them to move among variable environments to enhance their fitness [5]. This flexibility should increase species' persistence in landscapes experiencing anthropogenic change, such as in areas subject to high fragmentation. However, much of our perception

of how wide-ranging species respond to these landscape-scale processes is speculative, especially in peripheral populations where both occurrences and their detection probability are often limited. This shortcoming is especially relevant because as landscapes continue to be altered by anthropogenic disturbance, many species are faced with declines in range size [6]. An improved understanding of the effects of habitat loss and fragmentation on species occurrence patterns will enhance our understanding of how these processes may impact species distributions.

Habitat loss and fragmentation are separate processes whereby habitat loss is an overall reduction in the amount of suitable habitat resulting in a decline of patch size and habitat fragmentation is the breaking apart of habitat, independent of habitat loss [7]. While the effects of habitat loss on species are consistently negative, the effects of habitat fragmentation are less well understood, as few studies measure fragmentation independently of habitat loss [7]. While habitat fragmentation can have both weakly positive and weakly negative effects on biodiversity and population size, the impact of these effects is often far less important than the effects of habitat loss [7–9]. There is some evidence that the effects of habitat fragmentation depend upon the amount of habitat that is available in a landscape. The 'threshold hypothesis' predicts that individuals will be more affected by habitat fragmentation when the amount of habitat on the landscape is limiting (i.e. less than 30% habitat), and small and isolated patches become more numerous, than when habitat is abundant and patches are larger and more continuous [10,11]. Habitat fragmentation may compound the effects of habitat loss due to changes in patch size and landscape configuration, implying that fragmentation may have a greater effect at the range periphery, where habitat is often limiting [2]. This hypothesis has been supported by several studies examining population size and presence of birds and small mammals with habitat thresholds ranging from 10–30% [10,12–14]. In contrast, the 'flexibility hypothesis' suggests that individuals may alter their habitat selection patterns, permitting them to inhabit variable environments that would otherwise be unsuitable due to habitat fragmentation [5,15].

Canada lynx (*Lynx canadensis*) occur across the boreal forest of North America, where their primary prey is snowshoe hare (*Lepus americanus*). Since lynx are dependent upon snowshoe hares, they select forested habitat based on high hare abundance or where they are most easily depredated [16–18], whereas hares select young coniferous forests where both food and cover are adequate [19,20]. In the southern periphery of the lynx range, forest composition is more heterogeneous and hare densities are naturally lower, leading to reduced abundance and restricted distribution of lynx [21], which require densities between 1 to 1.5 hares per hectare to persist [22].Because habitat for both lynx and hare has become both reduced and fragmented due to anthropogenic activities in their southern ranges, the distribution and abundance of both species is now restricted [23,24]. This has reduced genetic diversity in southern populations of both hare [25] and lynx [26]. Additionally, the southern range of lynx in Ontario has contracted by over 175 km since 1970 [26]. Although the mechanisms ultimately limiting lynx populations at the southern range periphery remain to be fully understood, this may be due to sensitivity to habitat fragmentation [27], with habitat loss and climate change as other important factors [26]. Several other felid species are also reported to be sensitive to habitat fragmentation (e.g. Iberian lynx (*Lynx pardinus*) [28], bobcat (*Lynx rufus*) and cougar (*Puma concolor*) [29]). However, whether these species express any flexibility in selection patterns in relation to the amount of habitat on a landscape or whether these patterns hold true for habitat fragmentation, has not yet been explored.

We examined the occurrence patterns of Canada lynx across the 2 regions in the southern geographic range of the species in Ontario to assess patterns of occurrence in relation to habitat loss and fragmentation. Given that lynx are prey specialists, requiring areas within a narrow range of suitable conditions to meet prey and habitat requirements [30] as well as connectivity requirements [31], we predicted that lynx would be sensitive to habitat loss when habitat was widely available, and sensitive to both habitat loss and fragmentation when suitable habitat was less than 30%; this would support the 'threshold hypothesis' [10,11]. These patterns may be expressed more strongly near the southern range periphery, due to increased levels of habitat loss and reduced habitat quality [26], leading us to speculate that any sensitivity to habitat fragmentation would be most apparent there. Alternatively, the 'flexibility hypothesis' suggests that lynx will have tolerance to both habitat loss and fragmentation, such that their occurrence patterns may

not correlate with either process, indicating flexibility in habitat selection. We developed a habitat suitability model for lynx and tested the above predictions using patterns of track occurrence across the species' southern range periphery. We compared two regions each with three similar levels of suitable land cover as determined by the habitat suitability model, to examine if occurrence patterns differ across landscapes with varying amounts of suitable land cover. Observations of lynx tracks in areas with limited suitable land cover and increased fragmentation would imply that lynx are not sensitive to habitat fragmentation, or that the importance of suitable habitat on occurrence patterns at the range periphery are less critical than previously understood.

Methods

Ethics Statement

The Trent University Research Ethics Board approved the study (reference #21083). In the introduction of the study, participants were explicitly told that informed consent was implied if they submitted their survey data. The field component consisted of non-invasive track surveys conducted on public land, so no access permits or animal care protocols were required. Canada lynx are considered not at risk under provincial and federal guidelines.

Study Area

The study area encompassed 200 000 km² in central Ontario (Figure 1A), across the southern boreal forest and the Great Lakes St. Lawrence forest, a transition zone from boreal to deciduous forest, encompassing the southern range limit of lynx occurrence in the region [32]. The area is largely comprised of boreal forest, with spruce (*Picea glauca, P. mariana*), balsam fir (*Abies balsamea*), trembling poplar (*Populus tremuloides*) and white birch (*Betula papyrifera*) as dominant tree species. The southerly portions of the study area in the Great Lakes St. Lawrence region include pines (*Pinus resinosa, P. strobus*), eastern hemlock (*Tsuga canadensis*), yellow birch (*B. alleghaniensis*) and maples (*Acer saccharum, A. rubrum*). Habitat loss and fragmentation throughout the study area is caused primarily by forestry and associated road construction. Historically 1% of the entire region (approximately 2000 km²) was harvested annually [32], current levels are 0.04% or 800 km² (2000–2010 average; [33]). Other sources of habitat loss include populated areas, agriculture, and natural disturbance such as forest fire and pest infestations.

Habitat Suitability Model

In order to quantify lynx habitat suitability, we used the analytic hierarchy process, a decision-making procedure that is useful in the development of habitat suitability models for wide-ranging mammals (see [34,35] for description of methodology). We developed the survey design based on a literature review identifying important ecological factors affecting lynx occurrence, with an emphasis on the southern range periphery. The primary habitat characteristics were land cover attributes (e.g., [17,18]), forest age class (e.g., [18,36]), annual snowfall (e.g., [37]) and road density (e.g., [38]). We developed two separate models of habitat suitability, one based on expert-opinion, where we received 11 solicited responses from lynx researchers across North America, and the other using a literature-based approach with four 'naïve' participants with no previous knowledge of lynx ecology. Both experts and naïve participants received the same survey and the naïve participants also received four research papers providing a detailed description of the basic habitat requirements of lynx from across its range [17,18,38,39]. The survey consisted of five

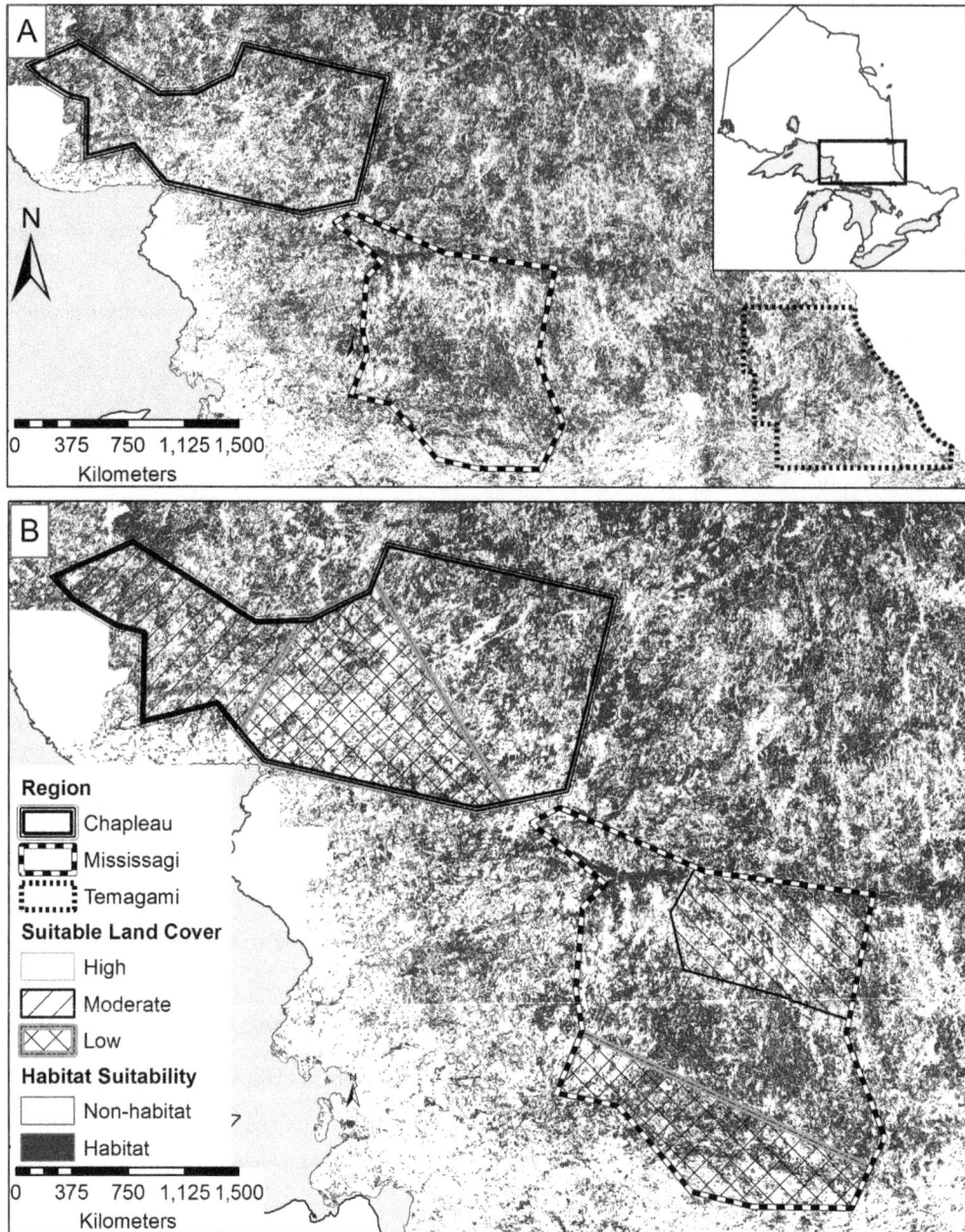

Figure 1. Habitat suitability map for Canada lynx in (A) central Ontario with Regions outlined and (B) suitable land cover levels within each region, as determined by the literature-based habitat suitability model.

separate pair-wise comparison matrices based on each of the features of interest (land cover, forest development stage, snowfall, and road density) and an overall comparison of the relative importance among all features. The overall ranking of features was used to weight parameters within the model and estimate the relative importance of factors affecting lynx habitat suitability, whereas weights within a feature determined the ranking for its attributes.

We used the Ontario Forest Resource Inventory to characterize land cover; these data provide a detailed description of species composition and forest stand age as determined by aerial photo interpretation. The study area included 41 provincial forest management units, and each unit was updated with forest fire

and harvest information up to and including 2008. Standardized forest units were combined to create six generalized land cover types (coniferous forest, deciduous forest, mixedwood forest, developed land, wetland, and open areas) and five forest development stages (presapling, sapling, immature, mature and old; [40]), which improved the accuracy of the dataset [41]. We converted the land cover map to a geospatial raster for analysis; all GIS analyses were conducted in ArcGIS 9.2 (ESRI, Redlands, CA, USA).

We evaluated the lynx habitat model in a portion of the study area near the North Bay - Temagami region of northeastern Ontario, Canada (47.01°N, 79.97°W; see Figure 1A). The Temagami region is approximately 8,000 km^2 and was selected because it is located within the southern range periphery of lynx in Ontario and the transition zone of boreal forest with the northern Great Lakes-St. Lawrence forest. Between January and March 2009, we surveyed lynx occurrence at 48 randomly selected sites that represented a gradient in available land cover types [38]. We assessed lynx presence by snowtracking triangular transects around the centroid of the cell (dimensions 0.5 km per side, [38]). Additional lynx tracks that were encountered opportunistically while travelling within the landscape were also considered as lynx presence. We calculated habitat suitability at the centre of each transect and each opportunistic track, using both models. We used receiver operating characteristic plots and the Area Under the Curve (AUC) as an independent measure of model accuracy via the program ROC/AUC [42]. AUC provides a measure of model accuracy, where values >0.7 indicate good model fit. We selected P_{fair}, the value where specificity and sensitivity are equal, as the threshold habitat suitable for lynx occurrence.

Lynx Occurrence Sampling

Two regions were selected to document lynx occurrence (estimated by track identification) in landscapes across a gradient of habitat fragmentation. Each region fell within the larger study area which encompassed the southern boreal forest and Great Lakes-St. Lawrence Forest, and was divided into three landscapes based on the amount of suitable land cover (high, moderate, and low) as determined by the habitat suitability model (Figure 1B). The Chapleau region was 12 900 km^2, located primarily in the boreal forest. The western portion of the region had the highest amount of suitable land cover and is the least fragmented landscape in this region. The central area of the Chapleau region is highly fragmented with the most habitat loss due to forestry, roads, and human settlements. The easternmost portion of this region has a moderate amount of suitable land cover and a moderate level of fragmentation due to forestry roads (Table 1). The Mississagi region was 12 800 km^2 located primarily in the Great Lakes St. Lawrence forest. The northern portion of this region had moderate amounts of suitable land cover, but was fragmented due to forestry roads; the central portion had the highest amount of suitable land cover and was least fragmented, and the southernmost landscape had the least amount of suitable land cover in this region, with habitat loss due to forestry, human settlements and roads. These regions were surveyed for occurrence of lynx tracks from January to March 2010 and each identified track point was recorded as a lynx occurrence. All forest access roads, trails, hydro-electric line corridors, cutovers and riparian areas were sampled via snowmobile, totalling 9 320 km of survey lines in both landscapes. All lynx track locations were documented; Chapleau had 104 track points and Mississagi had 89 tracks points (see Figure S1 in Information S1).

Roads in these two regions were limited to 1 or 2 highways, < 20 secondary roads, and forestry roads. To test whether there was bias arising from track proximity to surveyed roads, we randomly selected 100 points from roads (including highways, primary, secondary and tertiary roads, and snowmobile trails) and the surrounding landscape (not bisected by roads) and compared them at five spatial scales (10 km^2, 25 km^2, 50 km^2, 75 km^2, and 100 km^2) to assess any differences in habitat quality in each region. We found that there was no difference in the amount of lynx habitat (as defined by the suitability model) in any landscape, regardless of spatial scale and distance to roads (M. Hornseth, unpublished data, but see [38]). Accordingly, we deemed that proximity of locations to roads was not relevant to our particular analysis.

True absences are difficult to detect using typical survey methods, especially without repeated visits. We randomly selected points (equal to the number of lynx locations) from survey logs to represent pseudo-absences in Chapleau and Mississagi. These locations were at least 1 km apart and at least 2.5 km from the nearest lynx location. To examine the effect of spatial scale, and to encompass overall selection patterns, we buffered both observed lynx tracks and pseudo-absences with radii of 2.82 km and 5.61 km to create areas of 25 km^2 and 100 km^2 (from published home range size estimates), to assess the role of spatial scale on occurrence patterns (see [18,39]).

Habitat Amount and Fragmentation

Landscape connectivity can be considered across a variety of spatial and ecological scales, and for our analysis the metrics of interest included estimates of: (i) structural connectivity, which represents the spatial configuration of suitable patches; and (ii) functional connectivity, which includes animal response to patches [43]. We created a binary landscape of habitat quality using the literature based habitat suitability model and a critical threshold of habitat suitability value of 52 (threshold tuned by balancing the error rate between false positives and false negatives [42]). We quantified the percentage of habitat within each lynx and pseudo-absence area to estimate habitat amount. To avoid confusion of working at multiple scales, we used the term *suitable land cover* to describe the output of the habitat suitability model at a landscape-level and *suitable habitat* to describe this output at a finer spatial scale (25 km^2 and 100 km^2 areas).

We used PatchMorph [44] and the habitat suitability model to estimate a 'functionally' connected landscape for lynx from: (1) a critical threshold of habitat suitability value of 52, (2) a minimum patch size of 5 ha (the minimum mappable forest stand (Ontario Ministry of Natural Resources, unpublished data)), and (3) a crossing distance of 200 m (M. Hornseth, unpublished data). Note that crossing distance is defined as the distance that lynx will travel in unsuitable habitat; the minimum for this metric is two raster pixels and parameters were set conservatively as per published observations of lynx habitat use patterns (see [17,45]). Although we acknowledge that actual functional connectivity requirements for lynx are just beginning to be understood (see [46]), we consider our selected values as being within the range of those that are plausible, with minor deviations likely affecting our results only qualitatively. Additionally, we did a sensitivity analysis with crossing distances of 200 to 1000 m in 400 m increments to determine the effect of this parameter on our estimates of connectivity.

Effective mesh size can be defined as the average area potentially accessed by an animal on a given landscape without having to cross defined borders or low quality habitat, so larger values indicate that the landscape is more connected and smaller values indicate the landscape is more fragmented [44,47]. We used effective mesh size (M_{eff}) as our measure of habitat fragmentation in ArcMap 10.1 [48]. M_{eff} is calculated by:

$$M_{eff} = \frac{1}{A_t} \sum_{i=1}^{n} A_i^2,$$

where A is the area of a single patch and A_t can be either the total area of the polygon or the total amount of suitable habitat (i.e., the sum of all patch areas). In order to remove correlation between habitat amount and effective mesh size, we used the total amount of suitable habitat as the denominator (L. Fahrig, pers. comm.). Since correlations were still high (0.63–0.86), we regressed M_{eff}

Table 1. Summary of the amount of suitable land cover and habitat fragmentation across two regions in the southern boreal forest in Ontario, Canada.

Region	Land Cover[a] Level	Area (km^2)	Percentage of Suitable Land Cover	M$_{eff}$ (km^2)
Chapleau	High	5 085.7	41.88	87.31
	Moderate	3 162.8	34.95	22.41
	Low	4 639.7	20.64	5.68
Mississagi	High	7 873.2	42.84	258.61
	Moderate	3 016.8	31.85	23.14
	Low	2 356.4	25.5	18.55

[a]Land cover is the amount of suitable land cover measured at the landscape level as determined by the habitat suitability model.

against habitat amount and used the residuals as our estimate of habitat fragmentation (M$_{eff.r}$).

Data Analysis

We aimed to determine whether lynx are limited by habitat amount, fragmentation, or both processes, by contrasting patterns on landscapes with different amounts of suitable land cover. We hypothesized that lynx habitat requirements would restrict their occurrence to highly-connected areas in each landscape. We used one-sided unpaired t-tests to examine whether habitat amount and fragmentation were greater in presence areas than pseudo-absence areas at each spatial scale among landscapes with high, moderate, and low amounts of habitat amount in each region. We examined any correlations between these two within each region and landscape.

We tested 3 *a priori* hypotheses to explain lynx occurrence; i) lynx occurrence is limited only by fragmentation, ii) lynx occurrence is limited only by habitat loss, and iii) lynx occurrence is limited by both habitat amount and fragmentation. We used logistic regression and standard model selection procedures to determine which hypothesis best explained lynx occurrence in landscapes across the levels of suitable land cover. We used Akaike's information criterion to evaluate the candidate models for each lynx and pseudo-absence area and landscape within each region. We considered ΔAIC >2 to indicate a significant difference in model likelihood [49]. AIC does not assess model performance, and only models that performed well were considered plausible for the AIC model selection, so we used the Logistic Regression χ^2 model likelihood ratio test to determine model fit.

Results

Habitat Suitability Model

Both the expert-opinion and literature-based models suggested that coniferous forest land cover, and forest in a sapling developmental stage, provided the most suitable habitat for lynx. However, models differed with respect to the relative importance of overall features, with the literature-based model suggesting that land cover was only slightly (1.04 times) more important than development stage whereas expert opinion suggesting that development stage was substantially (1.20 times) more important than land cover type. We omitted annual snowfall and road density from the final habitat suitability models due to low overall importance in both models (see Table S1 in Information S1).

We detected lynx at 19% ($n = 48$) of the sites within the Temagami landscape; we also included 14 more lynx track occurrences that we encountered opportunistically within the

study site, increasing the total number of validation locations to 62. The literature-based model had a good overall fit (AUC: 0.912, $p<0.001$) and correctly predicted 83.9% of all sites ($n = 62$) and 82.6% of lynx occurrences ($n = 23$). The expert opinion model had a comparable fit (AUC: 0.855 $p<0.001$), correctly classifying 82.3% of all sites and 78.2% of lynx occurrences. Although both models performed well, the literature-based model surpassed the expert-opinion model in every comparison (see Table S1 in Information S1) and was selected for the remaining analyses (see Table S2 in Information S1).

Landscape Characteristics

The landscapes within both regions had similar amounts of suitable land cover (Table 1), but different levels of habitat fragmentation. The high-cover landscape in Chapleau consisted of 41.9% suitable land cover with an effective mesh size of 87.3 km^2. In Mississagi, the high-cover landscape had approximately the same amount of suitable land cover (42.8%), but a much larger mesh size of 258.6 km^2. The landscapes with a moderate amount of suitable land cover in the Chapleau and Mississagi regions had similar amounts of suitable land cover (35.0% and 31.9%, respectively) and mesh sizes (22.4 km^2 and 23.1 km^2, respectively). The low-cover landscapes had similar amounts of suitable land cover (20.6% in Chapleau, 25.5% in Mississagi), however, the landscape in the Chapleau region was substantially more fragmented (M$_{eff}$ 5.7 km^2) in comparison to the matched landscape in the Mississagi region (M$_{eff}$ 18.6 km^2). This indicated that although the two landscapes had similar amounts of suitable land cover, generally the Chapleau landscape was more fragmented.

Lynx Occurrence

Where possible, lynx selected areas with higher amounts of high quality habitat (structural connectivity) at the 25 km^2 spatial scale (Table 2). There was a positive correlation between the amount of suitable habitat and lynx occurrence areas in both high- and moderate-levels of suitable land cover in the Chapleau region, and in the landscape with a moderate-level of suitable land cover in the Mississagi region at the 25 km^2 area (Figure 2). In both regions, on landscapes with high- and moderate-levels of land cover, lynx consistently occurred in areas with at least 50% habitat and avoided areas with <30% habitat (Figure 3). However, in the low-cover landscapes, approximately half of lynx occurrences had less than 30% habitat at a spatial scale of 25 km^2. These trends were consistent across both regions. At a spatial scale of 100 km^2, there were no correlations between the amount of suitable habitat and lynx occurrence at any level of suitable land cover (Table 3). Once

Table 2. Summary of the differences in connectivity measures of Canada lynx occurrence and pseudo-absences in Ontario, Canada; all t-tests were one-sided with p-values<0.05 in bold and p-values<0.1 in italics.

Region	Area (km²)	Land Cover[a] Level	Variable	Present	Pseudo-absent	t-test	p-value
Mississagi	25	High	Habitat[b]	56.56	54.90	0.53	0.300
			$M_{eff,r}$[c]	0.17	−0.20	0.76	0.223
		Moderate	**Habitat**	**48.18**	**33.8**	**3.55**	**<0.001**
			$M_{eff,r}$	−0.77	0.45	−2.30	0.987
		Low	Habitat	26.4	29.98	−0.76	0.772
			M_{eff} residuals	−0.16	0.06	−0.51	0.691
	100	High	Habitat	47.98	47.72	0.10	0.461
			$M_{eff,r}$	−0.08	0.09	−0.26	0.601
		Moderate	*Habitat*	*35.35*	*31.11*	*1.33*	*0.095*
			$M_{eff,r}$	−1.12	0.66	−1.92	0.970
		Low	Habitat	23.22	27.64	−0.97	0.829
			$M_{eff,r}$	−0.24	0.10	−0.43	0.665
Chapleau	25	High	**Habitat**	**52.79**	**46.39**	**1.96**	**0.027**
			$M_{eff,r}$	−0.27	0.18	−0.76	0.774
		Moderate	**Habitat**	**51.88**	**43.31**	**3.01**	**0.002**
			$M_{eff,r}$	−0.36	0.44	−1.98	0.975
		Low	Habitat	29.87	29.88	−0.01	0.502
			$M_{eff,r}$	0.04	−0.03	0.31	0.380
	100	High	Habitat	41.16	39.88	0.52	0.300
			$M_{eff,r}$	−0.74	0.51	−1.29	0.899
		Moderate	**Habitat**	**39.78**	**34.75**	**2.33**	**0.011**
			$M_{eff,r}$	−0.41	0.50	−1.30	0.901
		Low	Habitat	22.22	21.89	0.14	0.442
			$M_{eff,r}$	−0.12	0.07	−0.96	0.829

[a]Land cover is the amount of suitable land cover measured at the landscape level as determined by the habitat suitability model.
[b]Habitat is the proportion of suitable habitat within each lynx- and pseudo-absence area based on the habitat suitability model.
[c]$M_{eff,r}$ is the residual of habitat regressed against mesh size (km²; see text), a measure of functional connectivity, within each lynx- and pseudo-absence area.

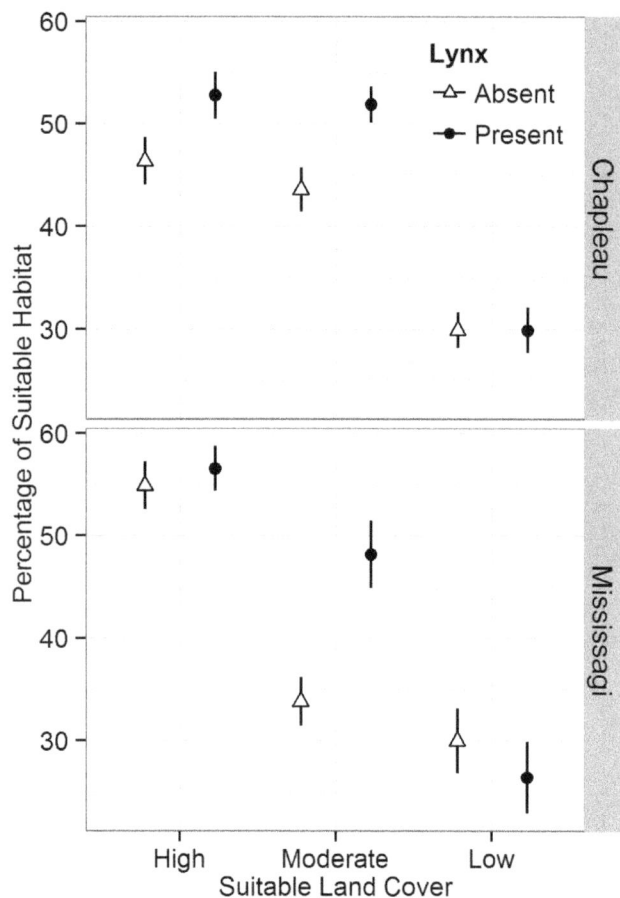

Figure 2. Mean percentage of suitable habitat (with standard errors) for lynx presences compared to pseudo-absences at the 25 km² scale in the regions of Chapleau and Mississagi with three levels of suitable land cover.

the effect of habitat amount was removed, there was no correlation between habitat fragmentation ($M_{eff.r}$) and lynx occurrence on any landscape, at either spatial scale (Table 2).

Lynx occurrence patterns differed across landscapes, but the trends were consistent across regions. In the landscapes with moderate levels of suitable land cover, the top model included both the proportion of suitable habitat and habitat fragmentation lynx occurrence. However, only the proportion of suitable habitat had a positive association on lynx occurrence, $M_{eff.r}$ had a negative correlation with lynx occurrence indicating that lynx selected areas with higher amounts of fragmentation (Figure 4; Table 3). In the high- and low-cover landscapes in both regions, there was no significant correlation between lynx occurrence patterns and proportion of suitable habitat or effective mesh size (Table 3).

Sensitivity Analysis

We examined 3 crossing distances in the PatchMorph output to determine if crossing distance was either underestimated or strongly influential on lynx occurrence. We tested crossing distances of 200 m, 600 m, and 1000 m, and used standardized regression coefficients from single variable logistic regressions to determine the level of influence. Effective mesh size coefficient estimates ranged from −0.02 to 0.04, with no visible trend; none of the coefficients were significant (p values ranged from 0.228–

0.589). Increasing the estimated crossing distance did not affect model fit.

Discussion

Our results confirm that lynx are not sensitive to habitat fragmentation at low levels of suitable habitat, and also suggest that lynx display considerable flexibility in habitat selection patterns, supporting the 'flexibility hypothesis'. We showed that in landscapes with moderate and high amounts of suitable land cover (30–35% and >40%, respectively), lynx occurred in areas with at least 30% available habitat and largely avoided areas below that threshold, while being unaffected by habitat fragmentation. Although this finding is consistent with the 'threshold hypothesis', this hypothesis also predicts that lynx would be more sensitive to habitat fragmentation on landscapes where suitable land cover was low. However, our results showed that on landscapes where suitable land cover was limited (<30%), lynx did not select areas with concentrated habitat and lynx occurrence patterns were not well correlated with either habitat amount or habitat fragmentation, instead supporting the 'flexibility hypothesis'. Overall, we detected a threshold at which lynx occurrence patterns changed, but instead of being more sensitive to habitat fragmentation at low levels of suitable habitat, lynx displayed more flexibility in habitat selection on these landscapes. This indicates that lynx habitat choice is complex and either involves factors beyond mere resource preference, or selection of different land cover types in these areas.

Patterns of Occurrence

As predicted by the literature-based habitat suitability model, lynx were most likely to occur in sapling-stage coniferous forest. These results are consistent with other literature on lynx habitat ecology [17,18] and also describes snowshoe hare habitat preferences [19,20]. Road density and annual snowfall were not important for describing lynx occurrence in Ontario. This finding contrasts with previous work (e.g., [37,38]) but is consistent with a companion occupancy model within our study area [46], suggesting that these factors differentially affect lynx occurrence across their range and may be threshold-dependent. We surmise that low variation in snowfall patterns and low abundance of major highways as well as low road density in our study site may have accounted for the disparate results. Lynx occurrence, as determined by snow tracks across the study area, also supported this model, signifying that our model is generally robust. We recommend the use of this habitat suitability model as a tool to evaluate future forest condition on resource availability for Canada lynx in Ontario.

Flexibility in Response to Habitat Loss

Our results suggest that when approximately 30–35% of the landscape consists of suitable land cover, there is a strong correlation between the amount of suitable habitat and lynx occurrence. While this trend was not significant at higher levels of land cover at a landscape scale, in landscapes with both high and moderate amounts of suitable land cover, lynx occurrence patterns suggest that lynx preferred areas with at least 50% suitable land cover. While lynx will occur in some areas with less than 50% available suitable land cover, lynx consistently avoided areas with less than 30% suitable habitat when suitable land cover was abundant at a landscape level. This is consistent with previous work on small mammals and birds showing that habitat occupancy dynamics are determined by species-specific tolerance thresholds [7,11,13].

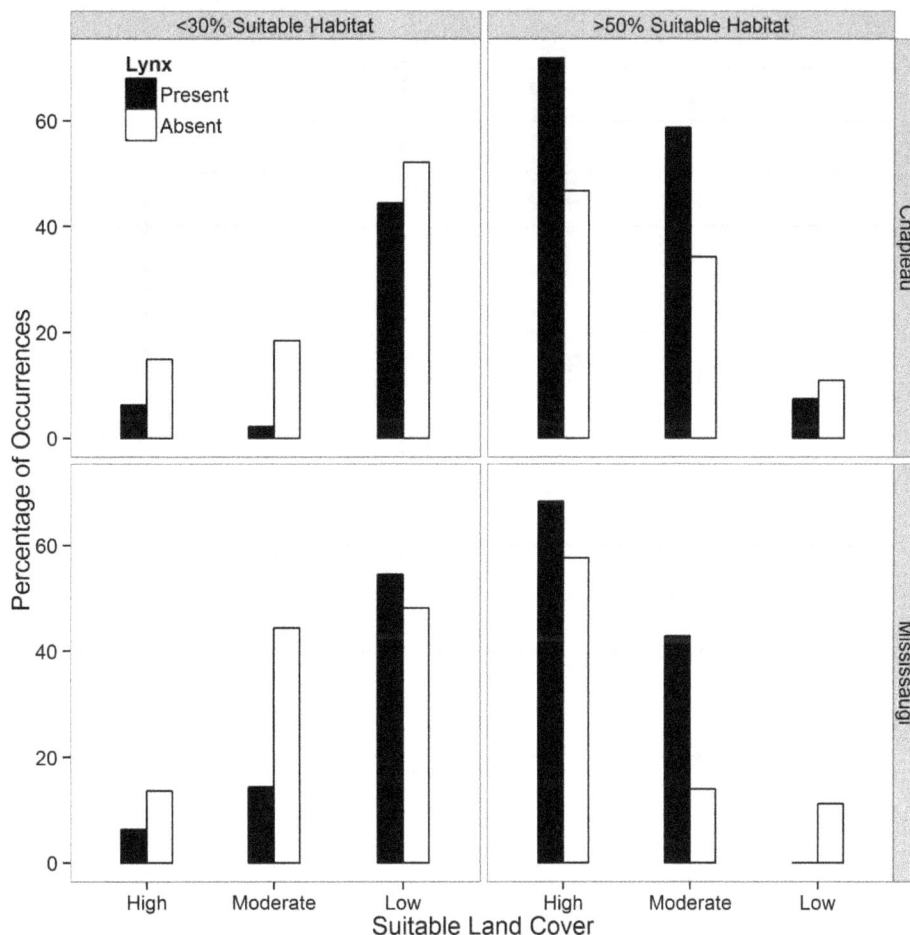

Figure 3. Distribution of lynx occurrences and pseudo-absences in relation to the amount of suitable habitat at the 25 km² scale in the regions of Chapleau and Mississagi with three levels of suitable land cover.

When suitable land cover comprised only 20–25% of the landscape, our results showed that there was no correlation between lynx occurrence and habitat amount, indicating some flexibility in habitat requirements on these landscapes. In contrast, when suitable habitat was limited, lynx did not avoid areas with less than 30% land cover and were not associated with areas with more than 50% suitable habitat, despite the local availability of these areas. It is possible that when suitable habitat is scarce, lynx can survive provided that hares, or suitable alternate prey, remain available on the landscape. This speculation is supported by observations of resident snowshoe hares occupying small patches <10 ha in fragmented landscapes [50,51] and the ability of lynx to include alternate prey items when hares are limited [31,52]. This pattern of labile specialization has been recently documented in birds, where the most specialized species tend to generalize their habitat selection pattern following disturbance [53]. However, the results of our study contrast with previous work by Swihart et al. [5,50], who showed that some species have greater sensitivity to habitat change at range margins. This suggests that there is a wide range of responses to habitat alteration and that further work is necessary to clarify the impact of landscape change on lynx.

Habitat Fragmentation

Our results show that there is no correlation between lynx occurrence patterns and habitat fragmentation ($M_{eff.r}$). $M_{eff.r}$ (mesh size) measures the connectivity of a landscape, independent of habitat loss, so a negative coefficient indicates a positive relationship with habitat fragmentation. Our results suggest that there is a weakly negative relationship with $M_{eff.r}$ at moderate levels of suitable land cover, which is the opposite of what we predicted. In addition, the results from our sensitivity analysis suggest that increasing crossing distance does not improve the measure of habitat fragmentation for lynx. While some studies have suggested that habitat fragmentation may only be important when habitat amount is below 30% [9–11], our results do not support this hypothesis. At low levels of suitable land cover there was no relationship between habitat fragmentation and lynx occurrence, which is consistent with studies showing that the effects of habitat loss are generally far greater than the effects of habitat fragmentation [7,11]. Our results concerning habitat loss and habitat fragmentation are especially applicable to forestry-dominated landscapes, where silvicultural practices can result in marked shifts in habitat features for a variety of species, including higher densities of prey species such as snowshoe hares [54]. Therefore, we recommend that planning decisions regarding lynx consider the amount of total available habitat, which should generally improve chances of population persistence, while also benefitting overall landscape structure and function. This point is especially relevant at the southern range periphery of lynx, where

Table 3. Model selection of 3 *a priori* hypotheses proposed to explain lynx occurrence patterns across 3 landscapes differing in the amount of suitable landscape-level land cover in 2 regions (Chapleau and Mississagi) within an area of 25 km^2 for each lynx track and pseudo-absence.

Chapleau	Coefficients		AIC	ΔAIC	Weight	χ²	p-value
	$M_{eff.r}$[a]	Habitat[b]					
High Land Cover							
Habitat Only	-	*0.031**	*107.2*	*0*	*0.57*	*3.61*	*0.057*
Habitat+$M_{eff.r}$	−0.079	0.032†	108.3	1.1	0.28	4.36	0.113
M_{eff} Only	−0.072	-	110.2	3.0	0.15	0.61	0.434
Moderate Land Cover							
Habitat+$M_{eff.r}$	*−0.253**	*0.086**	*108.7*	*0*	*0.71*	*13.00*	*0.002*
Habitat Only	-	*0.053**	*110.8*	*2.1*	*0.26*	*8.85*	*0.003*
M_{eff} Only	−0.244†	-	115.7	6.0	0.03	3.98	0.049
Low Land Cover							
$M_{eff.r}$ Only	0.098	-	100.3	0	0.43	0.06	0.802
Habitat Only	-	0.0001	100.4	0.1	0.41	0.00	0.996
Habitat+$M_{eff.r}$	0.098	−0.0002	102.4	2.1	0.15	0.11	0.945
Mississagi							
High Land Cover							
$M_{eff.r}$ Only	0.056	-	166.6	0	0.47	0.79	0.150
Habitat Only	-	0.006	166.9	0.5	0.36	0.28	0.596
Habitat+$M_{eff.r}$	0.056	−0.006	168.4	2.0	0.17	0.87	0.329
Moderate Land Cover							
Habitat+$M_{eff.r}$	*−0.388**	*0.073**	*63.6*	*0*	*0.87*	*17.50*	*<0.001*
Habitat Only	-	*0.063**	*67.8*	*4.2*	*0.12*	*11.27*	*0.008*
$M_{eff.r}$ Only	−0.390*	-	73.3	9.7	0.01	5.68	0.017
Low Land Cover							
$M_{eff.r}$ Only	−0.203	-	49.6	0	0.47	0.36	0.356
Habitat Only	-	−0.016	49.7	0.1	0.39	0.45	0.511
Habitat+$M_{eff.r}$	−0.234	0.004	51.6	2.0	0.14	0.86	0.651

Asterisk (*) indicates significant coefficients at $p < 0.05$,
† indicates significance at $p < 0.1$.
[a] $M_{eff.r}$ is the residual of habitat regressed against mesh size (km^2; see text), a measure of functional connectivity, within each lynx- and pseudo-absence area.
[b] Habitat is the proportion of suitable habitat within each lynx- and pseudo-absence area based on the habitat suitability model.

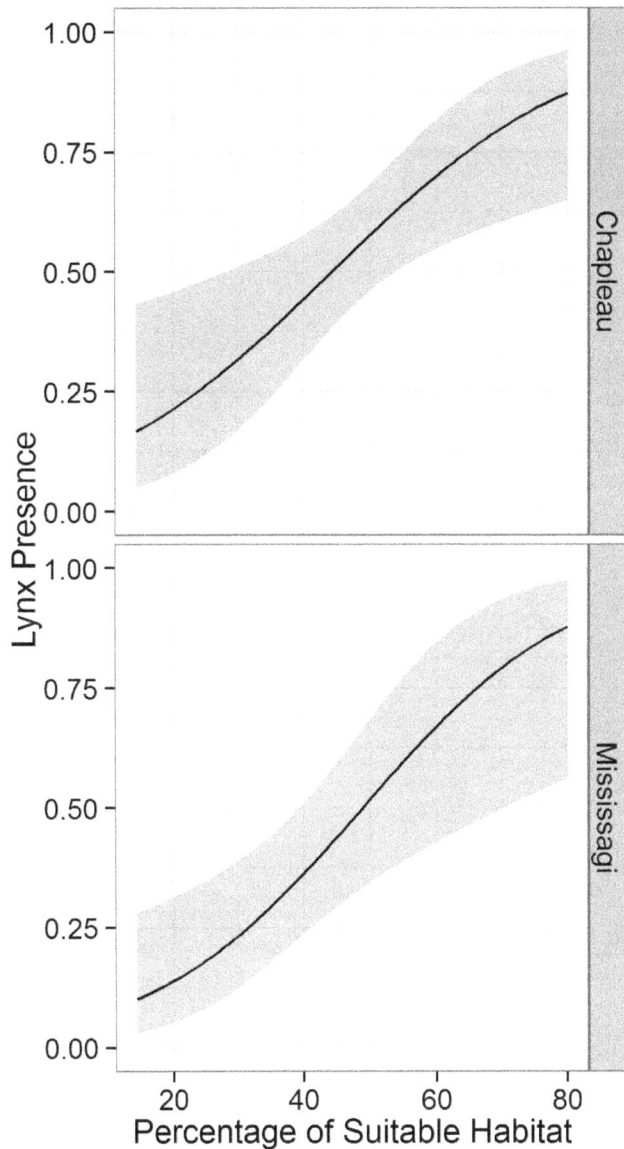

Figure 4. Regression plots for logistic models of Canada lynx occurrence in relation to the proportion of suitable habitat at the 25 km² spatial scale in the Mississagi and Chapleau regions with moderate levels of fragmentation. The shaded area indicates the standard error.

habitat loss is contributing to the northward regression of the species' distribution [26].

Conclusion

Our results highlight the importance of examining habitat fragmentation independently of habitat loss to isolate and understand the impacts of each process [7,8]. While previous research suggests that closely related species, such as bobcats and Iberian lynx, are sensitive to habitat fragmentation [28,29], our results show that habitat loss, not fragmentation, drives occurrence patterns for Canada lynx. The effects of habitat loss and fragmentation may be species-specific, so we recommend that this hypothesis be further evaluated in both specialist and generalist species to improve our understanding of the impacts of these wide-spread processes. This is especially necessary for carnivores, which are considered to be sensitive to both habitat loss and fragmentation [55]. Ultimately, as rates of habitat loss and fragmentation continue to increase on a global scale, this and additional research can improve conservation efforts by ensuring that recovery strategies focus on the appropriate management action.

Supporting Information

Information S1   Comparison of expert option and literature-based models. Table S1. Performance metrics for the expert-opinion and literature based habitat suitability models for Canada lynx occurrence in Ontario, Canada. Receiver operating characteristic was based on 62 presence/absence locations near Temagami, Ontario. Bold text indicates better model performance. **Table S2.** Expert-opinion and literature based model weights for all variables used in the development of the habitat suitability model for Canada lynx in Ontario, Canada. Models were based on a survey using the analytic hierarchy decision-making process to rate the importance of different variables. The expert-opinion model is based on the replies of nine lynx researchers; the literature based model is based on the responses of 4 unbiased observers after having reviewed four research papers on lynx habitat selection. **Figure S1.** Distribution of Canada lynx occurrence across within three landscapes differing in the amount of suitable land cover as determined by a literature-based habitat suitability model in the (A) Chapleau and (B) Mississagi Regions.

Acknowledgments

We thank K. Downing, L. Fahrig, E. Koen, K. Middel, and B. Pond for valuable input and feedback. We also thank all those who participated in the expert-opinion and literature-based surveys. Also thanks to T. Copeland, N. Woodhouse, E. Smith, A. Wilson, D. Ballak, and many volunteers for field support.

Author Contributions

Conceived and designed the experiments: MLH LW DLM JR MJF JB. Performed the experiments: MLH AAW. Analyzed the data: MLH DLM. Contributed reagents/materials/analysis tools: MLH JB AAW LW DLM. Wrote the paper: MLH.

References

1. Lawton J (1993) Range, population abundance and conservation. Trends in Ecology and Evolution 8: 409–413.
2. Gaston KJ (2003) The Structure and Dynamics of Geographic Ranges. New York: Oxford University Press. 266–266 p.
3. Grant MC, Antonovics J (1978) Biology of ecologically marginal populations of Anthozanthum odoratum. I. Phenetics and dynamics. Evolution 32: 822–838.
4. Yackulic CB, Sanderson EW, Uriarte M (2011) Anthropogenic and environmental drivers of modern range loss in large mammals. Proceedings of the National Academy of Sciences of the United States of America 108: 4024–4029.
5. Holt RD, Barfield M (2008) Habitat selection and niche conservatism. Israel Journal of Ecology and Evolution 54: 295–309.
6. Channell R, Lomolino MV (2000) Dynamic biogeography and conservation of endangered species. Nature 403: 84–86.
7. Fahrig L (2003) Effects of habitat fragmentation on biodiversity. Annual Review of Ecology, Evolution, and Systematics 34: 487–515.
8. Fahrig L (2002) Relative Effects of Habitat Loss and Fragmentation on Population Extinction. Journal of Wildlife Management 61: 603–610.
9. Flather CH, Bevers M (2002) Patchy reaction-diffusion and population abundance: The relative importance of habitat amount and arrangement. American Naturalist 159: 40–56.

10. Andrén H (1994) Effects of habitat fragmentation on birds and mammals in landscapes with different proportions of suitable habitat: a review. Oikos 71: 355–366.

11. Swift TL, Hannon SJ (2010) Critical thresholds associated with habitat loss: a review of the concepts, evidence, and applications. Biological reviews of the Cambridge Philosophical Society 85: 35–53.

12. Radford JQ, Bennett AF (2004) Thresholds in landscape parameters: occurrence of the white-browed treecreeper Climacteris affinis in Victoria, Australia. Biological Conservation 117: 375–391.

13. Reunanen P, Mönkkönen M, Nikula A, Hurme E, Nivala V (2004) Assessing landscape thresholds for the Siberian flying squirrel. Ecological Bulletins 51: 277–286.

14. Saunders DA, Hobbs RJ, Margules CR (1991) Biological consequences of ecosystem fragmentation - a review. Conservation Biology 5: 18–32.

15. Swihart RK, Lusk JJ, Duchamp JE, Rizkalla CE, Moore JE (2006) The roles of landscape context, niche breadth, and range boundaries in predicting species responses to habitat alteration. Diversity and Distributions 12: 277–287.

16. Fuller AK, Harrison DJ, Vashon JH (2007) Winter habitat selection by Canada lynx in Maine: prey abundance or accessibility? Journal of Wildlife Management 71: 1980–1986.

17. Murray DL, Boutin S, O'Donoghue M (1994) Winter habitat selection by lynx and coyotes in relation to snowshoe hare abundance. Canadian Journal of Zoology 72: 1444–1451.

18. Vashon JH, Meehan AL, Jakubas WJ, Organ JF, Vashon AD, et al. (2008) Spatial ecology of a Canada lynx population in northern Maine. Journal of Wildlife Management 72: 1479–1487.

19. Homyack JA, Harrison DJ, Litvaitis JA, William B (2006) Quantifying densities of snowshoe hares in Maine using pellet plots. Journal of Wildlife Management 34: 74–80.

20. Litvaitis JA, Sherburne JA, Bissonette JA (1985) Influence of understory characteristics on snowshoe hare habitat use and density. Journal of Wildlife Management 49: 866–873.

21. Aubry KB, Koehler GM, Squires JR (2000) Ecology of Canada lynx in southern boreal forests. In: Ruggiero LF, Aubry KB, Buskirk SW, Koehler GM, Krebs CJ et al., editors. Ecology and Conservation of Lynx in the United States: University Press of Colorado, Niwot, USA. pp. 373–396.

22. Steury T, Murray DL (2004) Modeling the reintroduction of lynx to the southern portion of its range. Biological Conservation 117: 127–141.

23. Murray DL (2000) A geographic analysis of snowshoe hare population demography. Canadian Journal of Zoology 78: 1207–1217.

24. Poole KG (2003) A review of the Canada lynx, Lynx canadensis, in Canada. Canadian Field Naturalist 117: 360–376.

25. Cheng E, Hodges KE, Melo-Ferreira J, Alves PC, Mills LS (2014) Conservation implications of the evolutionary history and genetic diversity hotspots of the snowshoe hare. Molecular Ecology 23: 2929–2942.

26. Koen EL, Bowman J, Murray DL, Wilson PJ (2014) Climate change reduces genetic diversity of Canada lynx at the trailing range edge. Ecography 37: 754–862.

27. Murray DL, Steury TD, Roth JD (2008) Assessment of Canada lynx research and conservation needs in the southern range: another kick at the cat. Journal of Wildlife Management 72: 1463–1472.

28. Ferreras P (2001) Landscape structure and asymmetrical inter-patch connectivity in a metapopulation of the endangered Iberian lynx. Biological Conservation 100: 125–136.

29. Crooks KR (2002) Relative sensitivities of mammalian carnivores to habitat fragmentation. Conservation Biology 16: 488–502.

30. Peers MJL, Thornton DH, Murray DL (2012) Reconsidering the specialist-generalist paradigm in niche breadth dynamics: resource gradient selection by Canada lynx and bobcat. PloS one 7: e51488–e51488.

31. Peers MJL, Wehtje M, Thornton DH, Murray DL (2014) Prey switching as a means of enhancing persistence in predators at the trailing southern edge. Global Change Biology 20: 1126–1135.

32. Perera AH, Balwin DJB (2000) Spatial patterns in the managed forest landscape of Ontario. In: Perera AH, Euler DL, Thomson ID, editors. Vancouver, BC: UBC Press. pp. 74–100.

33. OMNR (2011) Annual Report on Forest Management 2009/10. 106–106 p.

34. Clevenger AP, Wierzchowski J, Chruszcz B, Gunson K (2002) Identifying wildlife habitat linkages and planning mitigation passages. Conservation Biology 16: 503–514.

35. LaRue MA, Nielsen CK (2008) Modelling potential dispersal corridors for cougars in midwestern North America using least-cost path methods. Ecological Modelling 212: 372–381.

36. Mowat G, Slough B (2003) Habitat preference of Canada lynx through a cycle in snowshoe hare abundance. Canadian Journal of Zoology 81: 1736–1745.

37. Hoving CL, Harrison DJ, Krohn WB, Joseph RA, O'Brien M (2005) Broad-scale predictors of Canada lynx occurrence in eastern North America. Journal of Wildlife Management 69: 739–751.

38. Bayne EM, Boutin S, Moses RA (2008) Ecological factors influencing the spatial pattern of Canada lynx relative to its southern range edge in Alberta, Canada. Canadian Journal of Zoology 86: 1189–1197.

39. Burdett CL, Moen RA, Niemi GJ, Mech LD (2007) Defining space use and movements of Canada lynx with Global Positioning System telemetry. Journal of Mammalogy 88: 457–467.

40. Holloway GL, Naylor BJ, Watt WR (2004) Habitat relationships of wildlife in Ontario. Revised habitat suitability models for the Great Lakes-St. Lawrence and Boreal East forests. 110p.–110p. p.

41. Maxie AJ, Hussey KF, Lowe SJ, Middel KR, Pond BA, et al. (2010) A comparison of forest resource inventory, provincial land cover maps and field surveys for wildlife habitat analysis in the Great Lakes – St. Lawrence forest. The Forestry Chronicle 86: 77–86.

42. Bonn A, Schroder B (2001) Habitat models and their transfer for single and multi species groups: a case study of carabids in an alluvial forest. Ecography 24: 483–496.

43. Tischendorf L, Fahrig L (2000) On the usage and measurement of landscape connectivity. Oikos 90: 7–19.

44. Girvetz EH, Greco SE (2007) How to define a patch: a spatial model for hierarchically delineating organism-specific habitat patches. Landscape Ecology 22: 1131–1142.

45. Mowat G, Poole KG, O'Donoghue M (2000) Ecology of lynx in northern Canada and Alaska. In: Ruggiero LF, Aubry KB, Buskirk SW, Koehler GM, Krebs CJ et al., editors: University Press of Colorado, Niwot, USA. pp. 265–306.

46. Walpole AA, Bowman J, Murray DL, Wilson PJ (2012) Functional connectivity of lynx at their southern range periphery in Ontario, Canada. Landscape Ecology 27: 761–773.

47. Jaeger JAG (2000) Landscape division, splitting index, and effective mesh size: new measures of landscape fragmentation. Landscape Ecology 15: 115–130.

48. Girvetz E, Thorne J, Berry A, Jaeger J (2008) Integration of landscape fragmentation analysis into regional planning: A statewide multi-scale case study from California, USA. Landscape and Urban Planning 86: 205–218.

49. Burnham KP, Anderson DR (2002) Model selection and multimodel inference: Springer, New York, USA.

50. Lewis CW, Hodges KE, Koehler GM, Mills LS (2011) Influence of stand and landscape features on snowshoe hare abundance in fragmented forests. Journal of Mammalogy 92: 561–567.

51. Wirsing AJ, Steury TD, Murray DL (2002) A demographic analysis of a southern snowshoe hare population in a fragmented habitat: evaluating the refugium model. Canadian Journal of Zoology 80: 169–177.

52. Roth JD, Marshall JD, Murray DL, Nickerson DM, Steury TD (2007) Geographical gradients in diet affect population dynamics of Canada lynx. Ecology 88: 2736–2743.

53. Barnagaud JY, Devictor V, Jiguet F, Archaux F (2011) When species become generalists: on-going large-scale changes in bird habitat specialization. Global Ecology and Biogeography 20: 630–640.

54. Allard-Duchene A, Pothier D, Dupuch A, Fortin D (2014) Temporal changes in habitat use by snowshoe hares and red squirrels during post-fire and post-logging forest succession. Forest Ecology and Management 313: 17–25.

55. Crooks KR, Burdett CL, Theobald DM, Rondinini C, Boitani L (2011) Global patterns of fragmentation and connectivity of mammalian carnivore habitat. Philosophical Transactions of the Royal Society B-Biological Sciences 366: 2642–2651.

19

Global Drivers and Tradeoffs of Three Urban Vegetation Ecosystem Services

Cynnamon Dobbs[1,2]*, Craig R. Nitschke[2], Dave Kendal[3]

1 School of Botany, The University of Melbourne, Melbourne, Australia, 2 School of Forest Science and Ecosystem, Melbourne School of Land and Environment, The University of Melbourne, Melbourne, Australia, 3 Australian Research Centre for Urban Ecology, Royal Botanic Gardens Melbourne, c/o School of Botany, The University of Melbourne, Melbourne, Australia

Abstract

Our world is increasingly urbanizing which is highlighting that sustainable cities are essential for maintaining human well-being. This research is one of the first attempts to globally synthesize the effects of urbanization on ecosystem services and how these relate to governance, social development and climate. Three urban vegetation ecosystem services (carbon storage, recreation potential and habitat potential) were quantified for a selection of a hundred cities. Estimates of ecosystem services were obtained from the analysis of satellite imagery and the use of well-known carbon and structural habitat models. We found relationships between ecosystem services, social development, climate and governance, however these varied according to the service studied. Recreation potential was positively related to democracy and negatively related to population. Carbon storage was weakly related to temperature and democracy, while habitat potential was negatively related to democracy. We found that cities under 1 million inhabitants tended to have higher levels of recreation potential than larger cities and that democratic countries have higher recreation potential, especially if located in a continental climate. Carbon storage was higher in full democracies, especially in a continental climate, while habitat potential tended to be higher in authoritarian and hybrid regimes. Similar to other regional or city studies we found that the combination of environment conditions, socioeconomics, demographics and politics determines the provision of ecosystem services. Results from this study showed the existence of environmental injustice in the developing world.

Editor: Zoe G. Davies, University of Kent, United Kingdom

Funding: These authors have no support or funding to report.

Competing Interests: The authors have declared no competing interests exist.

* Email: cdobbsbr@gmail.com

Introduction

Urban areas are dynamic and complex landscapes, where socio-ecological processes can deliver ecosystem services across multiple scales [1]. The ecosystem services concept provides a framework that integrates ecology with socioeconomics, creating a transdisciplinary approach for understanding the benefits that can be delivered by nature and the implications of these benefits on human wellbeing [2,3]. Population growth, consumption and governance can all influence the provision of ecosystem services which in turn affect human health, livelihood, culture and equity [4]. This concept is particularly relevant in urban systems where natural resources are under enormous pressure and where the demand for ecosystem services is increasing [5].

Cities differ in their governance, infrastructure, economy and social equity [6]. They also vary in their development, with some cities having high rates of urbanization and uncontrolled population growth while other cities are experiencing declines in population. The social, political and biophysical context of the city shapes how socio-ecological interactions affect the provision of environmental benefits [1]. Quantifying how urban ecosystem services are provided under these different socio-political-biophysical conditions provides a useful framework for understanding how socio-political-biophysical factors influence the provision of ecosystem services.

The structure and composition of urban vegetation influences the provision of ecosystem services. A number of regulating services (e.g. maintenance of air quality, climate regulation, maintenance of soil fertility), cultural services (e.g. aesthetics, sense of place and recreation) and supporting services (e.g. habitat for flora and fauna and space for reproduction) are linked to the patterns of urban vegetation [7–9]. The distribution of vegetation is a consequence of many factors including topography, climate, transportation infrastructure, plant dispersal mechanisms, real estate markets, planning, cultural practices and social preferences [1,9–12]. Kendal et al. [13] found that temperature influences the composition of cultivated trees in urban areas, while both education and income can influence local vegetation structure and composition [13–15]. The relationships between politics and urban vegetation have shown to be characterized by an inequitable distribution, often favouring urban elites over marginalized and deprived groups, either racial or socioeconomic [16–19]. However, a global analysis on how governance is related to the provision of ecosystem services is lacking, especially in relation to the national political context. The national political context influences the practice of participatory governance, local-level management and the prioritization of greening policies [20].

We know little about global patterns of ecosystem services, and research at larger scales has generally been restricted to a single country, region or rural landscape [21]. Urban ecosystem services research has focussed mainly on cities in the United States of America [9,12,22–24] with a few studies in other continents [25–29]. These studies mostly focus on the quantification of ecosystem services. We also know little about the tradeoffs and synergies that occur in the provision of ecosystem services [30], particularly in urban landscapes. When assessing services that represent different ecosystem functions, i.e. regulation, supporting and cultural [7], it is necessary to explore their synergies and tradeoffs [31]. Synergies occur when multiple services are simultaneously enhanced, while tradeoffs occur when one service is enhanced at the cost of reducing another [32]. For example, the provision of services such as recreation, spiritual enhancement and psychological benefits typically all increase when the amount and quality of green space available for urban dwellers increases [33–35]. In contrast, increasing tree cover in parks leads to increases in carbon storage and habitat provision, but could lead to a reduction in recreational services as the space available for sport fields decreases [36].

To our knowledge, no previous studies have explored the global drivers of urban ecosystem services. This research therefore represents one of the first attempts to quantify global urban ecosystem services and the existence of synergies and tradeoffs and their relation with development, climate and governance. The objectives of the study are 1) to test the effect of socio-political factors on the provision of ecosystem services to explore whether patterns previously found at local-levels scale up globally and 2) to explore whether common biophysical, demography and socioeconomic factors can explain the synergies and tradeoffs in ecosystem services. To achieve our objectives we quantified services that represent different ecosystem functions: carbon storage, recreation potential, and habitat provision. Carbon storage helps mitigate climate change at the global scale by offsetting the urban footprint [37], while at the regional and local scale it contributes to improving air quality [12,26]. Habitat provision in the urban landscape is strongly linked to biodiversity and to the well-being of urban inhabitants [23,38–43]. In comparison to carbon and biodiversity, recreation potential has a more local effect as it relates to the provision of space for leisure, contemplation and exercising which has been linked to improve public health [44].

Methods

Urban vegetation extraction

A sample of one hundred cities was selected to represent a diversity of biophysical, socioeconomic, demographic and cultural factors (Table 1; Figure S1; Table S1). Remotely sensed data were used to provide a standardized method to quantify ecosystem services and look for synergies and tradeoffs across a large number of cities. Cities were selected from a global pool where good quality satellite imagery (Landsat 5 TM) was available during the vegetation-growing season between years 2006 to 2011. Cities from tropical regions in Asia and Africa were not included because of cloud cover over the cities. Landsat images were of high resolution (30 m^2 multispectral pixels), which allowed for fine scale analysis. Landsat images are widely used in urban landscape studies [45,46].

Identifying city boundaries is a critical step in any analysis of urban landscapes, and one that is notoriously difficult at large scales [47,48]. The wide range of cities included in the study meant that standardised metadata (e.g. current administrative

boundaries) were not available for all cities, therefore a method that could be applied to all cities was required. Following Schneider and Woodcock [47], the limits of a city were defined as the first area where less than 5% impermeable surface was present in a 200 m wide buffer located at the periphery of the urban area. To test the accuracy of this approach, the discrepancy between the administrative and calculated boundary was calculated for 30% of the cities in the study where administrative data was available. Discrepancy varied from a few square meters to 50 km^2 and was independent of the geographic location of the city, highlighting that that the error associated with our approach was randomly distributed.

The normalized difference vegetation index (NDVI) is an index of living green vegetation [49], and was calculated from the Landsat image of each city, using the red and infrared bands. An unsupervised classification of vegetation and impermeable surface was conducted; however, only vegetation (green cover) was retained for further analysis. To extract the classes representing vegetation the spectral value of 50 vegetation pixels by city were obtained from the NDVI image. The accuracy assessment was obtained using 80 random points within the vegetation class and cross referencing to Google Earth imagery. The Kappa coefficient for the vegetation classification was 0.8, while the user's accuracy corresponded to 75% and the producer's accuracy to 85% [50].

Quantification of ecosystem services

Recreation potential is derived from the vegetated areas that provide space for physical and psychological enjoyment. It includes vegetation in woodlands, grasslands and street trees that occurs in parks, smaller patches of vegetation and/or along streets. The recreation potential service was calculated as the amount of vegetated area per capita [23,51]. The area of vegetation from the NDVI imagery classification within our calculated urban boundary was divided by the population of the city, obtained from the United Nations global report on human settlements [6]. Due to the lack of data on census limits we assumed that the calculated urban boundary is consistent with the census data limits; while the accuracy of this approach may lead to over or underestimations, the error is randomly distributed and standardised across all cities.

Carbon storage in vegetation was calculated using an existing model based on Landsat derived NDVI [52]. This model was built for urban vegetation and has been validated with field data. The method has been previously used for assessing urban forest carbon offsets and quantifying carbon stock across an urban rural gradient in several cities of the United States [53,54]. The model is spatially explicit and calculates carbon storage per pixel (30 m^2) using the function:

$$Carbon(tonnes/pixel) = 0.10702e^{NDVI*0.0194}$$

Habitat potential is a function of vegetation structure at the landscape scale. Vegetation cover was obtained from the NDVI analysis. We recognize that different types of vegetation provide different degrees of habitat quality for floral and faunal guilds however this level of detail was not considered in the study. Areas of structural connectivity were identified using a Morphological Spatial Pattern Analysis (MSPA) available in the free software package GUIDOS (http://forest.jrc.ec.europa.eu/download/software/guidos/). MSPA has been used in studies focussed on assessing the connectivity of ecological habitats in both forested and urban landscapes [38,55,56]. It uses a land cover map of

Table 1. Socioeconomic, political and climatic characteristics for the 100 cities included in this study.

Climate (Köppen classification)	Population	Human Development Index (HDI)	Democracy Index (DI)
Tropical moist (11)	<1 million habitants (11)	Very high HDI (45)	Full democracies (38)
Dry climate (13)	1 to 2 million habitants (32)	High HDI (19)	Flawed democracies (32)
Moist mid latitude with mild winters (60)	2 to 6 million habitants (39)	Medium HDI (22)	Hybrid regimes (14)
Moist mid latitude with cold winters (16)	>6 million habitants (18)	Low HDI (14)	Authoritarian regimes (16)

vegetated/non vegetated areas to classify structural patterns following mathematical morphology methods [57].

The morphological segmentation of binary patterns obtained from an image with vegetated and non-vegetated pixels that produces seven categories according to their size, shape and connectivity.

Our measure of habitat potential is the proportion of area of vegetation larger than 1.44 ha which was classified as 'core' habitat [38]. 'Core' habitat areas are the pixels in patches where the distance to the non-vegetated area was greater than 60 m (2 Landsat pixels). Our definition of core areas include forest, woodland, shrubland and meadow areas and is consistent with broader ecological theory that shows that larger areas with relatively fewer edges are likely to support a wider of species (such as woodland birds [58]). However, we acknowledge that our measure based on NDVI is not a perfect measure of habitat potential, as it will include large areas of mown turf, which may have low ecological value, and exclude narrow linear corridors that may have high habitat potential. To achieve spatial concordance among the calculated ecosystem services across all cities values were standardised to the mean carbon storage per hectare, mean recreation potential per capita and habitat potential as the proportion of the total urban area covered by 'core' patches [59].

Socioeconomic and climatic factors

To explore the relationship between globally quantified ecosystem services with social and climatic factors we selected three social indicators: total population, the Human Development Index (HDI) and the Democracy Index (DI), and three climatic indicators: mean annual temperature, mean annual precipitation and the annual Heat Moisture Index (HMI) [60].

The Human Development Index (HDI) is a measure of development [61]. It is a composite index that combines indicators of life expectancy, educational attainment and income using a geometric mean [61]. The Democracy Index (DI) is an index that combines five metrics of governance: electoral process and pluralism, civil liberties, the functioning of the government, political participation and political culture [62] (Table S1). Values for mean annual temperature and precipitation were obtained from the World Meteorological Organization (http://www.wmo.int/pages/index_en.html) for the period between 1976 and 2005. The Heat Moisture Index (HMI) was calculated for each city following Wang et al. [60] as a ratio between mean annual temperature and mean annual rainfall. The HMI is a measure of the evaporative demand of the atmosphere and represents the aridity of the environment when the interaction between temperature and precipitation is considered.

Statistical analysis

The data were tested for normality using the Kolmogorov-Smirnov statistic. Recreation potential was log transformed to meet the analyses' assumptions of normality. Moran's I was calculated and used to assess if spatial autocorrelation exists within the analysed data [63].

A principal component analysis (PCA) was applied to the standardized data of each ecosystem service and social-climatic variables included in this study [63]. This multivariate data technique uses orthogonal transformation to group sets of correlated variables into principle components which are sets of linearly uncorrelated variables [64]. A Varimax rotation was used to improve result interpretation. The number of components was selected to provide the most interpretable solution with eigenvalues greater than one.

Following the PCA a Bayesian regression was used to assess the effect size of the socio-climatic variables on each of the ecosystem services considered in the study. This method uses a Markov Chain Monte Carlo technique to fit generalized linear models. Because of the lack of well established relationships between services and indicators, non-informative priors were used. The posterior distribution for each service model was simulated using a Markov chain Monte Carlo method. For each model we simulated 10000 iterations and a burn-in size of 2000, thinning the results by a factor of 1, reaching convergence [63]. For the socio-climatic parameters we report the 2.5% and the 97.5% credible interval of simulated posterior values, which represents the likely range of parameter values. We can be confident of a significant effect where the credible interval does not overlap with zero [65].

Once relations between ecosystem services were established, ANOVAs were used to test for differences between the categories of cities shown in Table 1, using Tukey's HSD multi-comparison test [63]. These are commonly used categories for classifying cities [6,61,62,66]. Synergies and tradeoffs were identified using pairwise Spearman correlations for all ecosystem services for all the cities together and separated by bio-socio-political relevant factors. Significance of the correlations was assessed at $p < 0.001$, $p < 0.05$ and $p < 0.01$.

Results

Quantification of global ecosystem services

There was high variability in green cover. The city with the lowest amount of vegetation cover was Calcutta (0.4%), while Winnipeg had the highest (63.1%). The percent of green cover was normally distributed (D = 0.99, p<0.001), with a mean of 32.6% (±12.1) with the majority of cities having between 20% and 40% green cover. Moran's I showed no indication of spatial autocorrelation (Z = 0.44, $p = 0.65$). The average recreation potential reached 8.9 m^2 per capita with the highest value (44 m^2) for Winnipeg and the minimum value for Istanbul (0.4 m^2). Carbon storage averaged 39 tonnes per ha, with the lowest value in Khartoum (0.2 tonnes/ha) and the maximum value in Paris (161 tonnes/ha). The average core habitat area provided within city habitat patches was 53%; the highest habitat potential was found for Montreal (98.3%) and the lowest for Bombay (2.4%). Results by city are detailed in Table S2.

Relationships between ecosystem services and urban climatic and development characteristics

The PCA was able to detect relationships between urban ecosystem services and development, political and climatic characteristics for the one hundred cities (Figure 1). A two-component solution provided the most interpretable result. Two components explained ~47% of the variance within the data (Table S3). Principal component 1 explained 31.6% of the variance in the data. The variables that loaded strongly (>0.4) on the first component were recreation potential, temperature, HDI and DI (Figure 1). Habitat provision, rainfall, HMI and population loaded strongly (>0.4) on the second component; this eigenvector explained 14.9% of the variance (Figure 1).

The PCA identified relationships between ecosystem services and some urban characteristics; however, to quantify the effect of these characteristics on services Bayesian regression was used to explore the size and direction of the effect for each urban characteristic (Figure 2; Table S3; Table S4). The association between carbon storage, HDI, DI and temperature, suggests that carbon storage tends to increase in wealthy, educated and democratic cities from cooler climates (Figure 1). The Bayesian analysis resulted in no significant effects detected for carbon storage (Figure 2); however, there were non-significant trends consistent with the PCA results suggesting a weak relationship with temperature and DI.

Recreation potential had a negative relationship with temperature, suggesting that cities from warmer climates tend to have a lower provision of this service. Higher recreation potential tends to occur in more democratic and more highly developed cities. Bayesian analysis showed that DI was the largest (positive) predictor of recreation potential, with population a significant (negative) predictor (Figure 2). There was also a non-significant trend suggesting that temperature may also be negatively related to recreation potential.

Habitat provision was positively related to HMI, suggesting that cooler and wetter cities have a lower provision of this service.

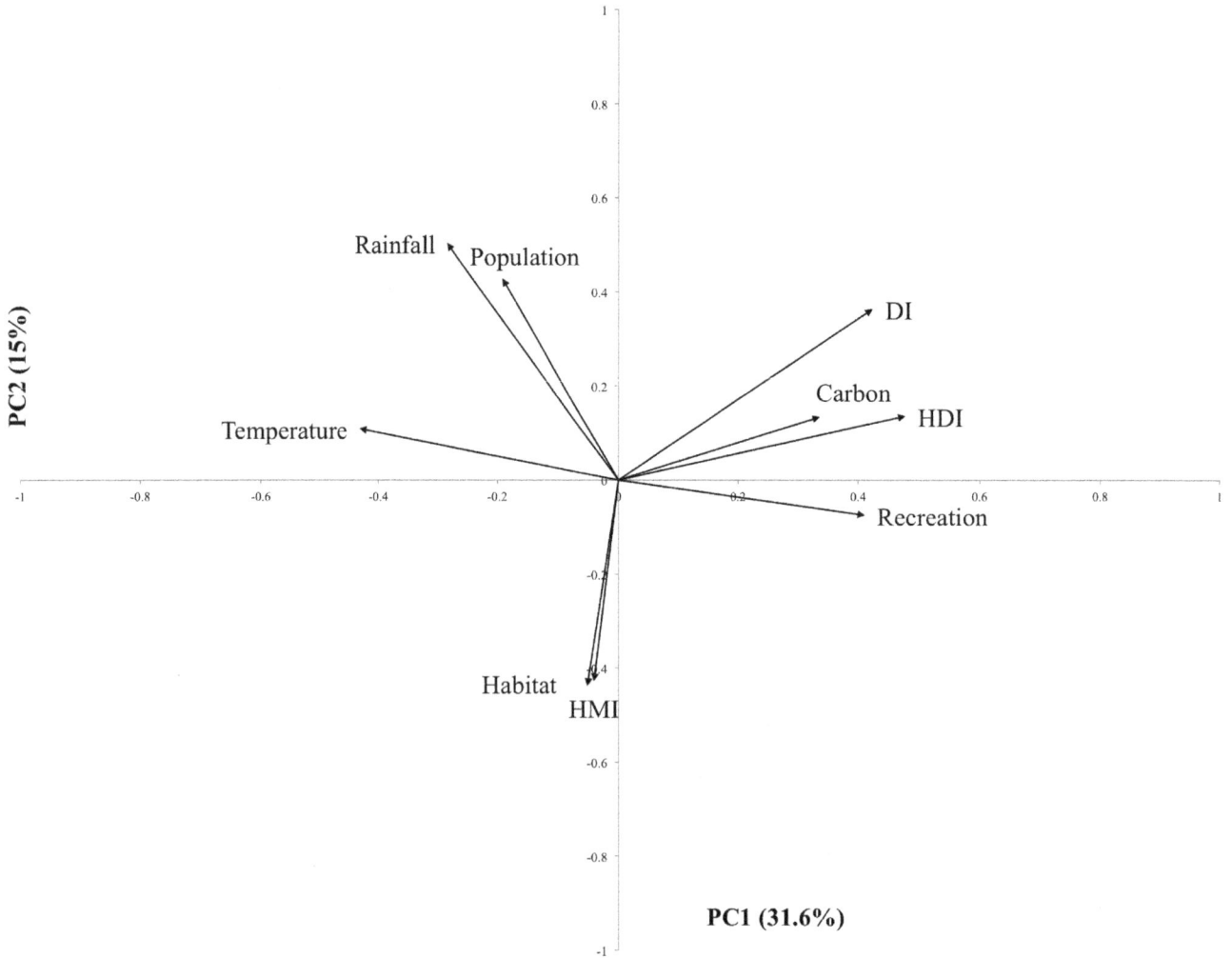

Figure 1. Ordination of the first and second standardized principal component for each ecosystem services and main drivers for 100 cities. The value of PC1 (Principal Component1) and PC2 (Principal Component 2) for the cities was standardised in order to more clearly show their location in the orthogonal space. The length of the arrow is an indication of the strength of the socio-political-climate variables and the ecosystem service in each PC. HMI: Heat Moisture Index, DI: Democracy Index, HDI: Human Development Index.

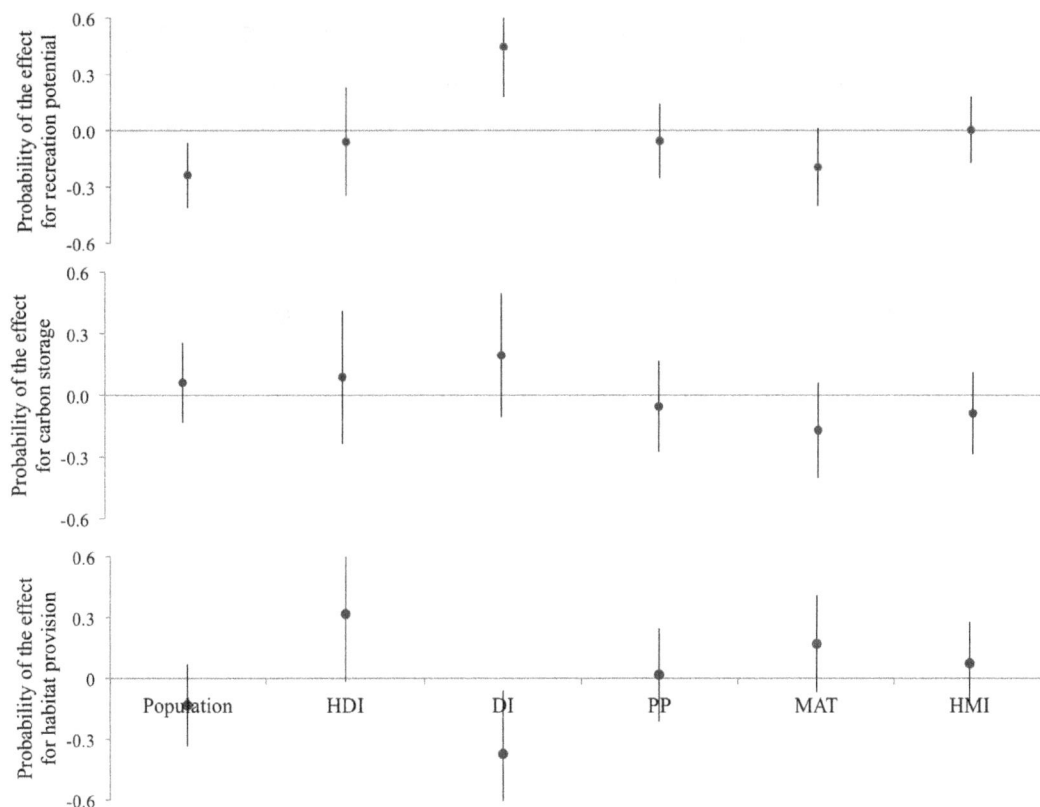

Figure 2. Bayesian models for three ecosystem services. Values overlapping zero imply a consistent effect of the bio-socio-political factor in the probability of having a positive or negative effect in the provision of ecosystem services. HDI: Human Development Index, DI: Democracy Index, PP: Annual precipitation, MAT: Mean Annual Temperature, HMI: Heat Moisture index.

Habitat provision had a negative relationship with population, suggesting that increases in population leads to fragmentation of urban vegetation (Figure 1). The Bayesian analysis showed that habitat potential was related (negatively) with DI (Figure 2), suggesting that more democratic cities tend to have lower habitat potential (more fragmented and less connected landscapes). A positive but non-significant trend was also detected between HDI and habitat potential.

The results from the ANOVAs and multiple comparison tests for the provision of ecosystem services by categories of city population, democratic regime and climate are summarised in Table 2. Carbon storage was higher in full than flawed democracies and authoritarian regimes. The analysis identified thresholds where the provision of ecosystem services declines. Cities with less than one million people tended to provide more recreation potential than larger cities, with megacities providing the least amount of recreational space. A higher provision of recreation service occurred in full democracies, and in continental cities over tropical and Mediterranean cities. Interestingly, authoritarian regimes tend to have a higher provision of habitat suggesting a higher degree of connectivity.

Synergies and tradeoffs of urban ecosystem services

There was a synergy between recreation potential and carbon storage across all cities (Table 3), which were positively and moderately correlated (Spearman's $\rho > 0.5$, $p < 0.001$). There were no significant correlations between habitat provision and recreation, or between carbon storage and habitat provision. Within city

categories, the synergy between recreation and carbon is consistent for all population sizes with cities over 1 million people showing stronger correlations (Spearman's $\rho \approx 0.6$) than cities under 1 million (Spearman's $\rho \approx 0.45$). This synergy is also consistent for different political regimes except for full democracies. The synergy is strongest in Desert and Mediterranean cities and does not hold for continental and tropical cities. Synergies between recreation potential and habitat provision are only present in cities with under 1 million inhabitants and in democracies. A weak but significant synergy ($p < 0.01$) was found in Mediterranean cities between carbon storage and habitat provision (Table 3).

Discussion

Three ecosystem services were characterized for one hundred cities, revealing that the global distribution of ecosystem services are shaped by both development factors and climate. By integrating NDVI based land cover information with development indicators we demonstrated that anthropogenic variables do influence the provisioning of ecosystem services within urban systems [32,67,68]. Population and political factors more directly influenced recreation potential and habitat potential, while carbon storage was influenced by political regime and temperature. Similar to other urban studies, ecosystem services were influenced by both the natural environment (climate) and by demographics, socioeconomics and governance [14,69–71]. The congruence with local and regional studies demonstrates that these aforementioned relationships scale up to the global scale and highlight the need to

Table 2. Significant differences (ANOVA) in the provision of ecosystem services using categories from significantly influential urban characteristics.

		Recreation potential (m^2 per capita)	Carbon storage (kg/ha)	Habitat provision (%)
	r^2	0.16	0.01	0.02
Population	<1 million	16.5^a (2.6–44)	39.5^a (0.7–81)	45^a (5–74)
	1 to 2 million	8.6^b (1.7–31)	42.3^a (0.7–151)	54^a (4–93)
	2 to 6 million	7.5^b (0.6–33)	38.4^a (0.2–149)	56.2^a (4–98)
	>6 million	3.2^b (0.4–7.6)	27.7^a (5–161)	49.3^a (2–93)
	p-value	0.0008	ns	ns
Democracy Index	r^2	0.29	0.14	0.04
	Authoritarian	3.4^a (0.8–6.5)	21.3^b (0.2–59)	63.4^a (7–93)
	Hybrid	3.3^a (0.4–8)	32^{ab} (3.1–114)	52.6^b (4–78)
	Flawed democracy	5.6^a (0.7–31)	31.5^b (0.7–161)	52.9^b (2–90)
	Full democracy	13.9^b (1–44)	55.5^a (3.4–151)	48.9^b (4–78)
	p-value	<0.0001	0.002	0.01
Climate	r^2	0.10	0.03	0.02
	Tropical	3.3^a (0.6–8)	25.5^a (0.7–94.1)	58.2^a (2–88)
	Desert	8.5^{ab} (0.6–29)	31.7^a (0.2–84)	61.5^a (7–90)
	Mediterranean	7.4^a (0.4–33)	40.8^a (1.9–161)	51.6^a (4–93)
	Continental	13.7^b (3–44)	46.5^a (0.7–103)	49.2^a (4–98)
	p-value	0.01	ns	ns

Values label with the different letter a imply significant differences among categories of analysis for Tukey's HSD comparison test.

consider socio-political-environmental dimensions when developing urban areas to achieve sustainable development goals.

Our study found that recreational potential is lower in cities with more than 1 million inhabitants, which is consistent with trends observed in some European cities [51]. Recreation potential also decreased with changes in governance as cities in countries governed by non-democratic regimes had a lower provision of this service. It is well understood that developing countries have social inequalities caused by urbanization [72], and this study highlights the existence of environmental inequalities that may be exacerbate by urbanization. Of the 100 cities studied, only 26% of the cities had more green cover person than World Health Organization (WHO) recommendations of 9 m^2 per capita [73] and only 12% of the cities met the green space per capita of 20 m^2 per capita [74,75]. The only cities that reached the WHO recommendation were mid-size cities (1 to 6 million people) that are predominantly located in North America and South Africa, with the exception of

Table 3. Spearman correlations between different ecosystem services for all the cities and by categories of population, democracy and climate.

		Recreation vs. Carbon	Recreation vs. Habitat	Carbon vs. Habitat
All cities		0.53***	0.09	0.02
Population	Less 1 million	0.46*	0.47*	0.009
	1 to 2 million	0.64 ***	0.1	0.05
	2 to 6 million	0.59***	0.24	0.11
	more than 6 million	0.62**	−0.12	−0.14
Democracy	Authoritarian regimes	0.47**	0.07	−0.04
	Hybrid regimes	0.66**	−0.11	−0.02
	Flawed Democracy	0.63***	0.25*	0.17
	Full Democracy	0.12	0.36*	0.15
Climate	Tropical	0.18	0.3	−0.19
	Desert	0.64**	0.009	−0.1
	Mediterranean	0.58***	0.12	0.19*
	Continental	0.27	0.24	0.05

Fisher significant test: *p-value<0.01, **p-value<0.05, ***p-value<0.001.

Bismarck, Oklahoma City and Winnipeg all located in North America.

Cities with higher HDI and under democratic regimes were found to provide more green space for their inhabitants, which may be due to the increased demand for environmental quality by residents [76]. This group included cities in Canada, which have strong environmental and urban forestry programs and policies at both the national and local level. These policies promote an increase in urban green space and street tree plantings and the maintenance of conservation areas [77]. At the opposite end of the spectrum are cities with low HDI, where the main policies at the country and local level are mainly related to socioeconomics; in addition, many of these countries have poor institutional capacity and insufficient budgets to deliver environmental policies [78,79].

Provision of the carbon storage service was found to be strongly influence by climate, though HDI and DI (in interaction with HDI) were also important factors. Cities with the highest provision of this service were mostly from continental biomes (Frankfurt, Paris, Prague), while the lowest provisioning of this service in its majority occurred in cities located in tropical to desert biomes (Ulaanbaatar, Sana'a, Pyonyang). Certain cities however were outliers, interestingly Mendoza, Phoenix and Las Vegas have been able to increase carbon storage despite being located in desert biomes [80,81]. Humans have greatly increased the number of trees within cities such as Phoenix, U.S.A. and Mendoza, Argentina, despite climatic limitations, thereby increasing the provision of this service. In general, more affluent and democratic cities are more likely to have a greater biomass of vegetation that can sequester more carbon, either by maintaining larger patches of vegetation and, or in the case of many continental cities, larger tree populations in streets.

Habitat potential was mainly related to HDI and governance, which is consistent with other studies (e.g. Schwarz 2010; Huang et al. 2007). Cities governed by flawed democracies, authoritarian and hybrid regimes from low-income countries tend to be more compact, with vegetation restricted to peri-urban areas and consequently have low levels of fragmentation. This might reflect the effect of motorization with developed countries with full democracies as they tend to be sprawling cites characterized by high levels of vehicle ownership and the associated transportation infrastructure required to facilitate commuting by vehicles [71]. In addition, control over land ownership under socialist or communist regimes tend to result in cities that are less fragmented [71], therefore maximizing habitat potential. When a city is governed within a full democratic regime, habitat potential is likely to be reduced. This is particularly apparent in cities where urban development happened between the 18th and 19th century under European colonization and where high rates of sprawling and dispersed urbanization are still prevalent (e.g. U.S.A. and Australia; [71]).

The delivery of ecosystem services varies according to city context. In general, megacities provide low levels of ecosystem services; mid-size cities provided average levels of ecosystem services, while cities with less than 1 million inhabitants typically had higher levels of recreation potential. Fuller and Gaston [51] found the same trend in European cities which highlights that a common signal exists at the global scale. The level of democracy within the country the city is located in affected both recreation potential and carbon storage services positively but had a negative effect on habitat provision. Overall this suggests that inequalities in the provision of ecosystem services exists between less and more developed countries which further supports our thesis that environmental inequalities found within cities maybe relevant at the global scale [17,83].

The cities that have the highest values for the three services combined in this study were Prague, Paris and Frankfurt, old European cities of very high development level under full democracy. While the lowest values provided for the combined services were Ulaanbaatar, Buenos Aires and Tegucigalpa with medium level of development within flawed and hybrid democracies.

The analysis revealed that synergies between cultural and regulation functions exist as do tradeoffs with some supporting services. Synergies and tradeoffs of this nature have also been found for a variety of other land uses [29,31]. Relationships between ecosystem services were not linear and varied according to the combination of socioeconomic and political characteristics of the urban ecosystem which is consistent with findings of global coastal ecosystem services [84]. Unsurprisingly, there was a moderately strong synergy between recreation and carbon across most cities as both services are related to the amount of vegetation in cities. However, the synergistic relation between recreation and carbon is very weak for full democracies, which might be due to the existence of parks with relatively few trees, and a relatively larger population of trees located along streets. This is confirmed by the positive relation between recreation and habitat and the weak correlation with habitat and carbon. The relation between recreation and carbon is also weaker in tropical and continental cities. Tropical cities have smaller vegetated areas that are primarily covered by trees, while continental cities have larger vegetated areas cover by fewer trees. There was a substantial variability in the habitat and carbon data however and the power of the analysis may be improved by increasing the sampled size of cities within each climate classification.

The analysis of ecosystem service provision for a hundred cities has its limitations, which need to be acknowledged. City boundaries were difficult to obtain for a wide range of cities; therefore, errors may exist in the estimation of the services. However, for the relationships between democracy, development and climate our methodology allows for a robust assessment. A better estimation of recreation potential could have been obtained if we had further details on the structure and composition of these spaces along with the use of these spaces by people. The resolution of Landsat imagery however was not fine enough to achieve this, and as finer resolution imagery is difficult to source for a large proportion of cities included in this study this level of analysis could not be conducted. The estimation of carbon stored through a model that uses NDVI is commonly used in natural areas and some urban areas [52–54], where the outcomes are biased by model performance and the calculation of NDVI; we at least had control over biases and errors in the latter. The habitat potential estimation was not intended to assess the functionality of each vegetation patch and therefore is a coarse metric of habitat. We feel however that the results from this study provide a level of precision that is consistent between cities and our results are consistent with the findings of other urban studies conducted at finer scales.

Further research should include the addition of indicators that can represent cultural background and the legacy of historic development such as the effects of colonization, wars, ethnic diversity, industrialization, planning regulation and infrastructure development, among others [71]. A temporal analysis may also be able to shed light on different urban morphological trajectories and their relationships with ecosystem services. The inclusion of other ecosystem services would reveal more about how the context of the city affects the provision of ecosystem services; however, finding available information for a large range of cities is problematic. Existing standardised global datasets, such as the one developed in this study, are useful to explore the tradeoffs or synergies between services, along with finer scale indicators of biodiversity and other services.

Conclusion

Cities are areas of human agglomeration that depend on natural resources for the maintenance of human wellbeing. This study has identified that a relationship exists between the bio-socio-political context and the provision of ecosystem services. Cities in countries with democratic systems and more developed economies tend to provide more ecosystem services to their inhabitants; in theory, this should promote improved human wellbeing. This relation becomes more evident for the cultural and provisioning services included in this study, while regulating services such as carbon storage are primarily driven by biophysical conditions followed by social context. The context of the city also influences the synergies and tradeoffs between ecosystem services. This highlights that improvements in economic conditions may not maximise and can hinder the provision of ecosystem services. As a global city scale analysis, this study was able to identify the existence of environmental inequalities according to political, economic and demographic context, which suggests further research should explore these relations within and across cities. Understanding the synergies between services and social and environmental context should ameliorate the development of environmental and social inequalities that are typical of urbanization. The relationship between ecosystem services and bio-socio-political context provides a key understanding of the influential factors that urban planning and policy making impinge upon and thus provide insights for creating liveable, sustainable, and resilient cities globally.

Supporting Information

Figure S1   Map of studied cities.

Table S1   List of cities included in this study. Details on population, Human Development Index and Democracy Index by each city are provided.

Table S2   Three urban forest ecosystem services for one hundred cities included in the study.

Table S3   Scores for principal components and their respective eigenvalues.

Table S4   Mean, 2.5% and 97.5% confidence intervals for estimated probabilities of the effect of socioeconomic, political and climate variable using Bayesian regression on each studied ecosystem services.

Acknowledgments

The authors would like to thank Jorge Aubad for early revisions of this manuscript and the two anonymous reviewers that greatly improve this manuscript.

Author Contributions

Conceived and designed the experiments: CD CN DK. Performed the experiments: CD. Analyzed the data: CD. Contributed reagents/materials/analysis tools: CD CN DK. Contributed to the writing of the manuscript: CD CN DK.

References

1. Grimm NB, Faeth SH, Golubiewski NE, Redman CL, Wu J, et al. (2008) Global change and the ecology of cities. Science 319: 756–760.
2. Millenium Ecosystem Assessment (2003) Introduction and Conceptual Framework. Ecosystems and Human Well-being: A framework for Assessment. Island Press. 26–48.
3. Daily G (1997) What are ecosystem services? Nature's Services: Societal Dependance on Natural Ecosystems. Washington D.C: Island Press.
4. Millenium Ecosystem Assessment (2003) Ecosystems and human well being. Ecosystems and Human Well-being: A framework for Assessment. Washington D.C: Island Press. 85–106.
5. Bastian O, Haase D, Grunewald K (2011) Ecosystem properties, potentials and services – The EPPS conceptual framework and an urban application example. Ecol Indic 21: 7–16.
6. UN-Habitat (2011) Cities and Climate Change: Global report on human settlements. London, Washington D.C: EarthScan.
7. De Groot RS, Wilson MA, Boumans RM (2002) A typology for the classification, description and valuation of ecosystem functions, goods and services. Ecol Econ 41: 393–408.
8. Nowak DJ, Walton JT, Stevens JC, Crane DE, Hoehn RE (2008) A ground-based method of assessing urban forest structure and ecosystem services. Arboric Urban For 34: 347–358.
9. Dobbs C, Escobedo FJ, Zipperer WC (2011) A framework for developing urban forest ecosystem services and goods indicators. Landsc Urban Plan 99: 196–206.
10. Alberti M, Marzluff JM, Sculenberger E, Bradley G, Ryan C, et al. (2003) Integrating humans into ecology: Opportunities and challenges for studying urban ecosystems. Bioscience 53: 1169–1179.
11. Andersson E (2006) Urban Landscapes and Sustainable Cities. Ecol Soc 11. Available: http://www.ecologyandsociety.org/vol11/iss1/art34/.
12. Escobedo FJ, Kroeger T, Wagner JE (2011) Urban forests and pollution mitigation: analyzing ecosystem services and disservices. Environ Pollut 159: 2078–2087.
13. Escobedo FJ, Nowak DJ, Wagner J, De la Maza CL, Rodríguez M, et al. (2006) The socioeconomics and management of Santiago de Chile's public urban forests. Urban For Urban Green 4: 105–114.
14. Kinzig AP, Warren PS, Martin CA, Hope D, Katti M (2005) The Effects of Human Socioeconomic Status and Cultural Characteristics on Urban Patterns of Biodiversity. Ecol Soc 10.
15. Szantoi Z, Escobedo FJ, Wagner J, Rodriguez JM, Smith S (2012) Socioeconomic Factors and Urban Tree Cover Policies in a Subtropical Urban Forest. GIScience Remote Sens 49: 428–449.
16. Heynen NC, Lindsey G (2003) Correlates of urban forest canopy cover: implications for local public works. Public Work Manag Policy 8: 33–47.
17. Heynen NC, Perkins HA, Roy P (2006) The political ecology of uneven urban green space: The impact of political economy on race and ethnicity in producing environmental inequality in Milwaukee. Urban Aff Rev 42: 3–25.
18. Kitchen L (2013) Are Trees Always "Good"? Urban Political Ecology and Environmental Justice in the Valleys of South Wales. Int J Urban Reg Res 37: 1968–1983.
19. Perkins H, Heynen NC, Wilson J (2004) Inequitable access to urban reforestation: the impact of urban political economy on housing tenure and urban forests. Cities 21: 291–299.
20. Wilkinson C, Sendstad M, Parnell S, Schewenius M (2013) Urbanization, Biodiversity and Ecosystem Services: Challenges and Opportunities. Elmqvist T, Fragkias M, Goodness J, Güneralp B, Marcotullio PJ, et al., editors Dordrecht: Springer Netherlands.
21. Eigenbrod F, Armsworth PR, Anderson BJ, Heinemeyer A, Gillings S, et al. (2010) The impact of proxy-based methods on mapping the distribution of ecosystem services. J Appl Ecol 47: 377–385.
22. Nowak DJ, Crane DE (2002) Carbon storage and sequestration by urban trees in the United States. Environ Pollut 116: 381–389.
23. Tratalos JA, Fuller RA, Warren PH, Davies RG, Gaston KJ (2007) Urban form, biodiversity potential and ecosystem services. Landsc Urban Plan 83: 308–317.
24. Edmondson JL, Davies ZG, McCormack SA, Gaston KJ, Leake JR (2014) Land-cover effects on soil organic carbon stocks in a European city. Sci Total Environ 472: 444–453.
25. Tratalos JA, Fuller RA, Evans KL, Davies RG, Newson SE, et al. (2007) Bird densities are associated with household densities. Glob Chang Biol 13: 1685–1695.
26. Davies ZG, Edmondson JL, Heinemeyer A, Leake JR, Gaston KJ (2011) Mapping an urban ecosystem service: quantifying above-ground carbon storage at a city-wide scale. J Appl Ecol 48: 1125–1134.
27. Cilliers S, Cilliers J, Lubbe R, Siebert S (2012) Ecosystem services of urban green spaces in African countries-perspectives and challenges. Urban Ecosyst 16: 681–702.

28. Ernstson H, Sörlin S, Elmqvist T (2009) Social Movements and Ecosystem Services – the Role of Social Network Structure in Protecting and Managing Urban Green Areas in Stockholm. Ecol Soc 13.

29. Haase D, Schwarz N, Strohbach MW, Kroll F, Seppelt R (2012) Synergies, Trade-offs, and Losses of Ecosystem Services in Urban Regions: an Integrated Multiscale Framework Applied to the Leipzig-Halle Region, Germany. Ecol Soc 17.

30. Millenium Ecosystem Assessment (2003) Drivers of change in ecosystems and their services. Washington D.C: Island Press.

31. Raudsepp-Hearne C, Peterson GD, Bennett EM (2010) Ecosystem service bundles for analyzing tradeoffs in diverse landscapes. Proc Natl Acad Sci U S A 107: 5242–5247.

32. Bennett EM, Peterson GD, Gordon LJ (2009) Understanding relationships among multiple ecosystem services. Ecol Lett 12: 1394–1404.

33. Fuller RA, Irvine KN, Devine-wright P, Warren PH, Gaston KJ (2007) Psychological benefits of greenspace increase with biodiversity. Biol Lett 3: 390–394.

34. Tzoulas K, Korpela K, Venn S, Yli-pelkonen V, Ka A, et al. (2007) Promoting ecosystem and human health in urban areas using Green Infrastructure: A literature review. Landsc Urban Plan 81: 167–178.

35. Jim CY, Chen WY (2009) Ecosystem services and valuation of urban forests in China. Cities 26: 187–194.

36. Bjerke T, Østdahl T, Thrane C, Strumse E (2006) Vegetation density of urban parks and perceived appropriateness for recreation. Urban For Urban Green 5: 35–44.

37. McDonald RI (2008) Global urbanization: can ecologists identify a sustainable way forward? Front Ecol Environ 6: 99–104.

38. Tannier C, Foltête J, Girardet X (2012) Assessing the capacity of different urban forms to preserve the connectivity of ecological habitats. Landsc Urban Plan 105: 128–139.

39. Aronson MFJ, La Sorte FA, Nilon CH, Katti M, Goddard MA, et al. (2014) A global analysis of the impacts of urbanization on bird and plant diversity reveals key anthropogenic drivers. Proc Biol Sci 281: 20133330.

40. Luck GW, Davidson P, Boxall D, Smallbone L (2011) Relations between Urban Bird and Plant Communities and Human Well-Being and Connection to Nature. Conserv Biol 25: 816–826.

41. Mayer FS, Ã CMF (2005) The connectedness to nature scale: A measure of individuals ' feeling in community with nature. J Environ Psychol 24: 503–515.

42. Keniger LE, Gaston KJ, Irvine KN, Fuller RA (2013) What are the benefits of interacting with nature? Int J Environ Res Public Health 10: 913–935.

43. Zhang JW, Howell RT, Iyer R (2014) Engagement with natural beauty moderates the positive relation between connectedness with nature and psychological well-being. J Environ Psychol 38: 55–63.

44. Pataki DE, Alig RJ, Fung AS, Golubiewski NE, Kennedy CA, et al. (2006) Urban ecosystems and the North American carbon cycle. Glob Chang Biol 12: 1–11.

45. Van de Voorde T, Vlaeminck J, Canters F (2008) Comparing Different Approaches for Mapping Urban Vegetation Cover from Landsat ETM+ Data: A Case Study on Brussels. Sensors 8: 3880–3902.

46. Zhu Z, Woodcock CE, Rogan J, Kellndorfer J (2011) Assessment of spectral, polarimetric, temporal, and spatial dimensions for urban and peri-urban land cover classification using Landsat and SAR data. Remote Sens Environ 117: 72–82.

47. Schneider A, Woodcock CE (2008) Compact, Dispersed, Fragmented, Extensive? A Comparison of Urban Growth in Twenty-five Global Cities using Remotely Sensed Data, Pattern Metrics and Census Information. Urban Stud 45: 659–692.

48. Potere D, Schneider A, Angel S, Civco L (2009) Mapping urban areas on a global scale: which of the eight maps now available is more accurate? Int J Remote Sens 30: 37–41.

49. Tucker CJ (1979) Red and photographic infrared linear combinations for monitoring vegetation. Remote Sens Environ 8: 127–150.

50. Congalton RG, Green K (2009) Assessing the Accuracy of Remotely Sensed Data: Principles and Practices. 2nd Editio. Boca Raton, FL: CRC/Taylor & Francis.

51. Fuller RA, Gaston KJ (2009) The scaling of green space coverage in European cities. Biol Lett 5: 352–355.

52. Myeong S, Nowak DJ, Duggin MJ (2006) A temporal analysis of urban forest carbon storage using remote sensing. Remote Sens Environ 101: 277–282.

53. Poudyal NC, Siry JP, Bowker JM (2010) Urban forests' potential to supply marketable carbon emission offsets: A survey of municipal governments in the United States. For Policy Econ 12: 432–438.

54. Hutyra LR, Yoon B, Alberti M (2011) Terrestrial carbon stocks across a gradient of urbanization: a study of the Seattle, WA region. Glob Chang Biol 17: 783–797.

55. Saura S, Vogt P, Velázquez J, Hernando A, Tejera R (2011) Key structural forest connectors can be identified by combining landscape spatial pattern and network analyses. For Ecol Manage 262: 150–160.

56. Wickham JD, Riitters KH, Wade TG, Vogt P (2010) A national assessment of green infrastructure and change for the conterminous United States using morphological image processing. Landsc Urban Plan 94: 186–195.

57. Vogt P, Riitters KH, Estreguil C, Kozak J, Wade TG, et al. (2006) Mapping Spatial Patterns with Morphological Image Processing. Landsc Ecol 22: 171–177.

58. Begon M, Townsend C, Harper J (2006) Ecology: from individuals to ecosystems. 4th Editio. Blackwell Publishing Ltd.

59. Naidoo R, Balmford a, Costanza R, Fisher B, Green RE, et al. (2008) Global mapping of ecosystem services and conservation priorities. Proc Natl Acad Sci U S A 105: 9495–9500.

60. Wang T, Hamann A, Yanchuk A, O'Neill GA, Aitken SN (2006) Use of response functions in selecting lodgepole pine populations for future climates. Glob Chang Biol 12: 2404–2416.

61. UNDP (2011) Human Development Report 2011 A Better Future for All. New York.

62. Economist Intelligence Unit (2011) Democracy index 2011: Democracy under stress. Available: www.eiu.com.

63. Fortin M-J, James PM a., MacKenzie A, Melles SJ, Rayfield B (2012) Spatial statistics, spatial regression, and graph theory in ecology. Spat Stat 1: 100–109.

64. Demšar U, Harris P, Brunsdon C, Fotheringham a S, McLoone S (2013) Principal Component Analysis on Spatial Data: An Overview. Ann Assoc Am Geogr 103: 106–128.

65. McMahon SM, Diez JM (2007) Scales of association: hierarchical linear models and the measurement of ecological systems. Ecol Lett 10: 437–452.

66. Bowler DE, Buyung-ali L, Knight TM, Pullin AS (2010) Urban greening to cool towns and cities: A systematic review of the empirical evidence. Landsc Urban Plan 97: 147–155.

67. Ellis EC, Ramankutty N (2008) Putting people in the map: anthropogenic biomes of the world. Front Ecol Environ 6: 439–447.

68. Mikkelson GM, Gonzalez A, Peterson GD (2007) Economic inequality predicts biodiversity loss. PLoS One 2: e444.

69. Angel S, Sheppard SC, Civco DL, Buckley R, Chabaeva A, et al. (2005) The Dynamics of Global Urban Expansion. Washington D.C: Transport and Urban Development Department, The World Bank.

70. Hope D, Gries C, Zhu W, Fagan WF, Redman CL, et al. (2003) Socioeconomics drive urban plant diversity. P Natl Acad Sci-Biol 100: 8788–8792.

71. Huang J, Lu XX, Sellers JM (2007) A global comparative analysis of urban form: Applying spatial metrics and remote sensing. Landsc Urban Plan 82: 184–197.

72. United Nations (2011) Are we building competitive and liveable cities? Guidelines for developing eco-efficient and socially inclusive infrastructure. United Nations.

73. Kuchelmeister G (1998) Urban forestry in the Asia-Pacific Region: Status and prospects. Rome.

74. Sukopp H, Numata M, Huber A (1995) Urban Ecology as the Basis of Urban Planning. The Hague: SPB Academic Pub.

75. Wang X-J (2009) Analysis of problems in urban green space system planning in China. J For Res 20: 79–82.

76. Farzin YH, Bond CA (2006) Democracy and environmental quality. J Dev Econ 81: 213–235.

77. Conway TM, Urbani L (2007) Variations in municipal urban forestry policies: A case study of Toronto, Canada. Urban For Urban Green 6: 181–192.

78. Camara de Comercio de Bogota (2009) Capítulo 3. Gestión del Espacio Público de Bogotá. Observatorio de la Gestion Urbana. Camara de Comercio de Bogota. 37–60.

79. Singh VS, Pandey DN, Chaudry P (2010) Urban forest and open green spaces: lessons for Jaipur, Rajasthan, India. Rajasthan, India.

80. Strohbach MW, Arnold E, Haase D (2012) The carbon footprint of urban green space-A life cycle approach. Landsc Urban Plan 104: 220–229.

81. Imhoff ML, Bounoua L, Defries R, Lawrence T, Stutzer D, et al. (2004) The consequences of urban land transformation on net primary productivity in the United States. Remote Sensing of. Remote Sens Environ 89: 434–443.

82. Schwarz N (2010) Urban form revisited-Selecting indicators for characterising European cities. Landsc Urban Plan 96: 29–47.

83. Pedlowski MA, Da Silva VAC, Adell JJC, Heynen NC (2002) Urban forest and environmental inequality in Campos dos Goytacazes, Rio de Janeiro, Brazil. Urban Ecosyst 6: 9–20.

84. Koch EW, Barbier EB, Silliman BR, Reed DJ, Perillo GME, et al. (2009) Non-linearity in ecosystem services: temporal and spatial variability in coastal protection. Front Ecol Environ 7: 29–37.

20

Fast Growing, Healthy and Resident Green Turtles (*Chelonia mydas*) at Two Neritic Sites in the Central and Northern Coast of Peru: Implications for Conservation

Ximena Velez-Zuazo[1,2], **Javier Quiñones**[3]*, **Aldo S. Pacheco**[2,4], **Luciana Klinge**[2], **Evelyn Paredes**[5], **Sixto Quispe**[3], **Shaleyla Kelez**[2]*

1 Department of Biology, University of Puerto Rico-Rio Piedras, San Juan, Puerto Rico, **2** ecOceánica, Lima, Perú, **3** Laboratorio Costero de Pisco, Instituto del Mar del Perú, Ica, Perú, **4** Instituto de Ciencias Naturales Alexander von Humboldt, CENSOR Laboratory, Universidad de Antofagasta, Antofagasta, Chile, **5** Unidad de Investigaciones en Depredadores Superiores, Instituto del Mar del Perú, Callao, Perú

Abstract

In order to enhance protection and conservation strategies for endangered green turtles (*Chelonia mydas*), the identification of neritic habitats where this species aggregates is mandatory. Herein, we present new information about the population parameters and residence time of two neritic aggregations from 2010 to 2013; one in an upwelling dominated site (Paracas ~14°S) and the other in an ecotone zone from upwelling to warm equatorial conditions (El Ñuro ~4°S) in the Southeast Pacific. We predicted proportionally more adult individuals would occur in the ecotone site; whereas in the site dominated by an upwelling juvenile individuals would predominate. At El Ñuro, the population was composed by (15.3%) of juveniles, (74.9%) sub-adults, and (9.8%) adults, with an adult sex ratio of 1.16 males per female. Times of residence in the area ranged between a minimum of 121 and a maximum of 1015 days (mean 331.1 days). At Paracas the population was composed by (72%) of juveniles and (28%) sub-adults, no adults were recorded, thus supporting the development habitat hypothesis stating that throughout the neritic distribution there are sites exclusively occupied by juveniles. Residence time ranged between a minimum of 65 days and a maximum of 680 days (mean 236.1). High growth rates and body condition index values were estimated suggesting healthy individuals at both study sites. The population traits recorded at both sites suggested that conditions found in Peruvian neritic waters may contribute to the recovery of South Pacific green turtles. However, both aggregations are still at jeopardy due to pollution, bycatch and illegal catch and thus require immediate enforcing of conservation measurements.

Editor: Elliott Lee Hazen, UC Santa Cruz Department of Ecology and Evolutionary Biology, United States of America

Funding: Green turtle research at El Ñuro was supported with funds from PADI Fundation, Idea Wild, Pro Natura Japan, Rufford Small Grants Foundation, Patagonia Footwear through 1% for the Planet, OAK Foundation through Duke Minigrants and Umanotera Foundation through Krilca Gifts. The Instituto del Mar del Perú provided funds for sea turtle research at Paracas. The funders had no role in study design, data collection and analysis, decision to publish, or preparation of the manuscript.

Competing Interests: The authors have declared that no competing interests exist.

* Email: javierantonioquinones@gmail.com (JQ); shaleyla.kelez@ecoceanica.org (SK)

Introduction

Green turtles (*Chelonia mydas*) experience ontogenic habitat changes during their life cycle, switching from a juvenile oceanic epipelagic phase where they drift and disperse for several years [1–3], to a juvenile, sub-adult and adult neritic phase, where they feed and grow to reach sexual maturity [4,5]. During the neritic phase, green turtles inhabit diverse ecosystems [5] such as sea grass beds [4,6], mixed-bottoms composed of algae, sea grass and mangroves [7–12], and coral reefs, but usually nearby sea grass or algae beds for foraging [13]. All of these ecosystems are considered important development habitats of juveniles until reaching sexual maturity [3]. These areas can also be shared with adults who use them for foraging in between reproductive periods [5,14]. Given the considerable time spent in these neritic areas, studies at these habitats represent unique opportunities to understand fundamental aspects of their life history, population dynamics [15–17],

feeding ecology [18,8], and functional role in the ecosystem [19]. Furthermore, the fact that green turtles are still endangered as a result of unsustainable practices (e.g., indiscriminate hunt for human consumption), habitat degradation and fisheries by-catch [20,21] studies in neritic habitats provide highly valuable information to propose conservation actions for this species such as preserving the habitat quality and the resources they feed on [22].

The Southeast Pacific holds a large aggregation of green turtles [23] recruiting from different nesting areas around the tropical-equatorial realm. In this region, this species may encounter different neritic conditions such as those imposed by the cold, nutrient-rich, upwelling coastal waters of the Humboldt Current System extending from central Chile (~35°S) to northern Peru (~6°S), and a transitional area (~6°–3°S) characterized by the convergence of northward upwelling waters with the warm, less-productive waters coming from the equator and warm tropical

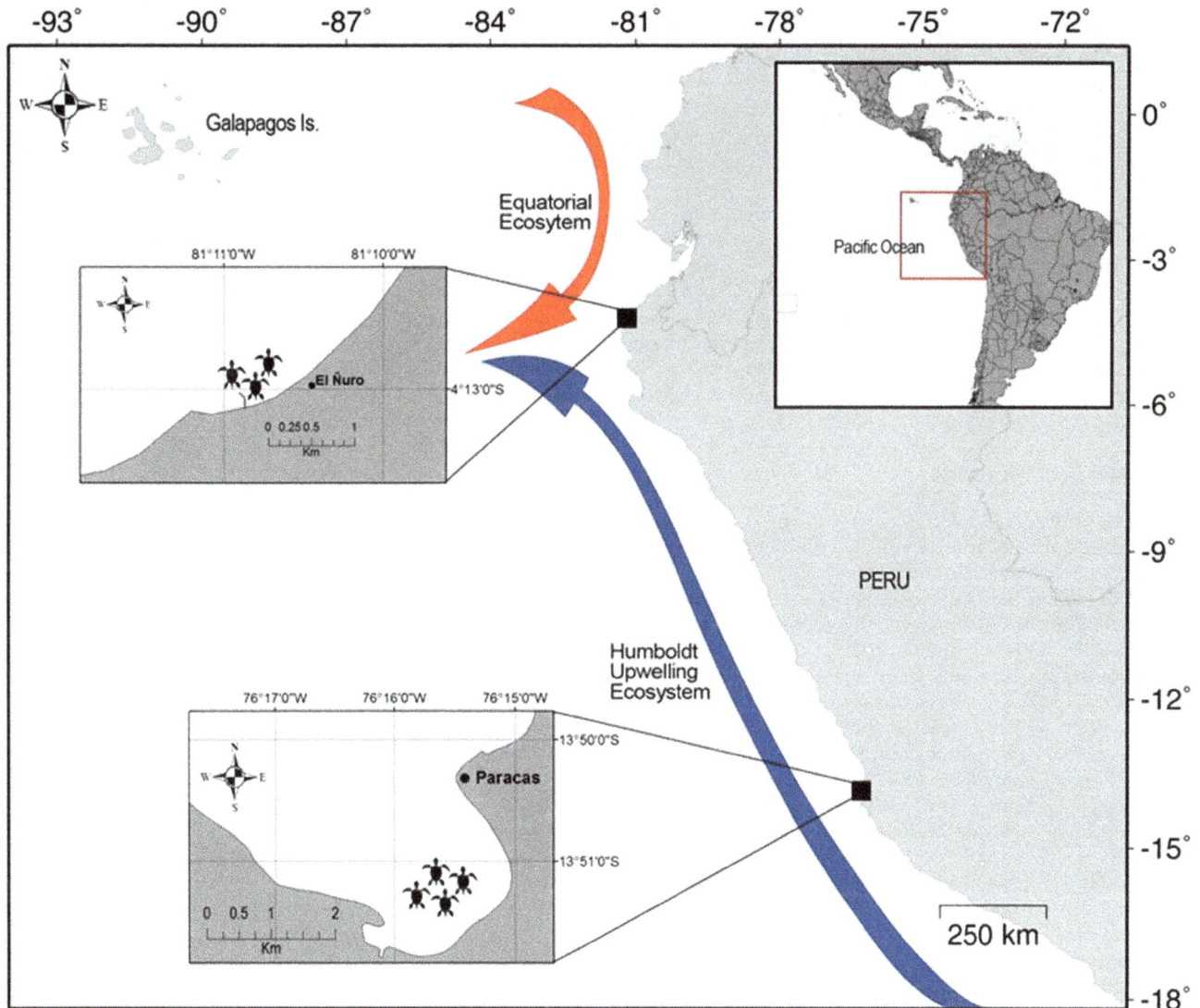

Figure 1. Green turtle (*Chelonia mydas*) neritic aggregations at two sites along the coast of Peru. (A) El Ñuro, at the northern part of the Peruvian coast, a region characterized by the convergence of the Humboldt upwelling ecosystem and the warm Equatorial ecosystem and (B) Paracas, characterized by permanent upwelling in the central coast of Peru. Map was created using Maptool available at www.seaturtle.org/maptool/.

waters that predominate in the coastal area of the northern tip of Peru towards Ecuador [24]. These conditions influence the existence and boundaries of different ecoregions [25] where green turtles may find a variety of foraging resources from distinct habitats ranging from coastal upwelling embayments to mangrove areas. Green turtles also have nesting rookeries located within the tropical realm [26,27].

Overall, the functionality of the different neritic habitats and their relationship with the distribution of each green turtle life stage within this region is still poorly understood and the information available is far from conclusive. A single study, based on the analyses of routes taken by post-nesting females from Galapagos, suggests sea surface temperature (SST) $\leq 25°C$ may act as a barrier to migration [28], but this might only be affecting adult females since juvenile and sub-adult green turtles are present in areas with lower SST. In central-south Peru, green turtles are abundant in warm neritic waters occurring during El Niño events [29]. Indeed, ca. 1000 tons of green turtles were captured and landed by artisanal fishermen during the El Niño 1987 [30]. However, landing data combines all turtle catch sites, thus precluding the assessment of habitat-life stage relationships. Green turtles can also be present in colder and temperate neritic habitats [8]. In Peru, green turtle aggregations have been registered inhabiting several areas along the coastline like Paracas bay ~14°S [29,30,31], Tambo de Mora ~13.3°S, Lobos de Afuera island ~7°S, Lobos de Tierra island ~6.5°S, Sechura bay ~5.7°S, Casitas y Punta Restin ~4.5°S and several areas along the coast of Tumbes (Punta Sal, Punta Mero, Bocapan, Puerto Pizarro) ~3.5– 4°S [29,31,32]. Yet, the roles of these different habitats in the population structure need to be evaluated.

Neritic habitats off Peru might constitute important feeding grounds for this species. Results from gut content studies suggest remarkable differences in the diet composition of green turtles from northern [33] and south-central sites along the coast of Peru [29,30,34]. These differences in food resources could reflect differences in the composition of the aggregation; thus, north versus south differences in life-stage distribution could also be expected. Furthermore, given that breeding areas are located in the equatorial realm, dispersal of post-hatchling individuals occurs towards southern locations of Peru thus southern neritic benthic habitats could be dominated by recruiting juveniles (i.e., developmental habitats) [16], while a mixed group composed by juveniles, sub-adults and adults could be most common in neritic habitats at the most northern locations in the coast of Peru.

Here we provide new information of two aggregations in an upwelling and an ecotone (Humboldt-Equatorial) site. The aims of this study were: (1) to estimate and compare population parameters at each site such as size frequencies, growth rates, and sex ratios in adults, and (2) to estimate and compare time of residency at each site. We predict that there will be proportionally more adult individuals in the ecotone site compared to the upwelling site where juvenile individuals would be relatively more abundant.

Materials and Methods

Ethics statements

According to the Ley Forestal y de Fauna Silvestre del Perú N° 29763 (Forestry and Wildlife Law of Peru) and its Reglamento de la Ley Forestal y de Fauna Silvestre through a Decreto Supremo DS No 014-2001-AG (Supreme Decree for regulation) requires permissions for conducting any research activity involving CITES (Convention of International Trade in Endangered Species of Wild Fauna and Flora) enlisted species present in the Peruvian territory outside of national protected natural areas. The Servicio Nacional Forestal y de Fauna Silvestre (National Forest and Wildlife Service) grants permission upon request and after a technical opinion from a specialist indicating that the research activities will not jeopardize the conservation status of endangered species. All research activities at El Ñuro were conducted under permits RD N° 0383-2010-AG-DGFFS-DGEFFS and RD N°0606-2011-AG-DGFFS-DGEFFS. The Servicio Nacional de Areas Naturales Protegidas (National Service for Natural Protected Areas) provided the permissions for conducting sea turtle research in Paracas National Reserve (N208-2013-SERNANP-RNP/J, N105-2013- SERNANP-RNP/J, N087-2011- SER-NANP-RNP/J). All permits allowed the manipulation and tagging of the animals as well as the collection of small tissue samples. All turtles were handled by trained personnel and following research and ethical protocols as in other sea turtle projects available in [35]. To minimize distress on captured turtles we kept them aboard for a maximum of 20 minutes, each individual was covered with wet towels to reduce overheating, and we released them in the same location were originally captured. In this study, no sea turtles were injured or killed during sampling manipulations.

Study locations

El Ñuro - the ecotone site. In-water surveys for green turtles were conducted in El Ñuro (4°13'S; 81°10'W, Fig. 1), a large sandy neritic area in the coast of the department of Piura, northern Peru. Green turtles were sampled in an area to the north-east of a pier in this area. As mentioned before, this area represents an ecotone between the cold upwelling waters of the Warm Temperate Southeastern Pacific Ecoregion and the warm and tropical conditions of the Tropical Eastern Pacific Ecoregion [25]. Sea surface temperature (SST) during the study period was on average 22.2°C (range: 19.3°–25.9°C). The seabed at this area is mainly composed by sandy sediment with scattered boulder reefs. The rocks and to some extent, the sediment substratum, are covered by the invasive green algae *Caulerpa* sp. (authors observations). The El Ñuro pier is used by ca. 350 artisanal fishermen mainly targeting demersal Peruvian hake (*Merluccius gayi*), large oceanic fish such as tuna (*Thunnus albacares*) and swordfish (*Xiphias glaudius*), and Humboldt squid (*Dosidiscus gigas*). Fishermen conduct day-fishing trips leaving between 3 and 4 am and returning from 11 am to 4 pm.

A nylon entanglement net (80 m×5 m, mesh size = 20.32 cm) was used to capture green turtles. Each in-water survey lasted 5 days, each day the net was deployed for around 5 hours usually starting at 8am and finishing at 1pm. The bottom part of the net was generally on top of the sea floor at ~5 m depth. In order to attract sea turtles to the net, some small pieces of Humboldt squid mantle were tied to the upper part of the net. Two to four free divers monitored the net every 30 minutes searching for sea turtles and helping to bring them on-board. A total of eight in-water surveys were conducted from the 31st of May 2010 to the 16th of August 2013, every 23.3 weeks on average (SD ±5.3).

Paracas – the upwelling site. In-water surveys were conducted at La Aguada a small inlet located in the South Eastern part of Bahia Paracas, in central Peru (13°51'S; 76°15'W, Fig. 1). These shallow waters are located within a cold temperate upwelling region, being one of the main feeding grounds in the Southeast Pacific [36]. SST during the study averaged 18.8°C (range: 15.1°–22.8°C). Green turtles were sampled within an estimated area of *ca.* 3.9 km². The seabed in this area is mostly sand covered by algae *Ulva* sp., *Enteromorpha* sp., *Chondracanthus* sp. [37] and *Caulerpa* sp. During the austral spring, summer and autumn, the pelagic medusae phase of the large jellyfish *Chrysaora plocamia* is very abundant [38]. This area is used by few fishermen targeting fish species such as mullets (*Mugil cephalus*) and silversides (*Odontesthes regia regia*).

A set of four entanglement nets (total length: 976 m×4 m, mesh size = 60–65 cm), traditionally used by local artisanal fisherman for catching sea turtles, was used. Each in-water survey lasted 2 days and the nets were deployed randomly within the area, usually starting at 5am and finishing at 1pm, around 8 hours on average. The net operates just below the water column surface with no bait, as traditionally used by local artisanal fishermen in this area. When a sea turtle was captured we approached the net sailing and carefully brought it on board for evaluation. A total of 21 in-water surveys were conducted every 8 weeks on average (SD±7.9) from the 4th of March 2010 to the 12th June 2013.

Collection of biological data at both sites

At both sites the following body measurements were taken from each turtle captured: the notch to tip Curved Carapace Length and Curved Carapace Width (CCLn-t, CCWn-t) and two measurements of the tail (Total Tail Length-TTL and Post-cloacal Tail Length) to 0.1 cm with a 100 cm soft measuring tape. In El Ñuro, notch to tip Straight Carapace Length and Straight Carapace Width (SCLn-t, SCWn-t) to 0.1 cm was obtained using a 100 cm Haglöf tree caliper. In Paracas, SCLn-t was not measured but estimated using the regression formula SCLn-t = 0.919CCLn-t+1.9645 (R^2 = 0.9845) estimated from turtles at El Ñuro. Green turtles were weighted using a 100 or 200 Kg scale (1 kg precision). For the largest turtles at El Ñuro, weight was not obtained using a scale, but estimated using the regression equation between size (CCLn-t) and weight (W): W = 1.916 (CCLn-t)–87.611, R^2 = 0.91; estimated using data of smaller individuals. In Paracas all turtles were weighed. The sex of the turtles was determined based on the secondary sexual characteristics i.e. length of the tail and nails. All captured green turtles were marked with external Inconel tags. At El Ñuro, tags were always applied in

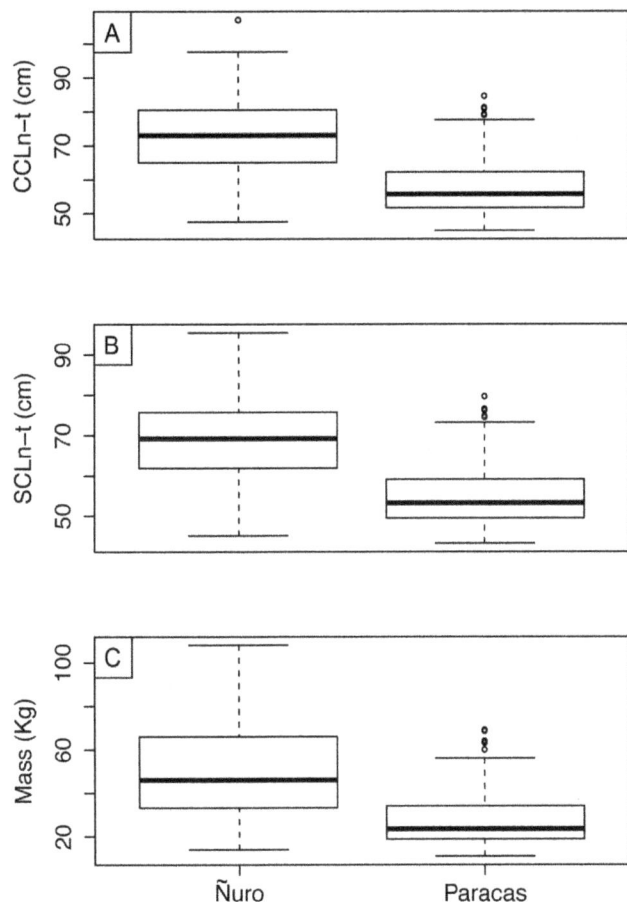

Figure 2. Box plots of (A) curve carapace length, (B) straight carapace length and (C) body weight of green turtles at both study sites.

both front flippers, except for nine individuals were tags were applied in only one of their flippers. In Paracas, Inconel tags were applied at both back flippers.

Population parameters, residency time, growth rate and body condition

Green turtles were arbitrarily classified in three different life-stages (i.e., juvenile, sub-adult and adult) using the minimum and mean CCLn-t size of nesting females in Galapagos (Patricia Zarate, personal communication). All individuals with CCLn-t, at first capture, under the minimum size of nesting females ($<$ 60.7 cm) were categorized as juveniles, all between the minimum and the mean size ($\geq$60.7 and $\leq$86.7 cm) as sub-adults, and all over the mean ($>$86.7 cm) as adults. Residence time was estimated as the time period elapsed between the first capture of the turtle and the last recapture and was expressed in days. Growth rates were estimated from the capture-recapture records as the increments in size (both for CCLn-t and SCLn-t) in cm per year. The body condition index (BCI) was calculated using the following formula: BCI = body mass/SCL [3]. This index is used as an indirect predictor of the nutritional status and/or health condition of the animal [39]. To calculate growth rates and BCI, only recapture intervals greater than 300 days were used, therefore reducing the effect of measurement error and seasonality effects [40,41,42].

Statistical analysis

Green turtle size (CCLn-t and SCLn-t) and weight were compared using a two- way PERMANOVA test considering site (Ñuro and Paracas) and sampling years (2010–2013) as fixed factors. PERMANOVA is a semi-parametric test based on Euclidian distance and calculates the significance using permutations. This test is robust enough in cases when data normality and variance homogeneity is not achieved such as in the case of our data set. This analysis also allows testing for significant effects of the interaction term (i.e. site×sampling years). Body condition index values were compared using PERMANCOVA test using a three-way design, considering CCLn-t as a co-variable and life stage (juveniles and subadults), site (Ñuro and Paracas), and years (2010–2013). Since adults were present only in el Ñuro, body condition index values of this life stage were not included in the statistical test. All statistical tests were performed using the PRIMER v6+PERMANOVA software [43].

Results

Number of capture and recapture individuals

At el Ñuro, a total of 228 green turtles were captured, with an average of 28.4 individuals per survey (range: 5–45). Of the total captured individuals, 154 were unique individuals while 51 were recaptures among surveys, 20 within surveys, and three captures from which data was not possible to get collected but it was used for CPUE estimates. For all subsequent analyses, within-survey recaptures were excluded. The mean number of new captures and recaptures was 19.5 new turtles per survey (range: 4–31, SD $\pm$10.4) and 6.4 turtles (range: 0–13, SD $\pm$4.5) respectively. One turtle, an adult female, had a tag from the Galapagos archipelago, applied in a nesting beach at Bahia Barahona, Isla Isabela, in 2010 (Macarena Parra personal communication). Also, two juvenile hawksbills (*Eretmochelys inmbricata*) were captured during the surveys. Two tagged green turtle from our project were found dead, one at El Ñuro and one at Los Organos, a beach 8 km to the north of El Ñuro. At Paracas, a total of 160 green turtles were captured, with an average of 5.3 individuals per survey (range: 1–27). Of the total captured turtles, 133 were unique individuals while 27 were recaptures among surveys. All turtles captured within surveys were excluded. The mean number of new captures and recaptures was 7.2 (range: 1–27, SD $\pm$5.7) and 0.7 turtles (range: 0–5, SD $\pm$1.3) respectively. One adult olive ridley sea turtle (*Lepidochelys olivacea*) was captured during surveys. In addition, one tagged turtle was found dead in the Pisco shores.

Size, weight, size classes and sex ratio

Overall, green turtles at El Ñuro were larger and heavier compared to the individuals captured at Paracas (Fig. 2, detailed body measurements are provided in Table S1). The size (CCLn-t) of captured green turtles at el Ñuro varied between 47.5 cm and 107 cm, with a mean size of 72.4 cm (SD $\pm$10.9), while at Paracas the size ranged from 44.9 to 84.5 cm, with an average size of 57.7 cm (SD $\pm$8.7). The PERMANOVA test indicated that differences for both CCLn-t (Pseudo-$F_{1, 356} = 196.9$, P$<$0.05) and SCLn-t (Pseudo-$F_{1, 356} = 192.9$, P$<$0.05) were significant. In addition, the interaction term time×year was also significant (CCLn-t; Pseudo-$F_{3, 356} = 3.1$, P$<$0.05 and SCLn-t; Pseudo-$F_{3, 356} = 2.9$, P$<$0.05) indicating that significant differences were persisting during all sampling years. At El Ñuro, green turtles weighed on average 50.2 kg (SD $\pm$21.4), ranging from 14 to 108 kg while at Paracas individuals weighted on average 27.7 kg (range: 11–69 kg, n = 160, SD $\pm$13.0). The PERMANOVA test also detected significant differences in weight (Pseudo-$F_{1,}$

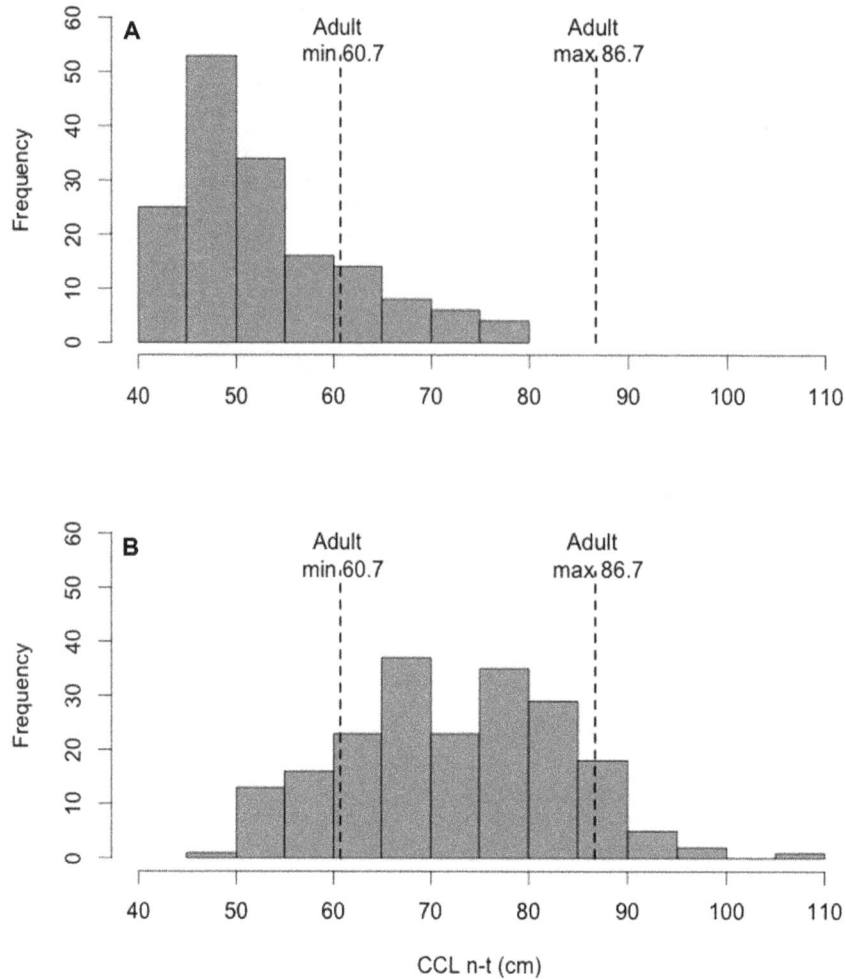

Figure 3. Size (CCLn-t) distribution of green turtles captured at (A) Paracas and (B) el Ñuro. Dotted lines are the minimum and mean size reported for adult green turtles in Galapagos Island [36], the largest green turtle rookery near Peru.

$_{356} = 146.2$, P<0.05) and the interaction term time×year (Pseudo-$F_{3,\ 356} = 3.3$, P<0.05).

At El Ñuro, the majority of individuals (81.7%; n = 166) were between 70 and 89 cm length (Figure 3) and 74.9% (n = 152) of the captured individuals were classified as sub-adults, 15.3% (n = 31) as juveniles and 9.9% (n = 20) as adults (Figure 3). The mean CCLn-t of adult individuals was 90.6 cm (range = 86.7–107, SD ±4.9). At Paracas, 73% (n = 117) individuals were classified as juveniles and 27% (n = 43) sub-adults; no adults were captured (Fig. 3). The mean CCLn-t for juveniles was 53.1 (range: 44.9–60.4 cm, SD ±3.7) and for sub-adults 69.5 (range: 61.3–84.5 cm, SD ±6.5). At El Ñuro, from all unique turtles captured, eight were identified as adult females and seven were identified as adult males. The mean TTL for adult males, including measurements obtained in subsequent recaptures, was 30.9 cm (range = 19–45 cm, SD ±7.8). These males ranged in size (CCLn-t) from 71.6 to 83.5 cm and averaged 79.4 cm. The sex ratio in captured adult sea turtles was 0.85 F: 1M. Since at Paracas no adults were recorded, the proportion of males and females of this aggregation remain unknown.

Time of residency, growth rates and body condition index

At El Ñuro, the mean recapture rate was 26%. In general, most of the individuals evaluated were captured only once (n = 115) but there were turtles captured more than once during these four years of surveys (two times: n = 31; three times: n = 15; four times: n = 3). There were more sub-adult turtles recaptured in comparison to the other life-stages (Fig. 4). The mean recapture time interval was 331.1 days (SD ±215.4), with a minimum of 121 days and a maximum recapture interval of 1015 days. At Paracas, the mean recapture rate was 12.5%. Most of the individuals evaluated were captured only once (n = 133), a few twice (n = 18) and only nine turtles were captured three times during these four years of surveys. There were more juvenile turtles recaptured in comparison to sub-adults (Fig. 4). The mean recapture interval was 277 days (SD ±236.2), with a minimum of 65 days and a maximum recapture interval of 680 days.

At El Ñuro, from all 45 recapture records (sub-adult = 44, adult = 1), we used 20 records with recapture intervals equal or higher than 300 days. The mean growth rate was estimated at 2.83 ± 1.51 cm year^{-1} (range: -0.1–5.58). At Paracas, three usable recaptures yielded a mean growth rate of 6.77 ± 1.75 cm

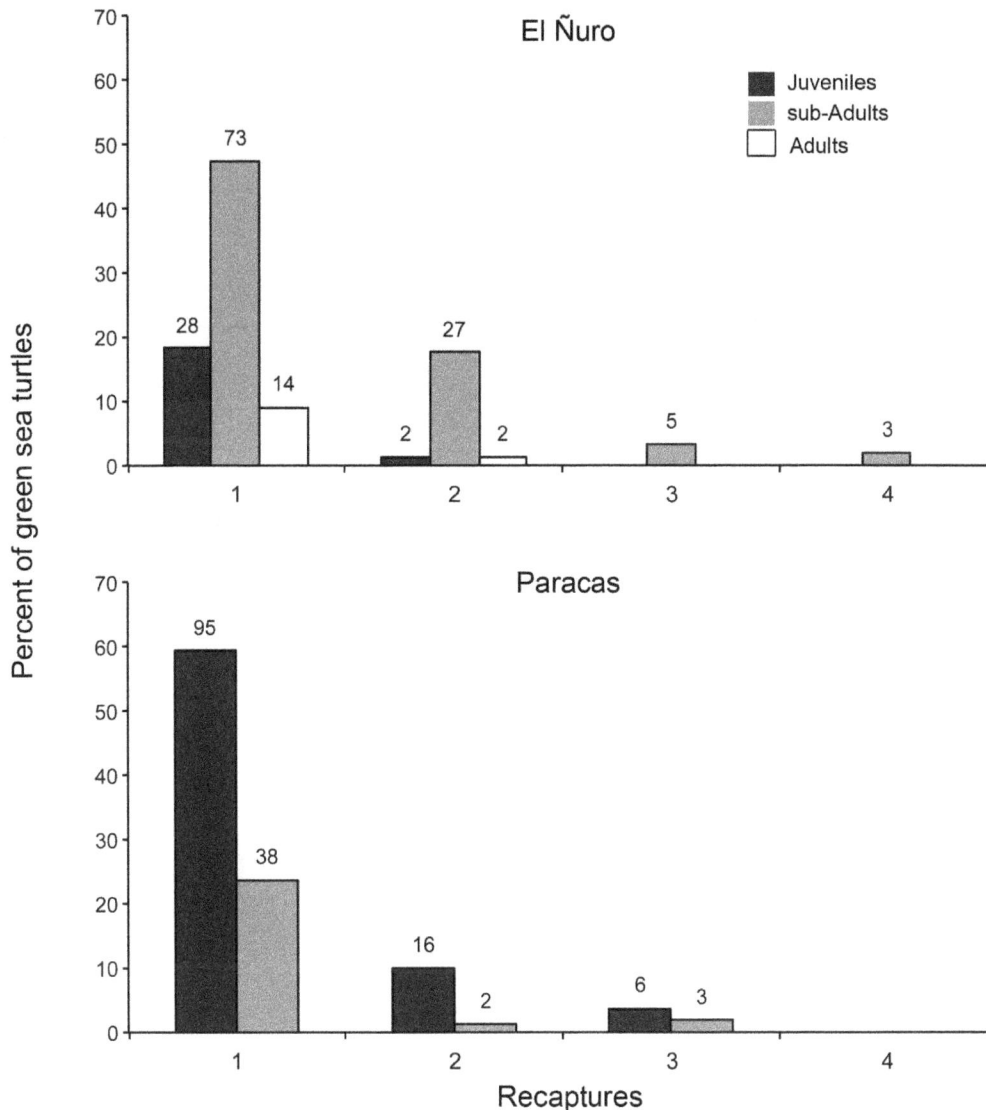

Figure 4. Number of times (%) that each turtle classified as juveniles (dark gray bar), sub-adult (light gray bar), or adult (white bar) was recaptured in El Ñuro and Paracas. The total number of turtles caught is indicated on top of each column.

year^{-1} (range: 4.77–8.07). Since we did not have equally and sufficient recaptures of individuals in all life stages, particularly from Paracas, no contrasting statistics were performed. Body condition index values at El Ñuro showed slight variation among life-stages and years. The lowest BCI exhibited per year and stage was for the juveniles in 2013 (BCI = 1.4) while adult green turtles exhibited the highest BCI in 2013 (1.6). In general, the mean BCI for each life stage (juveniles, sub-adults and adults) was 1.5. At Paracas, estimates of BCI were quite similar among life-stages and years (Fig. 5). Juvenile green turtles exhibited both, the lowest and maximum BCI (2010, BCI = 1.4 and 2013, BCI = 1.6 respectively) while adult green turtles exhibited similar values; the mean BCI for juveniles and sub-adults was the same (1.5). The PERMAN-COVA test comparing BCI values of juveniles and sub-adults (excluding adults since these where only present at El Ñuro) detected marginal differences for the interaction term co-variable (CCL)×year×site (Pseudo-$F_{3,\ 281}$ = 2.6814, P = 0.053).

Discussion

This is the first study comparing two neritic aggregations of green turtles in the Southeast Pacific. The presence of almost exclusively juvenile individuals in the upwelling site (La Aguada in Paracas) provides support to the development habitat hypothesis supporting the role of neritic habitats exclusively used by juveniles throughout the benthic life stage. Overall, the results of our study highlight the importance of neritic habitats for juveniles, sub-adults and adults green turtles in Peruvian waters and in the eastern Pacific in general. Preliminary growth rates and body index condition estimates suggest that green turtles at both sites are thriving in very favorable habitats exhibiting the highest values reported for foraging populations in the eastern Pacific.

The distribution of green turtles life-stages along these two sites suggests a spatial segregation pattern between sites; while the El Ñuro aggregation is composed of mainly sub-adults, with few juveniles and adults, for green turtles caught in Paracas the aggregation is dominated by juveniles and few sub-adults. Paracas

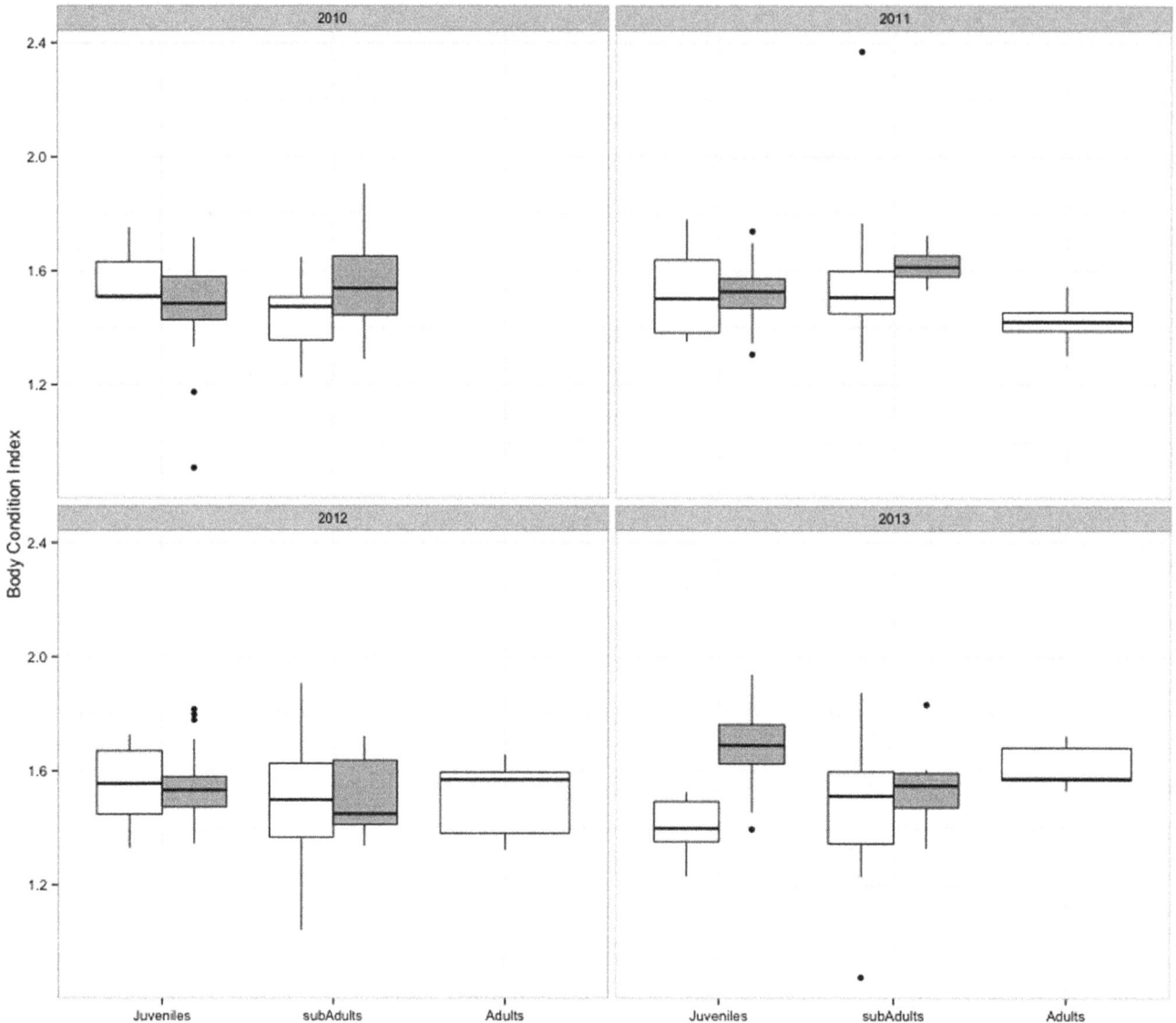

Figure 5. Box plots of body conditions values per green turtle life stages at the different sampling years at El Ñuro (grey boxes) and Paracas (white boxes).

is located within one of the most productive and permanent upwelling cells along the Peru-Chile coast [44]. In this area, early studies on the diet of this species suggested macroalgae as their principal prey item [29]. Recently, green turtles in this area have been reported preying intensively on the large jellyfish *Chrysaora plocamia* (up to 100 cm in bell diameter) [30,45,46] that can be very abundant in the water column, particularly during summer [38]. Upwelling is less intense and more seasonal in northern Peru, but still highly productive conditions exist there. At El Ñuro, the diversity and abundance of seaweeds is low; however, the algae *Caulerpa* sp. is very common and distributed in large patches along the seabed. Green turtles have been observed having pieces of this algae sticking out of their mouths (authors observations). However, we do not know whether green turtles really feed on this algae or this is unintentionally trapped in the mouth when disturbing the seabed searching for other benthic prey. Also, we cannot rule out the possibility of the influence of an external source of food (e.g., fish discard), particularly near El Ñuro pier.

Diet studies complemented with isotope data are necessary to understand the trophic ecology at El Ñuro.

Our estimates of growth rates suggest that the productivity may be playing a major role in the development of juveniles and sub-adults. Growth rate estimates obtained from green turtles at El Ñuro were the second highest estimates reported so far, after estimates obtained for the green turtle aggregation in Laguna Ojo de Liebre, in Mexico [47]. The green turtle aggregation from Paracas exhibited a high growth rate as well (6.8 cm/y ±1.8), but this estimate is preliminary and should be taken with caution. The growth rate estimates for Paracas were obtained from three recapture records, after filtering the dataset to comply with the minimum recapture interval established. Yet, these first growth rate estimates, at both sites, were high and individuals, particularly at Paracas, may be growing faster than individuals at El Ñuro. This difference is somehow expected since individuals at Paracas are significantly smaller individuals (mean CCLn-t = 57.7 cm) compared to green turtles at El Ñuro (mean CCLn-t = 72.42 cm)

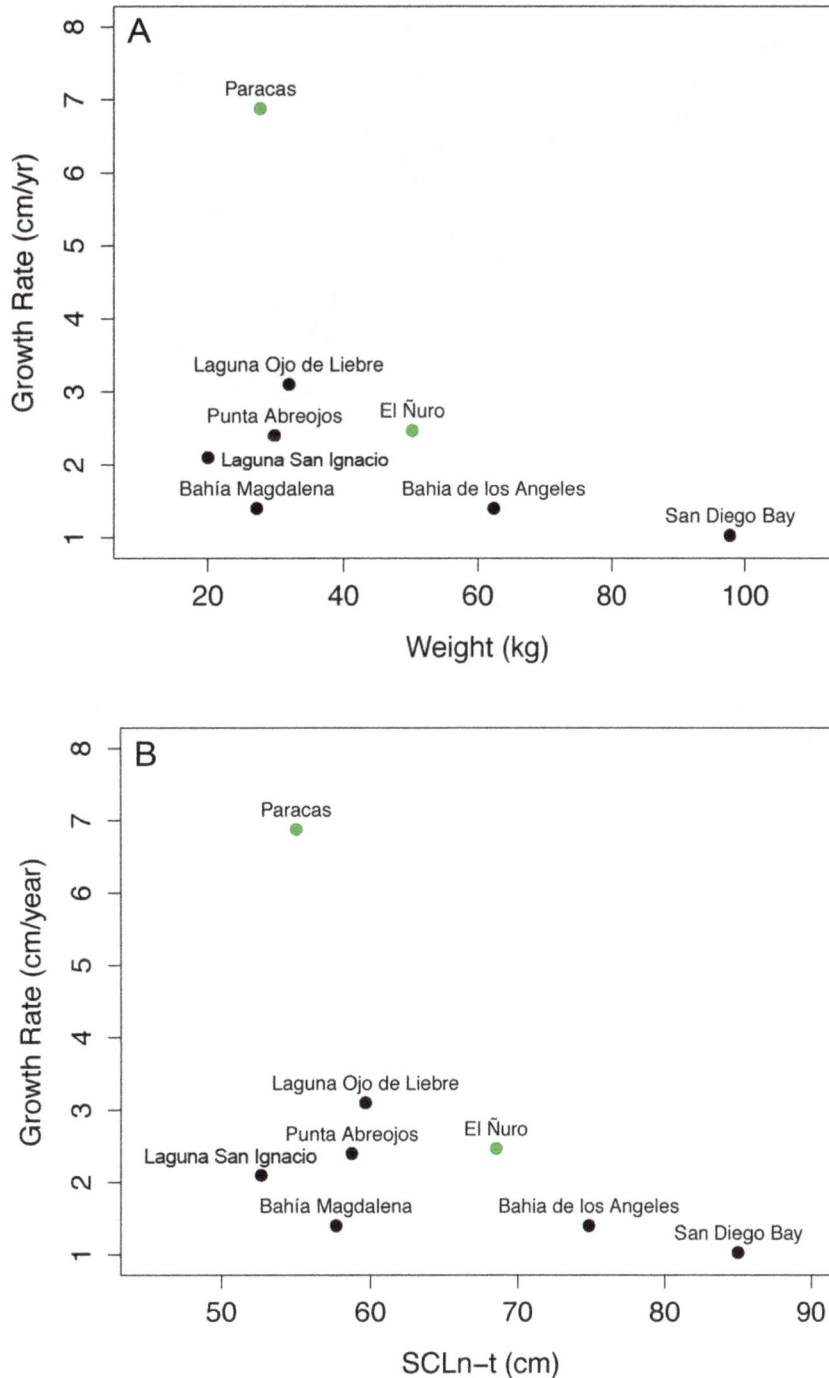

Figure 6. Relationship between weight and growth rate (A) and length (SCLn-t) and growth rate (B) for foraging aggregations of green sea turtles in the eastern and central Pacific (black dots), including our study (green dots).

and it is known that green turtles grow faster at smaller sizes e.g. [48,49,50]. Our results suggest that green turtles in Peru are growing faster than individuals at different locations in the central and northeast Pacific and this pattern is observed when comparing both weight and SCL versus growth rate (Fig. 6, Table 1). Even though the coastal region of Peru is located within tropical latitudes, the characteristics of the prevalent upwelling coastal system i.e. cold, low oxygen and nutrient-rich waters, resemble a more temperate realm extending northward close to the equator

(~4°S). Green turtles inhabiting temperate feeding grounds may boost their metabolism for growing faster compared to their counterparts living in warm habitats [51]. The combination of younger stages, low temperatures and high prey availability may yield fast growing green turtles.

In line with these results, BCI estimations for green sea turtles at both sites were high in comparison to other populations in the east Pacific. For example, in Mexico, mean estimates for different sites ranged from 1.27 to 1.42 while in Peru the mean estimates were

Table 1. Size, preliminary growth rates, and body condition index of green turtles (*Chelonia mydas*) at foraging grounds in the eastern Pacific.

Study site	N	Juv:Adt (% or ind)	CCL mean (±SD)	SCL mean (±SD)	Weight mean (±SD)	Growth rate mean (±SD)	BCI mean (±SD)	Habitat	Adult criteria	Reference
Palmyra Atoll	211	80:20	69.7 (16.1)	—	44.6 (29.7)	—	—	Reef with algae cover	CCL>85 cm (Gulf of Carpentaria) [57]	[58]
San Diego Bay	210	—	—	85 (17.3)	97.7 (52.1)	1.03	—	Power plant	—	[59,49]
Bahia de los Angeles	200	112:88 ind	80.9	74.8 (SE 0.7)	62.3	1.4 (0.93)	1.42 (0.015)	Benthic community dominated by macroalgae	MNS at Michoacan (SCL>77,3 cm) [60,61]	[8,51]
Laguna Ojo de Liebre	137	96:4	—	59.66 (10.39)	31.9 (18)	3.1 (2.2)	1.32 (0.16)	(Lagoon) *Zostera marina*, benthic macroalgae	MNS at Michoacan (SCL>77,3 cm)	[47]
Laguna San Ignacio	220	99:1	—	52.66 (7.71)	20 (10.9)	2.1 cm (1.3)	1.27 (0.23)	(Lagoon) seagrass beds and mangrove swamps	MNS at Michoacan (SCL>77,3 cm)	[47]
Punta Abreojos	604	94:6	—	58.74 (10.49)	29.8 (17.9)	2.4 (1.2)	1.34 (0.17)	(Lagoon) Mangrove forest, seagrass and algae	MNS at Michoacan (SCL>77,3 cm)	[47]
El Pardito	59	85:15	—	65.95 (11.10)	41.8 (22.4)	—	1.38 (0.13)	Islet within Gulf of California	MNS at Michoacan (SCL>77,3 cm)	[47]
Bahia Magdalena	169	97:3	—	57.68 (8.83)	27.2 (12.9)	1.4 (0.7)	1.32 (0.15)	(Lagoon) seagrass beds and mangrove	MNS at Michoacan (SCL>77,3 cm)	[47]
Gorgona National Park	86	83:3 ind	—	58.4 (7.8)	28.8 (10.7)	—	—	Soft and sandy bottoms, coral and soft corals areas	ND	[62]
Isla Plata	68	—	61.56 (5.69)	—	27 (7.58)	—	—	—	—	[63]
El Ñuro	203	90:10	72.4 (10.9)	68.5 (10.1)	50.2 (21.4)	2.8 (1.5)	1.5 (0.16)	Sandy bottom with patches of algae	MNS at Galapagos Island [3]	This study
Paracas	160	100:0	57.7 (8.7)	55 (8.0)	27.7 (13)	6.8 (1.8)	1.55 (0.14)	Sandy bottom with patches of algae	MNS at Galapagos Island	This study

Information includes site name, samples size (n), proportion of juveniles to adults (Juv: Adt) reported either as a percentage or as total count of samples individuals, mean and standard deviation (SD) of curve carapace length (CCL), straight carapace length (SCL) in cm, weight in kg, growth rate in cm year^{-1}, and body condition index (BCI). We include information of the habitat and the size criteria to distinguish juveniles from adults (mean nesting size-MNS of females).

1.49 and 1.55. These suggest that green turtles from El Ñuro and Paracas are thriving in neritic waters of central and northern Peru and may have found excellent habitat conditions for fast growth and development. Juveniles and sub-adult individuals may take advantage of such conditions before migrating as adults into relatively less productive waters off the nesting areas in the equatorial region.

Coastal and oceanic waters off Peru have been described as "a major sink for marine turtles in the Pacific" due to the high number of bycatch of all sea turtles species [23]. The highly mobile and migratory nature of sea turtles and the fact that the Peruvian coast may constitute an important migratory-displacement habitat could explain the occurrence of such large by-catch. However, our study emphasizes the fact that green turtles use coastal waters in this region to spend considerable amounts of time at localized sites in the neritic realm. At El Ñuro individuals monitored from 2010 to 2013 have stayed or departed and returned at intervals of up to 1015 days (2.7 years) while at Paracas the maximum residency time observed is 680 days. Green turtles may stay at the same area because the habitat may provide enough food resources and refuge. As stated before, prey availability is high, particularly in Paracas, while in the seabed at El Ñuro turtles use several caves and crevices within rocky reefs (Aldo S. Pacheco personal observations) for resting which may facilitate the localized residence at this site. Recapture rates were higher at El Ñuro in comparison to Paracas; however, strict comparisons between sites should be taken with caution since capture strategies and efforts were different. Sampling at Paracas was done using considerable larger nets compared to the one deployed at El Ñuro. Despite these differences, some plausible explanations can be drawn. Low recapture rates at Paracas may imply that the population is rather large and individuals are constantly moving throughout the area, likely exploiting actively natural resources as in the likely case of juveniles recruiting from the oceanic habitat [3,5]. In addition, we cannot rule out the possibility that the local conditions such as shallow depth of our sampling site is preferred by juveniles and that the adults may be distributed in deeper waters within the general area. The presence of almost exclusively juveniles provides support to the development habitat hypothesis, stating that habitats occupied only by juveniles may be found during the neritic stage as stated by Meylan et al. [16] who examined an extensive data set supporting this hypothesis for Caribbean green turtles. As an example of contradictory evidence they cite the work of Hays-Brown & Brown [29] reporting the presence of a mix of adult and juveniles individuals in Pisco, a few kilometers north of Paracas. However, Hays-Brown & Brown [29] present percentages of straight carapace length based on 416 individuals landed in the port of Pisco, but no indication is given regarding the localities where green turtles were caught. This precludes any assessment of size stage segregation between habitats in this whole area. Our data from La Aguada (Paracas) suggest that only juveniles may occur in specific sites. Nevertheless, the lack of adults can be also attributable to the effect of the former high fishery in Pisco-Paracas and the current illegal catch that may have depleted the number of adults in the area. Moreover, from 1976 larger animals were targeted by the legal fishery as a consequence of a ministerial decree which banned the capture of green turtles smaller than 80 cm of "total" length (not carapace length) until the complete closure of the fishery in 1995 [32]. In contrast, at El Ñuro sub-adults and adults were more common residents, possibly with less displacement movements given the differences of both areas. These turtles may use the area as a permanent residence habitat taking advantage of an additional food supply such as fish and squid discards released into the water during landing operations in the pier. At this location illegal catch is rare, and locals take care of the green turtles. Although this study did not explicitly examine the biotic and abiotic reasons explaining the residence times in a given habitat, these results highlight the importance of the use of neritic residences along the Southeast Pacific coast.

Outlook and conservation issues

In this study, we emphasize the presence of green turtle aggregations at two neritic sites of the Peruvian coast. Unlike most of the studies in Peru on biological aspects of sea turtles whose source comes from interactions with fisheries [23,30,52–55], from poaching [29,31,34] or from stranding [56], we present results from a less disturbed situation. Nevertheless, some aspects still need research attention such as the degree of connectivity between sites and it should be assessed with further satellite tracking and genetic research. In addition, isotopic studies may reveal important aspects of the trophic ecology of this species such as the role of different upwelling regimes in food supply. Giving the high density of humans living in coastal areas, green turtles in neritic habitats will be more exposed to negative human interactions and although the green turtle is under the protection of international and Peruvian laws, there are threats at both sites. At El Ñuro the aggregation occurs very close to an artisanal fishing pier, so green turtles are constantly at risk of interacting with human debris and the potential for vessel collision is high. Furthermore, snorkeling with green turtles has become a regular tourist activity that may have undesirable consequences on the feeding and migratory behavior of this species. In Paracas, a traditional sea turtle fishery existed [29], and although now illegal, it still occurs in significant numbers (Javier Quiñones, unpublished data). Collision with vessels is a constant threat to green turtles in this site due to high amount of recreational boats (particularly Jet Ski) navigating the bay. The biological traits examined in our study (e.g., growth rate, body condition) suggest that green turtles inhabiting the coast of Peru thrive under excellent habitat conditions which lately may cause a positive impact on the recovery of the whole population. However, if the aforementioned anthropogenic impacts are not mitigated the endangered situation of this species might remain. We strongly encourage enhancing conservation and educational programs for this species especially within artisanal fishermen communities.

Supporting Information

Table S1 Lengths, weight and growth rate and body condition index measurements of all individuals studied at el Ñuro.

Acknowledgments

We deeply thank Pacifico Adventures, Maurice Epstein, Gonzalo Villegas, Cathy Craig, Arla and Daryl Sinclair, Amy White, Rocio and Miguel Santander, Paola Nalvarte, Dagnia Nolasco, Carlos Kouri, Cynthia Céspedes, Enrique Basurto and many volunteers for providing lodging and logistic support during the surveys at El Ñuro. Capt. Pasache Jr., Segundo and Carlos Pizarro were crucial to conduct the in-water surveys. We deeply thank Cesar Mejia Villa, Cristobal Meca Andrade, Luis Delgado Arosti and several students from Universidad San Luis Gonzaga for their help during surveys at La Aguada.

Author Contributions

Conceived and designed the experiments: XVZ JQ SK ASP. Performed the experiments: XVZ JQ SK ASP LK EP SQ. Analyzed the data: XVZ JQ ASP. Contributed reagents/materials/analysis tools: XVZ JQ SK ASP LK EP SQ. Wrote the paper: XVZ ASP JQ SK.

References

1. Carr A, Meylan AB (1980) Evidence of passive migration of green turtle hatchlings in sargassum. Copeia 1980: 366–368.

2. Carr A (1987) New perspectives on the pelagic stage of sea turtle development. Conserv Biol 1: 103–121.

3. Luschi P, Hays GC, Papi F (2003) A review of long-distance movements by marine turtles, and the possible role of ocean currents. Oikos 103: 293–302.

4. Bjorndal KA (1980) Nutrition and grazing behavior of the green turtle *Chelonia mydas*. Mar Biol 56: 147–154.

5. Musick JA, Limpus CJ (1997) Habitat utilization and migration in juvenile sea turtles. In: Lutz PL, Musick JA, eds. The Biology of Sea Turtles. CRC Press, Boca Raton, Florida. pp 137–164.

6. Mortimer JA (1981) The feeding ecology of the West Caribbean green turtle (*Chelonia mydas*) in Nicaragua. Biotropica 13: 49–58.

7. Lopez-Mendilaharsu M, Gardner SC, Seminoff JA, Riosmena-Rodriguez R (2005) Identifying critical foraging habitats of the green turtle (*Chelonia mydas*) along the Pacific coast of the Baja California peninsula, Mexico. Aquat Conserv Mar Freshw Ecosyst 15: 259–269.

8. Seminoff JA, Resendiz A, Nichols WJ (2002) Diet of east Pacific green turtles (*Chelonia mydas*) in the central Gulf of California, México. J Herpetol 3: 447–453.

9. Arthur KE, McMahon KM, Limpus CJ, Dennison WC (2009) Feeding ecology of green turtles (*Chelonia mydas*) from Shoalwater Bay, Australia. Mar Turtle Newsletter 123: 6–12.

10. Carrion J, Zarate P, Seminoff J (2010) Feeding ecology of the green turtle (*Chelonia mydas*) in the Galapagos Islands. J Mar Biol Assoc UK 90: 1005–1013.

11. Senko J, Koch V, Megill W, Carthy R, Templeton R, et al. (2010) Fine scale daily movements and habitat use of East Pacific green turtles at a shallow coastal lagoon in Baja California Sur, Mexico. J Exp Mar Biol Ecol 391: 92–100.

12. Taquet C, Taquet M, Dempster T, Soria M, Ciccione S, et al. (2006) Foraging rhythms of the green sea turtle (*Chelonia mydas*) on seagrass beds in N'Gouja Bay, Mayotte (Indian Ocean), determined by acoustic transmitters and listening station. Mar Ecol Prog Ser 306: 295–302.

13. Chaloupka M, Limpus C, Miller J (2004) Green turtle somatic growth dynamics in a spatially disjunct Great Barrier Reef metapopulation. Coral Reefs 23: 325–335.

14. NMFS (1998) Recovery plan for East Pacific green turtle (*Chelonia mydas*). National Marine Fisheries Service, Silver Spring, MD.

15. Bjorndal KA, Bolten AB, Chaloupka MY (2003) Survival probability estimates for immature green turtles *Chelonia mydas* in the Bahamas. Mar Ecol Prog Ser 252: 273–281.

16. Meylan PA, Meylan AB, Gray JA (2011) The ecology and migrations of sea turtles 8. Tests of the developmental habitat hypothesis. Bull Am Mus Nat Hist 357: 1–70.

17. Patricio AR, Velez-Zuazo X, Diez CE, Van Dam R, Sabat AM (2011) Survival probability of immature green turtles in two foraging grounds at Culebra, Puerto Rico. Mar Ecol Prog Ser 440: 217–227.

18. Bjorndal KA (1997) Foraging ecology and nutrition of sea turtles. In: Lutz PL, Musick JA, eds. The Biology of Sea Turtles CRC Press, Boca Raton, Florida. pp 137–164.

19. Wabnitz CC, Balazs G, Beavers S, Bjorndal KA, Bolten AB, et al. (2010). Ecosystem structure and processes at Kaloko Honokhau, focusing on the role of herbivores, including the green sea turtle *Chelonia mydas*, in reef resilience. Mar Ecol Prog Ser 420: 27–44.

20. Wallace BP, Lewison RL, McDonald SL, McDonald RK, Kot CY, et al. (2010) Global patterns of marine turtle bycatch. Conserv Lett 3: 131–142.

21. Wallace BP, Kot CY, DiMatteo AD, Lee T, Crowder LB, et al. (2013) Impacts of fisheries bycatch on marine turtle populations worldwide: toward conservation and research priorities. Ecosphere 4: art40, doi:10.1890/ES12-00388.1

22. Christianen MJA, Herman PMJ, Bouma TJ, Lamers LPM, van Katwijk MM, et al. (2014) Habitat collapse due to overgrazing threatens turtle conservation in marine protected areas. Proc R Soc B 281, 20132890, doi:10.1098/rspb.2013.2890

23. Alfaro-Shigueto J, Mangel JC, Bernedo F, Dutton PH, Seminoff JA, et al. (2011) Small-scale fisheries of Peru: a major sink for turtles in the Pacific. J Appl Ecol 48: 1432–1440.

24. Swartzman G, Bertrand A, Gutíerrez M, Bertrand S, Vasquez L (2008) The relationship of anchovy and sardine to water masses in the Peruvian Humboldt Current System from1983 to 2005. Prog Oceanogr 79: 228–237.

25. Spalding M, Fox H, Allen G, Davidson N, Ferdaña Z, et al. (2007) Marine ecoregions of the world: a bioregionalization of coastal and shelf areas. Bioscience 57: 573–583.

26. Forsberg K, Casabonne F, Castillo J (2012) First evidence of green turtle nesting in Peru. Mar Turtle Newsletter 133: 9–11.

27. Wester J (2010) Turtle encounters on the coast of northern Peru, a global analysis of conservation. MSc thesis. Leiden University, Nederlands.

28. Seminoff J, Zárate P, Coyne M, Foley DG, Parker D, et al. (2008) Post-nesting migrations of Galápagos green turtles *Chelonia mydas* in relation to oceanographic conditions: integrating satellite telemetry with remotely sense ocean data. Endang Species Res 4: 57–72.

29. Hays-Brown C, Brown WM (1982) Status of sea turtles in the southeastern Pacific: emphasis on Peru. In: Bjorndal K, ed. Biology and Conservation of Sea Turtles. Smithsonian Institution Press, Washington DC. pp 235–240.

30. Quiñones J, González Carman V, Zeballos J, Purca S, Mianzan H (2010) Effects of El Niño-driven environmental variability on black turtle migration to Peruvian foraging grounds. Hydrobiologia 645: 69–79.

31. de Paz N, Reyes JC, Echegaray M (2002) Datos sobre captura, comercio y biología de tortugas marinas en el área de Pisco - Paracas. In: Mendo J, Wolff M, eds. I Jornada Científica: Bases ecológicas y socioeconomicas para el manejo de los recursos vivos de la Reserva Nacional de Paracas. Universidad Nacional Agraria La Molina, Perú. pp 125–129.

32. Aranda C, Chandler MW (1989) Las tortugas marinas del Perú y su situación actual. Bol de Lima 62: 77–86.

33. Santillan LA (2008) Análisis de la dieta de *Chelonia mydas agassizii* "Tortuga verde del Pacifico" en la Bahía de Sechura, Piura - Perú. Master Thesis. Universidad Nacional Agraria La Molina.

34. Paredes R (1969) Introducción al estudio biológico de *Chelonia mydas agassizii* en el perfil Pisco. Undergraduate Thesis. Universidad Nacional Federico Villareal, Perú. 83 pp.

35. Ehrhart LM, Ogren LH (1999) Studies in foraging habitats: capturing and handling turtles. In: Eckert KL, Bjorndal KA, Abreu-Grobois FA, Donnelly M, eds. Research and Management Techniques for the Conservation of Sea Turtles. IUCN/SSC Marine Turtle Specialist Group Publication No. 4. pp. 61–64.

36. Marquez R (1990) Sea turtles of the world. FAO Species Catalogue 125: 1–81.

37. Gil-Kodaka P, Mendo J, Fernandez E (2002) Diversidad de macroalgas del submareal en la Reserva Nacional de Paracas y notas sobre su uso potencial. In: Mendo J, Wolff M, eds. I Jornada Científica: Bases ecológicas y socioeconómicas para el manejo de los recursos vivos de la Reserva Nacional de Paracas. Universidad Nacional Agraria La Molina, Perú. pp 154–163.

38. Quiñones J (2008) *Chrysaora plocamia* Lesson, 1830 (Cnidaria, Scyphozoa), frente a Pisco, Perú. Informe Instituto del Mar del Perú 35: 221–230.

39. Bjorndal KA, Bolten AB, Chaloupka MY (2000) Green turtle somatic growth model: evidence for density dependence. Ecol Appl 10: 269–282.

40. Krueger BH, Chaloupka MY, Leighton PA, Dunn JA, Horrocks JA (2011) Somatic growth rates for a hawksbill turtle population in coral reef habitat around Barbados. Mar Ecol Prog Ser 432: 269–276.

41. Kubis S, Chaloupka M, Ehrhard L, Bresette M (2009) Growth rates of juvenile green turtles *Chelonia mydas* from three ecologically distinct foraging habitats along the east central coast of Florida, USA. Mar Ecol Prog Ser 289: 257–269.

42. McMichael E, Seminoff JA, Carthy R (2008) Growth rates of wild green turtles, *Chelonia mydas*, at a temperate foraging habitat in the northern Gulf of Mexico: assessing short-term effects of cold-stunning on growth. J Nat Hist 42: 2793–2807.

43. Anderson M, Gorley RN, Clarke RK (2008) PERMANOVA+ for PRIMER: Guide to Software and Statistical Methods.

44. Echevin V, Aumont O, Ledesma J, Flores G (2008) The seasonal cycle of surface chlorophyll in the Peruvian upwelling system: a modelling study. Prog Oceanogr 79: 167–176.

45. Mianzan HW, Cornelius PFS (1999) Cubomedusae and Scyphomedusae. In: Boltovskoy D, ed. South Atlantic zooplankton. Backhuys Publishers, Leiden. pp 513–559.

46. Riascos JM, Villegas V, Caceres I, Gonzalez JE, Pacheco AS (2013) Patterns of a novel association between the scyphomedusa *Chrysaora plocamia* and the parasitic anemone *Peachia chilensis*. J Mar Biol Assoc UK 93: 919–923.

47. López-Castro MC, Koch V, Marical-Loza A, Nichols WJ (2010) Long-term monitoring of black turtles *Chelonia mydas* at coastal foraging areas off the Baja California Peninsula. Endang Species Res 11: 35–45.

48. Bjorndal KA, Bolten AB (1988) Growth rates of immature green turtles, *Chelonia mydas*, on feeding grounds in the southern Bahamas. Copeia 1988: 555–564.

49. Eguchi T, Seminoff JA, LeRoux RA, Prosperi D, Dutton DL, et al. (2012) Morphology and growth rates of the green sea turtle (*Chelonia mydas*) in a northern-most temperate foraging ground. Herpetologica 68: 76–87.

50. Patricio R, Diez CE, van Dam RP (2014) Spatial and temporal variability of immature green turtle abundance and somatic growth in Puerto Rico. Endang Species Res 23: 51–62.

51. Seminoff JA, Jones TT, Resendiz A, Nichols WJ, Chaloupka MY (2003) Monitoring green turtles (*Chelonia mydas*) at a coastal foraging area in Baja California, Mexico: multiple indices describe population status. J Mar Biol Assoc UK 83: 1355–1362.

52. Kelez S, Velez-Zuazo X, Manrique C (2003) New evidence on the loggerhead sea turtle *Caretta caretta* (Linnaeus 1758) in Peru. Ecol apl 2: 141–142.

53. Castro J, de la Cruz J, Ramírez P, Quiñones J (2012). Sea turtles by-catch during El Niño 1997–1998, in northern Peru. Lat Am J Aquat Res 40: 970–979.

54. Alfaro-Shigueto J, Mangel JC, Caceres C, Seminoff JA, Gaos A, et al. (2010) Hawksbill turtles in Peruvian coastal fisheries. Mar Turtle Newsletter 129: 19–21.

55. Alfaro Shigueto J, Mangel JC, Seminoff JA, Dutton PH (2008) Demography of loggerhead turtles *Caretta caretta* in the southeastern Pacific Ocean:

fisheries-based observations and implications for management. Endang Species Res 5:129–135.

56. Rosales CA, Vera M, Llanos J (2010) Stranding and incidental catch of sea turtles in the coastal Tumbes, Peru. Rev Peru Biol 17: 293–301.

57. Hamann M, Schauble CS, Simon T, Evans S (2006) Demographic and health parameters of green sea turtles *Chelonia mydas* foraging in the Gulf of Carpentaria, Australia. Endang Species Res 2: 81–88.

58. Sterling EJ, McFadden KW, Holmes KE, Vintinner EC, Arengo F, et al. (2013) Ecology and conservation of marine turtles in a Central Pacific foraging ground. Chelonian Conserv Biol 12: 2–16.

59. Eguchi T, Seminoff JA, LeRoux RA, Dutton PH, Dutton DL (2010) Abundance and survival rates of green turtles in an urban environment: coexistence of humans and an endangered species. Mar Biol 157: 1869–1877.

60. Koch V, Brooks L, Nichols WJ (2007) Population ecology of the green/black turtle (*Chelonia mydas*) in Bahia Magdalena, Mexico. Mar Biol 153: 35–46.

61. Koch V, Nichols WJ, Peckham H, de La Toba V (2006) Estimates of sea turtle mortality from poaching and bycatch in Bahia Magdalena, Baja California Sur, Mexico. Biol Conserv 128: 327–334.

62. Amorocho D, Reina R (2007) Feeding ecology of the East Pacific green turtle (*Chelonia mydas agassizii*) at Gorgona National Park, Colombia. Endang Species Res 3: 43–51.

63. Muñoz Perez JP (2009) Identificación y estudio preliminar de los sitios críticos para anidación, forrajeo y descanso de las tortugas marinas en la costa centro y norte del Ecuador. Bachelor Thesis. Universidad San Francisco de Quito, Ecuador. 27 pp.

21

Evidence for Frozen-Niche Variation in a Cosmopolitan Parthenogenetic Soil Mite Species (Acari, Oribatida)

Helge von Saltzwedel*, Mark Maraun, Stefan Scheu, Ina Schaefer

Georg-August University, Johann-Friedrich-Blumenbach Institute of Zoology and Anthropology, Dept. Ecology, Göttingen, Germany

Abstract

Parthenogenetic lineages may colonize marginal areas of the range of related sexual species or coexist with sexual species in the same habitat. Frozen-Niche-Variation and General-Purpose-Genotype are two hypotheses suggesting that competition and interclonal selection result in parthenogenetic populations being either genetically diverse or rather homogeneous. The cosmopolitan parthenogenetic oribatid mite *Oppiella nova* has a broad ecological phenotype and is omnipresent in a variety of habitats. Morphological variation in body size is prominent in this species and suggests adaptation to distinct environmental conditions. We investigated genetic variance and body size of five independent forest - grassland ecotones. Forests and grasslands were inhabited by distinct genetic lineages with transitional habitats being colonized by both genetic lineages from forest and grassland. Notably, individuals of grasslands were significantly larger than individuals in forests. These differences indicate the presence of specialized genetic lineages specifically adapted to either forests or grasslands which coexist in transitional habitats. Molecular clock estimates suggest that forest and grassland lineages separated 16-6 million years ago, indicating long-term persistence of these lineages in their respective habitat. Long-term persistence, and morphological and genetic divergence imply that drift and environmental factors result in the evolution of distinct parthenogenetic lineages resembling evolution in sexual species. This suggests that parthenogenetic reproduction is not an evolutionary dead end.

Editor: Xiao-Yue Hong, Nanjing Agricultural University, China

Funding: The author(s) received no specific funding for this work.

Competing Interests: The authors have declared that no competing interests exist.

* Email: hsaltzw1@gwdg.de

Introduction

Parthenogenetic lineages often are successful colonizers of new or disturbed habitats. This success suggests that effective establishment of populations may occur without males and genetic exchange. In parthenogenetic species each individual represents a reproductive unit capable of founding a new population [1–3]. Thelytoky, the exclusive production of daughters from unfertilized eggs, also increases the number of reproductive individuals in a population and thereby population growth. In addition, genotypes that successfully establish in a new habitat are transmitted unchanged to the next generation whereas sexual reproduction potentially breaks up advantageous gene combinations every generation [4]. However, in the long-term, the lack of males and recombination is assumed to result in the accumulation of deleterious mutations [5], [6] and to limit adaptation to changing environments [7], [8]. Therefore, in the long-term parthenogenetic lineages are assumed, to be doomed to extinction due to mutational meltdown and competition with sexual sister-taxa.

Among the several hypotheses explaining the ecological and geographical distribution of parthenogenetic and sexual organisms [9] the Frozen-Niche-Variation (FNV) hypothesis [10–13] suggests that widespread parthenogenetic species consist of a number of locally adapted genotypes, each occupying a narrow niche. As parthenogenetic genomes are transmitted in full their genotypes are kept "frozen". In this model asexual individuals arise continuously from sexual populations resulting in genetically diverse populations. Evidence for such specialized genotypes supporting the FNV hypothesis have been found in fishes, frogs, spider mites, shrimps and water fleas [10], [14–19]. On the contrary, spatial and temporal variation of ecological niches may favor the evolution of parthenogenetic genotypes adapted to a wide range of ecological conditions, thereby representing a General-Purpose-Genotype (GPG) [2], [20] with only few parthenogenetic lineages dominating across habitats [21]. In these lineages mutations are the primary source of variation [22], [23] resulting in low genetic diversity within populations contrasting predictions of the FNV hypothesis. Evidence for GPG has been found in fishes, snails, ostracods, oribatid mites and ambrosia beetles [24–29], for a detailed list see [30].

The cosmopolitan thelytokous oribatid mite species *Oppiella nova* (Oudemans, 1902) lives in a variety of habitats including the soils of forests, grasslands, agricultural fields and suspended soils in tree canopies. It can reach high densities ($>20,000$ ind. m^{-2}) [31–34] and often co-occurs with sexual species of the same genus, such as *O. subpectinata* and *O. falcata* [35]. The existence of sexually reproducing congeneric species suggests that *O. nova* is a parthenogenetic offshoot of the predominantly sexual genus *Oppiella* [36]. However, phylogenetic relationships among *Oppiella* species are unresolved and the sexual sister-taxon of *O. nova* is unknown. The most prominent morphological variation in this

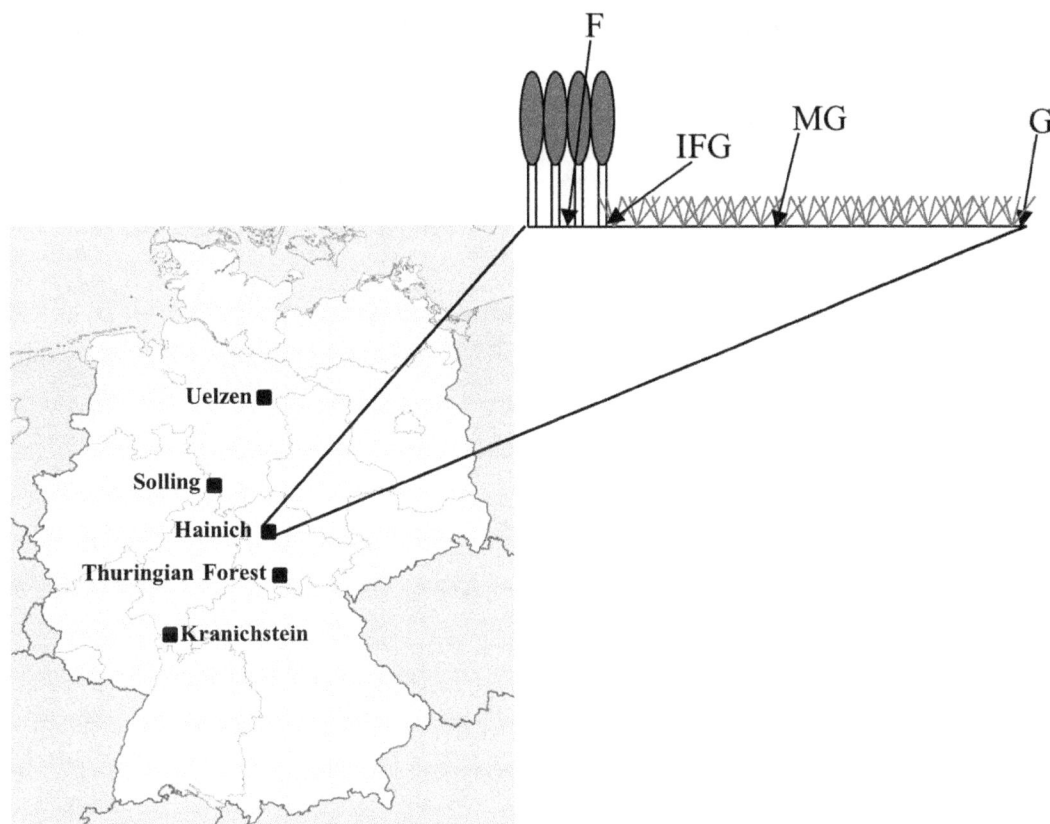

Figure 1. Sampling locations and sampling along a gradient from forest (F) to grassland (G) including transitional habitat types at the intersection between forest and grassland (IFG) and the margin of grassland (MG).

species is body size which ranges from 220 to 320 μm [37]. Due to morphological variation between habitats Woas [38] suggested *O. nova* to comprise different subspecies each adapted to a distinct habitat.

We analyzed the genetic and morphological variance of populations of *O. nova* from grassland and forest soils, forming two distinct soil habitats likely associated with distinct niches, to investigate whether the variation is driven by FNV or GPG processes. Grasslands and forests differ markedly in abiotic and biotic factors, including temperature, humidity, wind, soil structure and fungal community composition. Mites were sampled along a gradient from grassland to forest at five locations spaced at least 50 km from each other. The mitochondrial *COI* gene and the D3 region of the nuclear 28S rDNA were sequenced to identify genetic lineages; the D3 region also served as species marker [39], [40]. According to the FNV hypothesis we expected specimens of the same habitat to cluster together irrespective of sampling locations. In contrast, conform to the GPG hypothesis different habitats (and the associated niches) within the same location were expected to cluster together, i.e. to cluster according to distance. Although oribatid mites are generally poor dispersers, *O. nova* is able to migrate short distances and occasionally disperses long distances by wind [41]. To take dispersal into account, we tested for migration of genotypes between locations and between habitats, i.e. forest and grassland, within locations. Further, we investigated whether body size correlated with habitat type, genetic lineages or sampling location. Similar to haplotype distribution, we expected body size to correlate with habitat type

according to the FNV hypothesis but to correlate with distance of locations according to the GPG hypothesis.

Materials and Methods

Ethics statement

Permission for sampling at Kranichstein was given by the forestry office Darmstadt, permission at Hainich was issued by the state environmental office of Thüringen (§ 72 BbgNatSchG). All other sampling sites were outside Nature Reserve Areas and no permission for soil samples was required. The field study did not involve any endangered or protected species.

Sampling and study sites

A total of 147 individuals of the oribatid mite species *O. nova* were collected from five locations in Germany: Hainich (HA), Kranichstein (KW), Solling (SO), Thuringian Forest (TW) and Uelzen (UE) (**Table 1, Figure 1**). We restricted the analysis to the parthenogenetic species because the sexual sister-taxon is unknown. Individuals were sampled from soil and litter of adjacent grassland and forest along a gradient, including the habitat types forest (F) and grassland (G) and two transitional habitats, grassland margin (MG) and intersection of forest and grassland (IFG). MG was located in grassland but close to the forest edge which formed a sharp boundary, IGF samples were taken where tree litter and grassland vegetation mixed (**Figure 1**). The maximum distance between F and G sampling sites was 100 m, MG and IFG sampling sites were 15–20 m apart; sampling locations were 56–350 km apart. From each habitat three samples of 15×15 cm were

Table 1. Sampling locations, habitat type, number of collected individuals, number of sequences for *COI* and D3 and respective GenBank accession numbers of *Oppiella nova* analyzed in this study.

Location	Habitat type	n individuals	n sequences (*COI*)	GenBank acc. no.	n sequences (D3)	GenBank acc. no.
Kranichstein Forest, near Darmstadt (KW)	F	17	12	KF293419-26, 35–38,	14	KF293529, 33–41, 51–54
	G	23	19	KF293427-28, 39–55	21	KF293530, 42–44, 55–71
	MG	26	20	KF293415-16, 29–32, 56–69	24	KF293545-48, 72–90
	IFG	4	4	KF293417-18, 33–34	3	KF293532, 49–50
Solling Forest, near Neuhaus (SO)	F	12	7	KF293470-76	8	KF293591-98
	G	1	–	–	1	KF293599
	MG	7	4	KF293477-80	5	KF293600-04
Hainich Forest, near Weberstedt (HA)	F	4	4	KF293402-05	4	KF293514-17
	IFG	12	9	KF293406-14	11	KF293518-28
Thuringian Forest, near Ilmenau (TW)	F	3	3	KF293481-83	3	KF293605-07
	G	2	2	KF293484-85	1	KF293608
	IFG	1	–	–	1	KF293609
Uelzen Forest, near Uelzen (UE)	G	4	3	KF293486-88	3	KF293610-12
	MG	7	3	KF293489-91	7	KF293613-19
	IFG	24	20	KF293492-511	20	KF293620-39

Individuals were collected along a gradient from forest (F) to grassland (G), covering the intersection of forest and grassland (IFG) and margin of grassland (MG).

taken, including litter and the uppermost 5 cm of the soil. Invertebrates were extracted by heat [42] and collected in 75% EtOH. *O. nova* was separated using a dissecting microscope, and morphological identification was confirmed by light microscopy [37].

DNA extraction and PCR

Genomic DNA was extracted from single individuals using the DNeasy Blood and Tissue Kit (Qiagen; Hilden, Germany) following the manufacturer's protocol for animal tissue. Purified DNA was eluted in 30 µl buffer AE and stored at −20°C until further preparation. All PCR reactions for sequencing were performed in 25 µl volumes containing 12.5 µl HotStarTaq Mastermix (Qiagen; Hilden, Germany) with 1 µl of each primer (10 pM), 1 µl of MgCl$_2$ (25 mM) and variable volumes of template DNA (5 µl for D3 and 8 µl for *COI*) and H$_2$O (4.5 µl for D3 and 1.5 µl for *COI*). A 709 bp fragment of the *COI* gene was amplified using the primers LCO1490 (forward) 5′-GGT CAA CAA ATC ATA AAG ATA TTG G-3′ and HCO2198 (reverse) 5′-TAA ACT TCA GGG TGA CCA AAA AAT CA-3′ [43]. Amplification consisted of one initial activation step at 95°C for 15 min, followed by 35 amplification cycles of denaturation at 94°C for 30 s, annealing at 40°C for 60 s, elongation at 72°C for 60 s and a final elongation step at 72°C for 10 min. Amplification of the 356 bp fragment of the D3 region of the 28S rDNA was performed using the primers D3A (forward) 5′-GAC CCG TCT TGA AAC ACG GA-3′ and D3B (reverse) 5′-TCG GAA GGA ACC AGC TAC TA-3′ [44]. The PCR protocol for D3 consisted of an initial activation step at 95°C for 15 min, followed by 35 amplification cycles of denaturation at 94°C for 30 s, annealing at 54°C for 45 s, elongation at 72°C for 60 s and a final elongation

step at 72°C for 10 min. PCR products were purified with the QIAquick PCR Purification Kit (Qiagen; Hilden, Germany) following the manufacturer's protocol. Sequencing in both directions (forward and reverse strands) of *COI* fragments was done by Macrogen Inc. (Seoul, South Korea). The D3 fragments were sequenced at G2L (Institute for Microbiology and Genetics, Laboratory for Genomic Analyses, University of Göttingen). All nucleotide sequences are available at GenBank (www.ncbi.nlm.nih.gov/genbank; KF293402 - KF293513 for *COI* and KF293514 - KF293641 for D3).

Data analysis

Sequences were edited, ambiguous positions were corrected by hand, aided by the respective chromatograms, and nucleotide sequences were translated into amino acid sequences using the invertebrate mitochondrial code implemented in Sequencher 4.9 (Gene Codes Corporation, USA). Consensus sequences were assembled in BioEdit 7.0.1 [45] and aligned with ClustalX v1.81 [46] using multiple alignment parameters: 10.0 (gap opening) and 0.1 (gap extension) for the nucleotide and default settings for the amino acid dataset. In total, three different alignments were generated which included two individuals of *Berniniella hauseri* as outgroup. The D3 alignment included 126 individuals of *O. nova* (**Alignment S1**) and the *COI* alignment 110 individuals (**Alignment S2**). The nucleotide alignments were 356 bp (D3) and 709 bp (*COI*) long; the protein alignment of *COI* had 235 positions.

Phylogenetic trees were calculated with RAxML v8.0.2 [47], MrBayes v3.1.2 [48] and BEAST v1.7.4 [49]. Phylogenetic optimality criterion was maximum likelihood for RAxML, Bayesian inference for MrBayes and BEAST. The best fit model of sequence evolution was estimated with jModeltest 2.1.4 [50], [51], according to the AIC the best model was GTR+I+Γ [52], [53] for both nucleotide alignments. The MCMC chain was run for ten million generations and sampled every 1,000[th] generation in MrBayes. In BEAST, the MCMC chain ran for 100 million generations and sampled every 10,000[th] generation, the majority consensus trees were generated with a burnin value of 2,500 (25%). In RAxML 8,000 bootstrap replicates were calculated for statistical node support. A median-joining haplotype network for the nucleotide dataset of *COI* was generated with Network 4.6 (Fluxus Technology, Suffolk, Great Britain).

A strict molecular clock was performed with BEAST v1.7.4, BEAUti v1.7.4 and TreeAnnotator v1.7.4 [49] with a fixed substitution rate of 0.0115 which corresponds to the common invertebrate rate of *COI* of 0.023 substitutions per site per million years [54], [55] for the *COI* nucleotide alignment. The site model was GTR+I+Γ and as tree prior we used "Yule Process" [56] to allow higher rate variation among branches in this parthenogenetic species than coalescent tree priors do. The Yule.birth rate prior had uniform distribution; all priors were estimated by the software. The MCMC chain ran for 20,000,000 generations, every 2,000th generation was sampled and a burnin of 2,500 was applied and convergence of the MCMC was confirmed using Tracer v1.4 [57].

To test for potential migration between forest and grassland and between sampling locations, three models of migration were tested with grassland and forest specific *COI* haplotypes using Bayesian inference in MIGRATE-N 3.2.16 [58]. The three models included (1) panmixis among all locations (50 individuals from forest and grassland) assuming a single population, (2) migration between forest (26 individuals) and grassland (24 individuals), and (3) migration between the five sampling locations. This analyses included 4 individuals from HA, 31 from KW, 7 from SO, 5 from TW and 3 from UE. The models were tested in several independent runs. The following parameters deviated from default settings: 10,000 record steps in chain: heating set to on, static heating; 4 chains sampling at every 10th interval using the temperature scheme suggested with the character #; Theta prior distribution, uniform, 0 (minimum) 1 (maximum) 0.1 (delta); migration prior distribution, uniform, 0 (minimum) 10000 (maximum) 1000 (delta); running multiple replicates set to YES, 4 independent chains; number of long chains to run set to 2. To identify the best-fit model, marginal likelihoods of the three runs were compared by calculating the log Bayes Factor (LBF) and model probability (MP) by substracting the largest log likelihood from every other log likelihood, exponentiating the difference and summing up the results. The exponential elements were divided by this product; results indicate which model is most likely relative to the others [59].

Two independent analyses of molecular variance (AMOVA) were calculated in ARLEQUIN 3.5 [60] to investigate between and within population structure based on p-distances, selecting (1) habitat (forest and grassland) and (2) sampling locations as group. Populations represented by less than three individuals were excluded. According to the FNV hypothesis we expected higher variance within sampling locations (i.e., high variance between habitats and associated niches) than between sampling locations, whereas according to the GPG hypothesis we expected variance within habitats to be similar or lower than between sampling locations. Isolation by distance was tested by Mantel test implemented in ARLEQUIN using 10,000 permutations and straight-line (Euclidian) distances. Haplotype and nucleotide diversity were also calculated in ARLEQUIN. To distinguish between divergent selection and neutral drift the distribution of

synonymous and non-synonymous substitutions between locations and between habitat types was compared using the McDonald Kreitman test [61] in DnaSP v5.10.1 [62].

For morphological variation body length and width of 147 individuals were measured from dorsal pictures, taken with AxioCam HRm and processed with the image analyzing software AxioVision 4.8.2 (Zeiss, Göttingen, Germany) by quantifying pixels. Differences between mean values of body length of *O. nova* were analyzed in R 3.1 (R Development Core Team 2014) using the linear mixed effects model (nlme package) [63]. Locations were set as random variable and body-sizes were compared between habitat types (F, G, IFG, MG) and additionally between individuals with forest and grassland specific *COI* genotypes. *Post hoc* multiple comparisons of means were made using Tukey's honestly significant difference test (multicomp package) [64] with $p < 0.001$ as threshold for significance.

Results

Densities of *O. nova* varied between zero and 30 individuals per sample. To obtain equal numbers of individuals per habitat type, the three samples of each sampling location were pooled for further analysis. Numbers of individuals at the four habitat types were 36, 30, 40 and 41 for F, G, MG and IFG, respectively. Molecular variation of the D3 fragment was low; only nine positions of the 354 bp fragment varied in 126 analyzed individuals (positions 59–61, 114 and 120–124). Accordingly, the phylogenetic tree had no structure and habitats and locations were mixed (**Figure 2a, Figure S1–2**). Amino acid sequences of the *COI* fragment were almost identical in the 110 individuals sequenced. Only 16 specimens had one or two variable sites with non-synonymous substitutions (**Table S1**) and the overall genetic distances between protein sequences were low ($< 0.5\%$). In each of the phylogenetic trees *O. nova* was monophyletic and separated with high support from the outgroup taxon *B. hauseri*. Trees (MrBayes, BEAST and RAxML) based on the *COI* nucleotide alignment were similar (**Figure S3–4**) and *COI* haplotypes generally clustered according to habitat type, irrespective of sampling locations (**Figure 2b**). Applying a mitochondrial substitution rate of 2.3% per million years, F and G lineages diverged between 6 and 16 mya (**Figure 2b**).

The *COI* haplotype network also showed a strong habitat related structure (**Figure 3**). Individuals from several sampling locations had identical or closely related haplotypes. However, individuals from forest (F) and grassland (G) had distinct haplotypes, irrespective of sampling locations, i.e., individuals from F and G always clustered separately. Haplotype from IFG either clustered with individuals from G or F, individuals from MG either clustered with individuals from G or IFG (except for two individuals from Solling that formed an isolated clade).

In total, 37 haplotypes were sampled and one haplotype was very common with a total of 21 individuals, 11 from IFG, 9 from MG and one from G. Haplotypes from IFG commonly occurred in more than one habitat type, F and IFG shared five, G and MG shared four common haplotypes, whereas IFG and MG as well as G and IFG shared only one haplotype. In contrast, no haplotypes were shared between F and MG as well as between F and G. Haplotype (Hd) and nucleotide (Π) diversity showed similar patterns (**Table S2**). Within sampling locations Hd of the four habitat types was similar, being between 0.8–0.96. In HA, both Hd and Π were lowest, in SO haplotype diversity was highest (Hd = 0.95) but nucleotide diversity was only intermediate (Π = 0.1). Haplotypes in other locations and in F and G were more different from each other.

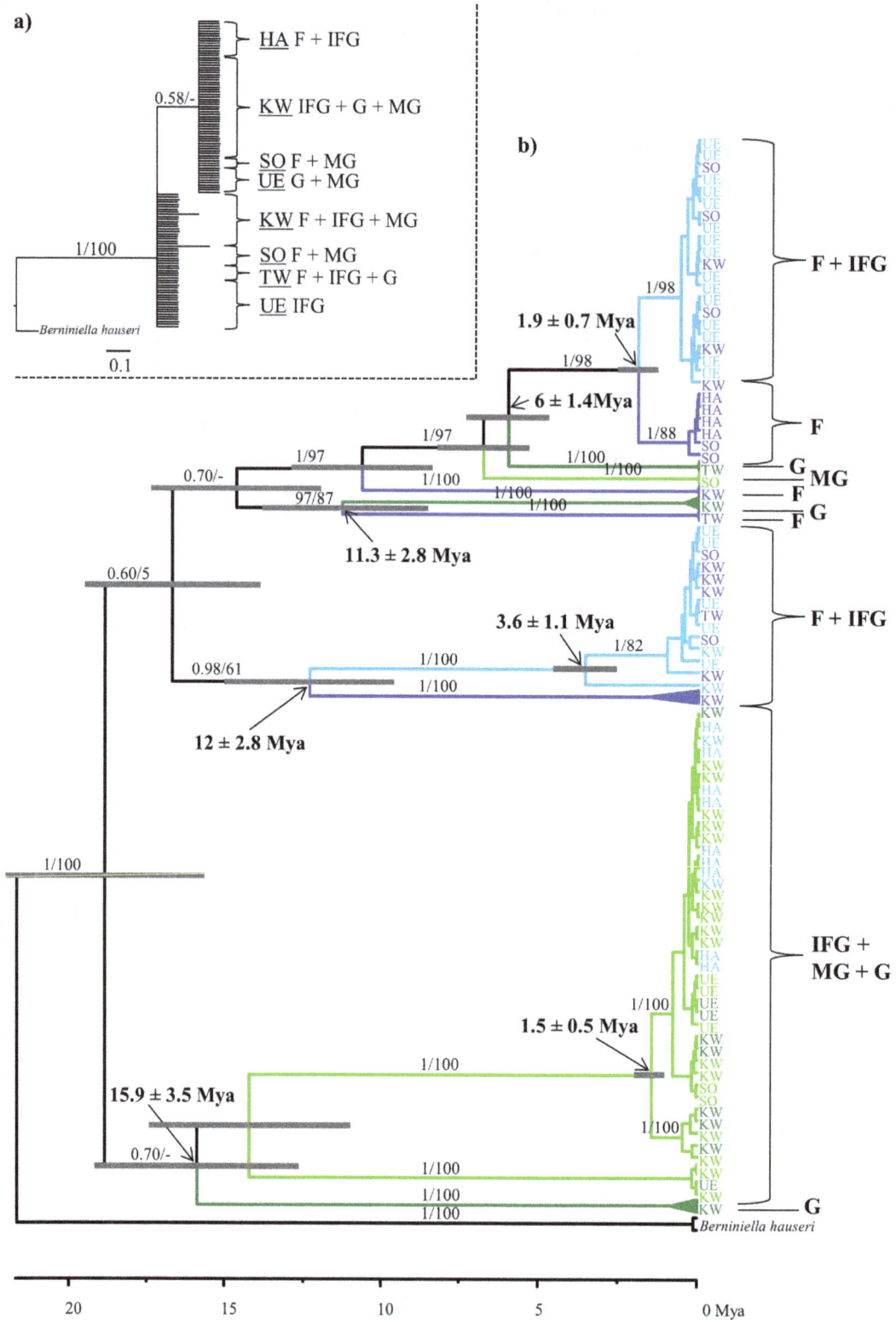

Figure 2. Bayesian phylogenetic trees showing the relatedness among individuals of *Oppiella nova* **from different habitats based on (a) nuclear (D3 region of the 28S rDNA, 126 individuals) and (b) mitochondrial markers (***COI***, 110 individuals).** Different colors indicate *COI* haplotypes of forest (F, blue), grassland (G, green), intersection of grassland and forest (IFG, light blue), and margin of grassland (MG, light green). Numbers on nodes represent posterior probabilities and bootstrap values, bold numbers are median estimated divergence times ±95% HPD calculated in BEAST using a strict molecular clock and grey bars on nodes indicate 95% HPD intervals. UE (Uelzen), SO (Solling), HA (Hainich), TW (Thuringian Forest) and KW (Kranichstein Forest) refer to the locations of the five forest – grassland gradients studied.

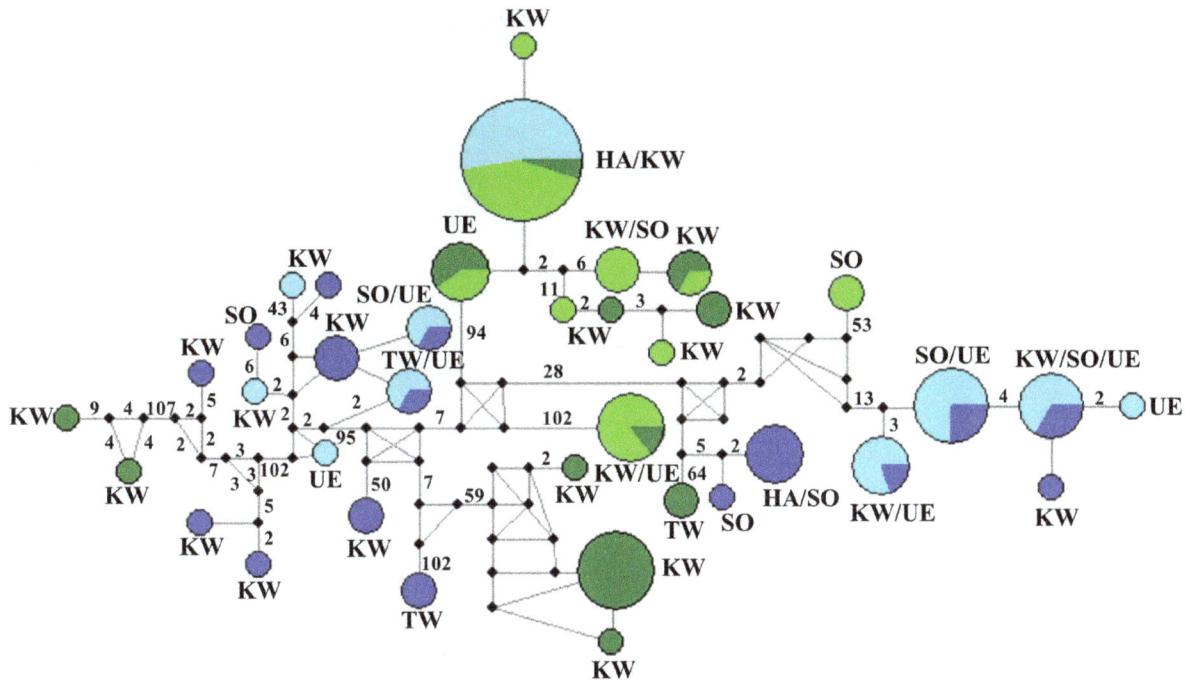

Figure 3. Median Joining Haplotype Network of 110 *COI* sequences of *Oppiella nova* collected from forest (blue), intersection of grassland and forest (light blue), margin of grassland (light green) and grassland (green) from five sampling locations (UE, Uelzen; SO, Solling; HA, Hainich; TW, Thuringian Forest; KW, Kranichstein Forest). The size of circles is proportional to the number of sequences per haplotype. Numbers on lines represent mutation steps separating the haplotypes; no number indicates a single mutation step. Haplotypes from forest and grassland are always separated by many mutation steps, but either haplotypes from forest or from grassland are closely related, even from distant sampling locations.

As indicated by AMOVA, genetic variance was generally high, being highest within samples (58%) and lower within locations (43%) and lowest within habitat types (35%) (**Table 2**). The negative variance component among locations resulted from low or nearly absent genetic structure. If the expectation of the estimator is zero, AMOVA can generate slightly negative variance components.

Among the three models tested with MIGRATE-N, migration between locations (model 3; log marginal likelihood = −5066, LBF = 0, MP = 1) was most likely. Substantially less likely were migrations between F and G (model 2; log marginal likelihood = −5317, LBF = −502, MP = 9.8E-110) and panmixis (model 1; log marginal likelihood = −5259, LBF = −386, MP = −1.5E-84). Isolation by distance was rejected being not significant (r(Y) = −0.24, P(rY) = 0.98). The McDonald Kreitman test was not significant for all comparisons as non-synonymous substitutions were not fixed within habitat types or locations.

Body length of *O. nova* (**Table S3**) in the different habitats ranged from an average of 251 to 275 μm with individuals from G being 24 μm longer than those from F ($F_{3,139}$ = 23.83, p<0.001 for habitat type; **Figure 4**). Accordingly, body length of individuals with forest and grassland specific genotypes differed significantly ($F_{1,22}$ = 22.06, p<0.001). Body size of individuals from IFG and F was similar; MG and IFG were in between that of individuals from F and G.

Discussion

The results indicate that *O. nova* differs both genetically and morphologically between forest and grassland. In agreement with the FNV hypothesis, haplotypes of forest and grassland were distinct and formed well-supported grassland and forest clades. Although individuals from both habitats were always distinct, some haplotypes also occurred in the transitional habitat types IFG and MG. This suggests niche-related environmental filtering between forest and grassland haplotypes with forest and grassland haplotypes coexisting in transitional habitats. Notably, forest and grassland haplotypes significantly differed in morphology with body size gradually increasing with distance from forest reaching a maximum in grassland specimens.

Considerable molecular variance was found in each of the locations and habitat types, suggesting independent colonization by different lineages rather than by a single locally adapted lineage. High molecular variance within sampling locations suggests that different lineages exist in neighboring habitat types at each sampling location. The results indicate that forest and grassland habitats are associated with certain niches selecting for specific genotypes with both niches being present in transitional habitats, which is consistent with the FNV model. Notably, haplotypes present in more than one habitat type also occurred at different locations. These widespread haplotypes predominantly colonized transitional habitats but haplotype diversity in these habitats was generally lower than in forests and grasslands. Environmental conditions in transitional habitats probably favor more generalist genotypes.

According to ecological niche theory, interspecific competition favors the evolution of species occupying separate niches. Species performance therefore is limited by environmental conditions and genetic adaptation, restricting geographic distribution. Similarly, intraspecific differentiation also can be linked to divergences in environmental conditions or resources. Niche differentiation typically is manifested in morphological differentiation, but may

Table 2. AMOVA of the *COI* gene of *Oppiella nova* on the variance among and within locations and among and within habitat types.

source of variation	d.f.	sum of squares	variance components	percentage of variation	fixiation indices	
Among locations	4	874	−0.40 Va	−1	Fct	−0.01
Among habitats	3	1,124	3.20 Va	6	Fct	0.06
Among samples within locations	8	1,609	22 Vb***	43	Fsc	0.43***
Among samples within habitats	9	1,308	19 Vb***	35	Fsc	0.39***
Within samples	97	2,898	30 Vc***	58	Fst	0.42***

Within samples variance was identical for both analyses; asterisks indicate significant differences at p<0.001; d.f. are degrees of freedom.

also be cryptic and only recognizable at physiological, genetic or transcriptomic levels [65–68]. Difference in body size is a common feature that separates individuals along a single resource dimension [69], whereas genetic differentiation is usually correlated with reproductive or geographic isolation [70–72].

In *O. nova* isolation by distance was not significant and differences in body size correlated with separation into forest and grassland, indicating that niche specific size-dimorphism is due to habitat specific adaptations rather than geographical differentiation. Variation in body size likely reflects niche differentiation, which often is induced by resource shifts and differential exposure to predators [73–75]. Stable isotope data from oppiid species indicate that *O. nova* lives as predator or scavenger [76], [77] and size dimorphism therefore may reflect adaptation to prey of different body size. However, differences in habitat structure and different predator communities in grasslands and forests may also

be responsible for the observed variations in body size. Adult oribatid mites typically are well protected from predation by morphological and chemical defenses [78–80]. However, Schneider and Maraun [81] demonstrated that gamasid mites, the most vigorous predators of soil microarthropods, prey heavily on *O. nova*. Gamasid mites preferentially prey on oribatid mite species of a body size of 200–300 μm [81], indicating that larger and smaller species live in size refuges. Large oribatid mites are heavily sclerotized while smaller ones and juveniles typically are weakly sclerotized but colonize pore space inaccessible for predators such as gamasid mites. For *O. nova*, which is small, weakly sclerotized and mobile, top-down control by gamasid mite predators is likely to be important with larger individuals suffering less from predation by gamasid mites than smaller ones. Differences in body size in forest and grassland therefore may reflect body size related differences in predation by gamasid mites. Unfortunately,

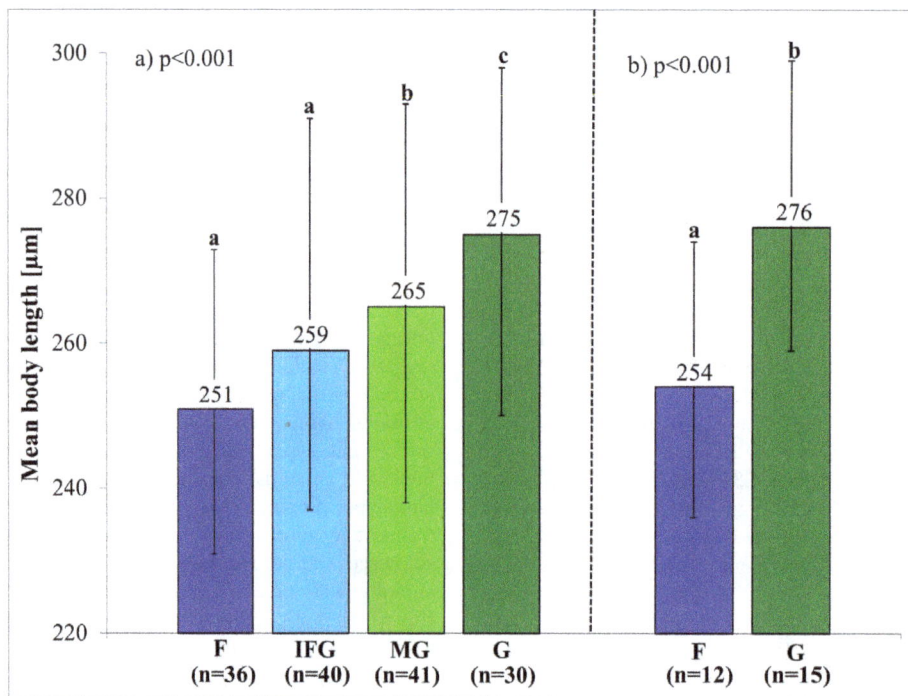

Figure 4. Body length of *Oppiella nova* from forests, grasslands and transitional habitat types between forests and grassland. (a) Differences between all collected individuals (147); forest (F), grassland (G), intersection of forest and grassland (IFG), and margin of grassland (MG). (b) Differences in body size between individuals with forest and grassland specific genotypes. Numbers in brackets refer to the number of individuals included in the analyses; error bars indicate standard deviation. P-values correspond to Tuckey's HSD test.

little is known on the control of oribatid mites by gamasid mite predators in the field and whether this differs between forest and grassland.

Overall, our data indicate ecological differentiation of a parthenogenetic lineage into discrete genetic and morphological entities. The gradual change in haplotype composition and body size between forest and grassland indicates adaptation to specific environmental conditions, i.e. a shift in ecological niches. Further, the results suggest that in addition to haplotypes from both forest and grassland, transitional habitats are colonized by widespread genotypes with lower haplotype diversity than forest and grassland. In contrast to forests and grassland, oribatid mites of transitional habitats may be less affected by predation but rather by abiotic forces due to more variable climatic conditions. Despite the distinctness of forest and grassland lineages and non-synonymous substitutions in the *COI* gene, no indications for divergent selection were found. This may be due to the large population size of *O. nova* as genetic drift and fixation probability of mutations decrease with increasing population size. *Oppiella nova* is among the most abundant oribatid mite species in grasslands and forests and can reach densities of thousands of individuals per square meter [82], [83], [34]. This suggests that extinction rates and bottlenecks are of minor importance explaining why genetic variance of the *COI* fragment within populations is high. Despite separation of shallow clades by long branches in the mitochondrial dataset, which may indicate a cryptic species complex, low D3 variance suggests that *O. nova* may best be treated as single (parthenogenetic) species. High intraspecific *COI* variance is common in arthropods [84], [85], especially in those living in soil [86–88], including parthenogenetic oribatid mites [28] and bdelloid rotifers [89].

In contrast to *O. nova*, haplotype diversity in the parthenogenetic oribatid mite *P. peltifer* suggested a general purpose genotype [28]. *Oppiella nova* is a fast reproducing [90] weakly sclerotized r-strategist [91] presumably feeding on living resources and therefore subject to co-evolutionary adaptations [76]. In contrast, *P. peltifer* reproduces slowly and is strongly sclerotized, characters typical for K-strategists. It predominantly feeds on dead organic matter suggesting that co-evolutionary processes between consumer and (dead) food resource are non-existing [91], [77], thereby facilitating more generalist genoptypes.

Our age estimations suggest that lineages of *O. nova* from grassland separated from those of forests during the Middle and Late Miocene (16-6 mya). The substitution rate of parthenogenetic species may differ from the general rate of *COI* established for arthropods. Still, age estimates and high genetic distances between forest and grassland lineages suggest long-term separation and persistence of lineages, contradicting the commonly held view that parthenogenetic lineages are short-lived evolutionary dead ends. Speciation of parthenogenetic lineages has been assumed to be responsible for the formation of large phylogenetic clusters in bdelloid rotifers [92–94] and certain groups of oribatid mites [40], [95]. The age of grassland lineages correlated well with the expansion of grasslands in the Miocene [96], [97] indicating long-standing adaptation to this habitat. Present day occurrence of grassland and forest lineages in managed European grasslands and forests, respectively, suggests recurrent establishment of lineages due to environmental filtering, i.e. grassland and forest lineages remained bound to the respective habitats.

Our results suggest that, as in sexual species, environmental filters and biotic interactions contribute to the evolution of parthenogenetic species. High genetic variability presumably is maintained by adaptation of certain genotypes to environmental settings as suggested by the FNV hypothesis. Habitat partitioning

and coexistence of parthenogenetic lineages at local scales suggest that speciation may occur sympatrically.

Supporting Information

Figure S1 Maximum Likelihood tree showing the relatedness among individuals of *Oppiella nova* from different habitats based on nuclear marker (D3 region of the 28S rDNA, 126 individuals). UE (Uelzen), SO (Solling), HA (Hainich), TW (Thuringian Forest) and KW (Kranichstein Forest) refer to the locations of the five forest – grassland gradients studied (see Figure 1).

Figure S2 Bayesian phylogenetic tree showing the relatedness among individuals of *Oppiella nova* from different habitats based on nuclear marker (D3 region of the 28S rDNA, 126 individuals). UE (Uelzen), SO (Solling), HA (Hainich), TW (Thuringian Forest) and KW (Kranichstein Forest) refer to the locations of the five forest – grassland gradients studied.

Figure S3 Maximum Likelihood tree showing the relatedness among individuals of *Oppiella nova* from different habitats based on mitochondrial marker (*COI*, 110 individuals). UE (Uelzen), SO (Solling), HA (Hainich), TW (Thuringian Forest) and KW (Kranichstein Forest) refer to the locations of the five forest – grassland gradients studied.

Figure S4 Bayesian phylogenetic tree showing the relatedness among individuals of *Oppiella nova* from different habitats based on mitochondrial marker (*COI*, 110 individuals UE (Uelzen), SO (Solling), HA (Hainich), TW (Thuringian Forest) and KW (Kranichstein Forest) refer to the locations of the five forest – grassland gradients studied.

Table S1 Non-synonymous amino acid substitutions among *COI* sequences of Oppiella nova from F, G, IFG and MG. Non-synonymous substitutions are highlighted in red and positions in the *COI* fragment are indicated; individuals affected are from Kranichstein Forest (KW), Thuringian Forest (TW) and Uelzen (UE).

Table S2 Number of individuals (n Ind), number of haplotypes (n haplo), haplotype (Hd) and nucleotide diversity (π) of habitats and locations of Oppiella nova.

Table S3 Body length [µm] of all individuals of *Oppiella nova* sampled for this study and are included in Fig. 4a.

Alignment S1 Alignment of the 28S rDNA (D3 region; 356 bp) including 126 individuals of *O. nova*.

Alignment S2 Alignment of the *COI* gene (709 bp) including 110 individuals of *O. nova*.

Acknowledgments

We thank Georgia Erdmann for help with species determination and Bernhard Klarner and Christoph Digel for help with statistical analyses.

Data Accessibility

DNA sequences: Genbank accession numbers KF293402 - KF293641.

References

1. Glesener RR, Tilman D (1978) Sexuality and the components of environmental uncertainty: clues from geographic parthenogenesis in terrestrial animals. Am Nat 112: 659–673.
2. Lynch M (1984) Destabilizing hybridization, general-purpose genotypes and geographical parthenogenesis. Q Rev Biol 59: 257–290.
3. Suomalainen E, Saura A, Lokki J (1987) Cytology and evolution in parthenogenesis. Bocca Raton: CRC Press.
4. Birdsell JA, Wills C (2003) The evolutionary origin and maintenance of sexual recombination: A review of contemporary models. Evol Biol 33: 27–137.
5. Muller HJ (1964) The relation of recombination to mutational advance. Mutat Res 1: 2–9.
6. Kondrashov AS (1988) Deleterious mutations and the evolution of sexual reproduction. Nature 336: 435–440.
7. Fisher RA (1930) The Genetical Theory of Natural Selection. In: Bennet JH, editor. The Genetical Theory of Natural Selection. Oxford: Oxford University Press.
8. Bell G (1982) The Masterpiece of Nature: The Evolution and Genetics of Sexuality. London: Guildford and King's Lynn.
9. Butlin R (2002) Evolution of sex: The costs and benefits of sex: new insights from old asexual lineages. Nature Rev Genet 3: 311–317.
10. Vrijenhoek RC. 1979 Factors affecting clonal diversity and coexistance. Amer Zool 19: 787–797.
11. Vrijenhoek RC (1984a) The evolution of clonal diversity in *Poeciliopsis*. In: Turner BJ, editor. Evolutionary Genetics of Fishes. New York: Plenum Press. 399–429.
12. Vrijenhoek RC (1984b) Ecological differentiation among clones: the Frozen-Niche Variation model. In: Wöhrmann K, Loeschke V, editors. Population Biology and Evolution. Heidelberg: Springer-Verlag. 217–231.
13. Wetherington JD, Schenck SA, Vrijenhoek RC (1989) The origins and ecological success of unisexual *Poeciliopsis*: the Frozen Niche-Variation model. In: Snelson FF Jr, editor. Ecology and evolution of livebearing fishes (Poeciliidae). New Jersey: Prentice Hall, Englewood Cliffs. 259–275.
14. Lima NRW (1998) Genetic analysis of predatory efficiency in natural and laboratory made hybrids of *Poeciliopsis* (Pisces: Poeciliidae). Behaviour 135: 83–98.
15. Gray MM, Weeks SC (2001) Niche breadth in clonal and sexual fish (*Poeciliopsis*): a test for the frozen niche variation model. Can J Fish Aquat Sci 58: 1313–1318.
16. Hotz H, Guex GD, Beerli P, Semlitsch RD, Pruvost NBM (2007) Hemiclone diversity in the hybridogenetic frog *Rana esculenta* outside the area of clone formation: the view from protein electrophoresis. J Zoolog Syst Evol Res 46: 56–62.
17. Groot TVM, Janssen A, Pallini A, Breeuwer JAJ (2005) Adaptation in the asexual false spider mite *Brevipalpus phoenicis*: evidence for frozen niche variation. Exp Appl Acarol 36: 165–176.
18. Browne RA, Hoopes CW (1990) Genotype diversity and selection in asexual brine shrimp. Evolution 33: 848–859.
19. Pantel JH, Juenger TE, Leibaold MA (2011) Environmental gradients structure *Daphnia pulex* x *pulicaria* clonal distribution. J Evolution Biol 24: 723–732.
20. Baker HG (1965) Characteristics and Modes of Origin of Weeds. In: Baker HG, Stebbins GL, editor. The Genetics of Colonizing Species. New York: Academic Press. 147–172.
21. Lynch M, Bürger R, Butcher D, Gabriel W (1993) The mutational meltdown in asexual populations. J Hered 84: 339–344.
22. Lynch M (1985) Spontaneous mutations for life history characters in an obligate parthenogen. Evolution 39: 804–818.
23. Lynch M, Gabriel W (1987) Environmental tolerance. Am Nat 129: 283–303.
24. Schlosser IJ, Doeringsfeld MR, Elder JF, Arzayus LF (1998) Niche relationships of clonal and sexual fish in a heterogeneous landscape. Ecology 79: 953–968.
25. Myers MJ, Meyer CP, Resh VH (2000) Neritid and thiarid gastropods from French Polynesian streams: how reproduction (sexual, parthenogenetic) and dispersal (active, passive) affect population structure. Freshw Biol 44: 535–545.
26. Van Doninck K, Schön I, De Bruyn L, Martens K (2002) A general purpose genotype in an ancient asexual. Oecologia 132: 205–212.
27. Van Doninck K, Schön I, Martens K, Backeljau T (2004) Clonal diversity in the ancient asexual ostracod *Darwinula stevensoni* assessed by RAPD-PCR. Heredity 93: 154–160.
28. Heethoff M, Domes K, Laumann M, Maraun M, Norton RA et al. (2007) High genetic divergences indicate ancient separation of parthenogenetic lineages of the oribatid mite *Platynothrus peltifer* (Acari, Oribatida). J Evolution Biol 2: 392–402.
29. Andersen FA, Jordal BH, Kambestad M, Kirkendall LR (2011) Improbable but true: the invasive inbreeding ambrosia beetle *Xylosandrus morigerus* has generalist genotypes. Ecol Evol 2: 247–257.
30. Schoen I, Martens K, van Dijk P (2009) *Lost Sex: The Evolutionary Biology of Parthenogenesis*. Heidelberg: Springer-Verlag.
31. Hutson BR (1980) Colonization of industrial reclamation sites by acari, collembola and other invertebrates. J Appl Ecol 17: 255–275.
32. Wanner M, Dunger W (2002) Primary immigration and succession of soil organisms on reclaimed opencast coal mining areas in Eastern Germany. Eur J Soil Biol 38: 137–143.
33. Lindberg N, Bengtsson J (2005) Population responses of oribatid mites and collembolans after drought. Appl Soil Ecol 28: 163–174.
34. Siira-Pietikäinen A, Penttinen R, Huhta V (2008) Oribatid mites (Acari: Oribatida) in boreal forest floor and decaying wood. Pedobiologia 52: 111–118.
35. Erdmann G, Scheu S, Maraun M (2012) Regional factors rather than forest type drive the community structure of soil living oribatid mites (Acari, Oribatida). Exp Appl Acarol 57: 159–167.
36. Cianciolo JM, Norton RA (2006) The ecological distribution of reproductive mode in oribatid mites, as related to biological complexity. Exp Appl Acarol 40: 1–25.
37. Weigmann G (2006) Hornmilben (Oribatida) In: *Die Tierwelt Deutschlands*. Keltern: Goecke & Evers.
38. Woas S (1986) Beitrag zur Revision der Oppioidea sensu Balogh, 1972 (Acari, Oribatei). Andrias 5: 21–224.
39. Cruickshank RH (2002) Molecular markers for the phylogenetics of mites and ticks. Syst Appl Acarol 7: 3–14.
40. Maraun M, Heethoff M, Schneider K, Scheu S, Weigmann G et al. (2004) Molecular phylogeny of oribatid mites (Oribatida, Acari): evidence for multiple radiations of parthenogenetic lineages. Exp Appl Acarol 33: 183–201.
41. Lehmitz R, Russell D, Hohberg K, Christian A, Xylander WER (2012) Active dispersal of oribative mites into young soils. Appl Soil Ecol 55: 10–19.
42. Kempson D, Lloyd M, Ghellardi R (1963). A new extractor for woodland litter. Pedobiologia 3: 1–21.
43. Folmer O, Black M, Hoeh W, Lutz R, Vrijenhoek R (1994) DNA primers for amplification of mitochondrial cytochrome c oxidase subunit I from diverse metazoan invertebrates. Mol Mar Biol Biotechnol 3: 294–299.
44. Litvaitis MK, Nunn G, Thomas WK, Kocher TD (1994) A molecular approach for the identification of meiofaunal turbellarians (Platyhelminthes, Turbellaria). Mar Biol 120: 437–442.
45. Hall TA (1999) BioEdit: a user-friendly biological sequence alignment editor and analysis program for Windows 95/98/NT. Nucleic Acids Symp 41: 95–98.
46. Thompson JD, Gibson TJ, Plewniak F, Jeanmougin F, Higgins DG (1997) The CLUSTAL_X windows interface: flexible strategies for multiple sequence alignment aided by quality analysis tools. Nucleic Acids Res 25: 4876–4882.
47. Stamatakis A (2014) RAxML Version 8: A tool for Phylogenetic analysis and Post-Analysis of large Phylogenetics. Bioinformatics.
48. Ronquist F, Huelsenbeck JP (2003) MrBayes 3: Bayesian phylogenetic inference under mixed models. Bioinformatics 19: 1572–1574.
49. Drummond AJ, Suchard A, Xie D, Rambaut A (2012) Bayesian Phylogenetics with BEAUti and the BEAST 1.7. Mol Biol Evol 29: 1969–1973.
50. Darriba D, Taboada GL, Doallo R, Posada D (2012). jModelTest 2: more models, new heuristics and parallel computing. Nat. Methods 9: 772.
51. Guindon S, Gascuel O (2003). A simple, fast and accurate method to estimate large phylogenies by maximum-likelihood. Syst Biol 52: 696–704.
52. Lanave C, Preparata G, Saccone C, Serio G (1984). A new method for calculating evolutionary substitution rates. J Mol Evol 20: 86–93.
53. Ziheng Y (1994). Maximum likelihood phylogenetic estimation from DNA sequences with variable rates over sites: Approximate methods. J Mol Evol 39: 306–314.
54. Brower AVZ (1994) Rapid morphological radiation and convergence among races of the butterfly *Heliconius erato* inferred from patterns of mitochondrial DNA evolution. Proc Natl Acad Sci 91: 6491–6495.
55. Avise JC (1994) Molecular Markers, Natural History and Evolution. New York: Chapman & Hall.
56. Gernhard T, Hartmann K, Steel M (2008) Stochastic properties of generalised Yule models, with biodiversity applications. J Math Biol 57: 713–735.
57. Drummond AJ, Rambaut A (2007) BEAST: Bayesian evolutionary analysis by sampling trees. BMC Evol Biol 7: 214.
58. Beerli P (2009) How to use MIGRATE or why are Markov chain Monte Carlo programs difficult to use? In: *Population Genetics for Animal Conservation* (ed. Bertorelle G, Bruford MW, Hauffe HC, Rizzoli A, Vernesi C), 42–79. Cambridge University Press, Cambridge, UK.
59. Beerli P, Palczewski M (2010) Unified framework to evaluate panmixia and migration direction among multiple sampling locations. Genetics 185: 313–326.
60. Excoffier LG, Laval G, Schneider S (2005) Arlequin ver. 3.0: An integrated software package for population genetics data analysis. Evol Bioinform Online 1: 47–50.

Author Contributions

Conceived and designed the experiments: IS. Performed the experiments: HvS. Analyzed the data: HvS. Contributed reagents/materials/analysis tools: SS. Contributed to the writing of the manuscript: HvS MM SS IS.

61. McDonald JH, Kreitman M (1991) Adaptive protein evolution at the Adh locus in *Drosophila*. Nature 351: 652–654.
62. Librado P, Rozas J (2009) DnaSP V5: A software for comprehensive analysis of DNA polymorphism data. Bioinformatics 25: 1451–1452.
63. Pinheiro J, Bates D, DebRoy S, Sarkar D and the R Development Core Team (2013) nlme: Linear and Nonlinear Mixed Effects Models. R Package version 3.1–111.
64. Hothorn T, Bretz F, Westfall T (2008) Simultaneous Inference in General Paramatric Models. Biomet J 40: 346–363.
65. Christensen B (1980) Constant differential distribution of genetic variants in polyploid parthenogenetic forms of *Lumbricillus lineatus* (Enchytraeidae, Oligochaeta). Hereditas 92: 193–198.
66. Posthuma L (1990) Genetic differentiation between populations of *Orchesella cincta* (Collembola) from heavy metal contaminated sites. J. Apll. Ecol. 27: 609–622.
67. Heethoff M, Etzold K, Scheu S (2004) Mitochondrial COII sequences indicate that the parthenogenetic earthworm *Octolasion tyrtaeum* (Savigny 1826) constitutes of two lineages differing in body size and genotype. Pedobiologia 48: 9–13.
68. Janssens TKS, Roelofs D, van Straalen NM (2009) Molecular mechanisms of heavy metal tolerance and evolution in invertebrates. Insect Sci 16: 3–18.
69. Hutchinson GE (1959) Homage to Santa Rosalia or why are there so many kinds of animals? Amer Nat 93: 145–159.
70. Hansen MM, Mensberg KLD (1998) Genetic differentiation and relationship between genetic and geographical distance in Danish sea trout (*Salmo trutta* L.) populations. Heredity 81: 493–504.
71. Whitaker RJ, Grogan DW, Taylor JW (2003) Geographic barriers isolate endemic populations of hyperthermophilic Archaea. Science 301: 976–978.
72. Ramachandran S, Deshpande O, Roseman CC (2005) Support from the relationship of genetic and geographic distance in human populations for a serial founder effect originating in Africa. Proc Natl Acad Sci 102: 15942–15947.
73. Brönmark C, Pettersson LB, Nilsson PA (1999) Predator-induced defense in crucian carp. In: Tollrian R, Harvell CD, editors. The Ecology and Evolution of Inducible Defenses. New Jersey: Princeton University Press. 203–217.
74. East TL, Havens KE, Rodusky AJ, Brady MA (1999) *Daphnia lumholtzi* and *Daphnia ambigua*: population comparisons of an exotic and native cladoceran in Lake Okeechobee, Florida. J Plankton Res 21: 1537–1551.
75. Cooper Jr WE, Stankowich T (2010) Prey or predator? Body size of an approaching animal affects decisions to attack or escape. Behav Ecol 21: 1278–1284.
76. Schneider K, Migge S, Norton RA, Scheu S, Langel R et al. (2004) Trophic niche differentiation in soil microathropods (Oribatida, Acari): evidence from stable isotope ratios (^{15}N/^{14}N). Soil Biol Biochem 36: 1769–1774.
77. Maraun M, Erdmann G, Fischer BM, Pollierer MM, Norton RA et al. (2011) Stable isotopes revisited: Their use and limits for oribatid mite trophic ecology. Soil Biol Biochem 43: 877–882.
78. Sanders FH, Norton RA (2004) Anatomy and function of the ptychoid defensive mechanism in the mite *Euphthiracarus cooki* (Acari: Oribatida). J Morphol 259: 119–154.
79. Peschel K, Norton RA, Scheu S, Maraun M (2006) Do oribatid mites live in enemy-free space? Evidence from feeding experiments with the predatory mite *Pergamasus septentrionalis*. Soil Biol Biochem 38: 2985–2989.
80. Heethoff M, Raspotnig G (2012) Expanding the 'enemy-free space' for oribatid mites: evidence for chemical defense of juvenile *Archegozetes longisetosus* against the rove beetle *Stenus juno*. Exp Appl Acarol 56: 93–97.
81. Schneider K, Maraun M (2009) Top-down control of soil microarthropods – Evidence from a laboratory experiment. Soil Biol Biochem 41: 170–175.
82. Maraun M, Scheu S (2000) The structure of oribatid mite communities (Acari, Oribatida): patterns, mechanisms and implications for future research. Ecography 23: 374–783.
83. Penttinen R, Siira-Pietikäinen A, Huhta V (2008) Oribatid mites in eleven different habitats in Finland. In: Bertrand M, Kreiter S, McCoy KD, Migeon A, Navajas M, Tixier MS, Vial L, editors. Integrative Acarology. Montpellier: Proceedings of the 6th European Congress of the EURAAC. 237–244.
84. Schäffer S, Koblmüller S, Pfingstl T, Sturmbauer C, Krisper G (2010) Contrasting mitochondrial DNA diversity estimates in Austrian *Scutovertex minutus* and *S. sculptus* (Acari, Oribatida, Brachypylina, Scutoverticidae). Pedobiologia 53: 203–211.
85. Edmands S (2001) Phylogeography of the intertidal copepod *Tigriopus californicus* reveals substantially reduced population differentiation at northern latitudes. Mol Ecol 10: 1743–50.
86. Rosenberger M, Maraun M, Scheu S, Schaefer I (2013) Pre- and post-glacial diversifications shape genetic complexity of soil-living microarthropod species. Pedobiologia 56: 79–87.
87. Torricelli G, Carapelli A, Convey P, Nardi F (2010) High divergence across the whole mitochondrial genome in the "pan-Antarctic" springtail *Friesea grisea*: Evidence for cryptic species? Gene 449: 30–40.
88. Boyer SL, Baker JM, Giribet G (2007) Deep genetic divergences in *Aoraki denticulata* (Arachnida, Opiliones, Cyphophthalmi): a widespread "mite harvestman" defies DNA taxonomy. Mol Ecol 16: 4999–5016.
89. Fontaneto D, Boschetti C, Ricci C (2008) Cryptic diversification in ancient asexuals: evidence from the bdelloid rotifer *Philodina flaviceps*. J Evolution Biol 21: 580–587.
90. Kaneko N (1988) Life history of *Oppiella nova* (Oudemans) (Oribatei) in cool temperate forest soils in Japan. Acarologia 29: 215–221.
91. Norton RA (1994) Evolutionary Aspects of Oribatid Mite Life Histories and Consequences for the Origin of the Astigmata. In: Houck MA, editor. Mites. Ecological and Evolutionary Analyses of Life-History Patterns. New York: Chapman & Hall. 99–135.
92. Mark Welch D, Meselson M (2000) Evidence for the evolution of bdelloid rotifers without sexual reproduction or genetic exchange. Science 288: 1211–1215.
93. Normark BB, Judson OP, Moran NA (2003) Genomic signatures of ancient asexual lineages. Biol J Linnean Soc 79: 69–84.
94. Birky Jr CW, Wolf C, Maughan H, Herbertson L. Henry E (2005) Speciation and selection without sex. Hydrobiologia 546: 29–45.
95. Laumann M, Norton RA, Weigmann G, Scheu S, Maraun M et al. (2007) Speciation in the parthenogenetic oribatid mite genus *Tectocepheus* (Acari, Oribatida) as indicated by molecular phylogeny. Pedobiologia 51: 111–122.
96. Retallack GJ (2001) Cenozoic expansion of grasslands and climatic cooling. J Geol 109: 407–426.
97. Osborne CP, Beerling DJ (2006) Nature's green revolution: the remarkable evolutionary rise of C$_4$ plants. Phil Trans R Soc A 361: 173–194.

22

Habitat Capacity for Cougar Recolonization in the Upper Great Lakes Region

Shawn T. O'Neil, Kasey C. Rahn¤, Joseph K. Bump*

School of Forest Resources and Environmental Science, Michigan Technological University, Houghton, Michigan, United States of America

Abstract

Background: Recent findings indicate that cougars (*Puma concolor*) are expanding their range into the midwestern United States. Confirmed reports of cougar in Michigan, Minnesota, and Wisconsin have increased dramatically in frequency during the last five years, leading to speculation that cougars may re-establish in the Upper Great Lakes (UGL) region, USA. Recent work showed favorable cougar habitat in northeastern Minnesota, suggesting that the northern forested regions of Michigan and Wisconsin may have similar potential. Recolonization of cougars in the UGL states would have important ecological, social, and political impacts that will require effective management.

Methodology/Principal Findings: Using Geographic Information Systems (GIS), we extended a cougar habitat model to Michigan and Wisconsin and incorporated primary prey densities to estimate the capacity of the region to support cougars. Results suggest that approximately 39% (>58,000 km^2) of the study area could support cougars, and that there is potential for a population of approximately 500 or more animals. An exploratory validation of this habitat model revealed strong association with 58 verified cougar locations occurring in the study area between 2008 and 2013.

Conclusions/Significance: Spatially explicit information derived from this study could potentially lead to estimation of a viable population, delineation of possible cougar-human conflict areas, and the targeting of site locations for current monitoring. Understanding predator-prey interactions, interspecific competition, and human-wildlife relationships is becoming increasingly critical as top carnivores continue to recolonize the UGL region.

Editor: Cédric Sueur, Institut Pluridisciplinaire Hubert Curien, France

Funding: This project was funded and supported by the School of Forest Resources and Environmental Science Ecosystem Science Center at Michigan Technological University. The funders had no role in study design, data collection and analysis, decision to publish, or preparation of the manuscript.

Competing Interests: The authors have declared that no competing interests exist.

* Email: jkbump@mtu.edu

¤ Current address: University of Montana, Missoula, Montana, United States of America

Introduction

Cougars (*Puma concolor*) once spanned North and South America, ranging from south of the boreal forests to Patagonia [1,2]. By the early twentieth century in the United States, human persecution, habitat degradation, and human expansion resulted in the extirpation of cougars from two-thirds of their historic range including eastern and midwestern America [3]. Cougars persisted only in the American west, where populations are increasing for the first time in nearly a century [4]. As a result of this increase, cougars are recolonizing portions of their former range. For example, natural recolonization, aided by changes in cougar protection and prey management, led to the return of viable populations in Wyoming and the Black Hills of South Dakota by 2000 [4]. Recolonization also occurred in the Badlands of North Dakota and in western Nebraska [5,6].

Cougars, particularly young males, will travel hundreds of kilometers in search of new territory [7–9]. In 2011, one individual traveled more than 1,700 km from Minnesota to Connecticut, and may have traveled a straight-line distance of 2,500 km from the Black Hills to the East Coast [10]. In recent years, reports of cougar presence have been more frequent and widespread in the northern Midwest and Great Lakes States. Verified photograph and video evidence from automatic cameras, human-cougar encounters, DNA samples (scat, hair and blood) and track records was recorded each year in Michigan, Wisconsin, and Minnesota between 2008 and 2013. DNA analysis of samples suggested that a minimum of 6 individual cougars visited Wisconsin during this time period [11]. Evidence of cougar presence was also verified in Ontario, Canada [12]. Biologists believe confirmed cougar occurrences are young dispersing males from western populations, as evidenced by the presence of radio collars originating from western-based research and monitoring programs [10, Michigan Department of Natural Resources (DNR) *unpublished data*]. This increase in occurrence is consistent with the expectation that as populations rise in the American west, long-distance dispersals become more frequent, particularly among males [9,13]. Cougars are now exploring much of the Midwest, ostensibly dispersing to expand to new territories, to increase mating opportunities, and to avoid overlap with existing home ranges [9,13,14].

The forests of northern Wisconsin and Michigan appear to exhibit high potential for cougar recolonization, due to favorable configurations of land cover, road density, abundant food resources, and relatively low human population densities [15,16].

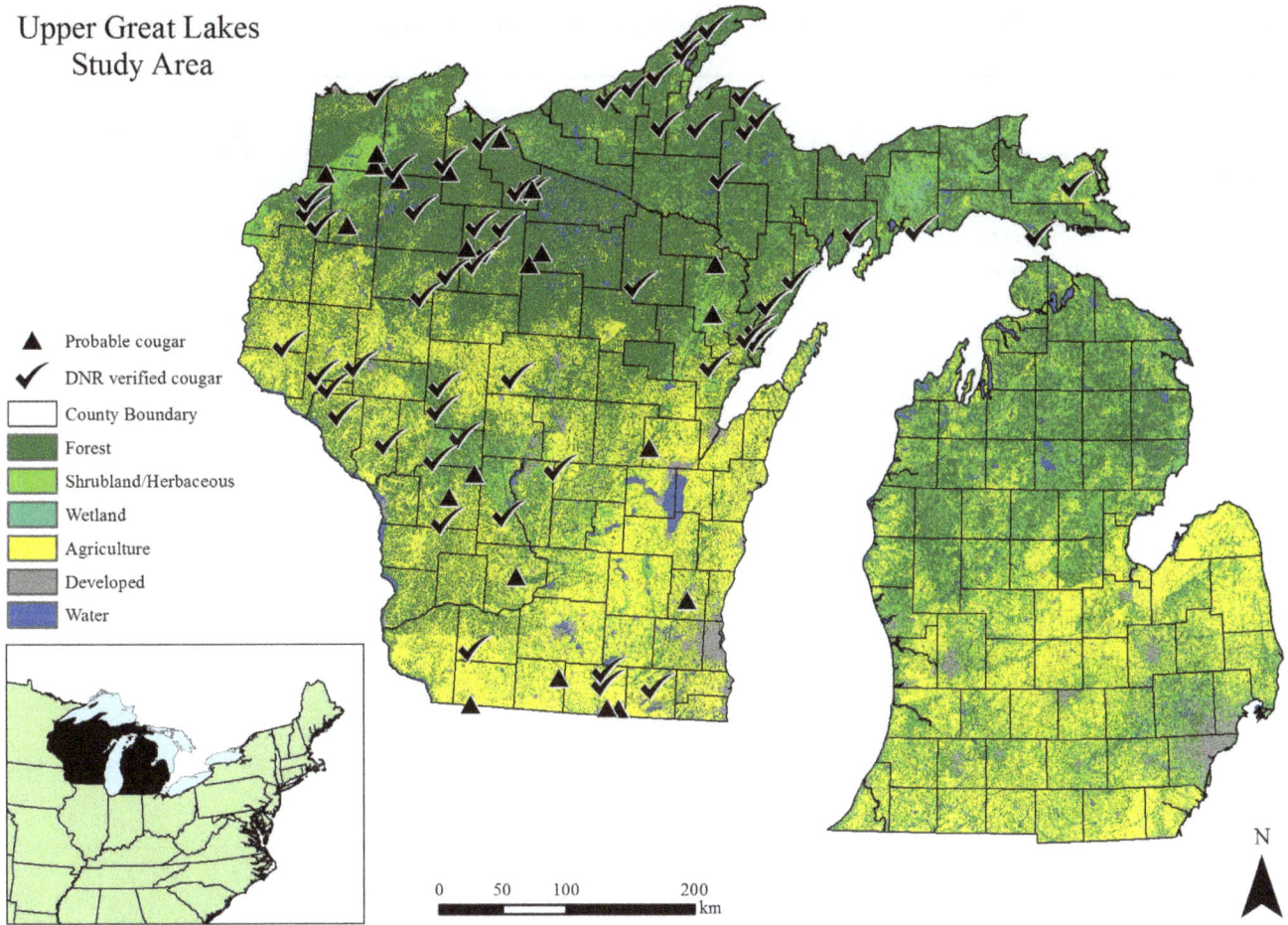

Figure 1. The study area for an analysis of cougar habitat and capacity for the Upper Great Lakes states, USA. Probable and verified cougar locations are represented from 1 January, 2008 to 1 June, 2013.

These same features have also shown to be important habitat features for gray wolves (*Canis lupus*) in this region [17–19]. Evidence of cougar range expansion has future conservation implications for the states of Michigan, Wisconsin, and Minnesota, which likely contain suitable habitat [20]. Despite little evidence of a breeding population [e.g. female cougars, kittens; but see [21,22], these states are preparing for the emergence of a reproducing cougar population at some point in the near future [23]).

Potential cougar range expansion to the Upper Great Lakes (UGL) region raises important ecological questions and may require novel strategies to manage the species. Current research and management needs include mapping the geographic extent of potential cougar habitat in the UGL and estimating its capacity to support cougars. Researchers have modeled potential cougar habitat based on life history requirements to address questions of habitat extent where accurate data on the species' distribution are lacking [16,20]. These models can be combined with prey densities to estimate an area's capacity for the species [24,25]. Here, we address the need for spatially explicit cougar habitat data in Michigan and Wisconsin by extending an expert-assisted GIS spatial model [16,20,26]. We also provide an exploratory validation of the model, and estimate cougar capacity based on prey resources (white-tailed deer [*Odocoileus virginianus*]) and favorable land cover characteristics.

Methods

Study area

Our study area encompassed the states of Michigan and Wisconsin where DNR-confirmed cougar locations were widespread since 2008 (Fig. 1). Land cover types in the northern regions of both states are forest-dominated, with agriculture and human development becoming more predominant in the south. Forested land occurred over approximately 47% of the entire study area. Forests were generally of northern hardwood association, with common types including maple (*Acer* spp.) – birch (*Betula spp.*) – hemlock (*Tsuga canadensis*), aspen (*Populus* spp.) – birch (*Betula* spp.), and oak (*Quercus spp.*). Additional forest types included spruce-fir (*Picea* spp., *Abies balsamea*), and pine (*Pinus banksiana, P. resinosa, P. strobus*). Agricultural land use occurred on approximately 31% of the study area and generally consisted of wheat, corn, and soybean crops. Livestock grazing and dairy farming were common in more southern regions, particularly in Wisconsin. Other land cover types included shrubland/herbaceous, wetlands, and urban/developed (Fig. 1). Elevation ranged from 141 to 602 m and terrain was generally flat to hilly with an overall median slope estimate of 1.07° (Interquartile Range [IQR] = 0.34°–2.73°); however, more rugged terrain was present throughout the region and slopes reached 81.5° in some of these areas. The study area was characterized by

Table 1. Weights for attributes within variables used to model potential habitat suitability for cougars in the Upper Great Lakes states (Adapted from [26]).

Variable	Attribute	Weight (±S.E.)	Percent importance from highest ranking variable
Land cover	Mixed forest	1.92 (0.51)	100
	Deciduous forest	1.61 (0.37)	84
	Evergreen forest	1.59 (0.62)	83
	Shrublands	1.12 (0.85)	58
	Wetlands	0.67 (0.29)	35
	Grasslands	0.61 (0.47)	32
	Agricultural	0.28 (0.17)	15
	Barren/developed/other	0.19 (0.05)	10
Distance to paved roads	Long (>5 km)	1.43 (0.71)	100
	Medium (0.3–5 km)	0.88 (0.34)	62
	Short (<0.3 km)	0.69 (0.73)	48
Distance to water	Short (<1 km)	1.57 (0.41)	100
	Medium (1–5 km)	0.92 (0.27)	59
	Long (>5 km)	0.52 (0.27)	33
Human density	Low (<5 persons/km^2)	2.28 (0.39)	100
	Medium-Low (6–10 persons/km^2)	1.00 (0.18)	44
	Medium-High (11–19 persons/km^2)	0.46 (0.27)	20
	High (>20 persons/km^2)	0.25 (0.07)	11
Slope	Steep (>15°)	1.17 (0.54)	100
	Moderate (5–15°)	1.17 (0.41)	100
	Gentle (<5°)	0.66 (0.53)	56

abundant lakes and streams, with distance to water not exceeding 11 km (Median = 0.35 km, IQR = 0.15–0.68 km). Human population density ≤5 persons/km^2 characterized much of the northern regions of the study area, as opposed to the southern regions where population centers were more common and population densities commonly exceeded 100 persons/km^2 in these areas.

Overall Modeling Approach

We applied an expert-assisted spatial habitat model for cougars [16,20,26] to our study area. As a means of validation, we summarized predicted habitat around confirmed locations, and investigated the association between model-predicted cougar habitat values and confirmed cougar locations during the study period. We used the results of the habitat model to eliminate areas deemed unsuitable for cougars, incorporated estimates of deer densities and associated deer biomass within the UGL to investigate potential prey biomass, and subsequently combined potential landscape habitat characteristics with prey resources to generate a range of cougar capacity estimates for the UGL region.

Cougar Habitat Model

We used previously established modeling framework and weights [16,20,26] to extend a cougar habitat model to the UGL. This model incorporated five, differently weighted parameters determined to be essential for cougar habitat: land cover, slope, human population density, distance to roads, and distance to water [26] (Table 1). All raster data generated by the habitat analysis were output to 30 m cell size in a GIS and we performed all spatial analysis in ArcGIS 10.1 (Environmental Systems Research Institute, Inc., Redlands, CA).

To represent roads we used Tiger/line shapefiles acquired from the U.S. Census Bureau. We excluded non-paved roads from the analyses; remaining features represented paved roads that presumably impede cougar movement and/or increase risk of mortality [27–29]. We calculated distance to roads and reclassified the distances into three categories for use in the habitat model (Table 1). We calculated human population density (persons/km^2) for each block group using 2010 U.S. Census Bureau data and grouped the results into four density classes (Table 1). We used the 1 arc-second (30 m spatial resolution) digital elevation model (DEM) from the National Elevation Dataset to calculate slopes (°) and reclassified the results into three categories (Table 1). We gathered rivers, streams, and lake features from the Michigan Geographic Data Library and from the ArcGIS Resource Center for Wisconsin. We retained all rivers, streams and water bodies assumed to hold water under normal, non-drought conditions (e.g. all lakes, rivers, and stream features including artificial paths, canals/ditches, intermittent streams, and perennial streams), calculated distance to these features, and reclassified distance to water into three categories (Table 1). We used the IFMAP/GAP Land Cover dataset [30] for Michigan and the National Land Cover Gap Analysis Project for Wisconsin. These datasets were comparable in that they allowed us to reclassify similar cover types into 8 final categories that were then weighted according to the initial habitat model [26] (Table 1).

We weighted each variable in the model according to expert-based variable rankings and weights published previously [16,26] (Table 1). Weights from this model were established via the Analytical Hierarchy Process (AHP) where experts reviewed all

Table 2. Weights for each variable used in the calculation of potential habitat suitability for cougars in the Upper Great Lakes states (Modified from [26]).

Variable	Weights (S.E.)	Percent importance from highest ranking variable (land cover)
Land cover	1.84 (0.59)	100
Human density	1.22 (0.82)	66
Distance to paved roads	0.86 (0.45)	47
Slope	0.61 (0.56)	33
Distance to water	0.47 (0.26)	26

Weights were based on an Analytical Hierarchy Process analysis [31] and represent the relative importance of each variable to cougar habitat as established by a survey of experts [20,26].

possible pairs of environmental attributes and assigned a rating based on each comparison [16,26,31]. Such analyses are useful for making *a priori* predictions in situations where empirical data are lacking, as is the case in our study area where until recently, cougars had not been confirmed present since the early 1900s [11,32]. Next, we completed a weighted summation using the five variables and their respective weights (Table 2). We divided all pixels in the weighted sum raster by the maximum value that a pixel could receive, which resulted in a relative ranking for habitat potential [16]. Resulting pixel values ranged from 0 to 1, with 1 representing greatest habitat potential. To eliminate noise associated with pixel-scale variation, we smoothed the habitat raster using the mean neighborhood statistic within a 13.75 km^2 circular moving window, corresponding to $1/4^{th}$ of the minimum home range size of a female cougar [9]. This size window retains a high level of resolution while also maintaining a realistic spatial scale for cougar perception of habitat, given that scale of selection is complex and likely varies depending on behaviors [33]. For ease of comparison, we scaled final habitat values into percentages and

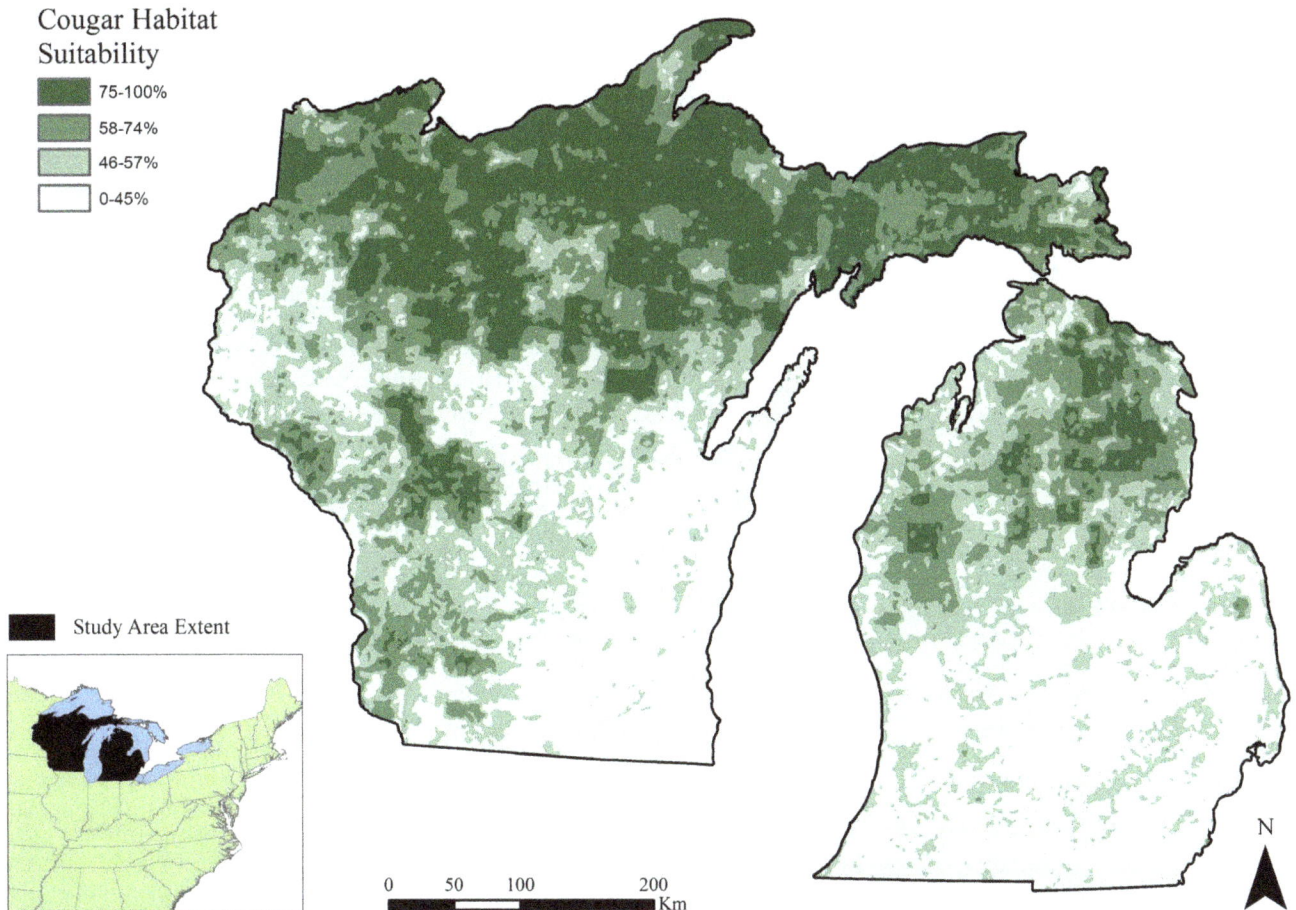

Figure 2. Cougar habitat rankings in the Upper Great Lakes region, USA. The cougar habitat model was generated based on expert-assisted variables and weights from LaRue [20], and LaRue and Nielsen [16,26].

Table 3. Confirmations of cougars by year, 1 January, 2008 to 1 June, 2013, in Michigan and Wisconsin, USA.

| Year | Cougar confirmations | | | |
	Michigan	Wisconsin	Probable sightings	Common methods[a]
2008	3	3	6	tracks, photo, DNA
2009	3	8	4	tracks, photo, observed (treed)
2010	1	6	5	photo, tracks, scat
2011	7	7	1	photo, tracks, scat (DNA), video
2012	6	13	9	photo, tracks, video, observed
2013[b]	0	1	NA	observed (treed)
Total	20	38	25	

[a]Ordered from most frequent to least frequent.
[b]Only includes confirmations reported through 1 June, 2013.
Confirmations are those verified by the Department of Natural Resources from either state as cougar (Michigan DNR, *unpublished data*; Wisconsin DNR, *unpublished reports*). Probable sightings were also reported in Wisconsin during the same time period.

reclassified them into the same categories as the original cougar habitat model [16] (Fig. 2).

Cougar Data and Model Validation

We reviewed current literature, archived reports, and press releases from the Michigan and Wisconsin DNRs, to document confirmed (i.e., verified evidence of) cougar occurrences in the study area from 1 January, 2008 to 1 June, 2013. Our review was supplemented by Rare Mammal Observations reported annually by the Wisconsin DNR [10,34–37] and a confirmation database from the Michigan DNR (Michigan DNR, *unpublished data*). These sources of information included all known cougar confirmations documented in the study area since 2008 (Fig. 1). Dates were chosen because the first cougar confirmed in these states since the 1900s occurred in 2008, despite statewide collection of cougar reports from Wisconsin since 1991 [38]. Numbers of verified occurrences increased after 2008 [10,34–37]. Verified evidence included tracks, photographs, scat, video, visual observation by wildlife officials, and DNA gathered from blood, scat, or hair. We recorded the date, township location, and type of evidence (photographs, tracks, scat, encounter, DNA) for each confirmation (Table 3, Fig. 1). Locations classified as "probable" occurrences by biologists based on available evidence [34] were included in Fig. 1 but not used for analysis.

We used the verified location data on the presence of cougars in Wisconsin and Michigan to assess the validity of the habitat model. We first summarized habitat model values around verified cougar locations at four spatial scales (i.e. buffers) to explore modeled habitat potential associated with known cougar space use. Four radial buffers were made around verified locations to represent the area potentially used by the observed animals within the time frame during which they were verified as present. Buffer radius distances were 1 km, 5 km, 10 km, and 25 km, respectively. These distances were chosen to cover a possible range of daily and weekly movement patterns, from smaller and more localized movements typically associated with resident cougars [28,39,40], to larger movements associated with long-distance dispersers [4,8,14]. We summarized mean modeled habitat values within these buffered areas. We then generated a random sample of "pseudo-absence" locations [41] to represent available habitat. We generated 100 random locations for every verified cougar locations over the study area, resulting in 5,800 availability samples [42]. Since no verifications occurred in the Lower Peninsula of Michigan, we

limited this sample to Wisconsin and the Upper Peninsula of Michigan. We distributed the locations equally by county to ensure a spatially-balanced sample [42]. Then, for each location, we generated buffers using the same radius distances as the cougar locations and summarized habitat values within each buffer using GIS. To compare habitat values within used buffers to habitat values within available buffers, we assigned a binary indicator variable to each location (1 for used, 0 for available) and fit a resource selection function (RSF; [43,44]) to these data via logistic regression [42]. Accordingly, the RSF assessed the influence that habitat values from our model had on the relative probability of cougar presence, allowing us to explore the validity of our habitat modeling approach. We performed the analysis for each buffer distance and thus fit 4 logistic regression models using 'glm' in R Version 3.0.0 (R Development Core Team, www.r-project.org, accessed 15 June 2013). We scaled the habitat model ranking to a percentage to simply model coefficient interpretation, assessed the resulting fit of models by comparing deviance residuals to the null model, and evaluated the ability of the habitat model to predict cougar occurrences using the Area Under Curve (AUC) statistic [45]. We calculated the AUC and its bootstrapped 95% confidence interval using the 'pROC' package in R [46].

Prey Biomass

Carnivore abundance and density depends on prey availability and biomass as well as available space and favorable land cover. Predicting carnivore density based on estimates of prey biomass has been applied in wolves [17,18], tigers (*Panthera tigris*) [47], and Canada lynx (*Lynx canadensis*) [24]. To estimate potential for cougar prey in the UGL, we used DNR estimates of deer density by deer management unit (DMU) in Michigan and Wisconsin [48,49]. We assumed that white-tailed deer would be the primary prey for cougars in the region [50,51]. To incorporate uncertainty in total deer biomass due to population sex/age structure across the UGL, we used three plausible sex/age structures (buck/doe/fawn) described for the northern Great Lakes and Ontario [52]: 40/30/30 (even/balanced), 25/40/35 (unbalanced, N. Wisconsin), and 15/50/35 (unbalanced, Michigan). We assumed the mean late-autumn deer size within these populations was 100 kg for bucks, 66 kg for does, and 33 kg for fawns [52–54]. Thus, deer biomass estimates (kg) for the three conceivable population structures were approximated within each DMU by the following:

Table 4. Modelled cougar habitat values (0–100%) association with 58 cougar occurrences in Michigan and Wisconsin, USA at 4 spatial scales between 2008 and 2013.

Buffer radius	Range	Mean (± SD)	β Habitat	Odds ratio (95% CI)	AUC (95% CI)
1 km	33–90	64±15	0.033	1.03 (1.02–1.05)	0.66 (0.59–0.72)
5 km	29–92	64±15	0.034	1.03 (1.02–1.05)	0.66 (0.59–0.72)
10 km	29–94	65±15	0.038	1.04 (1.02–1.06)	0.67 (0.60–0.73)
25 km	27–100	64±15	0.042	1.04 (1.02–1.06)	0.68 (0.62–0.74)

Coefficients, odds ratios, and area under curve (AUC) statistics were generated by logistic regression models linking modeled habitat values to verified occurrences.

$$M_{deer} = N_{deer} \times (M_{buck} \times N_{buck} + M_{doe} \times N_{doe} + M_{fawn} \times N_{fawn})\ (1)$$

where N_{deer} = estimated deer population/DMU, M_{buck} = presumed average mass of an UGL white-tailed buck, M_{doe} = presumed average mass of an UGL white-tailed doe, M_{fawn} = presumed average mass of an UGL fawn, and N values for buck/doe/fawn were the proportion of deer in the sex/age class given a specified population structure. To estimate capacity for cougars in the study area, we first assumed habitat patches scoring <0.75 in the cougar habitat model would not support a population [16]. We subset the habitat model to only include values ≥0.75, restricted the DMUs to this subset, calculated the area in 100 km^2 units for each DMU, and divided the deer biomass estimates by the area. Carbone and Gittleman [55] found that 10,000 kg/100 km^2 could result in 0.94 cougars/100 km^2; we multiplied deer biomass (10,000 kg/100 km^2) by 0.94 to achieve the potential cougar density for each DMU. Thus, potential cougars/100 km^2 depended on both deer density and available habitat based on favorable landscape characteristics.

Potential cougar density estimates (cougars/100 km^2) corresponding to the three aforementioned deer population structures were summed across the study area. To incorporate geographic variation, we averaged potential cougar density across all available habitats using a 100 km^2 circular assessment window. While densities of up to 13 cougars/100 km^2 have been reported in other locations [56], such densities are uncommon [33,57,58]. Given the potential for cougar-human conflicts leading to higher risk of mortality and competition with wolves and other predators for prey [59], we considered it unlikely for population densities to exceed 3 cougars/100 km^2 in the UGL. Thus, despite high estimates of prey biomass in some areas, we limited the maximum potential cougar density to 3/100 km^2. To incorporate additional uncertainty, we also carried out the analysis using maximum potential cougar densities of 2/100 km^2 and 1/100 km^2.

Results

Cougar Habitat Model

Our model indicated high potential for cougar habitat in Michigan and Wisconsin. Cougar habitat values ranged from 0.27 to 0.92 ($\bar{x} = 0.56$, SD = 0.15). Assuming the value ≥0.75 is a conservative estimate of suitable habitat [16] at least 39% (>58,000 km^2) of our total study area contained suitable cougar habitat. Habitat areas were generally contiguous and concentrated throughout the forested regions of northern Wisconsin, much of the Upper Peninsula of Michigan, and the northern Lower Peninsula of Michigan (Fig. 2). The largest contiguous patch of suitable habitat occurred in northern Wisconsin and the Upper Peninsula. This area contained 85% (49,216 km^2) of all habitat

receiving scores of 0.75 or higher. According to our model, patches appeared to decrease in frequency and size and were evidently more fragmented at southern latitudes. Less suitable cougar habitat (suitability values <0.75) often coincided with higher human population densities and agriculture-dominated landscapes; these areas were concentrated throughout the south and south-central regions of the study area (Fig. 2).

Cougar confirmations from the DNR in Michigan and Wisconsin totaled 58 from 1 January, 2008 to 1 June, 2013 (38 and 20 in Wisconsin and Michigan, respectively; Table 3). Model validation results suggested that the habitat model was effective in identifying areas of suitable cougar habitat. Habitat values summarized around cougar locations exhibited a wide range of habitat potentially used by cougars in the UGL, but were generally consistent with the expectation that verified locations would occur within higher-ranked vs. lower-ranked habitat. Observed habitat values within buffered areas ranged from 0.27 to 1.00 depending on buffer size (Table 4) and occurred within higher-ranked habitat than the mean cougar habitat ranking for the study area (0.56). Our cougar habitat model was a significant predictor of the relative probability of cougar occurrence at each buffer distance (Table 4). The RSF deviance residuals at each scale improved upon the null model and odds ratios associated with the habitat predictor were significantly greater than one (Table 4) indicating that the odds of an occurrence increased with increases in habitat ranking. The strength of this effect and the model's predictive performance (AUC) both increased slightly with larger buffer sizes (Table 4). At each scale, the odds of a cougar occurrence increased by 2–6% with each percent increase in cougar habitat ranking (Table 4).

Cougar Capacity

Deer density estimates per DMU ranged from 0.5 to >20 deer/km^2 ($\bar{x} = 11.1$, SD = 6.1; Fig. 3a). Using these estimates and three potential age/sex population structures, deer biomass per DMU could conceivably range from approximately 3,000 kg/100 km^2 (unbalanced age structure, lowest deer density) to over 200,000 kg/100 km^2 (balanced age structure, highest deer density). After restricting the DMUs to favorable cougar habitat, the mean deer biomass estimates per DMU within potential cougar habitat were approximately 10,000–165,000 kg/100 km^2 depending on the age/sex structure applied (Table 5). Thus, we estimated that prey biomass within potential cougar habitat was geographically variable and could support up to 15 cougars/100 km^2 (Fig. 3b). However, we also assumed that >3 cougars/100 km^2 anywhere within the study area was unrealistic despite high deer densities in some areas. Using three different maximum viable densities (1, 2, and 3 cougars/100 km^2) and allowing lower estimates to depend on approximations of deer biomass where

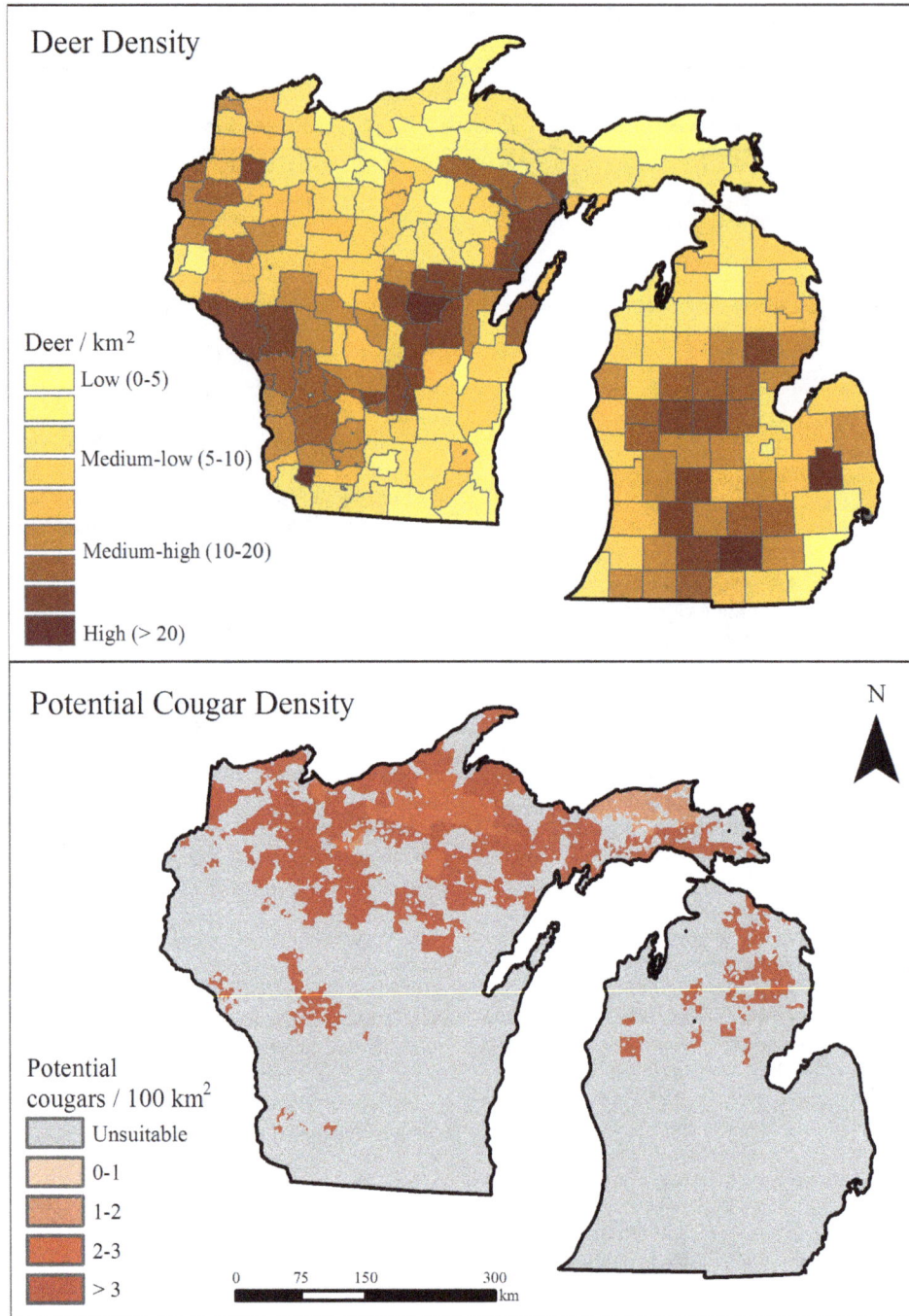

Figure 3. Prey densities and cougar capacity in the Upper Great Lakes region, USA. a) Whitetail deer density estimates were based on best available information from state natural resource management agencies [48,49]. **b)** Geographic variation in cougar capacity based on our most conservative potential deer biomass estimates. Deer biomass estimates were generally high enough to support >3 cougars/100 km^2 across much of the study area.

deer densities were low, we calculated that available resources could sustain 582 to 1,677 cougars within favorable habitat.

Discussion

Our extension of a habitat model for cougars [16,20,26] suggests that Michigan and Wisconsin contain >58,000 km^2 of potential cougar habitat. Based on potential deer biomass

estimates, more than 500 cougars could inhabit this overall area. For comparison, cougar range in Washington, USA covers 51% of the state (88,500 km^2) and consistently supports approximately 2,000 cougars [60], and estimates from several models in the northeastern U.S. suggested potential for 322–2,535 cougars depending on the area considered suitable [61]. The northern half of our study area is characterized by dense forest, low human population and road densities, abundant water resources, and

Table 5. Estimates of white-tailed deer biomass within deer management units for Michigan and Wisconsin (the Upper Great Lakes region, or UGL); estimates are based on three conceivable age/sex structures for white-tailed deer and are restricted to favorable habitat as indicated by cougar habitat model values ≥0.75.

Bucks/does/fawns	Range (kg/DMU)	Mean (kg/DMU)	SD (kg/DMU)	Total biomass (kg)
40/30/30	12,494–165,450	76,189	31,812	7.70×10^6
25/40/35	11,284–149,427	68,810	28,731	6.95×10^6
15/50/35	10,674–141,357	65,094	27,179	6.57×10^6

diverse topography. Large expanses of wild land exist that provide contiguous habitat characterized by forested landscapes which support high densities of white-tailed deer. Confirmed location data and associated radio-collar and DNA evidence suggest that multiple cougars occurred in this area between January 2008 and early 2013.

Our investigation of habitat capacity should not be viewed as an attempt to predict the precise distribution and/or exact size of a potential cougar population. Although it would be possible to speculate based on the models we presented (e.g. [18]), the nature of recolonization (i.e. breeding range selection, population size, geographic distribution) will likely depend on additional factors that we could not quantify such as future changes in management, public acceptance [62], the presence and density of competing predators, potential disease components, migratory behavior in deer [63] and other dynamics of prey populations. Both states have experienced similar conservation challenges as wolves have recolonized and recovered in the region [64,65]. Our model validation analysis is exploratory in nature, as the best available information for cougar presence in the UGL is primarily based on incidental, verified observations (Table 3); these location data are assumed to be associated with transient cougars from western populations [13]. Detection and verification probability may vary geographically, and this variation can bias information on cougar presence. Automatic cameras are not evenly spread across the landscape, and are likely underused on large blocks of public lands that contain some of the best cougar habitat in the region. In addition, resource selection by dispersing, transient animals may differ from that of resident individuals [26,66], perhaps with particular regard to human development and activity [15,33]. These reasons likely contributed to our habitat model's relatively low ability to predict cougar occurrences on our study site (model validation AUC scores between 0.59 and 0.74; Table 4). Without knowledge of the specific nature of detection probability across the study site, modeling cougar distribution (e.g. probability and/or likelihood of occurrence) based on presence-only data would require strong assumptions [67,68]. Similarly, it would not be appropriate to base estimates of cougar distribution entirely on locations associated from animals assumed to be non-resident individuals. As such, although our validation used the best available information to provide support for our habitat model, information on cougar habitat use in the UGL remains limited and may not be adequate for making predictions. Given that our objective was to model habitat potential for a future resident population, the expert-assisted model that we implemented is preferable because it only makes assumptions about the most general habitat requirements for cougars. We showed that verified cougar locations in the UGL were consistent with our model, which further suggests that areas of modeled suitable habitat may be able to support a viable cougar population in the future.

Recolonization and recovery of cougars to former ranges such as the UGL will likely require active and adaptive management at both state and federal levels. Natural recolonization is likely to occur eventually [69], providing states with favorable cougar habitat the opportunity to prepare. Public attitudes and their associated influences on public policy are important determinants of large carnivore population viability [62]. Recent evidence suggests that people may be more neutral and are consequently more impressionable in regions where cougars have previously been irrelevant [70], but that humans can be accepting and supportive of cougars, particularly if they have access to good information and are knowledgeable about cougar ecology [62,69,71]. These findings have implications for areas of potential cougar recolonization because there may be a relatively small window of opportunity for outreach and education programs to promote awareness and shape public opinion, as overcoming limited trust once it is instilled is difficult [70]. Large carnivore populations can be polarizing, in part because management tools are often controversial. The appropriateness and ethics of recreational hunting and population control of these animals are hotly debated [72,73] and understanding how public attitudes are influenced by these actions is complicated [74]. Consequently, a natural first step toward preparing for a cougar recolonization would be investigating social acceptance and potential human tolerance of cougars in the UGL states. Public education and outreach could positively shape public opinions and help to avoid agency mistrust [69]. The state of Wisconsin has drafted a management protocol in the event that a breeding population establishes [36]. Additionally, all confirmed cougar observations are publicly available [11]. Anticipating recolonization prior to its occurrence can improve future management and could lay the groundwork for a strong, collaborative cougar conservation program regionally.

In the event that a population becomes established in the UGL, state DNR agencies will need to address concerns of human safety, pet safety, and depredation of livestock [15,75], as well as develop long-term monitoring programs [76]. Strategies necessary to manage a cougar population will likely include plans for education and outreach associated with human safety concerns, potential compensation for livestock losses, mitigation strategies for conflicts with farmers/ranchers, and discussion of possible harvest scenarios [15,75]. Long-term research investigating cougar behavior, habitat use, prey selection, genetic structure of the population, competition with other predators, and impacts on white-tailed deer could eventually be warranted. The addition of another carnivore to the current predator guild in the UGL states could be politically challenging, yet long-term ecological benefits of a viable cougar population [77] could be realized under effective conservation planning scenarios.

Acknowledgments

We thank D. Beyer, Jr., A. Wydeven, B. Roell, J. Wiedenhoeft, S. Roeppke, and M. Hyslop for assistance in data collection. Thanks to the Michigan Department of Natural Resources and Wisconsin Department of Natural Resources for sharing information on cougar verifications. Thanks

to D. Beyer, Jr., D. Etter, and A. Wydeven for making improvements to the manuscript.

Author Contributions

Conceived and designed the experiments: STO JKB. Analyzed the data: STO KCR. Wrote the paper: STO KCR JKB.

References

1. Bolgiano C, Roberts J (2005) The eastern cougar: historic accounts, scientific investigations, and new evidence. Mechanicsburg, Pennsylvania, USA: Stackpole Books. 246 p.
2. Laundré JW, Hernández L (2010) What we know about pumas in Latin America. In: Hornocker M, Negri S, editors. Cougar: ecology and conservation. Chicago, Illinois, USA: University of Chicago Press. pp. 76–90.
3. Anderson CR, Lindzey FG, Knopff KH, Jalkotzy MG, Boyce MS (2010) Cougar management in North America. In: Hornocker M, Negri S, editors. Cougar: ecology and conservation. Chicago, Illinois, USA: University of Chicago Press. pp. 41–56.
4. Thompson DJ, Jenks JA (2010) Dispersal movements of subadult cougars from the Black Hills: the notions of range expansion and recolonization. Ecosphere 1: 1–11.
5. Fecske DM, Thompson DJ, Jenks JA, Oehler MW (2008) North Dakota Mountain Lion Status Report. In: Toweill DE, Nadeau S, Smith D, editors. Sun Valley, Idaho, USA. pp. 92–101.
6. Wilson S, Hoffman JD, Genoways HH (2010) Observations of reproduction in mountain lions from Nebraska. Western North American Naturalist 70: 238–240.
7. Sweanor LL, Logan KA, Hornocker MG (2000) Cougar dispersal patterns, metapopulation dynamics and conservation. Conservation Biology 14: 798–808.
8. Thompson DJ, Jenks JA (2005) Long-distance dispersal by a subadult male cougar from the Black Hills, South Dakota. Journal of Wildlife Management 69: 818–820.
9. Logan KA, Sweanor LL (2010) Behavior and social organization of a solitary carnivore. In: Hornocker M, Negri S, editors. Cougar: ecology and conservation. Chicago, Ilinois, USA: University of Chicago Press. pp. 105–118.
10. Wiedenhoeft JE, Wydeven AP, Bruner J (2012) Rare mammal observations 2011. 122–135 p.
11. Wisconsin Department of Natural Resources (2013) Cougars in Wisconsin. Available: http://dnr.wi.gov/topic/wildlifehabitat/cougar.html. Accessed 2013 June 1.
12. Rosatte R (2011) Evidence confirms the presence of cougars (Puma concolor) in Ontario, Canada. Canadian Field-Naturalist 125: 116–125.
13. LaRue MA, Nielsen CK, Dowling M, Miller K, Wilson B, et al. (2012) Cougars are recolonizing the midwest: Analysis of cougar confirmations during 1990–2008. Journal of Wildlife Management 76: 1364–1369.
14. Henaux V, Powell LA, Hobson KA, Nielsen CK, LaRue MA (2011) Tracking large carnivore dispersal using isotopic clues in claws: an application to cougars across the Great Plains. Methods in Ecology and Evolution 2: 489–499.
15. Sweanor LL, Logan KA (2010) Cougar-human interactions. In: Hornocker M, Negri S, editors. Cougar: ecology and conservation. Chicago, Illinois, USA: University of Chicago Press. pp. 190–205.
16. LaRue MA, Nielsen CK (2011) Modelling potential habitat for cougars in midwestern North America. Ecological Modelling 222: 897–900.
17. Mladenoff DJ, Haight RG, Sickley TA, Wydeven AP (1997) Causes and implications of species restoration in altered ecosystems. Bioscience 47: 21–31.
18. Mladenoff DJ, Sickley TA (1998) Assessing potential gray wolf restoration in the northeastern United States: A spatial prediction of favorable habitat and potential population levels. Journal of Wildlife Management 62: 1–10.
19. Potvin MJ, Drummer TD, Vucetich JA, Beyer DE, Peterson RO, et al. (2005) Monitoring and habitat analysis for wolves in upper Michigan. Journal of Wildlife Management 69: 1660–1669.
20. LaRue MA (2007) Predicting potential habitat and dispersal corridors for cougars in midwestern North America: Southern Illinois University.
21. Clark DW, White SC, Bowers AK, Lucio LD, Heidt GA (2002) A survey of recent accounts of the mountain lion (Puma concolor) in Arkansas. Southeastern Naturalist 1: 269–278.
22. Johnson K (2002) The mountain lions of Michigan. Endangered Species UPDATE 19: 27–31.
23. Wisconsin Cougar Working Group (2010) Draft Wisconsin Cougar Response Protocol. Available: http://www.legis.state.wi.us/senate/sen17/news/Press/2010/documents/CougarResponseProtocolforWI2010a.pdf. Accessed 2010 Sept 21.
24. Linden DW, Campa H, Roloff GJ, Beyer DE, Millenbah KF (2011) Modeling habitat potential for Canada lynx in Michigan. Wildlife Society Bulletin 35: 20–26.
25. Laundré JW (2013) The feasibility of the north-eastern USA supporting the return of the cougar Puma concolor. Oryx 47: 96–104.
26. LaRue MA, Nielsen CK (2008) Modelling potential dispersal corridors for cougars in midwestern North America using least-cost path methods. Ecological Modelling 212: 372–381.
27. Dickson BG, Jenness JS, Beier P (2005) Influence of vegetation, topography, and roads on cougar movement in southern California. Journal of Wildlife Management 69: 264–276.
28. Kertson BN, Spencer RD, Marzluff JM, Hepinstall-Cymerman J, Grue CE (2011) Cougar space use and movements in the wildland-urban landscape of western Washington. Ecological Applications 21: 2866–2881.
29. Teichman KJ, Cristescu B, Nielsen SE (2013) Does Sex Matter? Temporal and Spatial Patterns of Cougar-Human Conflict in British Columbia. Plos One 8.
30. Donovan ML, Nesslage GM, Skillen JJ, Maurer BA (2004) The Michigan GAP analysis final report. Lansing, MI: Wildlife Division, Michigan Department of Natural Resources.
31. Saaty TL (1980) The analytic hierarchy process: planning, priority setting, resources allocation. New York, New York, USA: McGraw-Hill International Book Company.
32. Wisconsin Cougar Working Group (2010) Draft Wisconsin Cougar Response Protocol. Available: http://www.legis.state.wi.us/senate/sen17/news/Press/2010/documents/CougarResponseProtocolforWI2010a.pdf. Accessed 2010 Sept 21.
33. Wilmers CC, Wang Y, Nickel B, Houghtaling P, Shakeri Y, et al. (2013) Scale Dependent Behavioral Responses to Human Development by a Large Predator, the Puma. Plos One 8: e60590.
34. Wiedenhoeft JE, Wydeven AP (2009) Rare mammal observations 2008. Wisconsin Wildlife Surveys April 2009, 19: 83–95.
35. Wiedenhoeft JE, Wydeven AP (2010) Rare mammal observations 2009. Wisconsin Wildlife Surveys 20: 103–113.
36. Wiedenhoeft JE, Wydeven AP, Bruner J (2013) Rare carnivore observations 2012. Wisconsin Wildlife Surveys 23: 93–102.
37. Wiedenhoeft JE, Wydeven AP, Bruner J (2011) Rare mammal observations 2010. Wisconsin Wildlife Surveys 21: 99–113.
38. Anderson EM, Wydeven AP, Holsman R (2006) Distribution of cougar observations in Wisconsin 1994–2003. In: McGrinnis HJ, Tischendorf JW, Ropski SJ, editors. Proceedings of the Eastern Cougar Conference, 28-April 2004 -1 May 2004, Morgantown, West Virginia, USA.
39. Beier P (1995) Dispersal of juvenile cougars in fragmented habitat. The Journal of Wildlife Management: 228–237.
40. Beier P, Choate D, Barrett RH (1995) Movement patterns of mountain lions during different behaviors. Journal of Mammalogy: 1056–1070.
41. McDonald L, Manly B, Huettmann F, Thogmartin W (2013) Location-only and use-availability data: analysis methods converge. Journal of Animal Ecology 82: 1120–1124.
42. Northrup JM, Hooten MB, Anderson CR, Wittemyer G (2013) Practical guidance on characterizing availability in resource selection functions under a use-availability design. Ecology 94: 1456–1463.
43. Boyce MS, McDonald LL (1999) Relating populations to habitats using resource selection functions. Trends in Ecology & Evolution 14: 268–272.
44. Johnson CJ, Nielsen SE, Merrill EH, McDonald TL, Boyce MS (2006) Resource selection functions based on use-availability data: Theoretical motivation and evaluation methods. Journal of Wildlife Management 70: 347–357.
45. Conkin JA, Alisauskas RT (2013) Modeling probability of waterfowl encounters from satellite imagery of habitat in the central Canadian arctic. Journal of Wildlife Management 77: 931–946.
46. Robin X, Turck N, Hainard A, Tiberti N, Lisacek F, et al. (2011) pROC: an open-source package for R and S plus to analyze and compare ROC curves. Bmc Bioinformatics 12.
47. Karanth KU, Nichols JD, Kumar NS, Link WA, Hines JE (2004) Tigers and their prey: predicting carnivore densities from prey abundance. Proceedings of the National Academy of Sciences of the United States of America 101: 4854–4858.
48. Wisconsin Department of Natural Resources (2013) Estimated Fall Deer Population, 2012. Madison, Wisconsin, USA: Wisconsin Department of Natural Resources, GIS Services Section. Available: http://dnr.wi.gov/topic/hunt/documents/fallabund.pdf. Accessed 2014 June 1.
49. Michigan Department of Natural Resources (2013) Clickable Map of Proposed 2006–2010 Deer Population Goals. Available: http://www.michigan.gov/dnr/0,1607,7-153-10363_10856_10905-129948-542,00.html. Accessed 2014 June 1.
50. Murphy K, Ruth TK (2010) Diet and prey selection of a perfect predator. In: Hornocker M, Negri S, editors. Cougar: ecology and conservation. Chicago, Illinois, USA: University of Chicago Press. pp. 118–137.
51. DelGiudice GD, McCaffery KR, Beyer Jr DE, Nelson ME (2009) Prey of wolves in the Great Lakes region. In: Wydeven AP, Deelen TR, Heske EJ, editors. Recovery of gray wolves in the Great Lakes Region of the United States: Springer New York. pp. 155–173.

52. Halls LK (1984) White-tailed deer: ecology and management. Harrisburg, Pennsylvania, USA: Stackpole Books.

53. Geist V (1998) Deer of the world: their evolution, behaviour and ecology. Harrisburg, Pennsylvania, USA: Stackpole Books.

54. Verme LJ (1989) Maternal investment in white-tailed deer. Journal of Mammalogy 70: 438–442.

55. Carbone C, Gittleman JL (2002) A common rule for the scaling of carnivore density. Science 295: 2273–2276.

56. Smallwood KS (1997) Interpreting puma (Puma concolor) population estimates for theory and management. Environmental Conservation 24: 283–289.

57. Lambert CMS, Wielgus RB, Robinson HS, Katnik DD, Cruickshank HS, et al. (2006) Cougar population dynamics and viability in the Pacific Northwest. Journal of Wildlife Management 70: 246–254.

58. Quigley H, Hornocker M (2010) Cougar population dynamics. In: Hornocker M, Negri S, editors. Cougar: ecology and conservation. Chicago, Illinois, USA: University of Chicago Press. pp. 59–75.

59. Ruth TK, Murphy K (2010) Competition with other carnivores for prey. In: Hornocker M, Negri S, editors. Cougar: ecology and conservation. Chicago, Illinois, USA: University of Chicago Press. pp. 163–172.

60. Washington Department of Fish and Wildlife (2008) Washington Department of Fish and Wildlife 2009–2015 Game Management Plan. Olympia, Washington, USA: Washington Department of Fish and Wildlife.

61. Glick HB (2014) Modeling cougar habitat in the Northeastern United States. Ecological Modelling 285: 78–89.

62. Smith JB, Nielsen CK, Hellgren EC (2014) Illinois resident attitudes toward recolonizing large carnivores. Journal of Wildlife Management 78: 930–943.

63. Murray BD, Webster CR, Bump JK (2013) Broadening the ecological context of ungulate-ecosystem interactions: the importance of space, seasonality, and nitrogen. Ecology 94: 1317–1326.

64. Beyer DE, Peterson RO, Vucetich JA, Hammill JH (2009) Wolf population changes in Michigan. In: Wydeven AP, Deelen TR, Heske EJ, editors. Recovery of gray wolves in the Great Lakes Region of the United States: an endangered species success story. : Springer New York. pp. 65–85.

65. Wydeven AP, Wiedenhoeft JE, Schultz RN, Thiel RP, Jurewicz RL, et al. (2009) History, population growth, and management of wolves in Wisconsin. In: Wydeven AP, Deelen TR, Heske EJ, editors. Recovery of gray wolves in the Great Lakes Region of the United States: an endangered species success story. : Springer New York. pp. 87–105.

66. Palomares F, Delibes M, Ferreras P, Fedriani JM, Calzada J, et al. (2000) Iberian lynx in a fragmented landscape: predispersal, dispersal, and postdispersal habitats. Conservation Biology 14: 809–818.

67. Pearce JL, Boyce MS (2006) Modelling distribution and abundance with presence-only data. Journal of Applied Ecology 43: 405–412.

68. Hastie T, Fithian W (2013) Inference from presence-only data; the ongoing controversy. Ecography 36: 864–867.

69. Beier P (2010) A focal species for conservation planning. In: Hornocker M, Negri S, editors. Cougar: ecology and conservation. Chicago, Illinois, USA: University of Chicago Press. pp. 177–189.

70. Davenport MA, Nielsen CK, Mangun JC (2010) Attitudes toward mountain lion management in the Midwest: implications for a potentially recolonizing large predator. Human Dimensions of Wildlife 15: 373–388.

71. Corona Research I (2006) Public opinions and perceptions of mountain lion issues. Denver, Colorado, USA: Corona Research, Inc.

72. Treves A (2009) Hunting for large carnivore conservation. Journal of Applied Ecology 46: 1350–1356.

73. Vucetich JA, Nelson MP (2014) Wolf Hunting and the Ethics of Predator Control. In: Kalof L, editor. The Oxford Handbook of Animal Studies. Oxford Handbooks Online: Oxford University Press, Oxford, UK.

74. Treves A, Naughton-Treves L, Shelley V (2013) Longitudinal Analysis of Attitudes Toward Wolves. Conservation Biology 27: 315–323.

75. Cougar Management Guidelines Working Group (2005) Cougar management guidelines. Wild Futures, Bainbridge Island, Washington, USA.

76. U.S. Fish and Wildlife Service (1982) Eastern cougar recovery plan. Atlanta, Georgia, USA: U.S. Fish and Wildlife Service.

77. Ripple WJ, Estes JA, Beschta RL, Wilmers CC, Ritchie EG, et al. (2014) Status and ecological effects of the world's largest carnivores. Science 343: 1241484.

PERMISSIONS

All chapters in this book were first published in PLOS ONE, by The Public Library of Science; hereby published with permission under the Creative Commons Attribution License or equivalent. Every chapter published in this book has been scrutinized by our experts. Their significance has been extensively debated. The topics covered herein carry significant findings which will fuel the growth of the discipline. They may even be implemented as practical applications or may be referred to as a beginning point for another development.

The contributors of this book come from diverse backgrounds, making this book a truly international effort. This book will bring forth new frontiers with its revolutionizing research information and detailed analysis of the nascent developments around the world.

We would like to thank all the contributing authors for lending their expertise to make the book truly unique. They have played a crucial role in the development of this book. Without their invaluable contributions this book wouldn't have been possible. They have made vital efforts to compile up to date information on the varied aspects of this subject to make this book a valuable addition to the collection of many professionals and students.

This book was conceptualized with the vision of imparting up-to-date information and advanced data in this field. To ensure the same, a matchless editorial board was set up. Every individual on the board went through rigorous rounds of assessment to prove their worth. After which they invested a large part of their time researching and compiling the most relevant data for our readers.

The editorial board has been involved in producing this book since its inception. They have spent rigorous hours researching and exploring the diverse topics which have resulted in the successful publishing of this book. They have passed on their knowledge of decades through this book. To expedite this challenging task, the publisher supported the team at every step. A small team of assistant editors was also appointed to further simplify the editing procedure and attain best results for the readers.

Apart from the editorial board, the designing team has also invested a significant amount of their time in understanding the subject and creating the most relevant covers. They scrutinized every image to scout for the most suitable representation of the subject and create an appropriate cover for the book.

The publishing team has been an ardent support to the editorial, designing and production team. Their endless efforts to recruit the best for this project, has resulted in the accomplishment of this book. They are a veteran in the field of academics and their pool of knowledge is as vast as their experience in printing. Their expertise and guidance has proved useful at every step. Their uncompromising quality standards have made this book an exceptional effort. Their encouragement from time to time has been an inspiration for everyone.

The publisher and the editorial board hope that this book will prove to be a valuable piece of knowledge for researchers, students, practitioners and scholars across the globe.

LIST OF CONTRIBUTORS

Yohei Nakamura and Kosaku Yamaoka
Graduate School of Kuroshio Science, Kochi University, Nankoku, Kochi, Japan

Masaru Kanda
Graduate School of Kuroshio Science, Kochi University, Nankoku, Kochi, Japan
Kuroshio Zikkan Center, Otsuki, Kochi, Japan

David A. Feary
School of the Environment, University of Technology, Sydney, New South Wales, Australia

Seth M. Harju, Chad V. Olson, Matthew R. Dzialak, James P. Mudd and Jeff B. Winstead
Hayden-Wing Associates LLC, Natural Resource Consultants, Laramie, Wyoming, United States of America

Aril Slotte, Cecilie Kvamme and Richard D. M. Nash
Institute of Marine Research, Bergen, Norway

Florian Eggers
Institute of Marine Research, Bergen, Norway
Department of Biology, University of Bergen, Bergen, Norway

Arne Johannessen
Department of Biology, University of Bergen, Bergen, Norway

Lísa Anne Libungan
Department of Life and Environmental Sciences, University of Iceland, Reykjavík, Iceland

Even Moland
Institute of Marine Research, Flødevigen, Norway

Esben M. Olsen
Institute of Marine Research, Flødevigen, Norway
Centre for Ecological and Evolutionary Synthesis (CEES), Department of Biosciences, University of Oslo, Oslo, Norway
Department of Natural Sciences, Faculty of Science and Engineering, University of Agder, Kristiansand, Norwa

Remo Freimann
Institute of Molecular Health Sciences, Professorship of Genetics, ETH Zurich, Zurich, Switzerland
Department of Aquatic Ecology, Swiss Federal Institute of Aquatic
Integrative Biology, ETH-Zurich, Zurich, Switzerland

Christopher T. Robinson
Department of Aquatic Ecology, Swiss Federal Institute of Aquatic
Science and Technology, Eawag, Dübendorf, Switzerland and Institute of Integrative Biology, ETH-Zurich, Zurich, Switzerland

Helmut Bürgmann
Department of Surface Waters – Research and Management, Swiss Federal Institute of Aquatic Science and Technology, Eawag, Kastanienbaum, Switzerland

Stuart E. G. Findlay
Cary Institute of Ecosystem Studies, Millbrook, New York, United States of America

Tony Chang, Andrew J. Hansen and Nathan Piekielek
Department of Ecology, Montana State University, Bozeman, Montana, United States of America

Yoko Nozawa, Che-Hung Lin and Ai-Chi Chung
Biodiversity Research Center, Academia Sinica, Taipei, Taiwan, ROC

Chris J. Brauer and Luciano B. Beheregaray
Molecular Ecology Laboratory, School of Biological Sciences, Flinders University, Adelaide, South Australia, Australia

Peter J. Unmack
Institute for Applied Ecology and Collaborative Research Network for Murray-Darling Basin Futures, University of Canberra, Canberra, Australian Capital Territory, Australia

Michael P. Hammer
School of Earth and Environmental Sciences, University of Adelaide, South Australia, Australia
Curator of Fishes, Museum and Art Gallery of the Northern Territory, Darwin, Northern Territory, Australia
Evolutionary Biology Unit, South Australian Museum, Adelaide, South Australia, Australia

Mark Adams
School of Earth and Environmental Sciences, University of Adelaide, South Australia, Australia
Evolutionary Biology Unit, South Australian Museum, Adelaide, South Australia, Australia

Stephen G. Hamilton and Andrew E. Derocher
Department of Biological Sciences, University of Alberta, Edmonton, AB, T6G 2E9 Canada

Laura Castro de la Guardia
Department of Biological Sciences, University of Alberta, Edmonton, AB, T6G 2E9 Canada
Department of Earth and Atmospheric Sciences, University of Alberta, Edmonton, AB, T6G 2E9 Canada

Vicki Sahanatien
Department of Biological Sciences, University of Alberta, Edmonton, AB, T6G 2E9 Canada
World Wildlife Fund Canada, PO Box 1750, Iqaluit, NU, X0A 0H0 Canada

Bruno Tremblay
Atmospheric and Oceanic Sciences, McGill University, Room 823, Burnside Hall 805 Sherbrooke Street West, Montreal, QC H3A 0B9 Canada

David Huard
David Huard Solutions, Québec, G1W 4G8 Canada

Mark W. Schwartz
John Muir Institute of the Environment, University of California Davis, Davis, California, United States of America
Department of Environmental Science & Policy, University of California Davis, Davis, California, United States of America

Lacy M. Smith
Department of Environmental Science & Policy, University of California Davis, Davis, California, United States of America

Zachary L. Steel
Graduate Group in Ecology, University of California Davis, Davis, California, United States of America

Claire D. Stevenson-Holt
Centre for Wildlife Conservation, University of Cumbria, Ambleside, Cumbria, United Kingdom

Kevin Watts
Centre for Ecosystems, Society and Biosecurity, Forest Research, Farnham, Surrey, United Kingdom

Chloe C. Bellamy
Centre for Ecosystems, Society and Biosecurity, Forest Research, Roslin, Midlothian, United Kingdom

Owen T. Nevin
School of Medical and Applied Sciences, Central Queensland University, Gladstone, Queensland, Australia

Andrew D. Ramsey
School of Biological and Forensic Sciences, University of Derby, Derby, Derbyshire, United Kingdom

Daniel Ayllón, Benigno Elvira, Irene Parra and Ana Almodóvar
Department of Zoology and Physical Anthropology, Faculty of Biology, Complutense University of Madrid, Madrid, Spain

Graciela G. Nicola
Department of Environmental Sciences, University of Castilla-La Mancha, Toledo, Spain

Nicolas Chemidlin Prévost-Bouré and Philippe Lemanceau
Uniteé Mixte de Recherche 1347 Agroécologie, Institut National de la Recherche Agronomique-AgroSup Dijon-Universitéde Bourgogne, Dijon, France

Lionel Ranjard
UnitéMixte de Recherche 1347 Agroécologie, Institut National de la Recherche Agronomique-AgroSup Dijon-Universitéde Bourgogne, Dijon, France
UnitéMixte de Recherche 1347 Agroécologie-Plateforme GenoSol, Institut National de la Recherche Agronomique-AgroSup Dijon-Universitéde Bourgogne, Dijon, France

Samuel Dequiedt, Mélanie Leliévre and Pierre Plassart
UnitéMixte de Recherche 1347 Agroécologie-Plateforme GenoSol, Institut National de la Recherche Agronomique-AgroSup Dijon-Universitéde Bourgogne, Dijon, France

Jean Thioulouse
Unité Mixte de Recherche 555 Laboratoire de Biométrie et Biologie Evolutive, UniversitéLyon 1-Centre National de la Recherche Scientifique, Villeurbanne, France

Nicolas P. A. Saby, Claudy Jolivet and Dominique Arrouays
Unitéde Services 1106 InfoSol, Institut National de la Recherche Agronomique, Orléans, France

Ariadna Rangel-Negrín, Alejandro Coyohua-Fuentes, Domingo Canales-Espinosa and Pedro Américo D. Dias
Instituto de Neuroetología, Universidad Veracruzana, Xalapa, Veracruz, Mexico

Roberto Chavira
Instituto de Ciencias Médicas y Nutricioón Salvador Zubirán, México D.F., Mexico

Barrie M. Forrest, Lauren M. Fletcher, Javier Atalah and Grant A. Hopkins
Coastal and Freshwater Group, Cawthron Institute, Nelson, New Zealand

Richard F. Piola
Coastal and Freshwater Group, Cawthron Institute, Nelson, New Zealand
Maritime Division, Defence Science and Technology Organisation, Melbourne, Victoria, Australia

Rebecca Kormos
Department of Integrative Biology, University of California, Berkeley, California, United States of America

Cyril F. Kormos
The WILD Foundation, Berkeley, California, United States of America

Tatyana Humle
Durrell Institute of Conservation and Ecology, School of Anthropology and Conservation, University of Kent, Canterbury, United Kingdom

Annette Lanjouw
Strategic Initiatives and Great Apes Program, The Arcus Foundation, Cambridge, United Kingdom

Helga Rainer
Conservation Program, The Arcus Foundation, Cambridge, United Kingdom

Ray Victurine
BusinessConservation Initiative and Conservation Finance, Wildlife Conservation Society, Bronx, New York, United States of America

Russell A. Mittermeier
Conservation International, Arlington, Virginia, United States of America

Mamadou S. Diallo
Guinée Écologie, Conakry, Republic of Guinea

Anthony B. Rylands
Conservation International, Arlington, Virginia, United States of America

Elizabeth A. Williamson
Scottish Primate Research Group, School of Natural Sciences, University of Stirling, Scotland, United Kingdom

Miguel A. Munguía-Rosas and Salvador Montiel
Departamento de Ecología Humana, Centro de Investigación y de Estudios Avanzados del Instituto Politécnico Nacional (CINVESTAV), Mérida, Yucatán, México

Haiyang Wang
Institute of Landscape Ecology of Montane Horticulture, Southwest University, Chongqing, 400716, China

Hui Feng and Yanru Zhang
Department of Botany, College of Horticulture and Landscape Architecture, Southwest University, Chongqing, 400716, China

Hong Chen
Institute of Landscape Ecology of Montane Horticulture, Southwest University, Chongqing, 400716, China
Department of Botany, College of Horticulture and Landscape Architecture, Southwest University, Chongqing, 400716, China

Megan L. Hornseth
Environmental and Life Sciences, Trent University, Peterborough, Canada

Aaron A. Walpole and Jeff Bowman
Wildlife Research & Monitoring Section, Ontario Ministry of Natural Resources and Forestry, Peterborough, Canada

Lyle R. Walton
Regional Operations Division, Ontario Ministry of Natural Resources, South Porcupine, Canada

Justina C. Ray
Wildlife Conservation Society Canada, Toronto, Canada

Marie-Josée Fortin
Department of Ecology and Evolutionary Biology, University of Toronto, Toronto, Canada

Dennis L. Murray
Department of Biology, Trent University, Peterborough, Canada

Cynnamon Dobbs
School of Botany, The University of Melbourne, Melbourne, Australia
School of Forest Science and Ecosystem, Melbourne School of Land and Environment, The University of Melbourne, Melbourne, Australia

Craig R. Nitschke
School of Forest Science and Ecosystem, Melbourne School of Land and Environment, The University of Melbourne, Melbourne, Australia

Dave Kendal
Australian Research Centre for Urban Ecology, Royal Botanic Gardens Melbourne, c/o School of Botany, The University of Melbourne, Melbourne, Australia

Ximena Velez-Zuazo
Department of Biology, University of Puerto Rico-Rio Piedras, San Juan, Puerto RicoecOceánica, Lima, Perú

Luciana Klinge and Shaleyla Kelez
ecOceánica, Lima, Perú

avier Quiñones and Sixto Quispe
Laboratorio Costero de Pisco, Instituto del Mar del Perú, Ica, Peru

Aldo S. Pacheco
ecOceánica, Lima, Perú
Instituto de Ciencias Naturales Alexander von Humboldt, CENSOR Laboratory, Universidad de Antofagasta, Antofagasta, Chile

Evelyn Paredes
Unidad de Investigaciones en Depredadores Superiores, Instituto del Mar del Perú, Callao, Peru

Helge von Saltzwedel, Mark Maraun, Stefan Scheu and Ina Schaefer
Georg-August University, Johann-Friedrich-Blumenbach Institute of Zoology and Anthropology, Dept. Ecology, Göttingen, Germany

Shawn T. O'Neil, Kasey C. Rahn and Joseph K. Bump
School of Forest Resources and Environmental Science, Michigan Technological University, Houghton, Michigan, United States of America

Index

A
Anoxic Conditions, 20, 24
Atlantic Herring, 20, 26-27, 31-32

B
Benthic Predation, 132, 135, 139-140
Biodiversity, 8, 53, 55, 62, 82, 89-91, 100, 112, 114, 121-122, 124, 130-132, 142, 144-145, 147, 150-159, 163, 173, 181, 184, 190-191, 212
Black Howler Monkeys, 124-130
Blue Tits, 124, 130
Broad-leaved Forest, 165-166, 169, 171
Brown Trout, 102-105, 109-112

C
Canada Lynx, 172-174, 177-178, 181-182, 218, 222
Canadian Arctic Archipelago, 75, 79
Carbon Storage, 170, 183-191
Carrying Capacity, 102-106, 108-112
Climate Change, 1, 8, 39, 41, 46-47, 53-55, 60, 63, 70, 72-73, 75, 77, 80, 103, 111-112, 173, 182, 184, 190
Climax Species, 165-167, 169-170
Cobble Habitats, 132, 135-136, 139
Community Dynamics, 122, 165
Conservation Funds, 82
Constant-capture Effort, 103
Coral Recruits, 55-58, 60-61
Cougars, 182, 214-222

D
Distance-decay Relationship, 114, 116-117, 121

E
Electrofishing, 103
Environmental Gradients, 55, 212
Environmentally-induced Recession, 33

F
Fecal Glucocorticoid Metabolite, 124, 129
Fishery Managers, 20
Fjord, 20-28, 32
Flexibility Hypothesis, 172-173, 178
Forest Fragments, 125, 128, 131, 158, 162

French Soil Quality Monitoring Network, 113
Frozen-niche-variation, 204

G
General Circulation Model, 43, 46, 76
General-purpose-genotype, 204
Geographical Information System, 91
Glacial Alpine Landscapes, 33
Glacial Water, 33-34, 37-38
Glaciated Alpine Floodplains, 33, 37
Global Positioning System, 9, 92, 182
Glucocorticoid Hormones, 124
Gorillas, 144-145, 149-152, 155-156
Greater Yellowstone Area, 41-43
Green Turtles, 192-203
Grey Squirrel, 91-93, 95, 97-101
Grizzly Bear, 41

H
Habitat Degradation, 192, 214
Habitat Fragmentation, 62-63, 67-71, 90, 130-131, 158, 163-164, 172-173, 175-176, 178-179, 181-182
Habitat Loss, 50, 75, 78, 80, 87, 89, 124-125, 144, 149, 164, 172-173, 175-176, 178-179, 181-182
Habitat Suitability Models, 91, 173, 176, 181-182
Heterotrophic Bacteria, 33
High Human Density, 82, 87-89
Hybrid Regimes, 183, 185, 188-189

I
Interbreeding, 20-21, 27
International Union for Conservation of Nature, 73, 125

L
Land Cost, 82
Least-cost Models, 18, 91-92, 97, 100
Long-lived Pioneers, 165-166, 169-170

M
Marine Ecosystems, 8, 75, 141
Maxent, 91-93, 95, 101
Molecular Clock, 204, 207-208
Mortality Events, 55, 60

Mule Deer, 10, 18

N
Natural Flow Regimes, 62
Natural Hydrology, 62, 70-71
Non-indigenous Species, 132

O
Obligate Corallivores, 1

P
Parthenogenetic Lineages, 204, 211-212
Patch Size, 48, 50, 52, 158-159, 161-164, 172, 175
Peripheral Populations, 172
Plant Population Performance, 82-83
Plant Regeneration, 165, 170
Polar Bears, 75-81

Q
Quaternary Geological Formations, 159

R
Red Squirrel, 41, 92-93, 100-101
Reef-building, 1
Representative Concentration Pathways, 41, 43
Resource Selection Functions, 9-10, 18, 222
Riverine Distance, 62

S
Sage-grouse, 9-14, 16-19

Sagebrush Steppe Conservation, 10
Sargassum, 1, 5, 8, 141, 202
Sea Ice, 75-81
Seagrass Habitats, 132
Shallow Coral Communities, 55
Soil Microbial Community Structure, 113
Spatial Scaling, 101, 113, 122
Structural Habitat Models, 183

T
Tana Mangabeys, 124
Taxa-area Relationship, 113
Thelytokous Oribatid Mite, 204
Threshold Hypothesis, 172-173, 178
Tosa Bay, 1-3, 5-7

U
Upper Great Lakes, 214-217, 220-221
Urbanization, 82, 88, 90, 142, 183, 188-191

V
Vessel Hulls, 132

W
Whitebark Pine, 41, 53-54

Y
Yarra Pygmy Perch, 62, 73
Young-of-the-year, 102-104

www.ingramcontent.com/pod-product-compliance
Lightning Source LLC
Chambersburg PA
CBHW061251190326
41458CB00011B/3644